Fundamentals of Materials Modelling for Metals Processing Technologies

THEORIES AND APPLICATIONS

Fundamentals of Materials Modelling for Metals Processing Technologies

THEORIES AND APPLICATIONS

Jianguo Lin

Imperial College London, UK

Imperial College Press

Published by

Imperial College Press
57 Shelton Street
Covent Garden
London WC2H 9HE

Distributed by

World Scientific Publishing Co. Pte. Ltd.

5 Toh Tuck Link, Singapore 596224

USA office: 27 Warren Street, Suite 401-402, Hackensack, NJ 07601

UK office: 57 Shelton Street, Covent Garden, London WC2H 9HE

Library of Congress Cataloging-in-Publication Data
Lin, Jianguo, 1958–
    Fundamentals of materials modelling for metals processing technologies : theories and applications / Jianguo Lin, Imperial College London, UK.
        pages cm
    Includes bibliographical references and index.
    ISBN 978-1-78326-496-4 (hardcover : alk. paper) -- ISBN 978-1-78326-497-1 (pbk. : alk. paper)
    1. Metals--Formability. 2. Materials science--Data processing. 3. Materials--Mathematical models. I. Title.
    TS205.L556 2015
    620.1'6--dc23
                                          2014049077

British Library Cataloguing-in-Publication Data
A catalogue record for this book is available from the British Library.

Preface

It is a particular pleasure for me to write the preface for this extraordinary book. Professor Lin has devoted many years to the development of numerical simulations of metal forming processes, and their verification experimentally. Prior to this work metal forming modelling was a mechanics exercise in which simple material models were used. The materials inputs were limited to elastic-plastic stress–strain curves with work hardening but no consideration of microstructure. Here, the subject has been taken forward by including the details of the evolution of the microstructure to achieve particular properties in the finished product. This is a major advance which is reflected in the large scale support the work has received in recent years. It is also a considerable intellectual achievement requiring a sustained effort over many years. I believe the book will be seen as a milestone in manufacturing and plasticity technologies.

Professor Gordon Williams
FREng, FRS
Imperial College London, August 2014

Introduction

For modern engineering applications, formed metal components are required to be manufactured with specified microstructures, mechanical and physical properties so that their performance characteristics and useful lifetime are assured. Material and process models, which characterise microstructural evolution in addition to process mechanics, have been developed to facilitate this requirement. Efforts have been made over many years to advance classical mechanics of materials models to achieve the goal. These advances have resulted in unified materials descriptions, which enable mechanical and physical phenomena arising in work-pieces during large deformation metal forming processes to be modelled.

The main purpose of this book is to present the state of the art in material and process modelling in a manner that makes it useable for development engineers to improve production processes and quality of formed metal parts. It is designed for readers who already have some basic knowledge of solid mechanics, the theory of plasticity and metal forming processes and are familiar with numerical methods and computing techniques. It is suitable for undergraduates, postgraduates and industrial engineers. Also, it is intended to be useful to academic supervisors and project managers.

The main focus here is the development of unified materials modelling theories for analysing mechanics and microstructural evolution problems encountered in metal forming processes. However, to cover a more general theme, the chosen title of the book is "Fundamentals of Materials Modelling for Metals Processing Technologies – Theories and Applications". Many metal processing operations do not involve bulk large plastic deformation, such as friction stir welding (FSW), inertia welding, heat treatment of materials (e.g. solution heat treatment followed by quenching), laser joining and cutting. In these cases small scale plastic deformation takes place due to mechanical and thermal loading and microstructure evolution occurs at high temperature under complex stress and strain conditions. Process mechanics and dynamic microstructural evolution phenomena can be modelled for these processes also, using the unified theories introduced in this book. Some of the examples given to guide the reader in the use of models and techniques appropriate to particular processes, relate to work carried out in the Metal Forming and Materials Modelling Group at Imperial College London and earlier at the University of Birmingham. For example, unified viscoplastic constitutive equations have been created to model the mechanics and microstructural evolution in inertia welding of RR1000 superalloys and bainitic steel (Shi, Doel, Lin, *et al.,* 2010).

A wide field of knowledge is covered, dealing with fundamentals of materials modelling, particularly the formulation of unified constitutive equations for metal processing technologies from macro to micro lengthscales. Inclusion of the numerous subjects related to modelling requires explanations to be concise and for more details readers should refer to relevant publications, which are listed in the references.

The contents of the book comprise an introduction to unified materials modelling theories and related practical engineering applications. They include fundamental mechanics theories, basic material and metallurgical knowledge, deformation and failure mechanisms and explanations of how they may be used to construct materials and process models to be used in computer-based process simulation of metal forming processes. Also included are crystal plasticity theories which are used in micro-forming simulation. The book is divided into four distinct parts:

- *Fundamental knowledge of metal forming techniques, mechanics and materials.* This comprises Chapters 1 and 2 and introduces the basic knowledge required for understanding how to formulate and efficiently characterise materials and undertake analyses of manufacturing processes based on the plastic deformation of metals. It aims particularly to educate young students, researchers and engineers working within diverse fields of metal forming research and applications.

- *Formulation and application of unified constitutive equations.* The second part of the book includes Chapters 3, 4, 5, 6 and 7, and introduces fundamental techniques for formulating and determining unified constitutive equations. Application domains for plasticity (Chapter 4) and viscoplasticity (Chapter 5) have been defined for different processes. This is important to enable young researchers and industrial engineers to choose the right theory for their particular applications. Material deformation and failure mechanisms for various deformation conditions and the theories for modelling individual process phenomena are introduced in detail in Chapter 6. Finally, numerical techniques for solving and calibrating unified constitutive equations are introduced in Chapter 7. This chapter is mainly for advanced researchers at universities and research institutes.

- *Application examples.* Examples, based on results from projects sponsored by industry, of applying the theories to practical engineering processes, are introduced in Chapter 8. These case studies are designed to help students, academic researchers and engineers to gain insight of techniques for using the theories and applying them appropriately to particular processes.

- *Crystal plasticity and micro-forming applications.* Mode of deformation, size of product and grain size are very important factors influencing the outcomes of micro-forming processes, but macro-mechanics theories often are not applicable to micro-forming applications. Thus crystal plasticity theories, based on slip systems in lattice structures, are introduced. Chapter 9 is mainly for researchers working on micro-mechanics technologies, including micro-forming and micro-machining. Also it is designed to help researchers working on fracture mechanics. It is an aid to understanding basic plastic deformation mechanisms.

The book has been written with the support of many people. My wife, Yuqin, has provided every opportunity for me to work on the book during a busy time in which I was supervising various research projects. Without her support, it would not have been possible to complete the book. In addition, my daughter, Kelly, and her husband, Edward, have spent their valuable time to undertake proofreading, which was hard work for them, since they are economists.

Professor Trevor Dean of the University of Birmingham, who is a well-known expert in metal forming technologies, has provided constructive suggestions for the book and gave me much advice and encouragement. He also helped me by reading and correcting most of the chapters of the book. To enable me to concentrate on the book writing, Professor Dean helped supervise my PhD students by reading and correcting their theses. Professor Gordon Williams of Imperial College London, who is a well-known research leader on mechanics of materials, has also spent his valuable time to proofread and comment for many chapters for the book. Professor Anthony Atkins of University of Reading also helped. Their efforts and support are greatly appreciated.

I give thanks to the many members and former members in the Metal Forming and Materials Modelling Group who have helped in proofreading chapters, checking equations, drawing figures, providing photographs and dealing with the many small details necessary to produce a quality book. These researchers include (in alphabetical order): Dr Qian Bai, Dr Jian Cao, Mr Omer El Fakir, Mr Haoxiang Gao, Mr Jiaying Jiang, Miss Erofili Kardoulaki, Dr Morad Karimpour, Dr Michael Kaye, Mr Aaron Lam, Dr Nan Li, Dr Jun Liu, Dr Mohamed Mohamed, Dr Denis Politis, Mr Nicholas Politis, Mr Zhutao Shao, Dr Zhusheng Shi, Dr Shiwen Wang and Mr Kailun Zheng. Particularly, major editing work has been carried out by Dr Zhusheng Shi of the Group.

Professor Jianguo Lin
FREng, FIMechE, FIMMM, CEng, PhD
23 August 2014 at Imperial College London

Contents

Chapter 1

Metal Forming and Materials Modelling

The aim of this chapter is to introduce the concept of metal forming processes, their characteristics and applications. Typical features of metal forming processes are compared with those of other types of manufacturing processes. The basic modelling concept is then introduced, leading to the discussion of why *mechanics* and *modelling* are fundamentals for metal forming research and development.

1.1 Introduction to Metal Forming Processes

Metal elements and their alloys are versatile engineering materials. At a temperature below its melting temperature, T_m, a metal is solid and can sustain load. Increasing the load augments deformation of the metal. If the final load is sufficiently low before removal, the metal returns to the original shape, as shown in **Fig. 1.1** region (I). This temporary deformation is known as *elastic*. Elastic deformation is usually small, and the initial and the deformed shapes are little different.

As the load increases to beyond the metal's elastic limit, point A in **Fig. 1.1**, permanent deformation occurs (**Fig 1.1** region (II)) and the ultimate displacement is *fracture*, as shown in **Fig. 1.1**, region (III). For all displacements in region (II) in **Fig. 1.1**, large enough to overcome the material's capacity to deform elastically but not lead to fracture, metals are said to be in an *elastic-plastic state* (**Fig. 1.1(b)**). Metals in this state will retain most of their deformed displacement after the removal of the load (B^0), since elastic recovery of metals is normally very small. Metals in a plastic state can thus be deformed into complex shapes depending on their ductility (the amount of deformation before failure takes place). Rolling, forging, drawing and extrusion are examples of controlled plastic deformation for forming metal into specific shapes and are known as *metal forming processes*.

The aim of this section is to introduce commonly used metal forming processes, their deformation features and modelling requirements.

(a) Load and displacement

(b) Deformation State

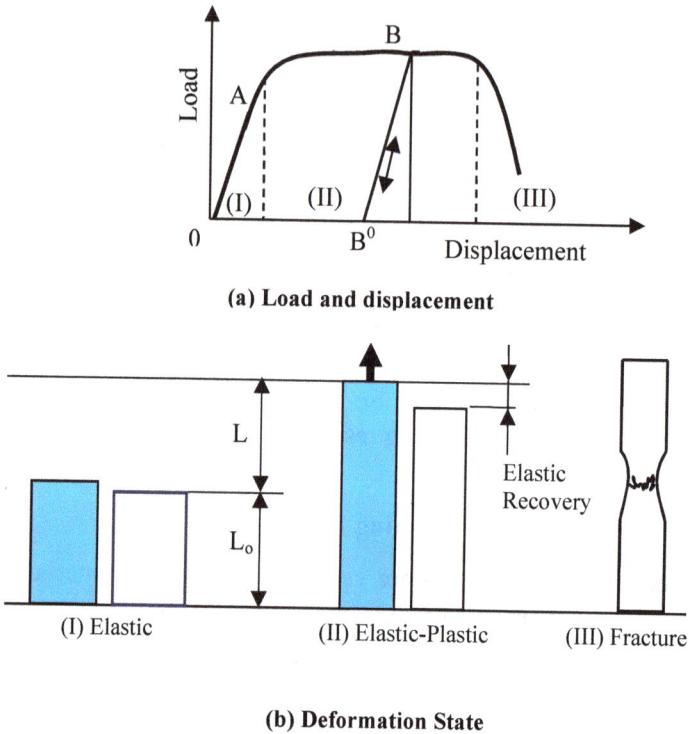

Fig. 1.1 Loading of a metal and the metal forming region (II).

1.1.1 *Compressive forming*

Compressive metal forming is the shaping of metals under compressive stresses. If the major compressive forces in three orthogonal directions are sufficiently high and differ sufficiently from each other (deviatoric stresses – discussed further in Chapter 2), it is believed that ductile metals can be formed into any shape. However, in practice it is sometimes very difficult to keep stresses compressive throughout a complete forming process, and fracture or cracking might take place in some locations.

a. Forging

Forging, as shown in **Fig. 1.2**, is the shaping of metal using localised compressive forces. Forging can produce an engineering artefact that is stronger than an equivalent cast or machined part. As the metal is shaped during the forging process, its internal grain structure deforms to follow the general shape of the part. As a result, the grain is continuous throughout the part, giving rise to a piece with improved strength characteristics. The flash is the excess material squeezed out from the die, which is necessary in closed die forging to ensure that the die cavity can be filled completely with a relatively low forging force.

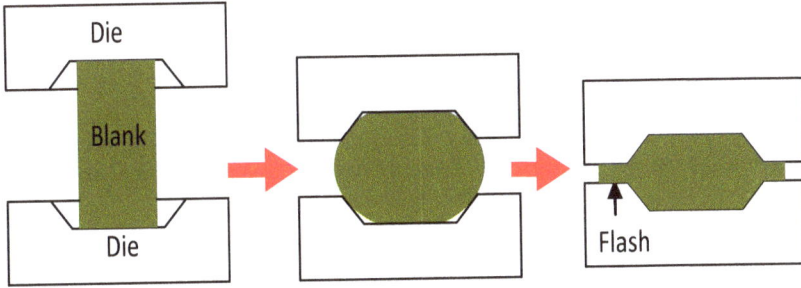

Fig. 1.2 Stages of impression die forging of a solid round billet. The flash is subsequently trimmed off after forging.

b. Extrusion and drawing

In the extrusion process, as shown in **Fig. 1.3**, a metal work-piece is forced (pushed) through a die. The die geometry remains the same throughout the operation, thus extruded products have a constant cross-section.

Depending on the material flow relative to the punch movement

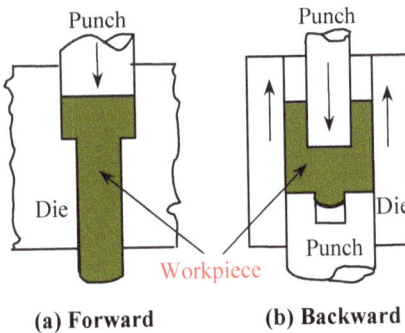

(a) Forward   (b) Backward

Fig. 1.3 Forward and backward extrusion processes.

direction, the extrusion process can be divided into forward extrusion (**Fig. 1.3(a)**) and backward extrusion (**Fig. 1.3(b)**).

In a drawing process, the material is pulled out from a die rather than pushed, as shown in **Fig. 1.4**. It is commonly used for the production of round bar or wire and the cross-section area is typically reduced. It is also used to shape tubes with or without a mandrel.

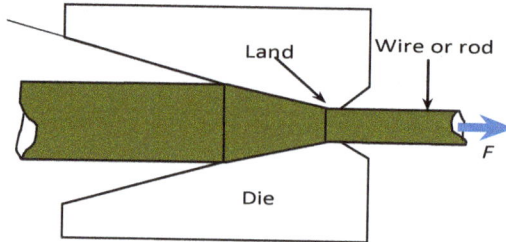

Fig. 1.4 A wire drawing process. The die angle, the reduction in cross-sectional area per pass, the speed of drawing, the temperature, and the friction, all affect the drawing force *F*.

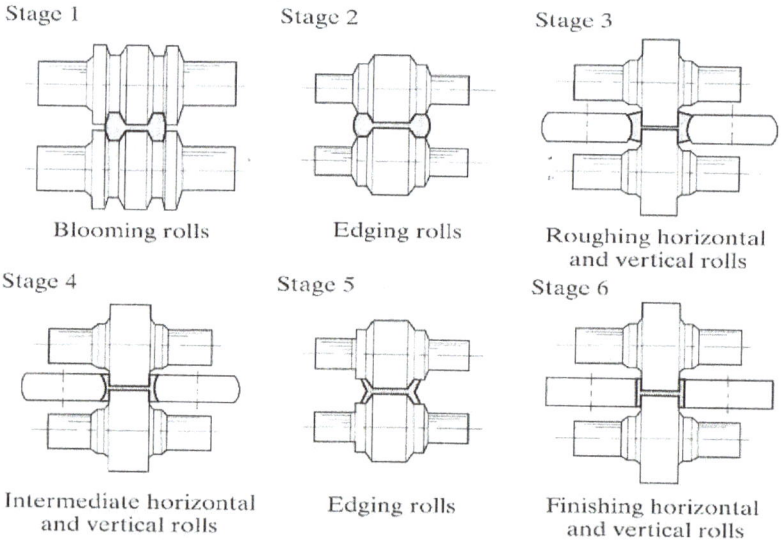

Stage 1

Stage 2

Stage 3

Blooming rolls

Edging rolls

Roughing horizontal and vertical rolls

Stage 4

Stage 5

Stage 6

Intermediate horizontal and vertical rolls

Edging rolls

Finishing horizontal and vertical rolls

Fig. 1.5 Stages in the shape rolling of an H-section part. Various other structure sections, such as channels and I-beams, are also rolled by this kind of process (Kalpakjian, Schmid, 2001).

c. *Rolling*

Rolling, as shown in **Fig. 1.5**, is a process in which metal stock is passed through a pair of rotating rolls. Rolling is classified according to the temperature of the metal work-piece. If it is above the recrystallisation temperature, the process is termed as hot rolling. There are many other types of rolling processes, including flat rolling, foil rolling, ring rolling, roll bending, roll forming, profile rolling and controlled rolling.

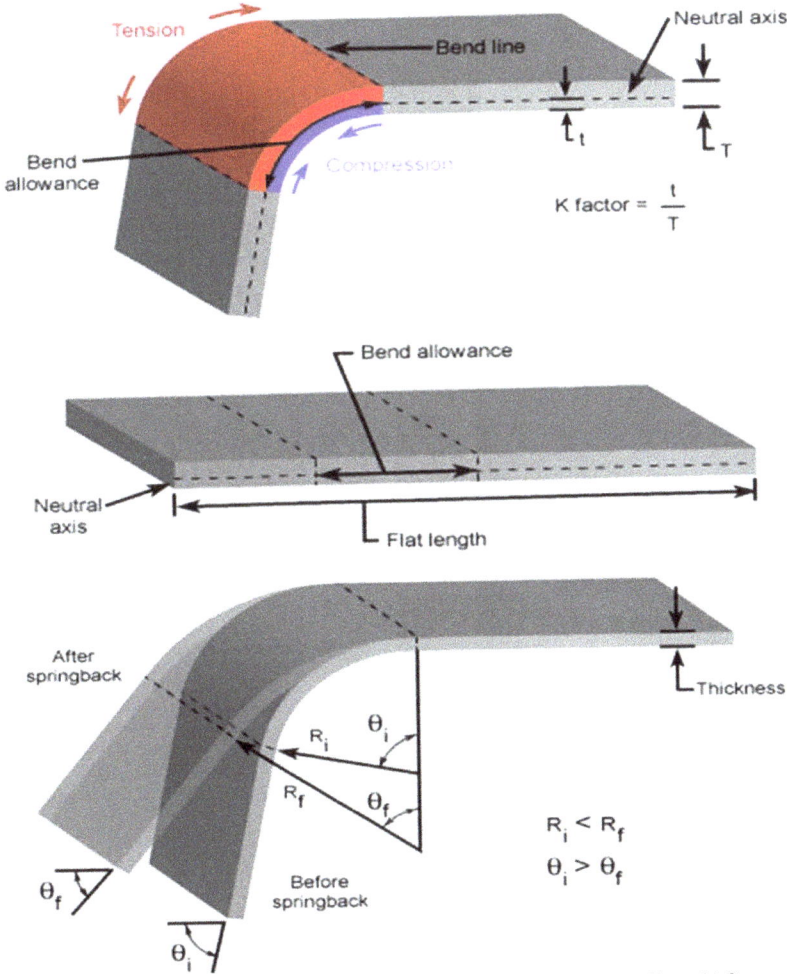

Fig. 1.6 Bending characteristics and springback in bending (After:
http://www.custompartnet.com/wu/sheet-metal-forming).

1.1.2 *Sheet metal forming*

A piece of sheet metal is plastically deformed by *tensile or bending stresses* into a two-dimensional or three-dimensional shape. This is known as sheet metal forming. Sheet metal forming processes include blanking, bending (**Figs. 1.6** and **1.7**), stretching-bending forming, stamping, drawing, superplastic forming and many other processes.

a. Bending

Figure 1.6 shows bending definition and springback in bending. The *neutral axis* or *neutral plane* is the location in the sheet where *strain is zero*, and thus remains at a constant length. The material on one side of the neutral plane is in tension and the other side is in compression, as shown in **Figs. 1.6** and **1.7**, which shows a few examples of bending operations in sheet metal forming. *Springback*, due to elastic recovery, is a very important feature and difficult to reduce or eliminate in bending. This is particularly important in bend forming materials with *low Young's modulus*, such as aluminium alloys, or high strength materials, such as ultra-high strength steels. In forming operations, springback is often compensated by over bending of the part. Another method is to increase plastic deformation locally at the bending radius region.

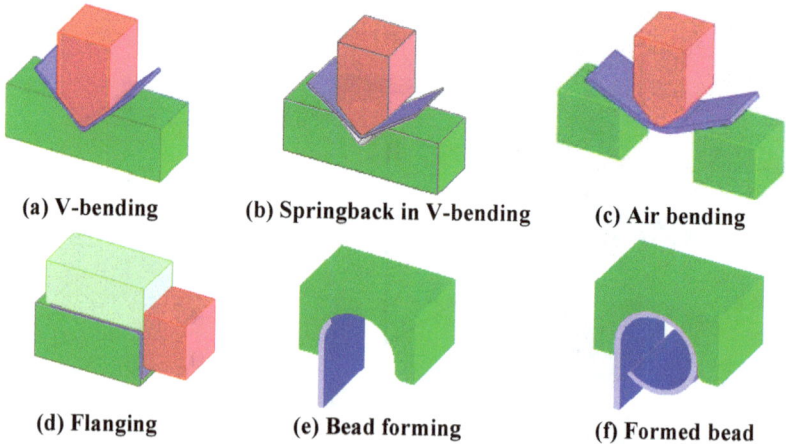

    (a) V-bending   (b) Springback in V-bending   (c) Air bending

    (d) Flanging   (e) Bead forming   (f) Formed bead

Fig. 1.7 Examples of bending processes. Springback is a big problem in bending processes, particularly in cold bending of high strength and low Young's modulus sheet metals.

b. Hydroforming

The *combined extrusion and hydroforming* process, as shown in **Fig. 1.8(a)**, enables more complex-shaped tubular components to be formed with less localised thinning. Numerous applications of hydroforming can be seen in exhaust manifolds, as shown in **Fig. 1.8(b)**. This could be sheet metal forming or tube forming depending on the definition and the thickness of the work-piece.

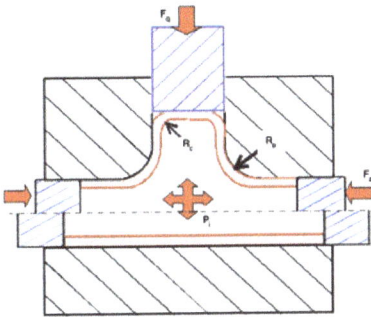

(a) Hydroforming   (b) Hydroformed parts

Fig. 1.8 A hydroforming process and examples of hydroformed parts (After: *http://www.thefabricator.com/article/hydroforming/the-basic-elements-of-tubular-hydroforming*).

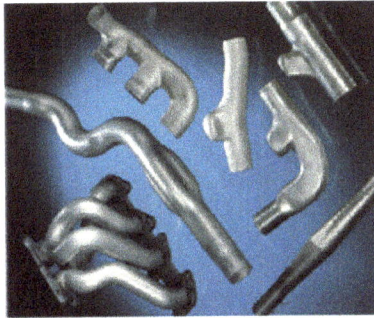

c. Drawing

A *deep drawing* process is shown in **Fig. 1.9**. In the deep drawing process the blank holding force should be sufficiently high to avoid wrinkling of the sheet but low enough to allow the material to be drawn into the die cavity without being fractured. Blank holding force is related to the friction coefficient of the sliding surfaces and the strength of the blank material. *Drawbeads* (**Fig. 1.10**) are often used to enhance

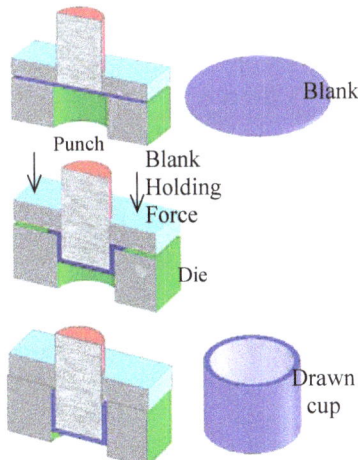

Fig. 1.9 A deep drawing process – drawing of a cup. Blank holding force is important to avoid wrinkling and tearing.

restraint on material flow into the die cavity, thus reducing or eliminating the possibility of buckling.

Ironing will reduce the wall thickness of a drawn cup or a tube, as shown in **Fig. 1.11**. Deep drawing and ironing are the key processes for producing beverage cans and typical operations are shown in **Fig. 1.12**. The production of beverage cans consists of seven operations from blanking (punching a material blank) to final seaming. Two deep drawing operations and one ironing operation are used to produce the main body structure of the can. The top neck is produced using a spin-forming process, which is outlined below.

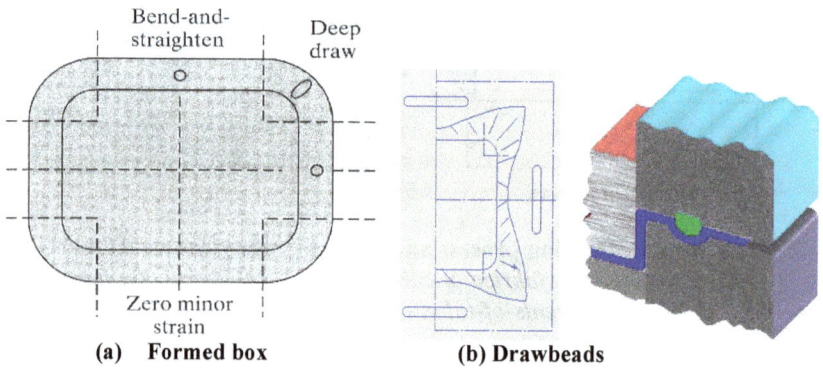

(a)   Formed box   (b) Drawbeads

Fig. 1.10 Drawbeads are used for drawing a box-component. (a) drawing process with drawbeads. (b) location of drawbeads.

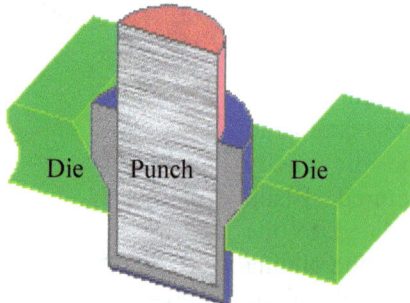

Fig. 1.11 Ironing of a cup – reducing the wall thickness of the work-piece.

Process	Process Illustration	Result

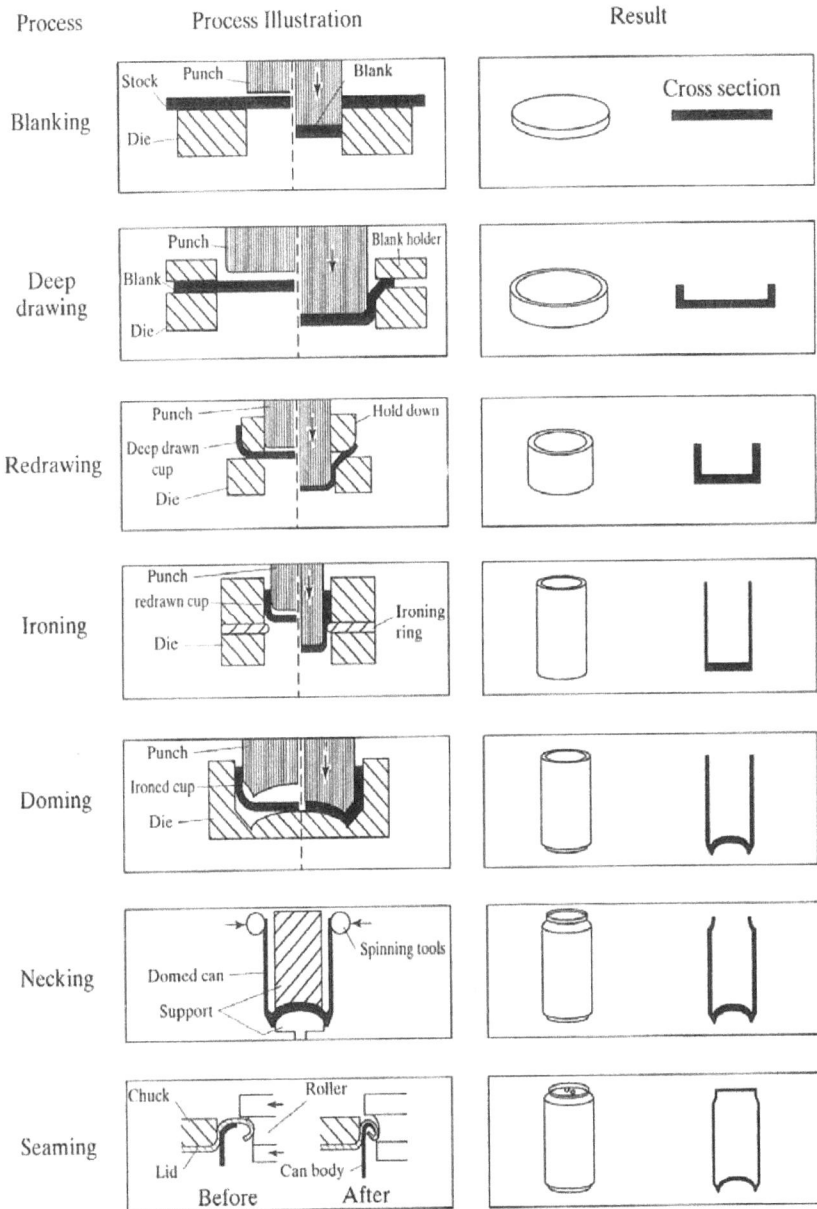

Fig. 1.12 Sheet metal forming processes involved in manufacturing a two-piece aluminium beverage can (Kalpakjian and Schmid, 2001).

d. Spin-forming

A *spin-forming* process, which is used to form rotationally symmetrical parts, is shown in **Fig. 1.13**. A sheet metal disc is rotated while rollers press the sheet against a tool, called a *mandrel*, to form the shape of the desired part. Spun metal parts have hollow shapes, such as cylinders, cones or hemispheres. Examples include cookware, hubcaps, satellite dishes, rocket nose cones and musical instruments.

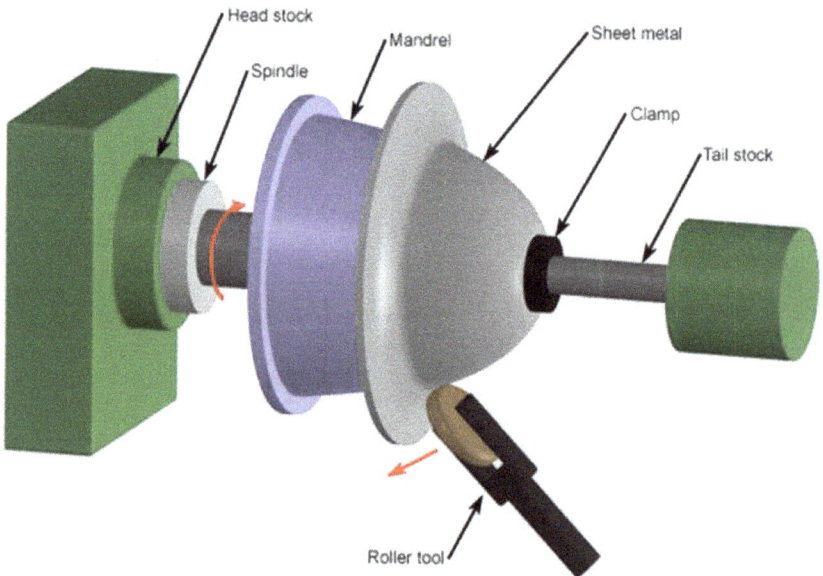

Fig. 1.13 A spin-forming process (After:
http://www.custompartnet.com/wu/images/sheet-metal/spinning.png).

1.1.3 *Advantages of forming*

The major advantages of metal forming are briefly outlined in this section.

a. High productivity for mass production

As shown in **Fig. 1.12**, the manufacturing route for aluminium beverage cans contains a number of forming processes, and expensive tooling, automatic handling and control facilities are required. However, once the production line has been set up, productivity can be very high. This results in the low unit price of the beverage cans. Metal forming processes are most suitable for high volume production, although many processes have been developed for low volume market, such as incremental forming (Wong, Dean and Lin, 2003), superplastic forming (Lin, 2003) and creep age forming (Zhan, Lin and Dean, 2011).

b. Increased materials yield

Forming is used to deform a simple shape metal blank into a component with a more complex shape and greater geometric definition through plastic deformation. This is different from cutting processes and no significant material removal is intended. Precision forming and net-shape forming techniques have been developed and the waste of the materials can be minimised or eliminated. **Figure 1.14** shows the tool and set-up for net-shape forging of gear tooth. There is no requirement for machining of the teeth profiles, which increase the strength and fatigue lives.

Fig. 1.14 Net-shape forging of gear teeth. Error of the gear teeth profile can be controlled within 5 μm. This meets the requirement of automotive gear box gears (Provided by Professor T.A. Dean of University of Birmingham).

c. Reducing defects and increasing strength

Defects within materials, as shown in **Fig. 1.15(a)**, such as voids and short cracks from casting, can be eliminated or reduced using hot compressive forming processes due to diffusion bonding or welding. In particular, hot rolling can reduce or eliminate defects within cast metal ingots and also deform the initial coarse grains, which on recrystallisation become finer and equiaxed, resulting in enhanced and uniform mechanical properties (**Fig. 1.15(c)**).

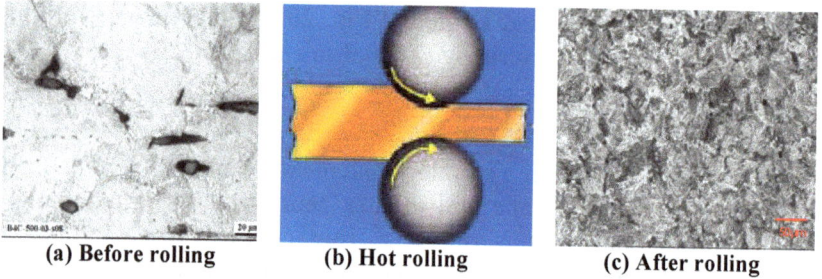

| (a) Before rolling | (b) Hot rolling | (c) After rolling |

Fig. 1.15 Defects (a) in the material can be reduced or eliminated (c) via a hot rolling process (b). The rolling process (c) can also result in recrystallisation, which reduces the grain size, thus increasing the strength, ductility and toughness of materials.

d. Increased structural integrity

Figure 1.16 shows the textures of the bars manufactured through different processes. Directional alignment through the extrusion process (a) has been deliberately oriented in a direction requiring maximum strength. This also yields ductility and resistance to impact and fatigue.

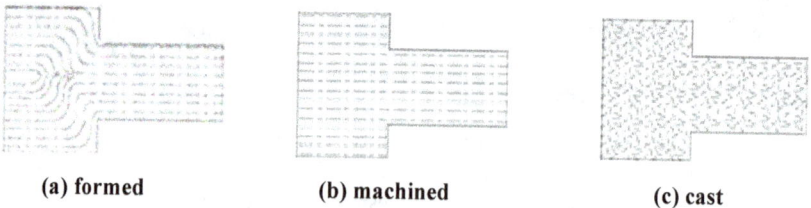

| (a) formed | (b) machined | (c) cast |

Fig. 1.16 Textures for: formed (a), machined (b) and cast (c) bars.

1.2 Modification of Materials Microstructure through Forming

This section introduces the mechanical and physical behaviour of materials that can be modified through forming processes.

1.2.1 *Grain size*

a. *Crystallised materials*

Metals have a crystalline structure (**Fig. 1.17**). When a metal solidifies from a molten state, tiny crystals start to grow and the size of crystals, known as grains in solid state, is dependent on chemical composition and cooling rate. The crystals can be dissolved and recrystallised during hot forming.

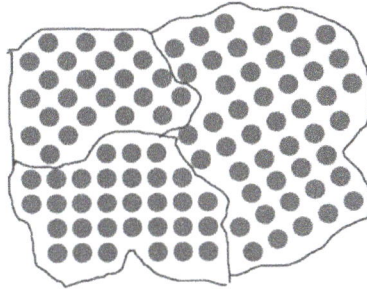

Fig. 1.17 Crystals in metals, with atoms and grain boundaries.

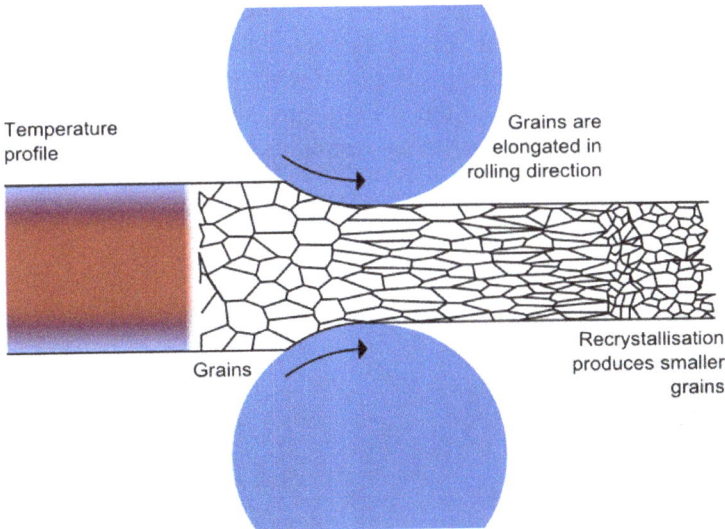

Temperature profile

Grains are elongated in rolling direction

Grains

Recrystallisation produces smaller grains

Fig. 1.18 Deformation, recrystallisation, and grain growth in hot rolling (Kaye, 2012).

b. *Recrystallisation and grain growth*

Plastic deformation in metal forming increases the dislocation density in a metal. Given time under hot forming conditions new crystals nucleate, normally at grain boundaries. This is known as *recrystallisation*, as shown in **Fig. 1.18**. If recrystallisation takes place during deformation, it is known as *dynamic recrystallisation. Static recrystallisation* takes place when a metal is not being deformed – either before or after a hot forming operation. Grain growth takes place during and after recrystallisation and grain size is related to temperature and time. As mechanical properties are dependent on grain characteristics, it is very important to predict recrystallisation and grain growth features in hot metal forming.

1.2.2 *Elimination of defects*

a. *Defects in metals*

The starting point of elemental metals and metal alloys is normally a casting process. Defects, such as gas porosity, shrinkage cavities, segregation etc. are difficult to avoid. Subsequent hot rolling reduces or eliminates such defects. **Fig. 1.19** shows the porosity type defects that can arise from casting processes, which should be removed using hot rolling operations due to the void closure and diffusion bonding at high temperature and high compressive stresses.

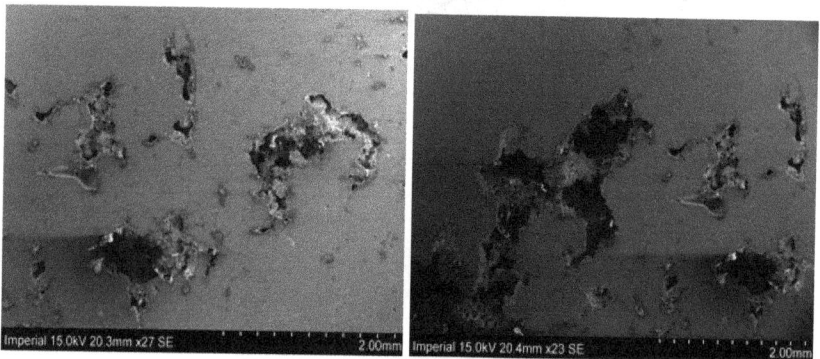

Fig. 1.19 SEM images of porosity in cast leaded free cutting steel blooms due to volume shrink in cooling (Afshan, 2013).

b. *Diffusion bonding*

Diffusion bonding is a solid-state process for joining many types of similar and dissimilar materials under compressive forces at high temperatures. Well known processes include diffusion bonding and superplastic forming for manufacturing complex-shaped titanium sheet metal parts, hot isostatic pressing (HIPing) and sintering of powders for producing high performance components. The diffusion bonding process is normally slow. However, during hot compressive forming processes, very high pressure is applied and the voids and defects of materials can be bound very quickly, as shown in **Fig. 1.20**. Thus in hot rolling, defects in cast can be reduced or eliminated and metals become more "solid".

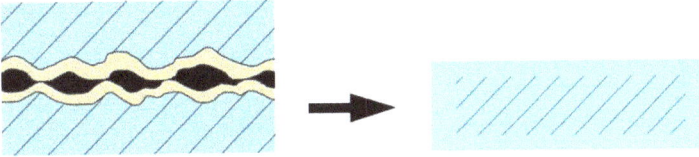

Fig. 1.20 Voids are healed during compressive forming processes.

1.2.3 *Hardening*

Work hardening, which is also known as *strain hardening*, is the strengthening of metals through plastic deformation. This is mainly due to accumulation of dislocations within the crystal structure of metals.

A typical strain hardening feature is shown in **Fig. 1.21 (a)** for a flow stress for 20 °C that increases with strain. A flow stress reducing with strain is known as strain softening, as shown for the curve for 300 °C in **(a)**. For metal deforming at a given strain rate, if the temperature is low, then less recovery of dislocations takes place, and thus more hardening can be observed. More uniform deformation of a material in forming can be observed for high strain hardening materials. This is particularly important in sheet metal forming when the possibility of localised necking and failure can be reduced for strain hardening metals. Strain hardening can be further divided into *isotropic hardening* and *kinematic hardening*, which will be detailed in Chapter 4.

Figure 1.21 (b) shows an example of *strain rate hardening*. In warm and hot metal forming processes the flow stress increases as the strain rate increases. This is known as *viscoplastic* behaviour and will be detailed in Chapter 5. From **Fig. 1.21 (b)**, it is also apparent that strain hardening becomes higher (e.g. the curve for strain rate of 1.0/s) as the strain rate increases. This is due to the fact that, for a given temperature, less time is available for annealing to take place at a higher deformation speed. Thus high densities of dislocations can be accumulated. The feature of strain rate hardening can also be used for forming a panel component with less localised thinning. This hardening mechanism has also been used in superplastic forming process to reduce localised necking (Lin, 2003).

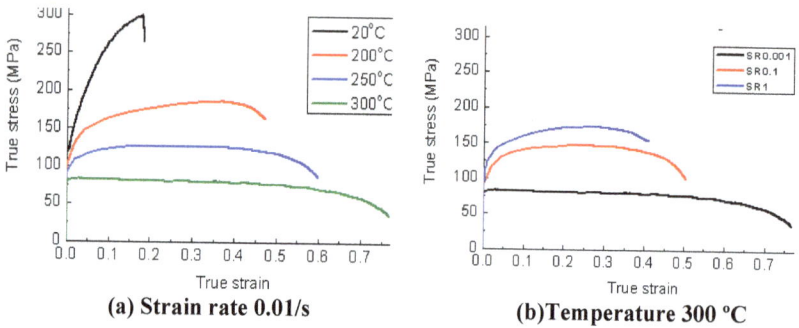

(a) Strain rate 0.01/s

(b)Temperature 300 °C

Fig. 1.21 Strain hardening (a) and strain rate hardening (b) in metal forming. The curves are obtained for AA5754 at different temperatures (a) and different strain rates (b).

If the strain hardening and high strain rate hardening of a metal can be effectively used in sheet metal forming (particularly in hot stamping/drawing, superplastic forming and hydroforming) localised thinning/necking/failure could be reduced or avoided. The mechanical characteristics of work-pieces during room temperature forming can be determined directly from stress–strain curves. However, for elevated temperature forming processes the characteristics are related to the material, temperature and deformation rate, which are very difficult to control.

1.2.4 *Failure of materials in forming*

In sheet metal forming processes, materials are subjected to tensile stresses and tearing often takes place, as shown in **Fig. 1.22**. It is important to monitor micro crack initiation and growth before macro failure takes place, so that sound components can be formed.

In bulk forming processes, such as cross wedge rolling shown in **Fig. 1.23**, tensile cyclic stresses occur in some regions and micro cracks can be formed, which reduces the quality of the parts. In many cases, micro cracks occur close to the axis of a work-piece and can only be detected if the formed part is sectioned for detailed investigation. This failure evolution has been studied using damage mechanics theories by Foster, Lin, Farrugia *et al.* (2011) and will be introduced in Chapter 6.

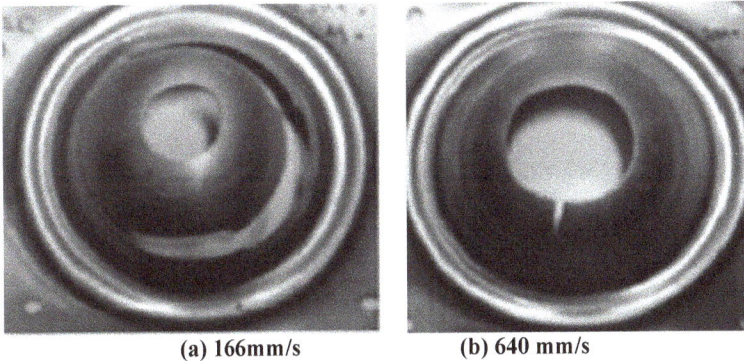

(a) 166mm/s   (b) 640 mm/s

Fig. 1.22 Tearing takes place for hot stamping and cold die quenching of a AA6082. Circumferencial crack is observed at the stroke speed of 166 mm/s, and cracks near the central hole occur at the forming speed of 640 mm/s (Mohamed, 2010).

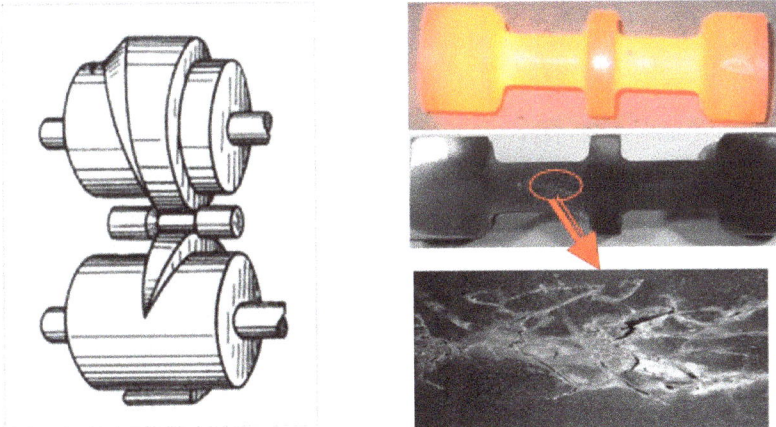

Fig. 1.23 Cross wedge rolling process and the micro cracks in the central of the cross-rolled part (Provided by Professor B. Wang of University Science Technology Beijing, China).

1.3 Methods of Materials Modelling

1.3.1 *Constitutive equations*

a. Stress–strain relationships

Tensile tests are normally used for the determination of mechanical properties, such as strength, ductility, elastic modulus and strain hardening. A standard tensile test specimen for sheet metal is shown in **Fig. 1.24 (a)**. The parallel part of this dog-bone type specimen is 120 mm and the original gauge length (L_0) is 100 mm. The original cross-sectional area of the specimen is A_0.

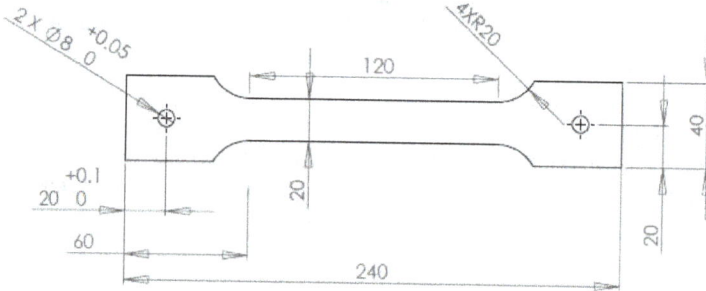

(a) **Geometry of a sheet tensile specimen.**

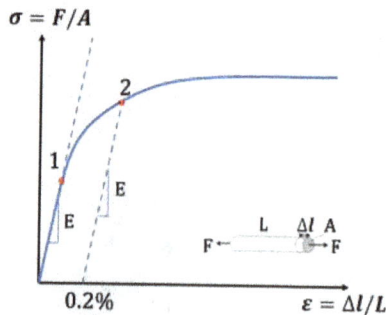

(b) **Stress–strain relationship.**

Fig. 1.24 A standard tensile specimen and typical stress–strain curve obtained from tensile tests.

A typical load-displacement curve for a tensile tests is shown in **Fig. 1.1(a)** and can be converted to stress (stress = load / cross-sectional area perpendicular to the load direction — force in unit area) and strain (strain = gauge-length elongation / specimen gauge-length — extension for unit length) relationships shown in **Fig. 1.24(b)**, where F is the force applied, A is the cross-sectional area of the specimen at the deformed state and L is the specimen length.

The *engineering stress*, or, *nominal stress*, is defined as the ratio of the applied load F to the original cross-sectional area A_0 of the specimen:

$$\text{engineering (nominal) stress:} \quad \sigma = \frac{F}{A_0} \ . \tag{1.1}$$

The engineering strain is defined as

$$e = \frac{\left(L - L_0\right)}{L_0} \ , \tag{1.2}$$

where L is the instantaneous length of the specimen gauge length.

As the load increases, the linear stress–strain relationship can be observed in **Fig. 1.24** (region I in **Fig. 1.1**). At a certain level of stress, the metal undergoes *permanent (plastic) deformation*. Beyond this level, stress and strain are no longer proportional, as they were in the *elastic* region. This stress level is known as the *yield stress*, σ_y, of the material. For soft and ductile metals (i.e. materials that deform extensively), it may not be easy to determine the exact location of the yield point since the slope of the strain (elastic) portion of the curve decreases slowly. Thus the yield point is often defined as the point on stress–strain curves that is *offset* by a strain of 0.002 (0.2%), as indicated in **Fig. 1.24**.

With further increases in load, the specimen will continue to deform and its cross-sectional area will decrease *permanently* and *uniformly* throughout its gauge length. The flow stress normally increases, known as *strain hardening*. The maximum stress point is known as the *tensile strength* (or ultimate tensile stress, σ_u) of the metal. It should be noted that the terms "yield strength" and "tensile strength" are relevant only for metals deforming at room temperature. At high temperatures metals behave viscoplastically and yield strength cannot be defined easily. Additionally the load often drops before necking taking place similar to that in superplastic forming processes.

The above stress and strain calculation assumes that the stress is evenly distributed over the entire cross-section of the specimen (**Fig.**

1.25). In practice, depending on how the specimen is attached at the ends and how it was manufactured, this assumption may not be valid. In that case, the value $\sigma = F/A_0$ will be only the average stress, called *engineering stress* or *nominal stress*. However, if the bar's length L is many times its diameter D, or width W, and it has no gross defects, then the stress can be assumed to be uniformly distributed over any cross-section that is more than a few multiples of D from both ends. This observation is known as the Saint-Venant principle (Adhémar Jean Claude Barré de Saint-Venant, French, 1797–1886). The length of the uniform cross-section part of a specimen necessary to avoid the stress concentration at its ends is dependent on the amount of deformation of the specimen.

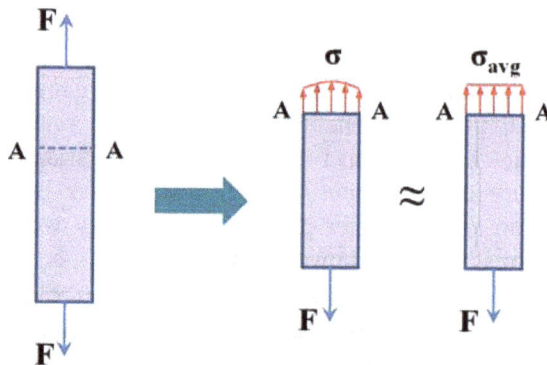

Fig. 1.25 Average stress measurement.

At the initial elastic deformation of a dog-bone type specimen, uniform stress can be assumed to exist on a cross-section $1.5D$ (or $1.5W$) away from the shank radius, D, (or width, W) of the specimen. Specimen design has a significant effect on the stress–strain distribution over the gauge length of the specimen, and this affects the accuracy of the measurement of stress and strain. Significant research has been carried out on high temperature creep and fatigue specimen designs, reported by Lin, Hayhurst and Dyson (1993a; 1993b), Lin, Dunne and Hayhurst (1999) and Kowalewski, Lin and Hayhurst (1994).

Ductility is a measure of the extent of deformation that a material undergoes before fracture. This is normally measured from uniaxial (cylindrical specimen) or plane stress (thin sheet specimen) tensile tests.

Two common measures are used to assess the ductility of a material. One is the total elongation, or the strain to failure, ε_f, of the specimen, as a percentage:

$$\text{total elongation, } \varepsilon_f = \frac{\left(L_f - L_0\right)}{L_0} \times 100\% \qquad (1.3)$$

where L_f is the gauge length of the specimen at fracture as shown in **Fig. 1.1(b)**. This is normally calculated based on the initial gauge length of the specimen. The other measure of ductility is the reduction of cross-sectional area:

$$\text{cross-sectional area reduction} = \frac{\left(A_0 - A_f\right)}{A_0} \times 100\% \qquad (1.4)$$

where A_f is the cross-sectional area of the specimen at failure. The reduction of cross-sectional area and elongation of a metal are generally related.

For large deformation, particularly for applications in metal forming, true stress and strain are generally used, which are defined below:

$$\text{true stress, } \sigma = \frac{F}{A}. \qquad (1.5)$$

For true strain, we considers the accumulation of the instantaneous elongation, dL, of a specimen at the current length L, as the gauge length of the specimen is deformed from the original gauge length L_0 to L:

$$\text{true strain, } \varepsilon = \int_{L_0}^{L} \frac{dL}{L} = \ln\left(\frac{L}{L_0}\right) = \ln\left(\frac{A_0}{A}\right) = 2\ln\left(\frac{D_0}{D}\right), \qquad (1.6)$$

where D_0 and D are the diameters (or widths) of the specimen at the original and deformed state respectively.

For small deformations, for which no necking arises, engineering strain and true strain are around the same value. However, at large deformations, they diverge rapidly. In the rest of the book, true stress and true strain definitions are used unless specifically stated, since metal forming is a large deformation process.

b. Simple constitutive equations

Equations, which are used to describe the deformation of a solid under forces, are known as *constitutive equations*. The simplest example of a constitutive equation is that describing linear elastic stress (σ) and strain (ε) relationship of metals, as shown in **Fig. 1.24** – the elastic region (region I):

$$\sigma = E\varepsilon.$$ (1.7)

The ratio of stress to strain in the elastic region (**Fig. 1.24**, region I) is known as *modulus of elasticity*, E, or, *Young's modulus* (Thomas Young, British, 1773–1829):

$$\text{modulus of elasticity,} \quad E = \frac{\sigma}{\varepsilon}.$$ (1.8)

This linear relationship is known as *Hooke's Law* (Robert Hooke, British, 1635–1703, British). The modulus of elasticity is a measure of the slope of the elastic portion of the curve, i.e. the stiffness of the material. E has the same unit as the stress, N/m^2. 1 N/m^2 is known as 1 Pascal (Blaise Pascal, French, 1623–1662). A metal with high Young's modulus requires large forces to deform it and thus has higher stiffness.

Constitutive equations, used to construct *material models*, become more complicated if the mechanical response of materials is considered under large elastic-plastic, elastic-viscoplastic deformation, or high-temperature conditions. Normally, a set of constitutive equations is required to accurately describe individual deformation mechanisms of materials forming under different temperatures and strain rates. Dominant deformation mechanisms are related to individual metals, their microstructure, and the deformation temperature and deformation rate.

c. Modelling of strain hardening

The true stress and true strain curve can be approximated by the following equation:

power-law for strain hardening: $\sigma = K\varepsilon^n$, (1.9)

where K is known as the strength coefficient and the n as the strain hardening (or work hardening) exponent. If we take the log for the equation, we can obtain

$$\log \sigma = \log K + n \cdot \log \varepsilon.$$ (1.10)

Thus on log-log axes, we find that the relation is a straight line, as shown in **Fig. 1.26**. The slope of the line is the strain hardening exponent, n. This shows that the flow stress increases with the strain. This phenomenon is known as strain hardening and can be rationalised with the parameter n. The

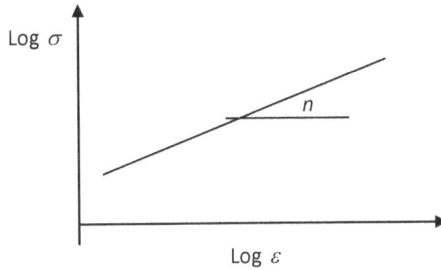

Fig. 1.26 Strain hardening.

greater the strain hardening of a metal is, the stronger it becomes as it is strained. A metal with high strain hardening can have a reduced tendency of necking during stretching or stamping processes, thus enabling a component to be formed with less localised thinning. In cold stamping processes, the n value can be used to judge if a metal can be stamped easily without much localised thinning.

On the physical level, hardening is due to an increase in the dislocation density; dislocations have a tendency to interlock and block each other. To a first approximation, on reloading, an increase in the elastic limit follows an increase in an above-the-yield point stress and it is this approximation which constitutes the theoretical basis of classical plasticity. Thus, for monotonic loading, the current limit of elasticity, also called the plasticity threshold or the yield stress, is equal to the highest values of stress previously attained. For a material with positive hardening, $d\sigma/d\varepsilon_p > 0$, the natural elastic limit σ_y is the smallest value of the yield stress, which is a function of the history of plastic deformation.

For high-temperature deformation processes, strain hardening is related to the strain rate (or deformation rate) and temperature. In general, if the strain rate is low and temperature is high, strain hardening is low, or can even be negative, i.e. strain softening, since recovery mechanisms may be involved during the deformation process. Thus necking or localised thinning can easily arise in warm or hot stamping processes since strain hardening of metals is low.

d. Modelling of strain rate hardening

If a specimen is deformed at high temperature, the flow stress of the material is related to the deformation rate (or strain rate $\dot{\varepsilon} = d\varepsilon/dt$, where t is time) and viscoplasticity of a material is observed. Ignoring the strain hardening, the flow stress can be expressed as:

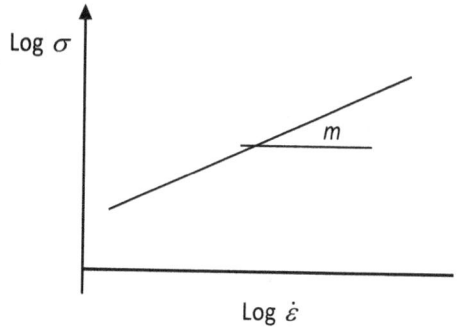

Fig. 1.27 Strain rate hardening.

power-law for strain rate hardening: $\sigma = A\dot{\varepsilon}^m$ (1.11)

where A is known as a coefficient and m as the *strain rate hardening exponent* (or *strain rate sensitivity parameter*). As for strain hardening, if we take the log for Eq. (1.11), we can obtain

$$\log \sigma = \log A + m \log \dot{\varepsilon}$$ (1.12)

Thus on log-log axes, we find that it is a straight line, as shown in **Fig. 1.27**. The slope of the line is the strain rate sensitivity parameter, m. This shows that the flow stress increases with the strain rate. This phenomenon is known as *strain rate hardening* and can be rationalised with the parameter m, which varies from 0 to 1. A material with high strain rate hardening can reduce the tendency of necking during stretching or stamping processes. In superplastic forming processes, the value of strain rate sensitivity parameter m is a key factor impacting the successful forming operation material as a metal with high m value can reduce localised thinning.

For hot stamping processes, for most metals it is often possible to choose values of temperature and strain rate to obtain high values of strain hardening and strain rate hardening. Thus both mechanisms can be used to form more complex-shaped panel components without localised failure. For superplastic forming processes, strain rate hardening is the main mechanism used for forming a complex-shaped part.

1.3.2 *Modelling of mechanical and physical properties of materials*

As stated above, the elastic-plastic, or, elastic-viscoplastic (flow stress) response of materials is related to microstructure, which changes dynamically during deformation, especially in high temperature forming conditions. Also, a change of microstructure may continue during an interval between two deformation operations. For example, recrystallisation may continue during the period between passes in hot rolling, although no deformation is happening. The physical properties of materials, which may change during forming, include:

a. Recrystallisation

This includes the nucleation and growth of recrystallised grains during static and dynamic conditions and recrystallisation cycles. Normally, recrystallisation takes place, if the following three conditions are met:

(i)   sufficient temperature – recrystallisation does not occur at room temperature for most metals;

(ii) sufficient dislocation density – if the dislocation density is too low, recrystallisation would not take place. Dislocations are generated due to plastic deformation and the amount depends on temperature and deformation rate; and,

(iii) sufficient time – incubation time is related to temperature and dislocation density.

If we use a variable X to represent the recrystallised volume fraction, which varies from 0 (initial state – no recrystallisation) to 1.0 (fully recrystallised), the evolution of recrystallisation can be a function such as:

$$\text{recrystallisation rate, } \dot{X} = \frac{dX}{dt} = f_X\left(T, \rho, ...\right) \tag{1.13}$$

where T is temperature and ρ is dislocation density, which is a function of plastic deformation, and possible recovery in hot forming conditions. The dislocation density accumulation rate is a function of

$$\text{dislocation accumulation rate, } \dot{\rho} = \frac{d\rho}{dt} = f_\rho\left(T, \dot{\varepsilon}, \rho, ...\right) \tag{1.14}$$

Within the equations "...." represents the many other factors that may be involved. They depend on metal composition, type of processing route and deformation conditions.

The introduction of *state variables* in equations to represent the evolution of individual physical phenomena of a metal during deformation, results in *phenomenal constitutive equations*, which have significant physical meaning. Sometimes, they are known as *physically-based constitutive equations*.

b. Grain size evolution

Grain refinement takes place due to recrystallisation. Static and dynamic grain growth occurs under hot forming conditions, such as superplastic forming (Lin, 2002). If the grain size is represented by d, we can write the following constitutive equation:

$$\text{grain size evolution, } \dot{d} = \frac{dd}{dt} = f_d\left(T, \dot{\varepsilon}, X, d, ...\right) \tag{1.15}$$

Grain growth rate increases significantly with temperature increase and reduces with increase in grain size.

c. Precipitate evolution

Precipitates are particularly important to the mechanical properties of light alloys and superalloys. Proper control of the shape, size and distribution of precipitates is important to achieve designed properties of an alloy. Static and dynamic dissolution of existing precipitates may take place during heating, soaking and deformation. Precipitates can be nucleated and grown during cooling. Precipitate dissolution, nucleation and growth are particularly important in assessing the mechanical properties of welded structures, since they are very difficult to control. State variables can be introduced to represent precipitate shape, size or distribution. Here, for convenience, S is used to present the average size of individual precipitates and the constitutive equation can be in the form:

$$\text{precipitate size evolution, } \dot{S} = \frac{dS}{dt} = f_S\left(T, \dot{T}, \rho, S, ...\right) \tag{1.16}$$

High dislocation density ρ encourages precipitates to be nucleated more quickly.

d. Phase transformation

This occurs in alloys and is particularly applicable to steels. As temperature increases, the phase can be changed to austenite and as temperature decreases, the austenite can be transformed to ferrite, pearlite, bainite and/or martensite depending on cooling rate and alloy composition. A phase transformation will cause a volume change of an alloy, thus creating extra strain. This is known as phase transformation plasticity and is particularly important for the prediction of residual stresses and deformation of post formed parts. This is a complicated process in the modelling of the phenomena, which is not discussed here but will be detailed in a later chapter, following an explanation of phases and transformation mechanisms.

1.3.3 Unified constitutive equations and state variables

a. Unified equations

As stated above, the mechanical response of a metal is related to its microstructure, which changes dynamically during deformation. In addition, a change in microstructure alters the dominant deformation mechanism, thus altering the mechanical response of a metal as signified by its stress–strain relationship. Thus a set of constitutive equations is required to represent the interactive response of a metal with the various process conditions imposed on it. For example, the following set of equations can be formulated:

plastic strain rate, $\quad \dot{\varepsilon}^{p} = f_{\varepsilon}\left(T,\rho,d,....\right)$

recrystallisation rate, $\quad \dot{X} = \dfrac{dX}{dt} = f_{X}\left(T,\rho,....\right)$

dislocation accumulation rate, $\dot{\rho} = \dfrac{d\rho}{dt} = f_{\rho}\left(T,\dot{\varepsilon}^{p},\rho,d,....\right)$ $\qquad$ (1.17)

grain size evolution, $\quad \dot{d} = \dfrac{dd}{dt} = f_{d}\left(T,\dot{\varepsilon}^{p},X,d,....\right)$

..... .

stress rate, $\quad \dot{\sigma} = E(\dot{\varepsilon}^{T} - \dot{\varepsilon}^{p})$, where $\dot{\varepsilon}^T$ and $\dot{\varepsilon}^p$ are total and plastic strain rates respectively. This set of constitutive equations is known as *unified constitutive equations*. These evolutionary equations are interdependent. For example, the plastic flow of the metal is related to grain size and dislocation density (strain hardening). The grain size is related to recrystallisation and others. Recrystallisation is related to dislocation density, which is related to plastic deformation rate and grain size as well. It can be seen that the interactive relation of the physical variables of the metal can modelled within the equation set. If more physical variables, such as precipitates for superalloys, phase transformation for steels etc. need to be considered, more equations can be included. Techniques for formulating constitutive equations are detailed in the rest of the book for different materials and applications.

It is assumed that all the physical phenomena evolve with time during a deformation/forming process. Thus the equations are a set of *ordinary differential equations* (ODE) in terms of time (t) and are normally solved using a numerical integration method; time is the unique scale for all the physical parameters. Thus by integrating the set of unified constitutive equations, particular physical problems can be solved. To solve the equation set, initial values for individual state variables must be given, known as the *initial value problems* in numerical integration. To simplify a problem, we suggest that only dominant or important physical phenomena be considered within the equation set, so that the number of equations can be minimised. The number of equations used depends on the particular problem to be solved.

b. State variables

Physical phenomena of materials under plastic deformation conditions can be described by a series of variables, as discussed above. These state variables can be divided into three types:

- *Observable variables*: including temperature T; the total strain from testing ε^T (the measured strain); physical properties of materials, such as recrystallised volume fraction X, grain size d; damage of materials ω. These evolve in the plastic or viscoplastic deformation of materials.

- *Internal variables*: including elastic strain ε^e and plastic strain ε^p (which cannot be measured from elastic-plastic (or viscoplastic deformation); hardening variables, such as isotropic hardening, kinematic hardening, which are determined from the fitting experimental data (to be discussed in more details in latter chapters).

- *Associated variables:* including stress tensors, σ, which are calculated from measured variables.

The state variables are summarised in **Table 1.1**. For some parameters, they are difficult to define. For example, micro damaging of material during plastic/viscoplastic deformation could be measured according to the void volume fraction, which is carried out in superplastic forming applications (Lin and Dean, 2005). However, for most cases, it is treated as an internal variable and determined from best fitting of experimental data, such as creep rupture problems (Lin, Hayhurst and Dyson, 1993a; 1993b), and calculated from evolutionary equations. Many of the physical variables can be defined as internal variables as well since real measurements are not often carried out during testing due to time constraints and high costs.

Table 1.1 Types of state variables.

Observable variables	Internal variables	Associated variables
Temperature; Total strain; Physical properties Damage.	Elastic strain; Plastic strain; Hardening variables; Possibly damage parameters.	True stresses; Deviatoric stresses.

The aim of this book is to introduce fundamental knowledge of the *mechanics and materials* necessary for the formulation of unified constitutive equations that enable the interaction between mechanical responses and physical properties of materials to be predicted for defined deformation/forming conditions. Techniques are also introduced for solving the equations and for determining the equations from experimental data.

Chapter 2

Mechanics of Metal Deformation

The aim of this chapter is to give a brief introduction to the physics, metallurgy and structure of metals, and the basic mechanics theories used to characterise metals for developing mathematical models. This is necessary for the fundamental understanding of deformation mechanisms of metals under different loading conditions, and thus for the formulation of mechanism-based constitutive equations. Firstly, structures of crystallised materials are introduced, and then dominant mechanisms for plastic deformation of metals are presented. Finally in this chapter, the basic concept of mechanics, including elastic and plastic theories, is introduced, which is related to various metal forming applications.

2.1 Crystallised Materials and Plastic Deformation

2.1.1 *Crystallised materials*

a. Atoms

Metals and their alloys are made of ordered arrangements of atoms held together by electromagnetic forces among the electrons of neighbouring atoms. As shown in **Fig. 2.1** the size or "radius" of an atom varies from 10^{-10} metres to 10^{-9} metres (0.1 nm–1 nm). Stable arrangements of atoms are determined by a minimum energy condition of the atom packing, which is a function of thermal activation. **Figure. 2.1**

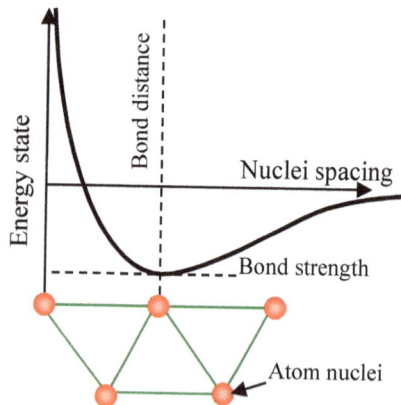

Fig. 2.1 Atoms joined together by electromagnetic forces.

shows that the energy state is a function of the distance between two atoms. The highest bonding strength exists at the minimum energy condition. In metals, the bonds between atoms are due to the sharing of electrons in their outer shells, (i.e. a "metallic bond").

At a microscopic scale, metals are found to consist of closely packed crystals. **Figure 2.2** shows a structural component in different lengthscales, from millimetre to micrometre to nanometre. Typical arrangements of atoms in crystals are described below.

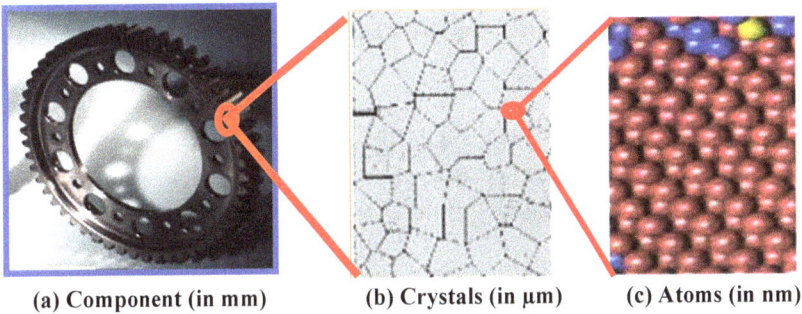

| (a) Component (in mm) | (b) Crystals (in μm) | (c) Atoms (in nm) |

Fig. 2.2 A metallic component shown in different lengthscales (From Professor H. Dong of Birmingham University).

b. Lattice structure and crystals

In solid-state metals, atoms arrange themselves into various orderly configurations known as crystals. The smallest group of atoms showing the characteristic *lattice structure* of a particular metal is known as a *unit cell*. A *single crystal*, also known as a *grain*, normally contains many unit cells. There are three types of basic atomic arrangements for unit cells:

- **Body centred cubic (BCC)** – This lattice structure is shown in **Fig. 2.3**, where each sphere represents an atom. The distance between the centre of atoms in the crystal structure is in the order of ($a =$) 0.1 nm. Each atom in the BCC crystal structure has eight neighbouring atoms. Alpha iron, chromium, tantalum and tungsten have the BCC structure.

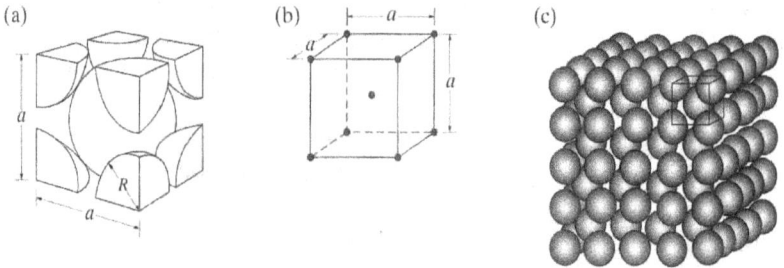

**Fig. 2.3 BCC crystal structure. (a) & (b) unit cell, and, (c) single crystal
with many unit cells (Moffatt, Pearsall and Wulff, 1976).**

- **Face centred cubic (FCC)** – This lattice structure is shown in
  Fig. 2.4 and consists of one atom on each plane. Gamma iron,
  aluminium, copper, nickel, lead, silver, gold and platinum have the
  FCC structure.

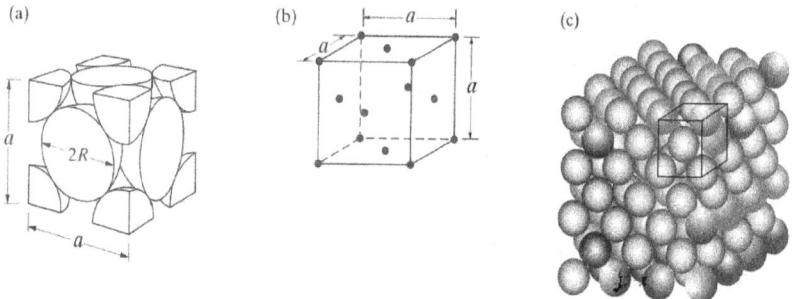

**Fig. 2.4 FCC crystal structure. (a) & (b) unit cell, and, (c) single crystal
with many unit cells (Moffatt, Pearsall and Wulff, 1976).**

- **Hexagonal close-packed (HCP)** – This lattice structure is shown
  in **Fig. 2.5**. In the HCP structure, the top and bottom planes are
  known as *basal planes*. Beryllium, cadmium, cobalt, magnesium,
  alpha-titanium, zinc and zirconium have the BCC structure.

The reason that metals form different crystal structures is to minimise
the energy required to fit together in a regular pattern. At different
temperatures the same metal may form different crystal structures due to
different energy requirement at a particular temperature. For example,

iron forms a BCC structure (alpha iron) below 912 °C but forms an FCC structure (gamma iron - austenite) between 912 °C to 1394 °C.

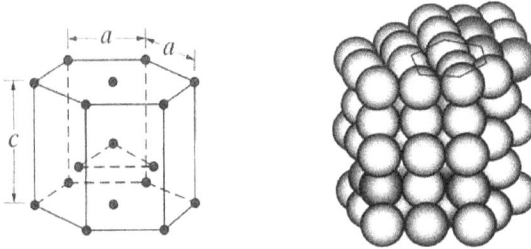

Fig. 2.5 HCP crystal structure. The unit cell, and, a single crystal with many unit cells (Moffatt, Pearsall and Wulff, 1976).

c. Lattice structure of steel

Steel is the most commonly used metal and its lattice structure is related to chemical compositions and temperatures. Commercially pure iron contains up to 0.008% carbon, steel contains up to 2.11% carbon, and cast iron contains up to 6.67% carbon. In this section the iron and carbon system is introduced and we will see how temperature and carbon affect the lattice structure of steel.

The iron–iron-carbide phase diagram is shown in **Fig. 2.6**. The diagram can be extended to the right to 100% carbon, which is pure graphite, but the range that is significant to engineering applications is up to 6.67%, since Fe_3C (cementite) is a stable phase.

- **Ferrite** – Alpha ferrite has *BCC crystal structure* and low solubility of carbon – up to 0.025% at 723 °C. α-ferrite, which exists at room temperature, is very important in engineering applications. δ-ferrite is stable only at very high temperatures; the maximum concentration of carbon in δ-ferrite is 0.09% at 1493 °C, as shown in **Fig. 2.6**. Ferrite is relatively soft and ductile – easy to form, since its solubility for carbon is very limited.

- **Austenite** – is known as γ-iron and has an *FCC structure* which has a solid solubility of up to 2.06% carbon at 1148 °C. Since the FCC structure has more interstitial positions, the solubility of austenite is much higher than that of ferrite, with the carbon atom occupying the interstitial positions, as shown in **Fig. 2.7(a)**. Austenite is an important phase in the heat treatment of steels. It is

denser than ferrite and its single phase of FCC structure is ductile
at elevated temperatures. Normally it possesses *excellent
formability*. Many other alloys can be dissolved in the FCC iron to
improve various properties of steel.

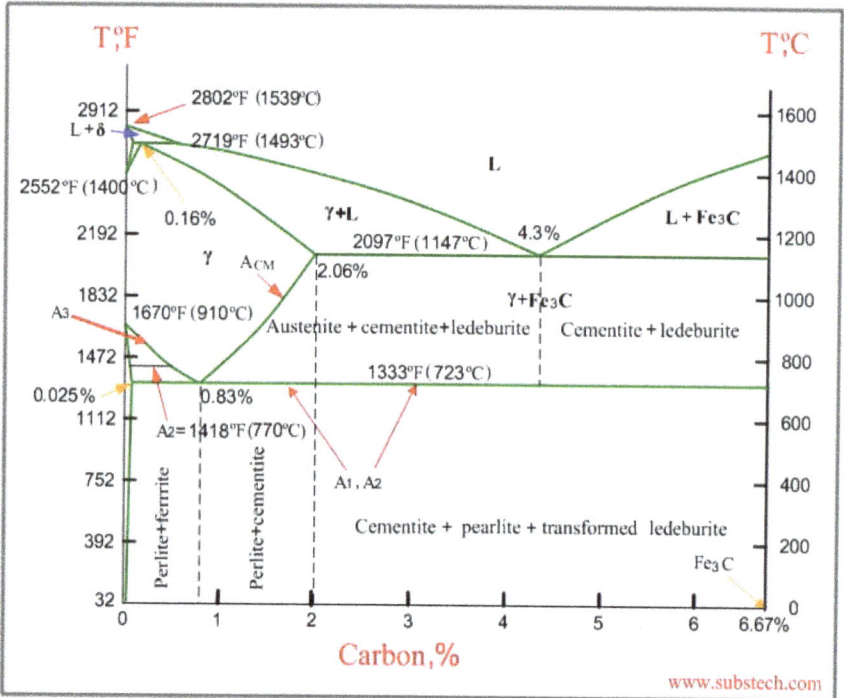

Fig. 2.6 Fe–Fe₃C Phase Diagram (After: www.substech.com/).

- **Martensite** – is formed when austenite is cooled at high rate, such
  as by quenching it in water; its FCC structure is transformed into a
  body centred tetragonal (BCT) structure. This structure can be
  described as a body centred rectangular prism which is slightly
  elongated along one of its principal axes in **Fig. 2.7(d)**. Martensite
  does not have as many slip systems (the concept of slip systems
  will be introduced latter) as BCC structure, thus it is very hard and
  brittle. Martensite transformation takes place almost instantly
  because it involves a slip mechanism (plastic deformation), not a
  time-dependent diffusion process. As austenite transforms to

martensite, its volume increases (its density decreases) by as much as 4%, which may cause quench cracking.

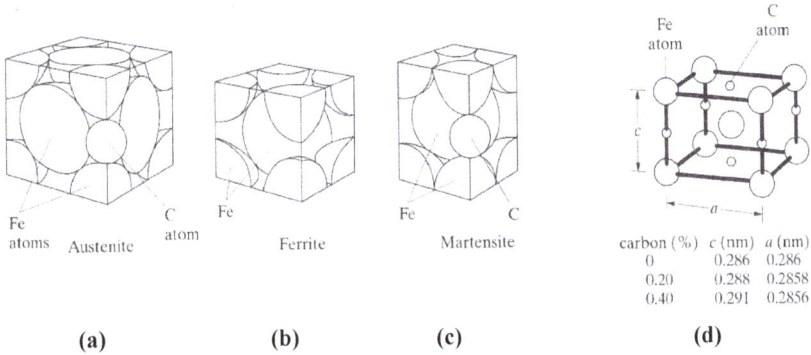

carbon (%)	c (nm)	a (nm)
0	0.286	0.286
0.20	0.288	0.2858
0.40	0.291	0.2856

(a)　　　　　(b)　　　　　(c)　　　　　(d)

Fig. 2.7 The lattice structure for (a) austenite, (b) ferrite, (c) martensite and (d) the effect of percentage of carbon (by weight) on the lattice structure on martensite. The dimension c increases with increasing carbon content. This effect causes the unit cell of martensite to be in the shape of a rectangular prism (Kalpakjian and Schmid, 2001).

- **Cementite** – is 100% iron carbide (Fe_3C), having a carbon content of 6.67%, as shown in **Fig. 2.6**. Cementite, also known as *carbide*, is a very hard and brittle inter-metallic compound, and has a significant influence on properties of steel. It may also include other alloying elements.

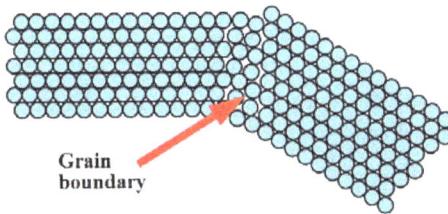

Fig. 2.8 Grains with grain boundary.

d. Crystal grains

It would be misleading to suppose that all the atoms' unit cells in a piece of metal are arranged in a regular way. Most commonly used metals are made up of a large number of randomly orientated *crystals*, which are also known as *grains*. Grains are regions of regular unit cells with a common orientation, as shown in **Fig. 2.8.** At grain boundaries, atoms are misaligned which creates a

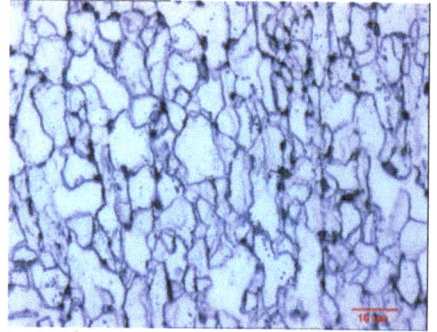

Fig. 2.9 Grain boundaries of a polycrystal structure for a pipeline steel.

high density of dislocations (the concept of dislocations will be discussed later). Thus we are dealing with metal structures that are not single crystals, but polycrystals in general, as shown in **Fig. 2.9.**

2.1.2 *Plastic deformation in single crystals*

a. The concept of elastic and plastic deformation

Elastic deformation. If a small stress is put onto the metal, the layers of atoms will start to roll over each other. If the stress is released again, they will fall back to their original positions. Under these circumstances the metal is said to be *elastic* as shown in **Fig. 2.10.**

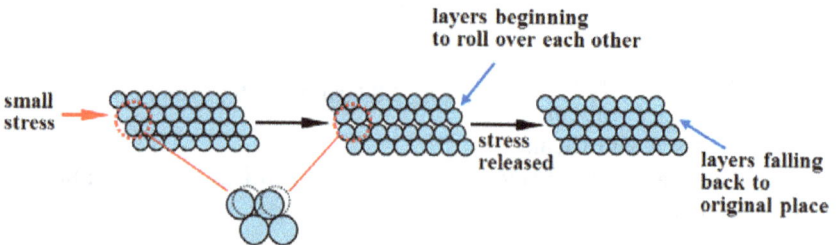

Fig. 2.10 Elastic deformation and elastic recovery.

Plastic deformation. If a larger stress is applied, the atoms roll over each other into a new position and the shape of the metal is permanently changed. This is known as *plastic* deformation and is shown in **Fig. 2.11**.

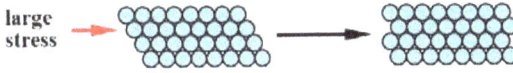

Fig. 2.11 Plastic (permanent) deformation.

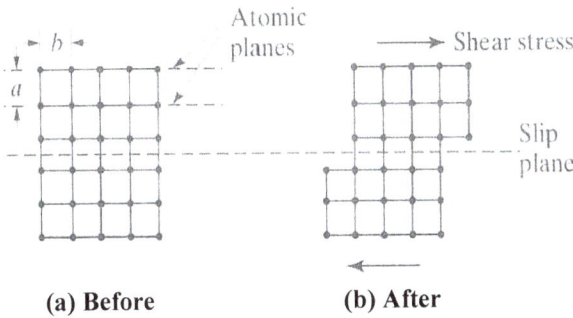

(a) Before   (b) After

Fig. 2.12 A single crystal (a) before and (b) after permanent deformation. (Kalpakjian and Schmid, 2001).

b. Slip systems

The slipping of one plane of atoms over an adjacent plane (slip plane) under a shear stress is shown in **Fig. 2.12**. The shear stress (the definition of shear stress is given in the next section) required to cause *slip* in single crystals is directly proportional to the ratio b/a in **Fig. 2.12**, where a is the spacing of the atomic plane, and b is inversely proportional to the atomic density in the atomic plane. As b/a decreases, the shear stress required to cause slip decreases. Thus, we state that slip in a crystal takes place along planes of maximum atomic density, or, in other words, that slip takes place in closely packed planes and in closely packed directions.

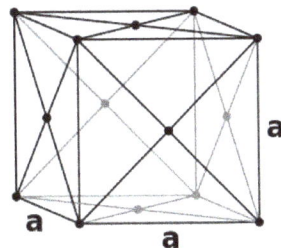

Fig. 2.13 12 slip systems for FCC.

Because the b/a ratio is different for different directions within the crystal, a single crystal has different properties in different directions, which is known as *anisotropy*.

For **FCC crystals** there are 12 slip systems on the 4 planes shown in **Fig. 2.13**. The probability of slip is moderate, but the required shear force to cause slip is low due to the relatively low value of b/a ratio (here $b = a$). These metals normally have moderate strength and good ductility, such as copper.

In **BCC crystals**, there are 48 possible slip systems. The probability of slip is high because the direction of an applied stress is likely to lie on one of the preferred slip directions and the externally applied shear stress operates on one of these systems and causes slip. Due to the relatively high b/a ratio in the crystals, however, the required shear stress is high. Metals with BCC structures generally have good strength and moderate ductility.

HCP crystals have three slip systems and they thus have a low probability of slip. However, more slip systems become active at elevated temperatures. Metals with HCP structures, for example magnesium alloys, are generally brittle at room temperature, thus the cold formability is very low.

c. Slip lines in single crystals

If a tensile stress is applied to a single crystal the maximum shear stress occurs on a plane at 45° to it. Deformation will be initiated on the crystal slip plane most closely oriented to the maximum shear stress direction. The deformation becomes plastic, if the tensile

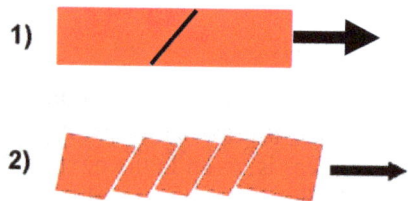

Fig 2.14 Slip in plastic deformation.

stress is sufficiently high. This phenomenon is shown in **Fig. 2.14**. It can also be seen that the portions of the single crystal that has slipped have rotated away from their original angular position toward the direction of the tensile force. Note that slip has taken place only along certain planes. With the use of a scanning electron microscope (SEM) it can be seen that what appears to be a single slip plane is actually a *slip band*, consisting of a number of slip planes (**Fig. 2.15**).

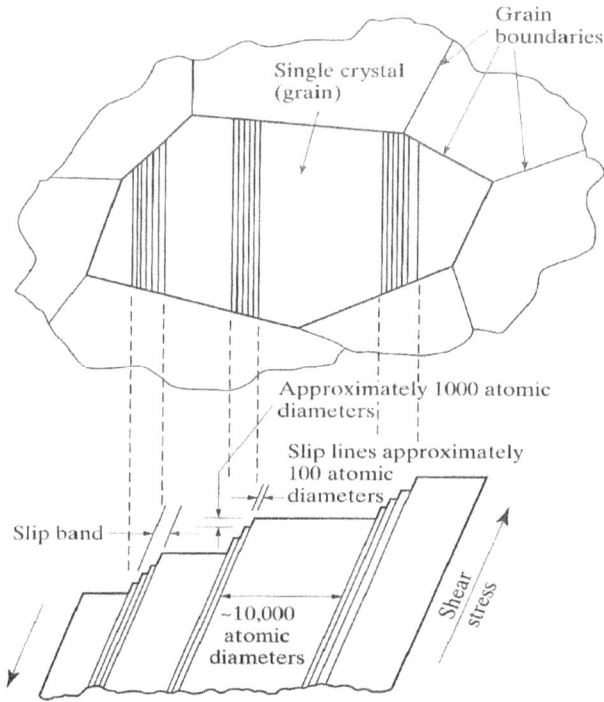

Fig. 2.15 Illustration of slip lines and slip bands in a single crystal (grain) subjected to a shear stress. The crystal shown in the top figure is an individual grain surrounded by many other grains (Kalpakjian and Schmid, 2001).

2.2 Dominant Deformation Mechanisms

The basic structures and deformation mechanisms in metals are introduced in Section 2.1. However, in metal forming processes, a metal with a particular microstructure, deforming under a particular process condition, deforms plastically with a specific dominant mechanism. This can lead to failure, which will be discussed in this section.

2.2.1 *Plastic deformation in cold forming*

Cold forming takes place at temperatures which are relatively low as compared to the material's absolute melting temperature in Kelvin, T_m, i.e. typically less than $0.3T_m$. Normally cold forming is carried out at room temperature.

a. Defects and Dislocations in metals

A crystalline material consists of a regular array of atoms, arranged into lattice planes, as discussed in the previous section. This is the ideal case. However, in reality, defects and imperfections exist in crystal structures. These cause the actual strength of metals to be one to two orders of magnitude lower than the theoretical intrinsic strength. Unlike the ideal models described earlier, actual metal crystals contain a large number of *defects* and *imperfections,* of which *dislocations* are the main kind. There are two types of dislocations, which are shown in **Fig. 2.16**.

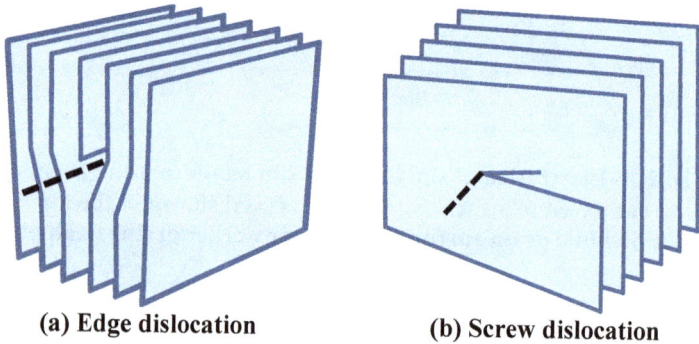

(a) Edge dislocation   (b) Screw dislocation

Fig. 2.16 Dislocation illustration.

An *edge dislocation*, shown in **Fig. 2.16(a)**, is a defect where an extra half-plane of atoms is introduced mid-way through a crystal, distorting nearby planes of atoms. When enough force is applied from one side of the crystal structure, this extra plane passes through planes of atoms breaking and joining bonds with them until it reaches the crystal boundary. The dislocation can be defined by two properties: a line direction, which is the direction running along the bottom of the extra half-plane, and the Burgers vector, **b**, which describes the magnitude and direction of distortion to the lattice. In an edge dislocation the Burgers vector is perpendicular to the line direction. The Burgers vector is shown in **Fig. 2.17**.

A *screw dislocation*, as shown in **Figs. 2.16 (b)** and **2.17(b)**, is much more difficult to visualise. It comprises a structure in which a helical path is traced around the linear defect (dislocation line) by the atomic planes in the crystal lattice. It may be envisaged that lattices on each side of the slip plane have been displaced relatively by a screwing action. In pure screw dislocations the Burgers vector is parallel to the line direction.

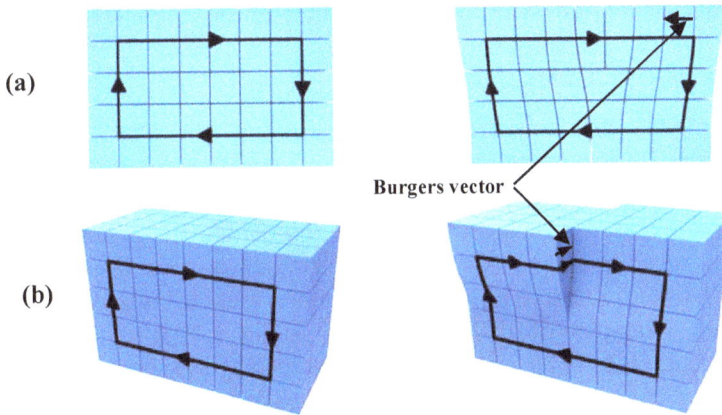

Fig. 2.17 Types of dislocations in single crystals: (a) edge dislocation and (b) screw dislocation.

Other defects in metals include a vacancy (missing atom), an interstitial atom (extra atom in lattice structure), impurity (alien atom introduced by imperfect control, usually in primary melting processes, to replace the atom of pure metal), voids or inclusions (nonmetallic elements, such as oxides, sulphides etc.), and grain boundary imperfections.

Dislocations are defects, inherent in crystals, that cause the discrepancy between the actual and theoretical strength of metals. A slip plane containing dislocations requires significantly less shear stress to allow slip than a plane in a perfect lattice.

Although the presence of dislocations decreases the shear stress required to cause slip, dislocations can:

- Become tangled and interfere with each other; and,
- Be impeded by barriers, such as grain boundaries, impurities, precipitates and inclusions.

Entanglement and the accumulation of dislocations around precipitates etc. increase the shear stress required for slip taking place. This increases the overall strength of the metal. This effect is known as *work hardening* or *strain hardening*. Large plastic deformation creates high dislocation density, thus increasing strength. Work hardening is used extensively for strengthening metals in cold metal forming processes, such as cold rolling, cold sheet metal forming and forging.

b. Plasticity in cold metal forming

When metals are cold formed, the dominant plastic deformation mechanism is dislocation movement due to high shear stress applied on the slip systems.

Accumulation of dislocations causes hardening of metals, thus the forming force increases during the forming process. Normally the strength of cold formed parts is increased due to work (or strain) hardening. Tangles of dislocations are found at the early stage of deformation and appear as non-well-defined grain boundaries. The process of *dynamic recovery* leads eventually to the formation of a cellular structure containing boundaries with misorientation lower than 15° (low angle grain boundaries).

An accumulation of dislocations can cause eventual failure of work-pieces during cold forming processes if deformation is too great. Thus the formability of metals is relatively low in cold forming. The accumulated dislocations can be removed by appropriate heat treatment (annealing), which promotes the *recovery* by *recrystallisation* of materials to gain the formability again. Therefore large amounts of deformation can be achieved by alternate forming and annealing operations.

2.2.2 Viscoplastic deformation in warm and hot metal forming

Warm metal forming takes place at temperatures which are above 0.3 T_m and below the recrystallisation temperature of materials. Hot metal forming takes place at temperatures above the solid-state recrystallisation temperature. To understand the dominant deformation mechanisms of a material under warm/hot forming processes, the following key features of metals need to be introduced first.

a. Grain boundary diffusion

At elevated temperatures the diffusion of atoms occurs in a material along grain boundaries where a high density of dislocation defects exists. Thus, plastic deformation takes place due to the diffusion of vacancies through their crystal lattice. Diffusion rate is sensitive to temperatures and stress conditions. Defects migrate to the crystal face along the direction of the maximum stress (Meyers and Chawla, 1999), as shown in **Fig. 2.18**. This is the typical deformation mechanism for high temperature creep of metals. This deformation mechanism causes grain boundary sliding to take place, resulting in plastic deformation. In hot forming processes, grain boundary diffusion normally is rapid.

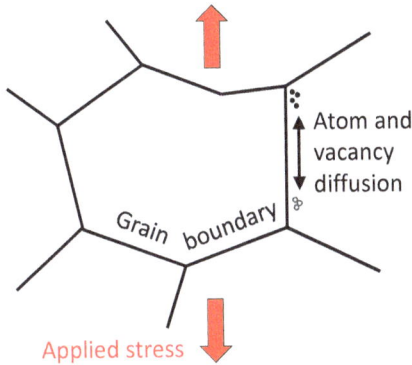

Fig. 2.18 Atom movement at grain boundaries.

b. Annihilation and rearrangement of dislocations

Each dislocation is associated with a strain field which contributes some small but finite amount to energy stored in the metal. When temperature is increased, dislocations become mobile and are able to glide, cross-slip and climb. If two dislocations of opposite directions (signs) meet, then they effectively cancel out and their contribution to the stored energy is removed. When annihilation is complete, dislocations of only one sign will remain.

After annihilation any remaining dislocations can align themselves into ordered arrays where their individual contribution to the stored energy is reduced by the overlapping of their strain fields. The simplest case is that of an array of edge dislocations of identical Burgers vector. The edge dislocations will rearrange themselves into tilt boundaries. Grain boundary theory predicts that an increase in boundary misorientation will increase the energy of the boundary but decrease the energy per dislocation. Thus, there is a driving force to produce fewer, more highly misoriented boundaries. The situation in highly deformed, polycrystalline materials is naturally more complex. Many dislocations with different Burgers vector can interact to form complex 2D networks.

c. *Recrystallisation, grain refinement and grain growth*

For elevated temperature forming conditions, a metal deforms and grain shape is distorted due to dislocation movement. The accumulated dislocations can be annealed dynamically and statically at high temperature, which reduces work hardening and increases ductility of metals. This mechanism, in which no significant change in microstructure occurs, is called recovery.

If accumulated dislocation density reaches a certain level and deformation temperature is sufficiently high, *recrystallisation* would take place given sufficient time, which enables new grains (small and dislocation-"free" as they are nucleated) to be created. This normally starts at grain boundaries where higher dislocation density exists.

Recrystallisation reduces dislocation density and may change the dominant deformation mechanism due to the grain refinement. Once recrystallisation is complete, the newly formed grains start to grow. The eventual size of grains is temperature- and time-dependent.

As shown in **Fig. 2.19**, the mechanical properties of a metal change with temperature and microstructure. The forming resistance reduces with temperature and ductility increases with temperature. Note that when temperature reaches a very high level, the ductility does not continue to increase, and thus the required forming load does not decrease further. This is due to the fact that grains grow very quickly at high temperature, and large grain size causes a change in deformation mechanisms from those described above and reduces the freedom of grain boundary sliding and grain rotation. This is discussed in detail in the following sections. Residual stress created in cold deformation reduces with temperature due to annealing (recovery) and recrystallisation.

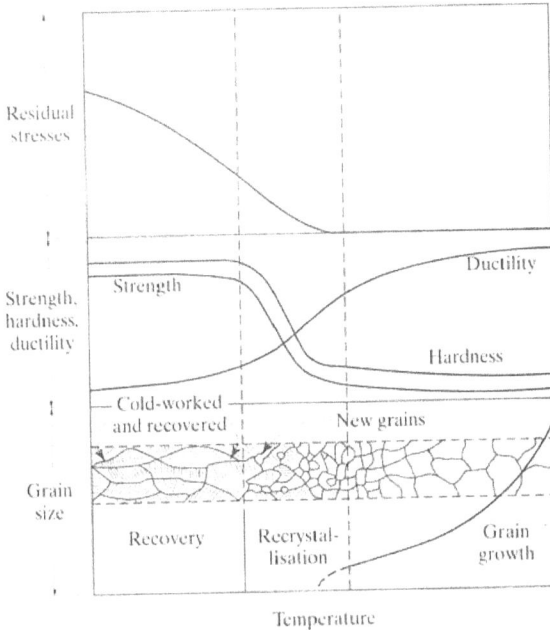

Fig. 2.19 Illustration of recovery, recrystallisation, grain refinement and grain growth and their effects on mechanical properties (Kalpakjian and Schmid, 2001).

d. Grain rotation and grain boundary sliding

In superplastic forming (SPF) a material with fine grain sizes is deformed at a slow rate in isothermal high temperature conditions. Grain boundary diffusion takes place quickly at the superplastic forming temperature. This favours grain boundary sliding (a creep-type deformation mechanism, which will be discussed later), but with a relatively high strain-rate. In addition, the fine-grained material promotes grain rotation due to the grain boundary diffusion. Thus, the dominant deformation mechanism for superplastic forming is grain boundary sliding and grain rotation.

To enable the grain boundary sliding and grain rotation mechanism to continue throughout the whole superplastic forming process, grain growth for the superplastic material under the forming temperature must be controlled. If grain growth takes place too quickly, the large grain size makes the grain rotation difficult. This may increase the necessary forming force/pressure during the forming process, or may cause premature failure of the work-piece. The acceptable forming rate can be increased by decreasing work-piece grain size.

e. Viscoplastic deformation

In warm/hot forming conditions there are many time-dependent mechanisms involved. For example, diffusion, recovery/annihilation, recrystallisation and grain growth are time-dependent parameters. These result in the force required for plastic deformation being time-dependent and for a given work-piece, deformation at elevated temperature is dependent on deformation speed (strain rate).

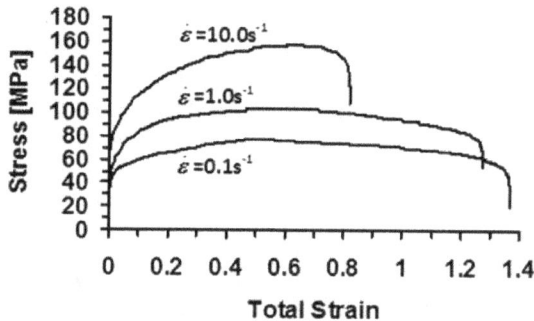

Fig. 2.20 Stress–strain relationships of a free cutting steel deformed at 1000 °C with strain rates of 0.1 s^{-1}, 1.0 s^{-1} and 10 s^{-1} (hot forming conditions).

Higher deformation rate is associated with the higher force, as shown in **Fig. 2.20**. This kind of plastic deformation is known as *viscoplasticity*, which is discussed in detail in Chapter 5.

f. Deformation mechanisms in warm/hot forming

As discussed above, deformation mechanisms arising in warm and hot metal forming are very complicated: they are dependent on the composition of a metal and microstructure, temperature and forming rate. Sometimes, the dominant deformation mechanism varies during a forming process since both temperature and microstructure may change.

For example, in a hot forming process, work-piece temperature may increase due to plastic deformation; in addition, the local work-piece temperature can decrease due to contact with cold dies. Also, dynamic recrystallisation may take place, which changes grain size.

The dominant deformation mechanisms in warm and hot forming are related to:

- *Temperature* – higher temperature facilitates grain boundary diffusion, thus causing more grain boundary sliding.

- *Deformation rate* – lower deformation rate allows more grain boundary diffusion to take place, thus causing more grain boundary sliding.

- *Grain size* – smaller grain size allows grain boundary sliding and promotes grain rotation.

Therefore, if the temperature is relatively low, deformation rate is high and grain size is large, then the dominant deformation mechanism is likely to be dislocation-based (similar to plastic deformation in cold forming conditions). It must be noted that in real warm/hot forming processes, the temperature and deformation rate usually vary from one location to another in a work-piece and change with time. This causes deformation mechanisms to vary throughout a work-piece, and during the processing time.

Note that at elevated temperatures, small-grain-sized material has lower strength and so can be deformed more easily. This is due to grain boundary diffusion, grain rotation and grain boundary sliding (such as the viscoplastic deformation in superplastic forming). However, at room temperature (cold forming conditions), small-grain-sized material possesses high strength and high ductility. This is due to the fact that there are more grain boundaries with high dislocations which make the deformation resistance high.

2.3 Stress and Strain Definitions

The uniaxial tension or compression test is the most common test for determining mechanical properties of metals, such as strength (yield stress and ultimate stress), ductility (strain to failure), elastic modulus, strain hardening etc. The tensile specimen can be designed in various shapes according to standards for individual applications. The basic information for stress and strain definitions are already given in

Chapter 1 (Section 1.3). **Figure 2.21** shows a typical stress–strain curve from a tensile test for a material.

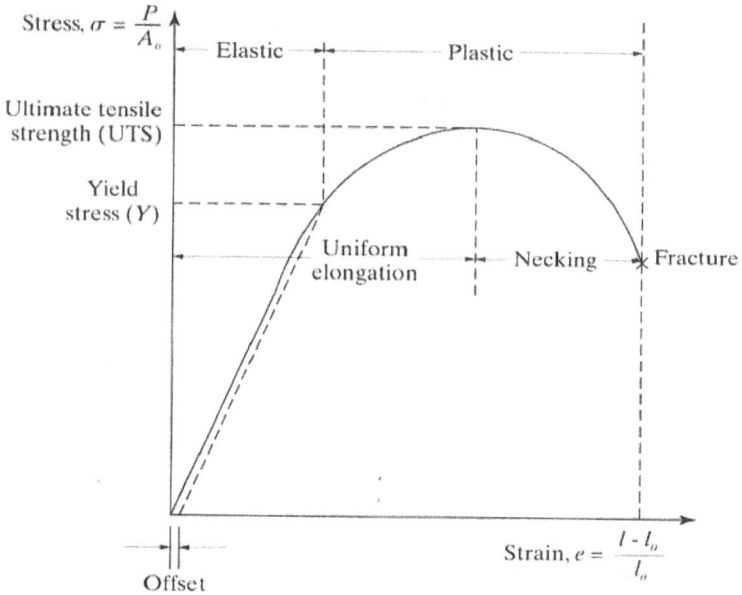

Fig. 2.21 A typical stress–strain curve showing different stages of deformation (Kalpakjian and Schmid, 2001).

When stress is first applied, the specimen elongates in proportion to the load. This is known as *linear elastic* behaviour. As the stress is increased to the level of Y (or represented as σ_y), known as the *yield stress (or tensile yield strength)*, the specimen begins to undergo *permanent (plastic) deformation*. Beyond that stress level, the stress and strain relationship is no longer proportional. As the limit of proportionality is often difficult to identify, yield stress is normally determined on the stress–strain curve by offsetting the linear region by a strain of 0.002 (or known as *proof stress*). This is shown in **Fig. 2.21**.

As deformation continues, stress increases nonlinearly due to the accumulation of dislocations, manifest as strain hardening, and its cross-sectional area decreases *permanently* and *uniformly* until the peak stress, known as *ultimate stress (ultimate tensile strength)*, is reached. If the specimen were unloaded below the yield stress it would return to its original shape. However, if the specimen is unloaded from the stress level higher than the yield stress, the curve follows a straight line downward and parallel to the original slope, as shown in **Fig. 1.1**. This is known as elastic recovery.

If the specimen is loaded beyond its tensile ultimate strength, it begins to *neck* and the cross-sectional area is no longer uniform along the gauge length of the specimen. As the deformation continues, fracture eventually takes place. The strain at this point is known as the strain to failure, ε_f. This is a measure of the ductility of the material.

Note that the above description is normally used for a material deforming at room temperature, i.e. cold deformation. At elevated temperatures, softening can take place at a very early stage of the deformation due to dynamic recrystallisation, microstructural evolution and static/dynamic recovery, and the cross-sectional area of the specimen remains uniform within the defined gauge length (uniform reduction of cross-sectional area of the specimen). The yield stress for warm/hot forming conditions is also very difficult to define and creep takes place at very low stress levels.

2.3.1 *Stresses*

a. 3D stress components

Figure 2.22 shows a body, assumed a continuum, deformed under forces F_1, F_2, F_3 etc. The 3D stresses acting on an infinitely small cube of material within the body are shown enlarged in the figure. In standard notation the first subscript numeral indicates the direction normal to the plane on which the force is acting and the second subscript numeral indicates in which direction the force is acting. Thus, σ_{11}, σ_{22} and σ_{33} are stresses parallel to the global orthogonal axes' (i.e. normal stresses) directions and the other stress components are shear stresses on corresponding planes. If the stresses related to the axis x_3 are zero, a *plane stress* state exists, i.e. the non-zero stresses are σ_{11}, σ_{12}, σ_{22} and σ_{21}, where the stresses $\sigma_{12} = \sigma_{21}$. Sheet metal forming processes are normally considered as plane stress deformation conditions since the stresses related to the thickness direction of a sheet are normally ignored.

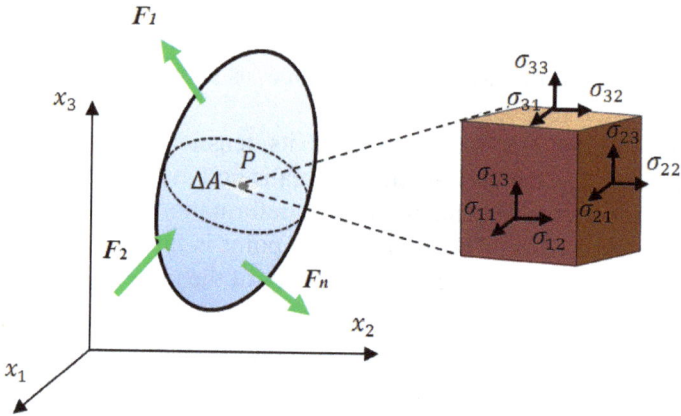

Fig. 2.22 3D Stress in a loaded deformable material body assumed as a continuum.

b. Stress and strain tensors

For a complex stress-state, stresses can be expressed as stress tensors, which can be written as:

$$\boldsymbol{\sigma} = \sigma_{ij} = \begin{bmatrix} \sigma_{11} & \sigma_{12} & \sigma_{13} \\ \sigma_{21} & \sigma_{22} & \sigma_{23} \\ \sigma_{31} & \sigma_{32} & \sigma_{33} \end{bmatrix}. \tag{2.1}$$

The stress tensor is also known as *Cauchy stress tensor (*Augustin-Louis Cauchy, French, 1789–1857), which is used for stress analysis of material bodies experiencing *small deformation*; it is a central concept in the *linear theory of elasticity*. For large deformation, also called *finite deformation*, other measures of stress are required, such as the *Piola–Kirchhoff stress tensor*, the *Biot stress tensor*, and the *Kirchhoff stress tensor*. These stress tensors are discussed later in this chapter.

Due to the symmetry of the matrices, Eq. (2.1) also can be written as

$$\sigma^T = [\sigma_{11}\ \sigma_{22}\ \sigma_{33}\ \tau_{12}\ \tau_{13}\ \tau_{23}],$$

where τ represents shear stresses.

Summation of stresses in a 3D (*x, y, z)* or (1, 2, 3) or (x_1, x_2, x_3) coordinate system:

$$x(1), \quad y(2), \quad z(3), \quad xy(12), \quad xz(13), \quad yz(23)$$

$$\sigma_{11}, \quad \sigma_{22}, \quad \sigma_{33}, \quad \sigma_{12}, \quad \sigma_{13}, \quad \sigma_{23}$$

$$\sigma_x, \quad \sigma_y, \quad \sigma_z, \quad \tau_{xy}, \quad \tau_{xz}, \quad \tau_{yz}$$

$$\tau_{12,} \quad \tau_{12}, \quad \tau_{12}$$

These notations are commonly used in mechanics books and technical literature.

c. *Principal stresses and invariant of stress tensors*

Principal stresses σ_1, σ_2 and σ_3 are defined in the planes where the shear stresses are zero, and are perpendicular to each other in a 3D isotropic continuum. An algebraic theorem states that the principal stresses can be solved from the following equations

$$\det\left(\mathbf{\sigma} - \sigma\delta_{ij}\right) = \left|\sigma_{ij} - \sigma\delta_{ij}\right| = 0 ,$$

where δ_{ij} is the Kronecker delta, which is defined as

$$\delta_{ij} = \begin{vmatrix} 1 & 0 & 0 \\ 0 & 1 & 0 \\ 0 & 0 & 1 \end{vmatrix} = \begin{cases} 1 & i = j \\ 0 & i \neq j \end{cases} .$$

Then the equation can be written as

$$\left|\sigma_{ij} - \sigma\delta_{ij}\right| = \begin{vmatrix} (\sigma_{11} - \sigma) & \sigma_{12} & \sigma_{13} \\ \sigma_{21} & (\sigma_{22} - \sigma) & \sigma_{23} \\ \sigma_{31} & \sigma_{32} & (\sigma_{33} - \sigma) \end{vmatrix} = 0 \qquad (2.2)$$

The value of the determinant of the 3×3 matrix is given by subtracting the products of the left hand downward diagonal terms from the right hand downward diagonal terms, thus the solution is

$$(\sigma_{11} - \sigma)(\sigma_{22} - \sigma)(\sigma_{33} - \sigma) + \sigma_{12}\sigma_{23}\sigma_{31} + \sigma_{13}\sigma_{21}\sigma_{32} - (\sigma_{11} - \sigma)\sigma_{23}^2$$
$$-(\sigma_{33} - \sigma)\sigma_{12}^2 - (\sigma_{22} - \sigma)\sigma_{13}^2 = 0 \qquad (2.3)$$

This is the cubic equation for which we are seeking, i.e. the *principal stress cubic*. After collecting terms and rearranging, this can be written as

$$\sigma^3 - (\sigma_{11} + \sigma_{22} + \sigma_{33})\sigma^2 + (\sigma_{11}\sigma_{22} + \sigma_{22}\sigma_{33} + \sigma_{33}\sigma_{11} - \sigma_{12}^{\ 2} - \sigma_{23}^{\ 2} - \sigma_{31}^{\ 2})\sigma$$
$$- (\sigma_{11}\sigma_{22}\sigma_{33} + 2\sigma_{12}\sigma_{23}\sigma_{31} - \sigma_{11}\sigma_{23}^{\ 2} - \sigma_{22}\sigma_{31}^{\ 2} - \sigma_{33}\sigma_{12}^{\ 2}) = 0$$

$$(2.4)$$

or

$$\sigma^3 - J_1\sigma^2 + J_2\sigma - J_3 = 0 \ , \tag{2.5}$$

where

$$J_1 = \sigma_{ii} = \sigma_{11} + \sigma_{22} + \sigma_{33} \tag{2.6a}$$

$$J_2 = \frac{1}{2}\left(\sigma_{ii}\sigma_{kk} - \sigma_{ik}\sigma_{ki}\right)$$
$$= \sigma_{11}\sigma_{22} + \sigma_{22}\sigma_{33} + \sigma_{33}\sigma_{11} - (\sigma_{12}^{\ 2} + \sigma_{23}^{\ 2} + \sigma_{31}^{\ 2}) \tag{2.6b}$$

$$J_3 = \frac{1}{6}\left(\sigma_{ii}\sigma_{jj}\sigma_{kk} + 2\sigma_{ij}\sigma_{jk}\sigma_{ki} - 3\sigma_{kk}\sigma_{ij}\sigma_{ji}\right)$$
$$= \sigma_{11}\sigma_{22}\sigma_{33} + 2\sigma_{12}\sigma_{23}\sigma_{31} - \sigma_{11}\sigma_{23}^{\ 2} - \sigma_{22}\sigma_{31}^{\ 2} - \sigma_{33}\sigma_{12}^{\ 2} \tag{2.6c}$$

Equation (2.5) has three real roots $\sigma = \sigma_1$, σ_2 and σ_3, where it is the convention to have $\sigma_1 > \sigma_2 > \sigma_3$. This is a very important equation since it carries the implication that the coefficients J_1, J_2 and J_3 must be *invariants*. That is to say, *for a given state of stress in a body,* however the original set of axes x_1, x_2 and x_3 are orientated in the body, the cubic equation must give the same values of σ_1, σ_2 and σ_3. Therefore, the values of the coefficients derived from any coordinate system must be the same. Thus, there are

$$J_1 = \sigma_{11} + \sigma_{22} + \sigma_{33} = \sigma_1 + \sigma_2 + \sigma_3$$

$$J_2 = \sigma_{11}\sigma_{22} + \sigma_{22}\sigma_{33} + \sigma_{33}\sigma_{11} - (\sigma_{12}^{\ 2} + \sigma_{23}^{\ 2} + \sigma_{31}^{\ 2})$$
$$= \sigma_1\sigma_2 + \sigma_2\sigma_3 + \sigma_3\sigma_1$$

$$J_3 = \sigma_{11}\sigma_{22}\sigma_{33} + 2\sigma_{12}\sigma_{23}\sigma_{31} - \sigma_{11}\sigma_{23}^{\ 2} - \sigma_{22}\sigma_{31}^{\ 2} - \sigma_{33}\sigma_{12}^{\ 2}$$
$$= \sigma_1\sigma_2\sigma_3$$

The shear stresses are zero on the principal stress planes. Methods of solving the principal stress cubic and the orientations of the principal stress planes will not be discussed in this book. However, these considerations will be raised again when discussing the mathematical theory of plasticity.

d. Hydrostatic stress

Hydrostatic stress is also known as *mean stress*. That is a stress which acts equally in all directions on an element and which will therefore cause only a change of volume, which, as yield cannot arise under a pure compressive hydrostatic stress state, is recoverable on removal of the stress. This is expressed as

$$\sigma_H = \frac{J_1(\sigma)}{3} = \frac{\sigma_{ii}}{3} = \frac{1}{3}(\sigma_{11} + \sigma_{22} + \sigma_{33})$$. (2.7)

For the case of simple *uniaxial tension*, we have

$$\sigma_{11} = \sigma_1 = \sigma_H/3 \text{, and } \sigma_{22} = \sigma_{33} = \sigma_2 = \sigma_3 = 0$$.

Hydrostatic stress has been used widely to assess the damage initiation and growth of a material within plastic/viscoplastic deformation conditions. For example, it could be assumed that when $\sigma_H < 0$, it is compressive forming and no micro-damage is initiated.

e. Deviatoric stresses

In writing 3D constitutive laws, the stress tensor can be represented by a summation of two other stress tensors: a mean volumetric stress tensor, and a stress deviator tensor **S**, i.e

$$\sigma_{ij} = S_{ij} + \frac{1}{3}\delta_{ij}\sigma_{kk}$$.

Thus, *stress deviators*, S_{ij}, are defined by

$$S_{ij} = \sigma_{ij} - \frac{1}{3}\delta_{ij}\sigma_{kk} \text{, or } \mathbf{S} = \mathbf{\sigma} - \frac{1}{3}J_1(\sigma)\delta_{ij} = \mathbf{\sigma} - \frac{1}{3}\mathbf{Tr}(\sigma)\delta_{ij}$$, (2.8)

or

$$S_{ij} = \sigma_{ij} - \frac{1}{3}(\sigma_{11} + \sigma_{22} + \sigma_{33})\delta_{ij}$$,

or

$$S_{ij} = \begin{bmatrix} \dfrac{2\sigma_{11}-\sigma_{22}-\sigma_{33}}{3} & \sigma_{12} & \sigma_{13} \\[2mm] \sigma_{21} & \dfrac{2\sigma_{22}-\sigma_{11}-\sigma_{33}}{3} & \\[2mm] \sigma_{31} & \sigma_{32} & \dfrac{2\sigma_{33}-\sigma_{22}-\sigma_{11}}{3} \end{bmatrix}.$$

The plastic deformation of a continuum is directly caused by the *deviatoric stresses*. Since S_{ij} is a second-order tensor, it has principal axes. The principal values of the stress deviators are the root of the cubic equation (similar to Eq. (2.5)) and the invariants of the deviatoric stress tensor can be expressed as

$$\begin{aligned} J_1' &= S_{ii} = S_{11} + S_{22} + S_{33} \\ &= (\sigma_{11}-\sigma_H)+(\sigma_{22}-\sigma_H)+(\sigma_{33}-\sigma_H)=0 \end{aligned} \tag{2.8a}$$

$$\begin{aligned} J_2' &= \frac{1}{2}\left(S_{ii}S_{kk}-S_{ik}S_{ki}\right) \\ &= S_{11}S_{22}+S_{22}S_{33}+S_{33}S_{11}-(\sigma_{12}^2+\sigma_{23}^2+\sigma_{31}^2) \\ &= \frac{1}{2}\left[(\sigma_{11}-\sigma_{22})^2+(\sigma_{22}-\sigma_{33})^2+(\sigma_{33}-\sigma_{11})^2+6(\sigma_{12}^2+\sigma_{23}^2+\sigma_{31}^2)\right] \\ &= \frac{1}{2}\left[(\sigma_1-\sigma_2)^2+(\sigma_2-\sigma_3)^2+(\sigma_3-\sigma_1)^2\right] \end{aligned} \tag{2.8b}$$

The third invariant, J_3', is the determinant of the deviatoric stress tensor. The relationship between the second invariants of stress tensor (J_2) and deviatoric stress tensor (J_2') is

$$J_2' = 3\sigma_H^2 - J_2.$$

For the uniaxial tension/compression case, the deviatoric stress tensor, which causes the plastic deformation of materials, is

$$\mathbf{S} = \begin{bmatrix} \dfrac{2}{3}\sigma_{11} & 0 & 0 \\[2mm] 0 & -\dfrac{1}{3}\sigma_{11} & 0 \\[2mm] 0 & 0 & -\dfrac{1}{3}\sigma_{11} \end{bmatrix},$$

where $S_{12} = \sigma_{12} = 0$, $S_{13} = \sigma_{13} = 0$ and $S_{23} = \sigma_{23} = 0$.

f. von-Mises stress

The *von-Mises stress*, a scalar, is also known as *equivalent stress* and defined as the second invariant of deviatoric stress tensor (J_2').

$$\sigma_e = \sqrt{J_2'(S)} = \left[\frac{3}{2} S_{ij} : S_{ij}\right]^{1/2}$$

$$= \sqrt{\frac{1}{2}\left[\left(\sigma_{11}-\sigma_{22}\right)^2 + \left(\sigma_{22}-\sigma_{33}\right)^2 + \left(\sigma_{33}-\sigma_{11}\right)^2 + 3(\sigma_{12}^{\,2}+\sigma_{23}^{\,2}+\sigma_{13}^{\,2})\right]}$$

$$= \sqrt{\frac{1}{2}\left[\left(\sigma_{1}-\sigma_{2}\right)^2 + \left(\sigma_{2}-\sigma_{3}\right)^2 + \left(\sigma_{3}-\sigma_{1}\right)^2\right]}$$

$$(2.9)$$

For the uniaxial tension/compression in "11" direction, we have

$$\sigma_e = \sigma_{11} = \sigma_1 \ .$$

The *von-Mises yield criterion* is often used to assess the start of plastic deformation of a material under multiaxial stress conditions. This is discussed in later chapters.

2.3.2 *Displacements, strains and strain rates*

Notations of displacement and strain components in 3D (*x, y, z*) or (1, 2, 3) or (x_1, x_2, x_3) coordinate systems:

$$x(1), \quad y(2), \quad z(3), \quad xy(12), \quad xz(13), \quad yz(23)$$

$$\varepsilon_{11}, \quad \varepsilon_{22}, \quad \varepsilon_{33}, \quad \gamma_{12}, \quad \gamma_{13}, \quad \gamma_{23}$$

$$\varepsilon_{x}, \quad \varepsilon_{y}, \quad \varepsilon_{z}, \quad \varepsilon_{12}, \quad \varepsilon_{13}, \quad \varepsilon_{23}$$

$$u \quad\quad v \quad\quad w$$

These notations are commonly used in mechanics books and in technical literature.

a. Strains and displacements

Experimentally measured strains have certain limitations and are not suitable for scientific analysis. In this section the general relationships between strains and displacements are introduced. Strains are not related to rigid body motion, but are related to the relative movement of positions in a continuum. This means that strains can be written as functions of displacements.

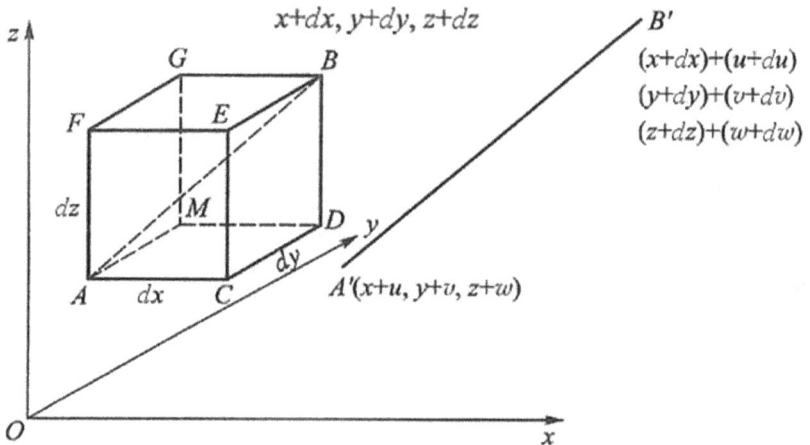

Fig. 2.23 Deformation of line AB to $A'B'$.

To demonstrate this, an element in an isotropic body can be selected, as shown in **Fig. 2.23**. AB is the diagonal line before deformation. The line $A'B'$ is the line AB after displacement and deformation. Assuming that the element (and hence the line) is deformed uniformly and the deformed line $A'B'$ is parallel to the original line AB. The new position of

A' is $(x+u, y+v, z+w)$, where u, v and w are the displacement from A to A' in the x-, y- and z-directions respectively.

The coordinates of position B are $(x+dx, y+dy, z+dz)$ and the deformed position B' $[(x+dx)+(u+du), (y+dy)+(v+dv), (z+dz)+(w+dw)]$, where du, dv and dw are increments of the line AB after deformation. If $u = f(x,y,z)$, then $u + du = f(x + dx, y + dy, z + dz)$. This, expanded using a Taylor series, becomes

$$u + du = f(x, y, z) + \frac{\partial f}{\partial x} dx + \frac{\partial f}{\partial y} dy + \frac{\partial f}{\partial z} dz + \dots \dots \quad (2.10)$$

Since $u = f(x,y,z)$ is a small value, the higher order differentials become vanishingly small, and can be eliminated. Thus, an increment in the x-direction (i.e. the increment of the line AC shown in **Fig. 2.23**) can be expressed as

$$du = \frac{\partial u}{\partial x} dx + \frac{\partial u}{\partial y} dy + \frac{\partial u}{\partial z} dz \quad (2.11a)$$

Similarly, for the y- and z-directions:

$$dv = \frac{\partial v}{\partial x} dx + \frac{\partial v}{\partial y} dy + \frac{\partial v}{\partial z} dz \quad (2.11b)$$

$$dw = \frac{\partial w}{\partial x} dx + \frac{\partial w}{\partial y} dy + \frac{\partial w}{\partial z} dz \quad (2.11c)$$

where $\frac{\partial u}{\partial x} dx$ represents the position change of the original length dx or AC in **Fig. 2.23** and also shown in **Fig. 2.24** in 2D. Thus the strain in the Ox-direction ε_x (or ε_{11}) can be expressed as

$$\varepsilon_x = \frac{\partial u}{\partial x} dx \Big/ dx = \frac{\partial u}{\partial x} \quad (2.12)$$

Similarly, the strains in the y- and z-directions can be expressed as

$\varepsilon_y = \frac{\partial v}{\partial y}$ and $\varepsilon_z = \frac{\partial w}{\partial z}$, respectively.

It can be seen in the xOz plane, shown in **Fig. 2.24**, that $\angle JA'C' = \angle F'A'M$. Thus

$$\angle JA'C' = \alpha_{zx} \approx \tan \alpha_{zx}$$

$$= \frac{C'J}{A'J} = \frac{\dfrac{\partial w}{\partial x} dx}{dx + \dfrac{\partial u}{\partial x} dx}$$

$$= \frac{\dfrac{\partial w}{\partial x}}{1 + \dfrac{\partial u}{\partial x}} \approx \frac{\partial w}{\partial x}$$

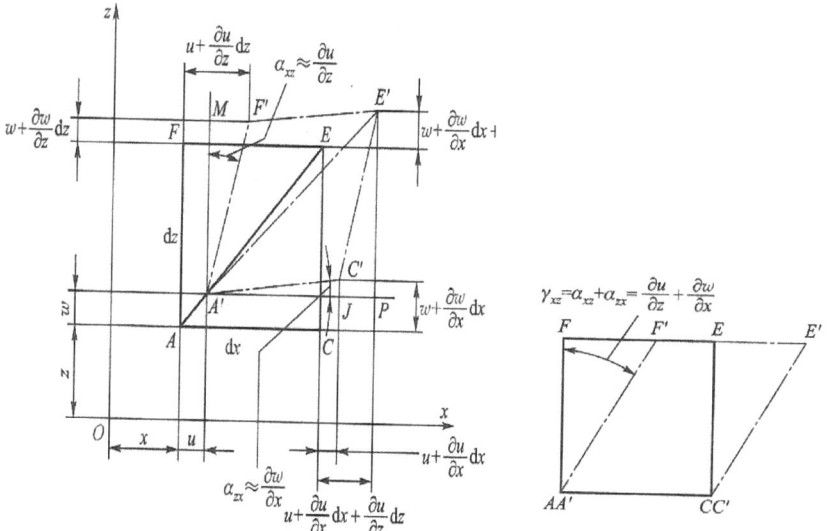

Fig. 2.24 Displacements and strains in an xOz plane (Ford and Alexander, 1977).

Similarly

$$\angle F'A'M = \alpha_{xz} \approx \tan \alpha_{xz} = \frac{F'M}{A'M} \approx \frac{\partial u}{\partial z}.$$

Thus the shear strain in xOz plane can be written

$$\gamma_{xz} = \frac{\partial w}{\partial x} + \frac{\partial u}{\partial z} \qquad (2.13a)$$

Similarly, in the other planes:

$$\gamma_{yz} = \frac{\partial w}{\partial y} + \frac{\partial v}{\partial z} \qquad (2.13b)$$

$$\gamma_{xy} = \frac{\partial v}{\partial x} + \frac{\partial u}{\partial y} \qquad (2.13c)$$

According to the deformation, two half shear strain values can represent the shear deformation. Thus $\frac{1}{2}\gamma_{xz}$ and $\frac{1}{2}\gamma_{zx}$ can be used to represent γ_{xz}.

The shear strain stain tensor can be written as

$$\varepsilon = \varepsilon_{ij} = \begin{bmatrix} \dfrac{\partial u}{\partial x} & \dfrac{1}{2}\left(\dfrac{\partial u}{\partial y}+\dfrac{\partial v}{\partial x}\right) & \dfrac{1}{2}\left(\dfrac{\partial u}{\partial z}+\dfrac{\partial w}{\partial x}\right) \\[2mm] \dfrac{1}{2}\left(\dfrac{\partial u}{\partial y}+\dfrac{\partial v}{\partial x}\right) & \dfrac{\partial v}{\partial y} & \dfrac{1}{2}\left(\dfrac{\partial v}{\partial z}+\dfrac{\partial w}{\partial y}\right) \\[2mm] \dfrac{1}{2}\left(\dfrac{\partial u}{\partial z}+\dfrac{\partial w}{\partial x}\right) & \dfrac{1}{2}\left(\dfrac{\partial v}{\partial z}+\dfrac{\partial w}{\partial y}\right) & \dfrac{\partial w}{\partial z} \end{bmatrix}$$

$$= \begin{bmatrix} \varepsilon_{11} & \dfrac{1}{2}\gamma_{12} & \dfrac{1}{2}\gamma_{13} \\[2mm] \dfrac{1}{2}\gamma_{21} & \varepsilon_{22} & \dfrac{1}{2}\gamma_{23} \\[2mm] \dfrac{1}{2}\gamma_{31} & \dfrac{1}{2}\gamma_{32} & \varepsilon_{33} \end{bmatrix} = \begin{bmatrix} \varepsilon_{11} & \varepsilon_{12} & \varepsilon_{13} \\ \varepsilon_{21} & \varepsilon_{22} & \varepsilon_{23} \\ \varepsilon_{31} & \varepsilon_{32} & \varepsilon_{33} \end{bmatrix} \qquad (2.14)$$

The strain tensor has the same characteristics as the stress tensor (reference Eq. 2.1). It determines the strain state for an element of a deformed body. For rigid plastic deformation, the volume does not change, thus this results in $\varepsilon_x + \varepsilon_y + \varepsilon_z = \varepsilon_{11} + \varepsilon_{22} + \varepsilon_{33} = 0$, and the hydrostatic strain $\varepsilon_H = (\varepsilon_x + \varepsilon_y + \varepsilon_z)/3 = 0$. The deviatoric strains are a strain tensor.

In elastic deformation, the volume of a solid does change. In general, the deformation of a solid involves a combination of volume and shape change. Similar to the definitions of invariants of stress tensors, the three principal strains can be calculated from the cubic equation

$$\varepsilon^3 - I_1\varepsilon^2 + I_2\varepsilon - I_3 = 0,$$

(2.15)

where

$$I_1 = \varepsilon_{ii} = \varepsilon_{11} + \varepsilon_{22} + \varepsilon_{33}$$

(2.15a)

$$I_2 = \frac{1}{2}\left(\varepsilon_{ii}\varepsilon_{kk} - \varepsilon_{ik}\varepsilon_{ki}\right)$$

$$= \varepsilon_{11}\varepsilon_{22} - \varepsilon_{22}\varepsilon_{33} - \varepsilon_{33}\varepsilon_{11} - \frac{1}{4}\left(\gamma_{12}^{\ 2} + \gamma_{23}^{\ 2} + \gamma_{31}^{\ 2}\right)$$

(2.15b)

$$I_3 = \varepsilon_{11}\varepsilon_{22}\varepsilon_{33} + \frac{1}{4}\gamma_{12}\gamma_{23}\gamma_{31} - \frac{1}{4}\left(\varepsilon_{11}\gamma_{23}^{\ 2} + \varepsilon_{22}\gamma_{31}^{\ 2} + \varepsilon_{33}\gamma_{12}^{\ 2}\right)$$

(2.15c)

Similar to stress analysis, the principal strains, $\varepsilon_1 > \varepsilon_2 > \varepsilon_3$, can be calculated from Eq. (2.15). The *maximum shear strain* can be expressed as

$$\gamma_{max} = \pm\frac{1}{2}\left(\varepsilon_1 - \varepsilon_2\right)$$

(2.16a)

The first invariant of the strain tensor can be used to express the volume change:

$$\Delta = I_1 = \varepsilon_{ii} = \varepsilon_{11} + \varepsilon_{22} + \varepsilon_{33}$$

(2.16b)

The mean strain or *hydrostatic strain* can be defined as

$$\varepsilon_H = \frac{I_1}{3} = \frac{\varepsilon_{ii}}{3} = \frac{\varepsilon_{11} + \varepsilon_{22} + \varepsilon_{33}}{3}$$

(2.16c)

The part of the strain tensor that is involved in shape change rather than the volume change is known as the *deviatoric strain tensor*, ε'. Similar to the definition of deviatoric stresses, the deviatoric strains can be expressed as

$$\varepsilon'_{ij} = \varepsilon_{ij} - \varepsilon_H \delta_{ij}$$

(2.16d)

or

$$
\varepsilon'_{ij} =
\begin{bmatrix}
\varepsilon_{11} - \varepsilon_H & \varepsilon_{12} & \varepsilon_{13} \\
\varepsilon_{21} & \varepsilon_{22} - \varepsilon_H & \varepsilon_{23} \\
\varepsilon_{31} & \varepsilon_{32} & \varepsilon_{33} - \varepsilon_H
\end{bmatrix}
$$

$$
=
\begin{bmatrix}
\dfrac{2\varepsilon_{11} - \varepsilon_{22} - \varepsilon_{33}}{3} & \varepsilon_{12} & \varepsilon_{13} \\[2ex]
\varepsilon_{21} & \dfrac{2\varepsilon_{22} - \varepsilon_{11} - \varepsilon_{33}}{3} & \varepsilon_{23} \\[2ex]
\varepsilon_{31} & \varepsilon_{32} & \dfrac{2\varepsilon_{33} - \varepsilon_{11} - \varepsilon_{22}}{3}
\end{bmatrix} .
$$

The *effective strain* is

$$
\varepsilon_e = \sqrt{\frac{2}{9}\left[\left(\varepsilon_1 - \varepsilon_2\right)^2 + \left(\varepsilon_2 - \varepsilon_3\right)^2 + \left(\varepsilon_3 - \varepsilon_1\right)^2\right]} \tag{2.16e}
$$

The effective strain, similar to the effective stress, is used to quantify complicated stress states using only principle strains. For the uniaxial case, $\varepsilon_e = \varepsilon_1 = \varepsilon_{11}$, where $\varepsilon_2 = \varepsilon_3 = -\varepsilon_1/2$.

b. Strain rate and strain rate tensor

The distance between two points within a deformable material changes during plastic deformation. The magnitude of the strain depends on the degree of distance change and the strain rate, which is related to the displacement velocities, $\dot{u}$, $\dot{v}$ and $\dot{w}$, defines the speed with which the change takes place. The displacement rates in the three directions can be expressed as

$$
\begin{aligned}
\dot{u} &= f_1\left(x, y, z, t\right) \\
\dot{v} &= f_2\left(x, y, z, t\right) \\
\dot{w} &= f_3\left(x, y, z, t\right)
\end{aligned} \tag{2.17}
$$

For small deformation, displacement rate can be expressed as

$$\dot{u} = \frac{\partial u}{\partial t}$$

$$\dot{v} = \frac{\partial v}{\partial t} \hspace{4cm} (2.18)$$

$$\dot{w} = \frac{\partial w}{\partial t}$$

Thus the strain rate can be expressed as displacement rate, such as:

$$\dot{\varepsilon}_x = \frac{\partial \dot{u}}{\partial x} = \frac{\partial}{\partial t}\left(\frac{\partial u}{\partial x}\right) = \frac{\partial \varepsilon_x}{\partial t}$$

$$\dot{\gamma}_{xy} = \frac{\partial \dot{u}}{\partial y} + \frac{\partial \dot{v}}{\partial x} = \frac{\partial}{\partial t}\left(\frac{\partial u}{\partial y} + \frac{\partial v}{\partial x}\right) = \frac{\partial \gamma_{xy}}{\partial t} \hspace{1.5cm} (2.19)$$

The other strain rate components can be obtained in the similar way. It should be noted that the unit for displacement rate, e.g. *m/s*, is different for strain rate, e.g. 1/*s*.

2.3.3 *Finite deformation and other strain definitions*

Consider a point in a material, *P*, with a position vector $\mathbf{X} = X_i\mathbf{I}_i$ for the undeformed body in the coordinate system X_i with the unit vector of $\mathbf{I}_i$, where $i = 1,2,3$ (**Fig. 2.25**). After displacement and deformation of the body, the new position of the material point, *p*, in the new coordinate system of x_i, with the unit vector of $\mathbf{e}_i$, is given by $\mathbf{x} = x_i\mathbf{e}_i$. The coordinate systems for the undeformed and deformed configurations can be superimposed for convenience, as shown in **Fig. 2.25**.

a. Finite deformation

Assume *Q* is a close neighbouring point to P (i.e. d**X**, is very small) with a position vector of

$$Q: \ \mathbf{X} + d\mathbf{X} = (X_i + dX_i)\mathbf{I}_i \ .$$

Q is displaced to the new position *q*, then

$$q: \ \mathbf{x} + d\mathbf{x} = \mathbf{X} + d\mathbf{X} + \mathbf{u}(\mathbf{X} + d\mathbf{X}) \ .$$

Thus, there is:

$$d\mathbf{x} = d\mathbf{X} + d\mathbf{u}, \tag{2.20}$$

Where $d\mathbf{u}$ is the vector, which represents the relative displacement of Q with respect to P in the deformed body. That is related to the quantity of deformation.

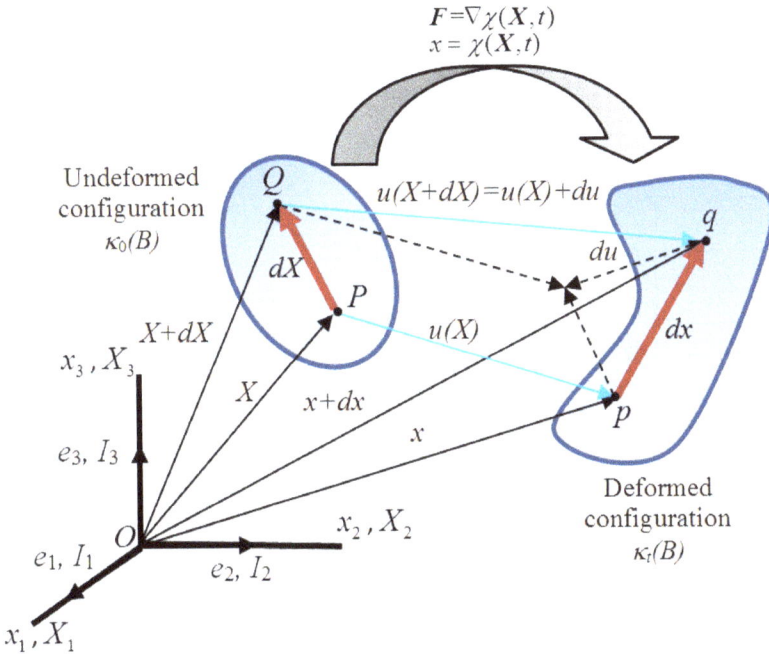

Fig. 2.25 Finite deformation of a continuum body.

Assuming continuity in the displacement field, for an infinitesimal element $d\mathbf{X}$, it is possible to use a Taylor series expansion around the material point P and neglect higher order terms to approximate the components of the relative displacement vector for the neighbouring point Q as

$$\mathbf{u}(\mathbf{X} + d\mathbf{X}) = \mathbf{u}(\mathbf{X}) + d\mathbf{u} \approx \mathbf{u}(\mathbf{X}) + \nabla_x \mathbf{u} \cdot d\mathbf{X},$$

or

$$u_i^q = u_i + du_i \approx u_i + \frac{\partial u_i}{\partial X_j} dX_j$$

b. Deformation gradient tensor

Equation (2.20) can be written as:

$$dx = dX + du = dX + \nabla_x \mathbf{u} \cdot dX = (\mathbf{I} + \nabla_x \mathbf{u})dX = \mathbf{F}dX$$

(2.21)

where,

$$\mathbf{F} = \frac{\partial \mathbf{x}}{\partial \mathbf{X}} = F_{ij} = \frac{\partial x_i}{\partial X_j}, \ i = 1,2,3 \text{ and } j = 1,2,3$$

(2.22)

That is,

$$\mathbf{F} = \frac{\partial x_i}{\partial X_j} = \begin{bmatrix} \dfrac{\partial x_1}{\partial X_1} & \dfrac{\partial x_1}{\partial X_2} & \dfrac{\partial x_1}{\partial X_3} \\[2mm] \dfrac{\partial x_2}{\partial X_1} & \dfrac{\partial x_2}{\partial X_2} & \dfrac{\partial x_2}{\partial X_3} \\[2mm] \dfrac{\partial x_3}{\partial X_1} & \dfrac{\partial x_3}{\partial X_2} & \dfrac{\partial x_3}{\partial X_3} \end{bmatrix}$$

This is known as a *deformation gradient tensor*, a second-order tensor that represents the gradient of the mapping functional relation. The tensor describes the motion of a continuum. The material deformation gradient tensor characterises local deformation at a point (at the neighbouring area) by transforming a line element from the original configuration to the current (or deformed) configuration. **F** is positively defined to enable one-to-one mapping (to ensure continuity). Thus we have

$$dx = \frac{\partial \mathbf{x}}{\partial \mathbf{X}} dX = \mathbf{F}dX,$$

or

$$dx_i = \frac{\partial x_i}{\partial X_j} dX_j = F_{ij} dX_j .$$

For a special case of rigid body translation **S**,

$$\mathbf{x} = \mathbf{X} + \mathbf{S},$$

the deformation gradient tensor reduces to the unit tensor **I**, e.g.

$$x_1 = X_1 + S_1$$
$$x_2 = X_2 + S_2$$
$$x_3 = X_3 + S_3$$

or

$$\mathbf{F} = \begin{bmatrix} \dfrac{\partial x_1}{\partial X_1} & \dfrac{\partial x_1}{\partial X_2} & \dfrac{\partial x_1}{\partial X_3} \\ \dfrac{\partial x_2}{\partial X_1} & \dfrac{\partial x_2}{\partial X_2} & \dfrac{\partial x_2}{\partial X_3} \\ \dfrac{\partial x_3}{\partial X_1} & \dfrac{\partial x_3}{\partial X_2} & \dfrac{\partial x_3}{\partial X_3} \end{bmatrix} = \begin{bmatrix} 1 & 0 & 0 \\ 0 & 1 & 0 \\ 0 & 0 & 1 \end{bmatrix}.$$

For rigid body rotation **R**,

$$\mathbf{x} = \mathbf{RX},$$

the deformation gradient tensor becomes

$$\mathbf{F} = \mathbf{R}.$$

A strain tensor vanishes in a case of rigid body motion without deformation.

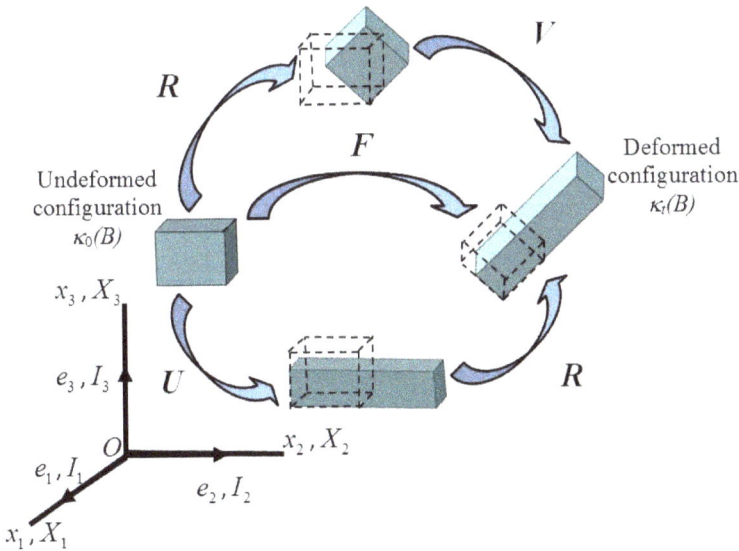

Fig. 2.26 Polar decomposition of deformation gradient for a continuum body.

c. Polar decomposition

The deformation gradient tensor, **F**, like any second-order tensor, can be decomposed, using the polar decomposition theorem, into a product of two second-order tensors: an *orthogonal tensor*, **R**, and a *positive definite symmetric tensor*, U, i.e.

$$\mathbf{F} = \mathbf{RU} = \mathbf{VR}, \tag{2.23}$$

where the orthogonal tensor, **R** has the properties of $\mathbf{R}^{-1} = \mathbf{R}^{T}$ and $|\mathbf{R}| = +1$, and represents a rotation of the body. The tensor **U** is the *right stretch tensor* and **V** the *left stretch tensor* referencing the rotation tensor **R** (**Fig. 2.26**). Both **U** and **V** are positive definite and second-order symmetric tensors, i.e. $\mathbf{U} = \mathbf{U}^{T}$ and $\mathbf{V} = \mathbf{V}^{T}$.

This decomposition indicates that the deformation of an undeformed line element $d\mathbf{X}$ into $d\mathbf{x}$ in the deformed state, i.e. $d\mathbf{x} = \mathbf{F}d\mathbf{X}$, can be obtained either by first stretching the element by U, followed by a rotation R (**Fig. 2.26**), i.e. $d\mathbf{x} = \mathbf{RU}d\mathbf{X}$; or, equivalently, by applying a rigid rotation R first, followed by a stretching V, i.e. $d\mathbf{x} = \mathbf{VR}d\mathbf{X}$ (**Fig. 2.26**). In this case, there is

$$\mathbf{V} = \mathbf{R} \cdot \mathbf{U} \cdot \mathbf{R}^{T}$$

So that **U** and **V** have the same eigenvalues or principal stretches.

Since pure rotation does not induce stress in a deformable body, it is often convenient to use rotation independent measures of deformation in continuum mechanics; a rotation followed by its inverse rotation leads to no change, i.e.

$$\mathbf{RR}^{T} = \mathbf{R}^{T}\mathbf{R} = 1.$$

d. Deformation tensor

The rotation can be excluded by multiplying *F* by its transpose. Green (George Green, British, 1793–1841) introduced a deformation tensor known as the *right Chauchy–Green deformation tensor* or *Green deformation tensor,* defined as

$$\mathbf{C} = \mathbf{F}^{T}\mathbf{F} = \mathbf{U} \tag{2.24}$$

or

$$C_{ij} = F_{ki}F_{kj} = \frac{\partial x_{k}}{\partial X_{i}} \frac{\partial x_{k}}{\partial X_{j}}.$$

In physics, the Cauchy–Green tensor gives us the square of local change in distance due to deformation, that is:

$$dx^2 = d\mathbf{X} \cdot \mathbf{C} d\mathbf{X}$$ (2.25)

Invariants of $\mathbf{C}$ are often used in the expression for *strain energy density functions*. The most commonly used invariants are:

$$I_1^C = Tr(\mathbf{C}) = C_{ii} = \lambda_1^2 + \lambda_2^2 + \lambda_3^2$$

$$I_2^C = \frac{1}{2}\left[(Tr\mathbf{C})^2 - Tr(\mathbf{C}^2) \right] = \frac{1}{2}\left[(C_{ii})^2 + C_{ik}C_{ki} \right] = \lambda_1^2\lambda_2^2 + \lambda_2^2\lambda_3^2 + \lambda_3^2\lambda_1^2$$

$$I_3^C = \det(\mathbf{C}) = \lambda_1^2\lambda_2^2\lambda_3^2$$

where λ_i is a stretch ratio for unit elements initially oriented along three coordinate axes.

Deformation of a material element can be defined using the tensors introduced above.

e. Other strain definitions

The Seth–Hill (B.R. Seth, Indian, 1907–1979; Rodney Hill, British, 1921–2011) family of strain measures can be expressed as:

- *Green-Lagrangian strain tensor and strain*

$$\varepsilon = \frac{1}{2}\left(\mathbf{U}^2 - \mathbf{I} \right) = \frac{1}{2}(\mathbf{C} - \mathbf{I}),$$

e.g. for uniaxial

$$\varepsilon_{11} = \frac{1}{2}\left(\lambda_1^2 - 1 \right)$$

- *Biot strain tensor (engineering strain)*

$$\varepsilon = \mathbf{U} - \mathbf{I} = \mathbf{C}^{1/2} - \mathbf{I},$$

e.g. for uniaxial

$$\varepsilon_{11} = \lambda_1 - 1$$

- *True (natural) strain or Hencky strain*

$$\varepsilon = \ln \mathbf{U} = \frac{1}{2}\ln \mathbf{C},$$

e.g. for uniaxial

$$\varepsilon_{11} = \ln \lambda_1,$$

where λ_1 is the *stretching/compression ratio* in the tensile/compression direction for an uniaxial tensile/compression test, i.e. $\lambda_1 = L/L_0$, where L_0 and L are the gauge lengths of the specimen before and after deformation. All strain definitions give similar values for small strains. However, as deformation increases, differences between values of the different strain measures become larger.

The Green–Lagrangian strain tensor is a symmetric tensor. Thus it is readily applied in large displacement *finite element analysis* for metal forming process simulation. For this application, incremental methods are used and deformation within one increment is normally small. Thus the accumulated strain from Green–Lagrangian strain definition eventually approaches the true strain value.

f. Stress measures in finite deformation

The most commonly used measure of stress is the *Cauchy stress tensor*, often called simply the *stress tensor* or "*true stress*", and is used for small deformations. For large deformations, particularly in continuum and computational mechanics other stress measures are often used. They are:

- The Kirchhoff stress ($\boldsymbol{\sigma}^K$).
- The Nominal stress ($\boldsymbol{\sigma}^N$).
- The first Piola–Kirchhoff stress ($\boldsymbol{\sigma}^P$). This stress tensor is the transpose of the nominal stress ($\boldsymbol{\sigma}^P = (\boldsymbol{\sigma}^N)^T$).
- The second Piola–Kirchhoff stress or PK2 stress ($\boldsymbol{\sigma}^S$).
- The Biot stress ($\boldsymbol{\sigma}^B$)

Some stress definitions can be given as:

- *Kirchhoff stress*

$$\boldsymbol{\sigma}^K = |\mathbf{F}| \boldsymbol{\sigma},$$

where $\mathbf{F}$ is the *deformation gradient tensor* and $\boldsymbol{\sigma}$ is the *Cauchy stress tensor* defined previously. The *Kirchhoff stress tensor* is used widely in numerical algorithms in *metal plasticity*, where there is no change in volume during plastic deformation.

- *First Piola–Kirchhoff stress tensor*

$$\mathbf{\sigma}^{P} = |\mathbf{F}| \mathbf{\sigma} \mathbf{F}^{-T}$$
.

The *first Piola–Kirchhoff stress tensor* expresses stress relative to the resulting material configuration. This is in contrast to the Cauchy stress tensor which expresses the stress relative to the present configuration. For infinitesimal deformations or rotations, the Cauchy and Piola–Kirchhoff tensors are identical.

- *Biot stress*

$$\mathbf{\sigma}^{B} = \frac{1}{1}\left(\mathbf{R}^{T} \cdot \mathbf{\sigma}^{P} + (\mathbf{\sigma}^{P})^{T} \cdot \mathbf{R}\right)$$
,

where $\mathbf{R}$ is the *rotation tensor* obtained from polar decomposition of the deformation gradient tensor. *Biot stress* is also called *Jaumann stress*. The quantity of $\mathbf{\sigma}^{B}$ does not have any physical interpretation.

2.3.4 Stress and strain relationships

a. Linear elasticity

To obtain linear elastic constitutive equations it is sufficient to choose a thermo-dynamic potential, which is the positive definition of a quadratic function of strain. The potential $\mathbf{\Psi}$ can be defined as

$$\psi = \frac{1}{2\rho}\mathbf{\alpha}:\mathbf{\varepsilon}:\mathbf{\varepsilon}\ , \qquad (2.26)$$

where ρ is the density of the material, $\mathbf{\alpha}$ is a fourth-order tensor and is composed of the elastic modulus E and Poisson ratio υ. By definition of the associated variable, the stress tensor $\mathbf{\sigma}$ is derived from the potential $\mathbf{\Psi}$ according to

$$\mathbf{\sigma} = \rho\left(\frac{\partial\psi}{\partial\varepsilon}\right) = \mathbf{\alpha}:\mathbf{\varepsilon} \qquad (2.27)$$
.

The tensor $\mathbf{\alpha}$ satisfies certain symmetric properties. Then we have

$$\sigma_{ij} = \rho\left(\frac{\partial\psi}{\partial\varepsilon_{ij}}\right) = \alpha_{ijkl}\varepsilon_{kl} \qquad (2.28)$$
.

This is the generalised Hooke's law for an isotropic linear elastic continuum.

b. 3D Linear isotropic elasticity

Equation (2.27) can also be written as

$$\sigma_{ij} = \lambda \varepsilon_{kk} \delta_{ij} + 2\mu \varepsilon_{ij} \,, \tag{2.29}$$

where λ and μ are two Lamé constants, which are defined below. This equation can be expressed in the expanded form:

$$\sigma_{ij} = \frac{vE}{(1+v)(1-2v)} \varepsilon_{kk} \delta_{ij} + \frac{E}{(1+v)} \varepsilon_{ij}.$$

Employing the dual form, the strain can be expressed as a function of stress:

$$\varepsilon = \frac{1-v}{E} \sigma - \frac{v}{E} \mathbf{Tr}(\sigma)\delta_{ij} \tag{2.30}$$

or

$$\varepsilon_{ij} = \frac{1-v}{E} \sigma_{ij} - \frac{v}{E} \sigma_{kk} \delta_{ij} \,,$$

where E is Young's modulus and v is Poisson's ratio, which are characteristics particular to metal composition. *Hydrostatic strain ε_H* and *deviatoric strain ε'_{ij}* are functions of corresponding stresses:

$$\varepsilon_H = \frac{1-2v}{E} \sigma_H$$

$$\varepsilon'_{ij} = \frac{1-v}{E} S_{ij} = \frac{1}{2G} S_{ij}.$$

The relationships between the commonly used parameters in elasticity are summarised below.

Lamé constants: $\lambda = \dfrac{vE}{(1+v)(1-2v)}$, $\mu = \dfrac{E}{2(1+v)}$

Young's Modulus: $E = \mu \dfrac{3\lambda + 2\mu}{\lambda + \mu}$

Poisson's ratio: $v = \dfrac{\lambda}{2(\lambda + \mu)}$

Shear modulus: $G = \mu = \dfrac{E}{2(1+v)}$

Bulk modulus: $K = \dfrac{3\lambda + 2\mu}{3}$

Young's modulus E is always positive. Also, to comply with theories of elasticity, a limitation of the range of Poisson's ratio, $-1 < v < 0.5$, is necessary.

c. Plane stress conditions

In *sheet metal forming*, or for thin shell components, a plane stress condition (stress in the thickness direction is zero) is often used to make analysis more efficient. That is:

$$\sigma_{33} = \sigma_{13} = \sigma_{23} = 0 .$$

Then Eqn. (2.23) can be written as

$$\begin{Bmatrix} \varepsilon_{11} \\ \varepsilon_{22} \\ \varepsilon_{12} \end{Bmatrix} = \begin{bmatrix} 1/E & -v/E & 0 \\ -v/E & 1/E & 0 \\ 0 & 0 & (1+v)/E \end{bmatrix} \begin{Bmatrix} \sigma_{11} \\ \sigma_{22} \\ \sigma_{12} \end{Bmatrix} \tag{2.31}$$

and

$$\varepsilon_{33} = -\dfrac{v}{1-v}\left(\varepsilon_{11} + \varepsilon_{22}\right); \ \varepsilon_{13} = \varepsilon_{23} = 0 .$$

d. Plane strain conditions

In some *rolling and forging processes*, the strain in a particular direction (for the case of rolling, it is normally in the lateral direction – parallel to roll axes), is close to zero, i.e.

$$\varepsilon_{33} = \varepsilon_{13} = \varepsilon_{23} = 0 .$$

Then

$$\begin{Bmatrix} \sigma_{11} \\ \sigma_{22} \\ \sigma_{12} \end{Bmatrix} = \begin{bmatrix} \lambda + 2\mu & \lambda & 0 \\ \lambda & \lambda + 2\mu & 0 \\ 0 & 0 & 2\mu \end{bmatrix} \begin{Bmatrix} \varepsilon_{11} \\ \varepsilon_{22} \\ \varepsilon_{12} \end{Bmatrix} \tag{2.32}$$

and

$$\sigma_{33} = -\frac{\lambda}{2(\lambda + \mu)}(\sigma_{11} + \sigma_{22}) = v(\sigma_{11} + \sigma_{22}); \quad \sigma_{13} = \sigma_{23} = 0.$$

e. Orthotropic (anisotropic) elasticity

Each orthogonal direction of an *orthotropic material* has different mechanical properties. Orthotropic materials are thus anisotropic; their properties depend on the direction from which they are being considered. An *isotropic material*, in contrast, has the same properties in every direction. An example of an orthotropic material is a metal which has been cold rolled to form a thin sheet; the properties in the *rolling direction* and each of the two *transverse directions* will be different due to the anisotropic structure that develops during rolling.

It is important to keep in mind that a material which is anisotropic on one lengthscale may be isotropic on another (usually larger) lengthscale. For instance, most metals are polycrystalline with very small grains. Each of the individual grains may be anisotropic, but if the material as a whole comprises many randomly oriented grains, then its measured mechanical properties will be an average of the properties over all possible orientations of the individual grains. This point will be discussed in detail in Chapter 9 (crystal plasticity modelling).

The matrix of an orthotropic material contains only nine independent material constants. With the principal axes of orthotropy as the reference axes, the elasticity law can be expressed in the following form:

$$\begin{Bmatrix} \varepsilon_{11} \\ \varepsilon_{22} \\ \varepsilon_{33} \\ \varepsilon_{23} \\ \varepsilon_{31} \\ \varepsilon_{12} \end{Bmatrix} = \begin{bmatrix} \frac{1}{E_1} & \frac{-v_{12}}{E_1} & \frac{-v_{13}}{E_1} & 0 & 0 & 0 \\ \frac{-v_{21}}{E_2} & \frac{1}{E_2} & \frac{-v_{23}}{E_2} & 0 & 0 & 0 \\ \frac{-v_{31}}{E_3} & \frac{-v_{32}}{E_3} & \frac{1}{E_3} & 0 & 0 & 0 \\ 0 & 0 & 0 & \frac{1}{2G_{23}} & 0 & 0 \\ 0 & 0 & 0 & 0 & \frac{1}{2G_{31}} & 0 \\ 0 & 0 & 0 & 0 & 0 & \frac{1}{2G_{12}} \end{bmatrix} \begin{Bmatrix} \sigma_{11} \\ \sigma_{22} \\ \sigma_{33} \\ \sigma_{23} \\ \sigma_{31} \\ \sigma_{12} \end{Bmatrix} \quad (2.33)$$

Symmetric conditions result in:

$$\frac{v_{12}}{E_1} = \frac{v_{21}}{E_2}, \ \frac{v_{13}}{E_1} = \frac{v_{31}}{E_3}, \ \frac{v_{23}}{E_2} = \frac{v_{32}}{E_3}.$$

The nine independent material constants are:

- three elastic moduli in the direction of orthotropy: E_1, E_2, E_3,
- three shear moduli: G_{12}, G_{23}, G_{31},
- three Poisson (contraction) ratios: v_{12}, v_{23}, v_{31}.

2.3.5 *Elastic strain energy*

All the work performed during elastic deformation is stored as energy within the solid and this energy is recovered on the release of the applied forces. This is known as *elastic strain energy, W*. **Figure 2.27** shows the linear relationship of the force, *F*, and displacement, *u*, under elastic loading. The energy (or work) can be calculated using:

$$W = \frac{1}{2}Fu \tag{2.34}$$

For a uniaxial case, the increment of elastic strain energy is given by

$$\partial W = \frac{1}{2}F\partial u = \frac{1}{2}\left(A\sigma_{11}\right)\left(\varepsilon_{11}\partial x\right) = \frac{1}{2}\left(\sigma_{11}\varepsilon_{11}\right)\left(A\partial x\right).$$

This equation is obtained according to the stress and strain definitions, $F = A\sigma_{11}$ and $\partial u = \varepsilon_{11}\partial x$ respectively, where A is the cross-sectional area of an element of the solid and $A\partial x$ is the volume of an element of the solid. Thus the *strain energy per unit volume* or *strain energy density* W_0 can be expressed as

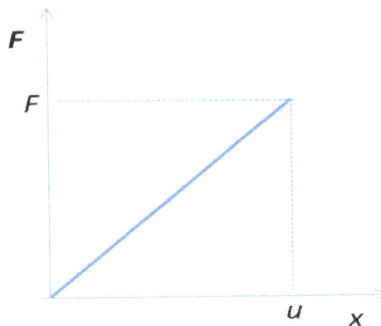

Fig. 2.27 **Work done by the load and displacement for an elastic solid.**

$$W_0 = \frac{\partial W}{A \partial x} = \frac{1}{2} \sigma_{11} \varepsilon_{11} \tag{2.35}$$

Through the use of the elastic stress–strain relationship for a uniaxial case, $\sigma = E\varepsilon$, the strain energy density can be expressed as stress or strain only:

$$W_0 = \frac{1}{2E} \sigma_{11}^2 \tag{2.35a}$$

$$W_0 = \frac{1}{2} E \varepsilon_{11}^2 \tag{2.35b}$$

The stress–strain relationships can be obtained by the differentiation of the equations, i.e.

$$\varepsilon_{11} = \frac{\partial W_0}{\partial \sigma_{11}} = \frac{\sigma_{11}}{E}$$

$$\sigma_{11} = \frac{\partial W_0}{\partial \varepsilon_{11}} = E\varepsilon_{11}$$

The use of the strain energy density equations to obtain stress–strain relationships is very useful in stress analysis.

Similar to the uniaxial tensile case, the strain energy density of an element subject to pure shear, can be obtained as

$$W_0 = \frac{1}{2} \tau_{12} \gamma_{12} = \frac{1}{2G} \tau_{12}^2 = \frac{1}{2} G \gamma_{12}^2,$$

where G is the shear modulus of the material.

The elastic energy in the general 3D stress state can be expressed as:

$$W_0 = \frac{1}{2} \sigma_{ik} \varepsilon_{ij} \tag{2.36}$$

or in expanded form:

$$W_0 = \frac{1}{2} \left(\sigma_{11}\varepsilon_{11} + \sigma_{22}\varepsilon_{22} + \sigma_{33}\varepsilon_{33} + \sigma_{12}\varepsilon_{12} + \sigma_{13}\varepsilon_{13} + \sigma_{23}\varepsilon_{23} \right).$$

Using the stress–strain relationships discussed above, by eliminating either stress or strain, elastic strain energy density can be described as a function of either stress or strain alone and expressed as either

$$W_0 = \frac{1}{2E}\left(\sigma_{11}^2 + \sigma_{22}^2 + \sigma_{33}^2\right) - \frac{v}{E}\left(\sigma_{11}\sigma_{22} + \sigma_{22}\sigma_{33} + \sigma_{11}\sigma_{33}\right)$$
$$+ \frac{1}{2G}\left(\sigma_{12}^2 + \sigma_{13}^2 + \sigma_{23}^2\right) \tag{2.36a}$$

or

$$W_0 = \frac{1}{2}\lambda\Delta^2 + G\left(\varepsilon_{11}^2 + \varepsilon_{22}^2 + \varepsilon_{33}^2\right) + \frac{1}{2}G\left(\gamma_{12}^2 + \gamma_{13}^2 + \gamma_{23}^2\right). \tag{2.36b}$$

As discussed above for the uniaxial tensile case, differentiation of strain energy density functions, results in stress–strain relationships, such as

$$\sigma_{ij} = \frac{\partial W_0}{\partial \varepsilon_{ij}} = \lambda\Delta + 2G\varepsilon_{ij}, \tag{2.37a}$$

where λ is the Lamé constant and Δ is volumetric strain. Also,

$$\varepsilon_{ij} = \partial W_0 / \partial \sigma_{ij}. \tag{2.37b}$$

The methods for obtaining stress–strain relationships using strain energy density functions are very useful in elasticity analysis.

2.4 Yield criteria

As discussed above, for a uniaxial tensile test, if the applied stress exceeds the yield stress of the material, plastic (permanent) deformation occurs. This is the mechanism used in metal forming – exploitation of permanent deformation of a metal to produce a specified shape. However, in real metal forming processes, the state of stress at any position within a work-piece is usually not uniaxial but is of a complicated multiaxial form and an important criterion necessary to analyse a process is the state of stress required for yielding to occur. The most commonly used yield criteria for 3D stress states are those of von-Mises (Richard von-Mises, American, 1883–1953) and Tresca (Henri Édouard Tresca, French, 1814–1885).

2.4.1 *von-Mises yield criterion*

The *von-Mises yield criterion* states that yielding of metals begins when the second stress invariant J_2' reaches a critical value. For this reason, it is sometimes called the J_2'-*plasticity* or J_2' *flow theory*. It is the part of plasticity theory that applies best to ductile metals.

This criterion was first formulated by Clerk-Maxwell in 1865 on mathematical grounds and generalised by von-Mises in 1913. In materials science and engineering, the von-Mises yield criterion can also be formulated in terms of the *von-Mises stress* or *equivalent tensile stress*, σ_e, a scalar stress value that can be computed from the Cauchy stress tensor. In this case, a metal is said to start yielding when the imposed von-Mises stress reaches a critical value k, which is related to the yield strength, Y (or σ_Y). The von-Mises stress is used to predict the yielding of materials under any loading condition and its value can be obtained from results of uniaxial tensile tests.

The von-Mises yield criterion can be expressed as (the second invariant of deviatoric stress tensor):

$$J_2' = k^2$$

(2.38)

where k is a constant. According to the definition of deviatoric stresses, there is

$$2J_2' = S_{ij} : S_{ij} = S_{11}^2 + S_{22}^2 + S_{33}^2 + 2(\sigma_{12}^2 + \sigma_{23}^2 + \sigma_{31}^2) = 2k^2$$

(2.38a)

Based on stress tensor, it can be expressed as

$$(\sigma_{11} - \sigma_{22})^2 + (\sigma_{22} - \sigma_{33})^2 + (\sigma_{33} - \sigma_{11})^2 + 6\sigma_{12}^2 + 6\sigma_{23}^2 + 6\sigma_{31}^2 = 6k^2$$

(2.38b)

With the use of principal stresses, it can be expressed as

$$\left(\sigma_1 - \sigma_2\right)^2 + \left(\sigma_2 - \sigma_3\right)^2 + \left(\sigma_3 - \sigma_1\right)^2 = 6k^2 .$$

(2.38c)

The constant k can be interpreted in terms of the yield stress Y (or σ_Y) in simple tension, where there is

$$\sigma_1 = \sigma_Y$$

$$\sigma_2 = \sigma_3 = 0 .$$

Then Eq. (3.43c) becomes

$$\left(\sigma_1 - 0\right)^2 + \left(0 - 0\right)^2 + \left(0 - \sigma_1\right)^2 = 6k^2$$

$$2\left(\sigma_1\right)^2 = 6k^2$$

and

$$2\left(\sigma_Y\right)^2 = 6k^2.$$

Then for the uniaxial case we have

$$k = \frac{\sigma_Y}{\sqrt{3}},$$

where $\sigma_Y/\sqrt{3}$ is the yield stress of the material in pure shear. Thus, a more important meaning for k can be found by considering pure shear and k can be interpreted as the shear yield stress of a metal. At the onset of yielding, the magnitude of the shear stress in pure shear is $1/\sqrt{3}$ times the tensile yield stress in the case of simple tension. Therefore

$$\sqrt{\frac{1}{2}\left[\left(\sigma_1 - \sigma_2\right)^2 + \left(\sigma_2 - \sigma_3\right)^2 + \left(\sigma_3 - \sigma_1\right)^2\right]} = \frac{\sigma_Y}{\sqrt{3}}$$

or

$$\left[\left(\sigma_1 - \sigma_2\right)^2 + \left(\sigma_2 - \sigma_3\right)^2 + \left(\sigma_3 - \sigma_1\right)^2\right]^{\frac{1}{2}} = \sqrt{\frac{2}{3}}\sigma_Y.$$

A pure shear is equivalent to two principal stresses, σ_1, $\sigma_2 = 0$, $\sigma_3 = -\sigma_1$. Therefore

$$\left(\sigma_1 - 0\right)^2 + \left(0 + \sigma_1\right)^2 + \left(-\sigma_1 - \sigma_1\right)^2 = 6k^2$$

and

$$\sigma_1 = k = -\sigma_3 = \frac{\sigma_Y}{\sqrt{3}}.$$

The shear stress in pure shear is numerically equal to the principal stresses σ_1 and σ_3, thus k is the yield stress in pure shear.

The von-Mises criterion implies that yielding is not dependent upon any one particular stress component or shear stress, but rather is dependent on a function of all three shear stresses resulting from different values of the three principal stresses. Thus in simple tension, the maximum shear stress is $\tau_{max} = \sigma_1/2$, i.e.

$$\frac{\sigma_1}{2} = \frac{\sigma_Y}{2} = \frac{\sqrt{3}k}{2} = \frac{k}{1.155}.$$

Mathematically, the von-Mises yield criterion is independent of hydrostatic stress. It is independent of the signs of the stresses (all squared terms), of their relative magnitudes and its geometric representation on the deviatoric plane is a circle about point O (Ref. **Fig. 2.30** in Section 2.4.3), radius $\sqrt{2}k = \sqrt{2/3}\sigma_Y$. This is discussed in detail in Section 2.4.3.

2.4.2 *Tresca*

The *Tresca yield criterion* is also known as the *maximum shear stress theory*. In terms of the principal stresses the Tresca criterion is expressed as

$$\frac{1}{2}\max\left\{|\sigma_1 - \sigma_2|, |\sigma_2 - \sigma_3|, |\sigma_1 - \sigma_3|\right\} = \tau_{max} = \frac{\sigma_Y}{\sqrt{3}}, \tag{2.39}$$

where τ_{max} is the yield strength in shear. For the principal stresses if we define $\sigma_1 > \sigma_2 > \sigma_3$, then, there is

$$\frac{1}{2}(\sigma_1 - \sigma_3) = \tau_{max}. \tag{2.40}$$

For a uniaxial tension, $\sigma_1 = \sigma_Y$, and, $\sigma_2 = \sigma_3 = 0$, according to the definition, there is

$$\tau_{max} = \frac{1}{2}(\sigma_1 - \sigma_3) = \frac{\sigma_Y}{\sqrt{3}}.$$

For a state of pure shear, $\sigma_1 = -\sigma_3 = k$, $\sigma_2 = 0$, the maximum shear stress predicts that yielding will occur when

$$\tau_{max} = \frac{1}{2}(\sigma_1 - \sigma_3) = \frac{1}{2}[k - (-k)] = k.$$

The Tresca yield criterion is less mathematically complicated than the von-Mises criterion and for this reason it is often used in engineering design. However, the Tresca yield criterion does not take into consideration the intermediate principal stress. In reality, it is often difficult to identify which of the principle stresses are the maximum and minimum when undertaking analysis.

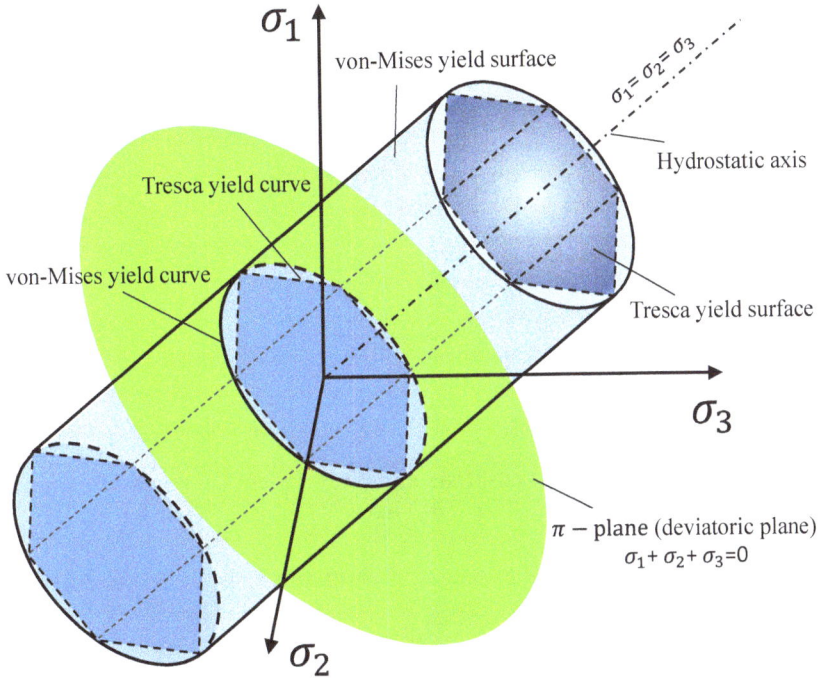

Fig. 2.28 Comparison of von-Mises and Tresca yield surfaces in 3D.

2.4.3 *Graphical illustration of von-Mises and Tresca yield surfaces*

Figure 2.28 graphically depicts both von-Mises and Tresca yield criteria, showing σ_1, σ_2 and σ_3 in 3D space. The radius of the von-Mises yield curve is $\sqrt{2/3}\sigma_Y$.

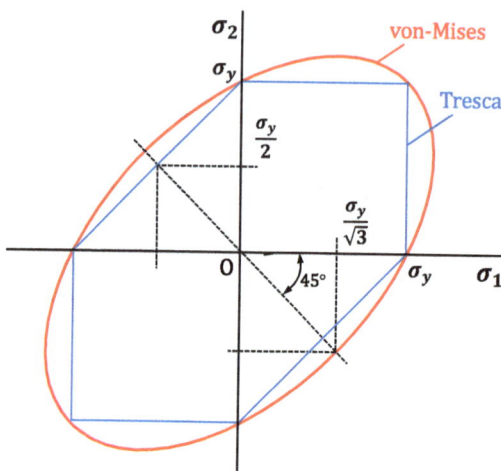

Fig. 2.29 Comparison of von-Mises and Tresca yield curves in 2D σ_1 and $\sigma_2(\sigma_3)$ axes.

In coordinates of principal stress, as shown in **Figs. 2.28** and **2.29**, the von-Mises yield surface is a cylinder with a central axis of $\sigma_1 = \sigma_2 = \sigma_3$. **Figures 2.28** and **2.29** also show the Tresca yield surface as a symmetric hexagonal cylinder with the same axis as that of the von-Mises surface. For certain stress states, von-Mises and Tresca yield criteria have the same values, but in general Tresca provides a more conservative prediction of the yield of a material compared with the von-Mises criterion. The two yield criteria are those most commonly used for metals.

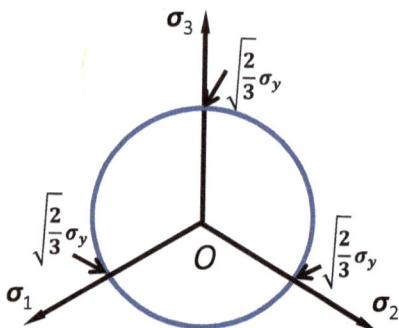

Fig. 2.30 Illustration of von-Mises yield criterion.

If the von-Mises yield surface, shown in **Fig. 2.28**, is projected to the plane perpendicular to its central axes, $\sigma_1 = \sigma_2 = \sigma_3$, it becomes a circle (**Fig. 2.30**). The angles between the axes are 120° and the radius of the circle is $\sqrt{2/3}\sigma_Y$.

2.4.4 *Other yield criteria*

The yield criteria introduced below are not commonly used in modern engineering analysis for metals and metal forming analysis. However, a brief introduction to the theories is given here to highlight that other formulations and assumptions are available to assess the yield of metals.

a. *Maximum principal stress theory*

This was introduced by W.J.M. Rankine (William John Macquorn Rankine, Scottish, 1820–1872). Yield occurs when the largest principal stress exceeds the uniaxial tensile yield strength. Although this criterion allows for a quick and easy comparison with experimental data it is rarely suitable for design purposes. However, this theory can be applicable to brittle materials.

$$\sigma_1 \leq \sigma_Y$$

b. *Total strain energy theory*

This theory assumes that the stored energy associated with elastic deformation at the point of yield is independent of the specific stress tensor. Thus yield occurs when the strain energy per unit volume is greater than the strain energy at the elastic limit in simple tension. For a 3D stress state this is given by

$$\sigma_1^2 + \sigma_2^2 + \sigma_3^2 - 2v\left(\sigma_1\sigma_2 + \sigma_2\sigma_3 + \sigma_1\sigma_3\right) \leq \sigma_Y^2 .$$

There are many other yield criteria for isotropic and anisotropic materials, which are not discussed here. The main purpose of the introduction to various yield criteria here is to demonstrate that most of them were proposed for metals, particularly steel and for room temperature design applications. However, at high temperatures and for many other new materials developed recently, such traditional yield criteria may not be suitable.

The yield stress of strain hardening materials increases due to plastic deformation and results in the expansion of the yield surfaces. For materials under cyclic loading, the yield surface loci also change. These materials will be discussed further in following chapters.

2.5 Stress States for Different Metal Forming Conditions

In metal forming processes, stress and strain states are complicated and vary with processes, materials and forming conditions. However, from a mechanics point of view, these can be divided into a number of typical types. In this section, the stress-states for different deformation conditions are analysed.

The *spherical stress tensor* is defined as the hydrostatic stress σ_H,

$$H_{ij} = \begin{bmatrix} \sigma_H & 0 & 0 \\ 0 & \sigma_H & 0 \\ 0 & 0 & \sigma_H \end{bmatrix},$$

and the *deviatoric stress tensor*, which is a symmetric tensor and the driving force for the plastic deformation of a work-piece, is

$$S_{ij} = \begin{bmatrix} \sigma_{11} - \sigma_H & \sigma_{12} & \sigma_{13} \\ \sigma_{21} & \sigma_{22} - \sigma_H & \sigma_{23} \\ \sigma_{31} & \sigma_{32} & \sigma_{33} - \sigma_H \end{bmatrix}.$$

If the von-Mises yield criterion is used, the value of J_2' can be used to judge if plastic deformation occurs.

If it is assumed that, $\sigma_{11} > \sigma_{22} > \sigma_{33}$ in the coordinate system chosen, the deviatoric stress components are:

- $S_{11} = \sigma_{11} - \sigma_H > 0$ — positive

- $S_{33} = \sigma_{33} - \sigma_H < 0$ — negative, and,

- $S_{22} = \sigma_{22} - \sigma_H < 0$ or > 0 — could be positive or negative.

This results in the third invariant:

- $J_3' = S_{11}S_{22}S_{33} > 0$ — tensile strain, i.e. the work-piece extended relative to the loading direction;

- $J_3' = 0$ — plane strain, i.e. strain in one direction is zero;

- $J_3' < 0$ — compressive deformation, i.e. the work-piece shortened.

It should be noted that for a forming process, the stress state can vary from one material point to another. In finite element forming process simulations, the values of J_3' for individual integration points (or elements) are obtainable. Based on the values of J_3', the deformation state of the material at a particular forming time can be understood.

2.5.1 *Simple deformation conditions*

a. *Simple tension*

Uniaxial tensile tests are the simplest and most commonly used tests for obtaining the basic properties of a material. The stress states are given in **Fig. 2.31** (the counter forces are not shown in the figure).

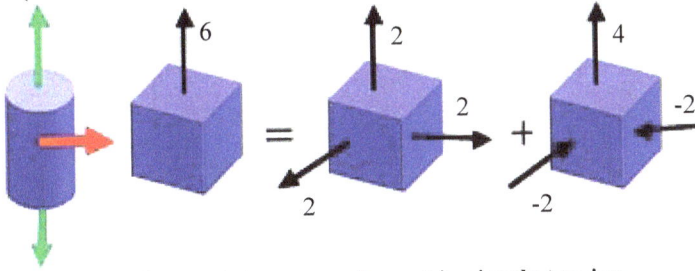

Fig. 2.31 Stress states on an element in simple tension (units are in MPa).

For example, the *stress tensor* for the uniaxial tension case ($\sigma_{11} = \sigma_1 = 6$ (MPa); other stress components are zero) can be decomposed into *spherical stress tensor*, i.e.

$$\sigma_H = \sigma_{ii}/3 = (\sigma_{11} + \sigma_{22} + \sigma_{33})/3 = 2 \text{ (MPa)}$$

and *deviatoric stress tensor*,

$$S_{11} = \sigma_{11} - \sigma_H = 6 - 2 = 4 \text{ (MPa)}$$

$$S_{22} = \sigma_{22} - \sigma_H = 0 - 2 = -2 \text{ (MPa)}$$

$$S_{33} = -2 \text{ (MPa)};$$

other stress components are zero. That is

$$\begin{bmatrix} 6 & 0 & 0 \\ 0 & 0 & 0 \\ 0 & 0 & 0 \end{bmatrix} = \begin{bmatrix} 2 & 0 & 0 \\ 0 & 2 & 0 \\ 0 & 0 & 2 \end{bmatrix} + \begin{bmatrix} 4 & 0 & 0 \\ 0 & -2 & 0 \\ 0 & 0 & -2 \end{bmatrix}$$

$$\text{Total} \qquad \text{Spherical} \qquad \text{Deviatoric}$$

Here, $J_3' = S_{11}S_{22}S_{33} = 16 > 0$; the maximum principal stress is tensile, i.e. the material has been extended. The hydrostatic stress is positive ($\sigma_H = 2\text{MPa}$), which indicates that the specimen is in tension in general.

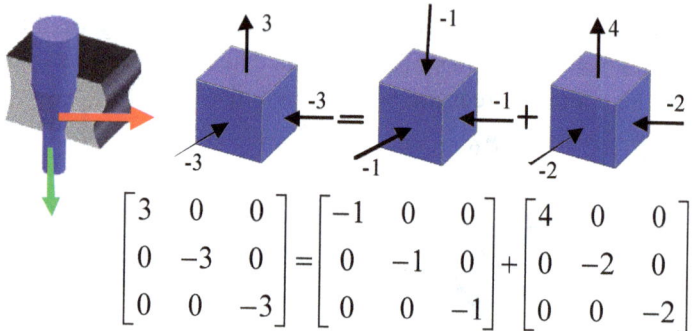

$$\begin{bmatrix} 3 & 0 & 0 \\ 0 & -3 & 0 \\ 0 & 0 & -3 \end{bmatrix} = \begin{bmatrix} -1 & 0 & 0 \\ 0 & -1 & 0 \\ 0 & 0 & -1 \end{bmatrix} + \begin{bmatrix} 4 & 0 & 0 \\ 0 & -2 & 0 \\ 0 & 0 & -2 \end{bmatrix}$$

Fig. 2.32   Stress states in drawing (units are in MPa).

b. Drawing

The stress states and stress tensors for a drawing process are given in **Fig. 2.32**. The stress tensor can be divided into spherical and deviatoric tensors, i.e. $\sigma_H = \sigma_{ii}/3 = (3 - 3 - 3)/3 = -1$ MPa; $S_{11} = \sigma_{11} - \sigma_H = 4$ MPa, $S_{22} = S_{33} = -2$ MPa. Note that this deviatoric tensor is the same as that in pure tension since the deformation mode is the same ($J_3' = S_{11}S_{22}S_{33} = 16 > 0$) – the material is extended. The plastic deformation is only related to deviatoric stresses.

In addition it can be seen that the hydrostatic stress is negative, i.e. $\sigma_H = -1$ MPa. This indicates that the material is in compressive in general at the material point.

c. Extrusion

The stress state and stress tensors for an extrusion process are shown in **Fig. 2.33**. According to the stress values given in the global coordinate system, the hydrostatic stress is $\sigma_H = \sigma_{ii}/3 = (-2 - 8 - 8)/3 = -6$ MPa, and the deviatoric stresses are $S_{11} = \sigma_{11} - \sigma_H = 4$ MPa, $S_{22} = S_{33} = -2$ MPa. The material is subjected to higher compressive stresses than for the drawing operation due to the high compressive value of the hydrostatic stress. Compared with the tension and drawing, higher forming forces are required for an extrusion process. However, the same deformation mode compared with the above two cases has been observed, since $J_3' = S_{11}S_{22}S_{33} = 16 > 0$ and they have the same stress components of the deviatoric stress tensor.

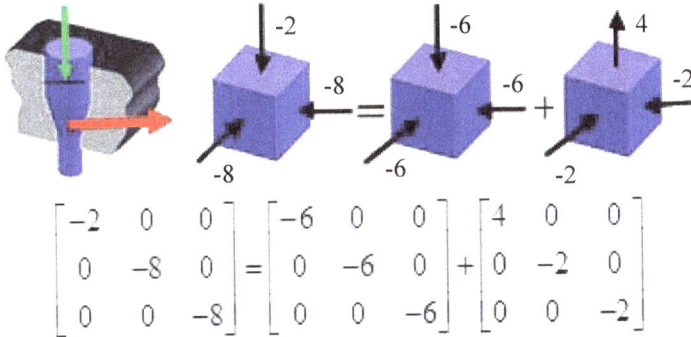

$$\begin{bmatrix} -2 & 0 & 0 \\ 0 & -8 & 0 \\ 0 & 0 & -8 \end{bmatrix} = \begin{bmatrix} -6 & 0 & 0 \\ 0 & -6 & 0 \\ 0 & 0 & -6 \end{bmatrix} + \begin{bmatrix} 4 & 0 & 0 \\ 0 & -2 & 0 \\ 0 & 0 & -2 \end{bmatrix}$$

Fig. 2.33   Stress states in extrusion (units are in MPa).

From the above three cases (simple tension, drawing and extrusion) it can be seen that their stress states are different (see the first stress tensors for individual cases) and hydrostatic stress values are also different (they vary from positive to negative). However, their deviatoric stress tensors are the same and $J_3' > 0$. This indicates that their deformation modes are similar; that is, the work-piece is extended in the axial direction and contracted in lateral directions. It should be noted that for forming complex-shaped components, stress states change with the deformation and the J_3' value varies both spatially and with forming time. An example of this feature has been given by Zeng, Yuan, Wang *et al.* (1997) for a simple test-piece.

2.5.2 *Stress states in forming processes*

A coordinate system is defined with the definition of $\sigma_{11} > \sigma_{22} > \sigma_{33}$. For simplicity of analysis it is also assumed that the directions of principal stress are the same as the chosen coordinate system, that is $\sigma_1 > \sigma_2 > \sigma_3$. A stress state diagram for various forming processes is shown in **Fig. 2.34**, which contains three vertical lines starting from "*extension*" to "*plane strain*" then to "*contraction*" of a work-piece. The three curves $\sigma_1 = 0$, $\sigma_2 = 0$ and $\sigma_3 = 0$, shown in **Fig. 2.34**, represent the three types of plane stress conditions, i.e. (i) two compressive stresses ($\sigma_1 = 0$ and the other two principal stresses are compressive); (ii) one tension and one compression ($\sigma_2 = 0$, which indicates $\sigma_1 > 0$ and $\sigma_3 < 0$); and (iii) tension in two directions ($\sigma_3 = 0$ and the other two principal stresses are tensile).

A parameter μ_σ is introduced to assess the stress state at different loading conditions. It is defined as

$$\mu_\sigma = \frac{2\sigma_2 - (\sigma_1 + \sigma_3)}{\sigma_1 - \sigma_3}.$$ (2.41)

For example,

- for simple tension, $\sigma_1 = \sigma_Y$ (or Y), $\sigma_2 = \sigma_3 = 0$, and $\mu_\sigma = -1$;

- for simple compression, $\sigma_3 = -\sigma_Y$ (or $-Y$), $\sigma_1 = \sigma_2 = 0$, and $\mu_\sigma = +1$;

- for plane strain, $\sigma_2 = (\sigma_1 + \sigma_3)/2$, and $\mu_\sigma = 0$.

These characteristic lines are shown in **Fig. 2.34**. The three vertical lines are:

- *extension line*: $\sigma_2 = \sigma_3$ (can be tension or compression or equal to 0) and $\mu_\sigma = -1$, which is equivalent to simple "tension" (strain state is similar to simple tension). The top of line (extension; $\mu_\sigma = -1$), $\sigma_2 = \sigma_3 = 0$ represents the pure tension as the values of compressive stresses σ_2 and σ_3 ($\sigma_2 = \sigma_3$) increase when moving to the lower end of the line, even if the same deformation modes are observed (extension). However, the force required to achieve the same deformation increases, as the values indicate in **Figs. 2.31, 2.32** and **2.33**.

- *plane strain line*: $\sigma_2 = (\sigma_1 + \sigma_3)/2$ and $\mu_\sigma = 0$ according to Eq. (2.46). This indicates that as a work-piece is extended in direction "1", the strain in direction "2" is zero. Compressive stress σ_2 is normally required to keep the strain in that direction is zero in compressive forming conditions, such as the forming cases shown in **Fig. 2.34**.

- *contraction line*: $\sigma_1 = \sigma_2$ and $\mu_\sigma = +1$. Under the compressive load in direction "3", the dimension of a work-piece in the direction reduced. This is opposite to the extension line discussed earlier. The values of compressive stresses σ_1 and σ_2 ($\sigma_1 = \sigma_2$) reduce from the bottom of the line to the top – the pure compression, which is $\sigma_1 = \sigma_2 = 0$.

Between the lines of "extension" and "plane strain" $\sigma_2 < (\sigma_1 + \sigma_3) / 2$, that is $\mu_\sigma < 0$, the "length" of the work-piece increases during deformation. However, between the lines of plane strain and contraction, $\sigma_2 > (\sigma_1 + \sigma_3) / 2$, that is $\mu_\sigma > 0$, the "length" of the work-piece decreases during deformation. Thus in the forming process, we can estimate the work-piece is extended or contracted under different stress state situations.

Based on the tension and compression forming conditions, four different stress states can exist, they are (**Fig. 2.34**):

- In the area below the line $\sigma_1 = 0$. That is $0 > \sigma_1 > \sigma_2 > \sigma_3$. The work-piece is under compressive forming in all three directions, but the "length" of the work-piece in direction "1" is extended for extrusion (also see **Fig. 2.33**). For an extrusion of a plate, the lateral strain increment in mid-section of the work-piece is zero, i.e. $d\varepsilon_2 = 0$. The length of the work-piece in direction "3" is compressed in the forging of a component in a die with a rubber surrounded, which provides compressive stresses around the work-piece. In the compressive notched bar tests, the material at the notch is subjected to compression in the axial direction and in the other two directions due to the constraint of the notch. In a rubber constraint forging process, all the stress components are compressive so the work-piece becomes shorter and thicker.

- In the area between the lines of $\sigma_1 = 0$ and $\sigma_2 = 0$. In this region $\sigma_1 > 0 > \sigma_2 > \sigma_3$. This means that σ_1 is tensile, and both σ_2 and σ_3 are compressive. In the wire drawing process (also see **Fig. 2.32**), the material is extended due to the tensile stress σ_1 and the other two stress and strain components are compressive; in a forging process, the material is constrain by the die in direction "2" and thus the strain increment in direction "2" is zero, $d\varepsilon_2 = 0$. The material is compressed in direction "1" and extended in direction "3".

- In the area between the lines of $\sigma_3 = 0$ and $\sigma_2 = 0$. In this region $\sigma_1 > \sigma_2 > 0 > \sigma_3$. Only the stress in direction "3" is compressive in the main deformation area. In many sheet metal forming processes, the material is subjected to stretching in the plane directions and the compressive stress in thickness direction, especially when back pressure is applied to reduce the damage growth in superplastic forming (If no back pressure is applied, the stress in the thicker direction is normally ignored as it is fairly small compared with the stresses in the plane. Then it becomes plane stress condition).

- In the area above the line of $\sigma_3 = 0$. The work-piece is subjected to tensile stresses in all three directions. For example, in the notched bar tensile tests, the material around the symmetric axis of the notch area is subjected to tension in all three directions due to the constraint of material in the parallel section.

It should be noted that the same deformation state (straining state) could correspond to different stress states. For example, both simple tension and extrusion would cause the extension of the work-piece as the values of μ_σ and J_3' are similar but their stress states are different as the values of σ_H are not the same. It has been mentioned previously that the hydrostatic stress does not affect plastic deformation of materials. Deviatoric stress tensors and J_3' values for simple tension, drawing and extrusion are the same type, thus their deformation states are the same as well.

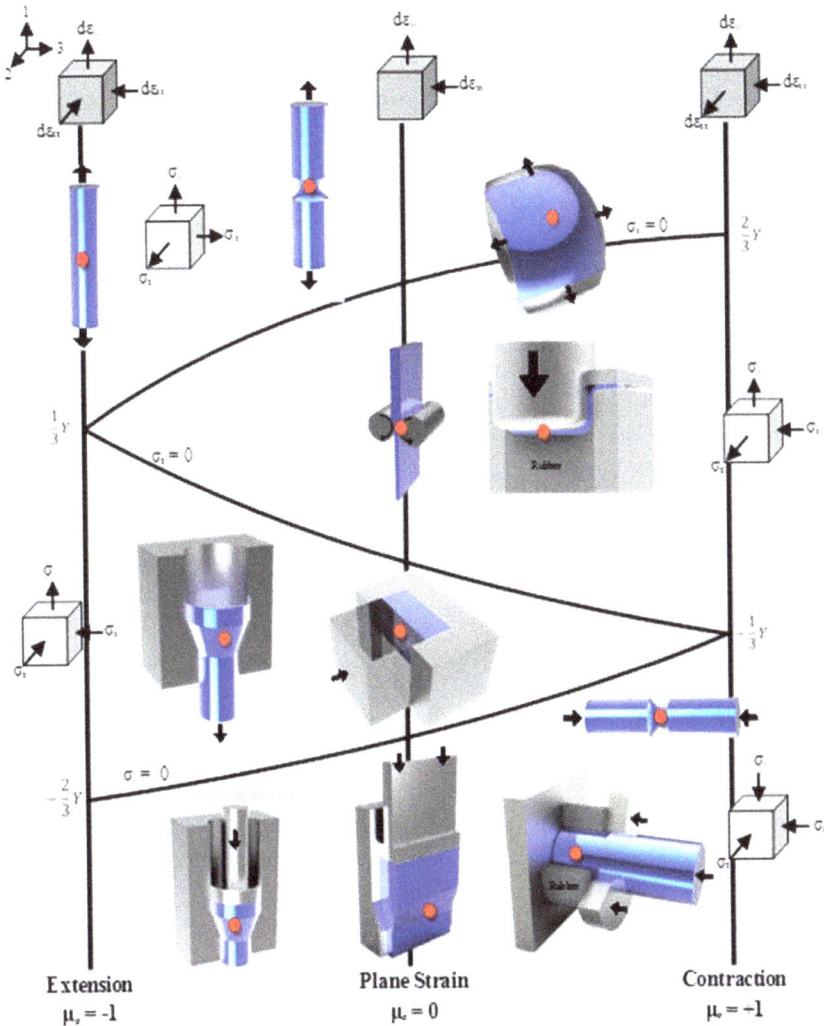

Fig. 2.34 Analysis of stress states in various forming processes. Stress states at red spot locations on blue work-pieces.

Based on the information given above, the following three aspects are now analysed.

a. Forming forces

The forming force decreases as the deformation states shown in **Fig. 2.34** move up from the bottom of the figure to achieve the same amount of deformation. This supports planning forming processes and tool designs, enabling complex-shape components to be formed using lower capacity presses.

At the bottom part of the figure, i.e. below the line of $\sigma_1 = 0$, all principal stresses are compressive. Large forces are required to achieve the same amount of deformation. For example, the applied force is higher in extrusion processes compared to that in wire drawing (cf. **Figs. 2.32** and **2.33**) for the same material using the same die. The formability in extrusion is higher due to the high compressive hydrostatic stress, which reduces the damage of the material in processing. This is discussed in Chapter 6.

At the top part of the figure, i.e. above the line of $\sigma_3 = 0$, all the principal stresses are positive and the material can be deformed readily with the applied forces. However, in this case, the material is subjected to tensile hydrostatic stress and failure is already taking place.

b. Uniformity of deformation

For pure extension and compression, such as uniaxial tension and up-setting processes, $\mu_\sigma = -1$ and $\mu_\sigma = +1$, the deformation should be uniform throughout the work-piece. The instability of materials is not considered here. However in real forming processes, friction occurs between the die and work-piece, which would introduce shear forces at interfaces and thus causes non-uniform deformation. This phenomenon can be often seen in up-setting forging processes. In addition, temperature gradients would also cause non-uniform plastic flow of materials in hot/warm forming processes.

c. Formability and potential failure

Forming forces are high at the forming stress states shown at the bottom part of **Fig. 2.34**. However, due to the high compressive value of σ_H ($\sigma_H < 0$), damage would not be nucleated easily, thus higher deformations could be achieved before failure or cracking occurring. That is why in compressive forging processes large deformations can be achieved and more complex-shaped components can be formed. In sheet metal forming, the material is mainly subjected to tensile stresses, in which $\sigma_H > 0$ and failure and tearing can take place more readily. In superplastic forming processes, back pressure is sometimes used to reduce the positive value of, σ_H and, thus to increase the formability of the material. Obviously, this would increase the forming pressure to overcome the back pressure forces.

The stress state varies spatially and with deformation. A forming process can be optimised to control the stress-state histories so that more complex-shaped components can be formed using lower forming forces. Formability of the material can be studied using damage mechanics theories detailed in Chapter 6.

Chapter 3

Unified Constitutive Modelling Techniques

In high temperature creep rupture analysis, scientists (such as Kachanov, 1958; Hayhurst 1972; Dyson and McLean, 1977; Leckie and Hayhurst, 1977) introduced internal state variables to represent damage evolution, which enables the damage and creep strain development of materials in service conditions to interact, so that lifetime and tertiary creep can be predicted. Lemaitre and Chaboche (1990) introduced isotropic hardening and kinematic hardening state variables, which were incorporated into elastic-plastic and elastic-viscoplastic flow laws to predict the hardening behaviour of materials under cyclic loading conditions. These represent the early stage development of unified constitutive equations and mainly concentrated on the modelling of stress–strain response, hardening and failure of materials under different loading conditions. The developments on formulating unified constitutive equations established the basis for the creation of unified theories for metal forming simulations.

In metal forming processes, particularly in hot/warm forming conditions, microstructure evolution, such as recrystallisation (Lin, Liu, Farrugia *et al.*, 2005), grain growth (Cheong, Lin and Ball, 2001; Lin and Dunne, 2001; Lin, 2003) and grain refinement, precipitate dissolution, formation and growth (Lin, Ho and Dean, 2006), become more important. That is because one of the advantages of metal forming processes is to make the material stronger and tougher, both of which are directly related to the material microstructures. In addition, micro-damage of materials during large deformation forming processes is also important, since the failure of materials for these processes should be predicted (Lin and Dean, 2005; Lin, Foster, Liu *et al.*, 2007) to avoid defects and control damage tolerance in formed parts. This is especially important for sheet metal forming processes, such as stamping (Lin, Mohamed, Dean *et al.*, 2014) and superplastic forming (Lin, Cheong and Yao, 2002).

In this chapter, the theories and commonly used state variables (mainly hardening state variables) for formulating unified constitutive equations will be introduced. To enable the concept of unified constitutive modelling methods to be applied easily, the basic theories of plasticity and viscoplasticity will also be briefly introduced. This includes the choice of basic constitutive laws and viscoplastic potentials.

3.1 Viscoplastic Potential and Basic Constitutive Laws

3.1.1 *Basic definitions for elastic-plastic problems*

a. General stress–strain definition

Figure 3.1 shows a piece of material loaded to the plastic deformation region with the initial Young's modulus of E. If the damage is not considered for the material during the first stage of elastic-plastic deformation, the unloading curve will be parallel to the elastic part of the initial loading curve. Consequently the value of the Young's modulus, E, does not change. If the material is

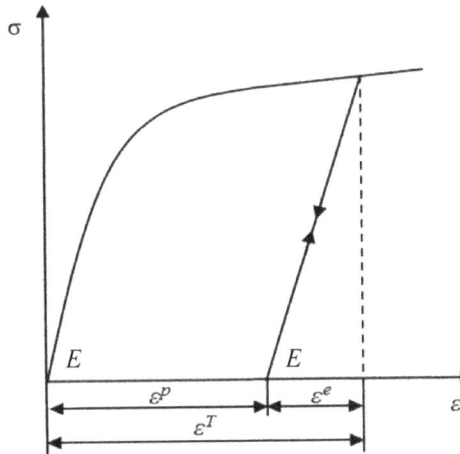

Fig 3.1 Stress–strain relationship.

loaded again, it will follow the unloading path and the yield stress will be higher due to plastic deformation (work hardening). In this figure, we have

$$\varepsilon^T = \varepsilon^P + \varepsilon^e \text{, or, } \varepsilon^e = \varepsilon^T - \varepsilon^P \text{,} \tag{3.1}$$

where ε^T is the *total strain*, which is the *measured strain* from a uniaxial tension/compression test. ε^P and ε^e are known as *plastic strain* and *elastic strain*, respectively.

Based on the elastic-plastic stress–strain relationship, for the uniaxial case we have

$$\sigma = E\varepsilon^e = E\left(\varepsilon^T - \varepsilon^P\right). \tag{3.2}$$

For general multiaxial cases, the above equation can be written as

$$\sigma_{ij} = D_{ijkl}\varepsilon_{ij}^e = D_{ijkl}\left(\varepsilon_{ij}^T - \varepsilon_{ij}^p\right). \tag{3.3}$$

Normally the total strain is the measured strain from experiments, i.e. nominal strain. In FE analysis, the total strain is obtained from nodal displacements, which are achieved by solving global system equations. In this case, to enable the stresses to be calculated, the calculation of plastic strains becomes of vital importance. Thus in materials modelling the key problem is how to calculate plastic strain accurately.

b. Power-law and strain hardening exponent

A typical uniaxial true stress-true strain curve shown in **Fig. 3.1** can be approximated by a power-law equation (without considering the material's damage and failure in the plastic deformation):

$$\sigma = K\varepsilon^n. \tag{3.4}$$

Note that this equation indicates neither the elastic region nor the yield point of the material. These quantities, however, are readily available from the stress–strain curve. K is known as *strength coefficient*. A high K-value indicates the material is strong – hard to deform. Note that K is the true stress at a true strain of unity. The constant, n, is the *strain-hardening exponent*. A material with high n-value indicates that the material has high strain hardening feature, which enables the material to be deformed in a more uniform manner. For most of metals, n-value reduces with increasing temperature. Taking the log of Eq. (3.4):

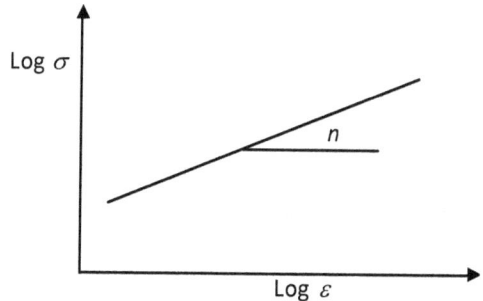

Fig. 3.2 **Stress–strain relationship in log-log scales.**

$$\log\sigma = \log K + n\log\varepsilon. \tag{3.5}$$

If the true stress-true strain curve is plotted on a log-log scale, a straight line is obtained and shown in **Fig. 3.2**: The slope n is the *strain-hardening exponent.*

Again this equation is only able to approximate the stress–strain curves before the stress drops (necking). However, the n-value is an important parameter in assessing the strain hardening feature of a material. This is especially important for sheet metal forming, such as stamping, hydroforming and superplastic forming.

c. Strain rate hardening

Equation (3.4) is mainly used for room temperature conditions. In hot/warm forming conditions, strain rate ($\dot{\varepsilon} = d\varepsilon/dt$) effects on flow stress become important. By the introduction of strain rate effect, Eq. (3.4) can be written as

$$\sigma = K\varepsilon^n\dot{\varepsilon}^m, \tag{3.6}$$

where m is the *strain rate hardening exponent.* Strain hardening normally reduces at high temperature forming conditions, such as hot stamping and superplastic forming, due to static and dynamic recovery of dislocations. This recovery takes place even more at low strain rate deformation process as more time is available for the recovery and annealing. For simplicity here the strain hardening exponent is assumed to be zero ($n = 0$). The above equation can therefore be written as:

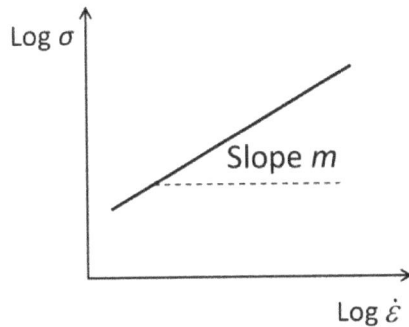

Fig. 3.3 Stress–strain rate relationship in log-log scales.

$$\sigma = K\dot{\varepsilon}^m. \tag{3.7}$$

Taking *log* of the above equation,

$$\log\sigma = \log K + m\log\dot{\varepsilon}. \tag{3.8}$$

It can be seen from the equation that the relationship between flow stress and strain rate is linear in *log-log* scales, and the slope of the straight line is the value of the strain rate hardening exponent *m*, as shown in **Fig. 3.3**.

Strain rate hardening is an important parameter in hot forming conditions such as superplastic forming. This is the mechanism to reduce localised thinning (Lin, 2003) in gas-blow forming of complex-shaped panel components.

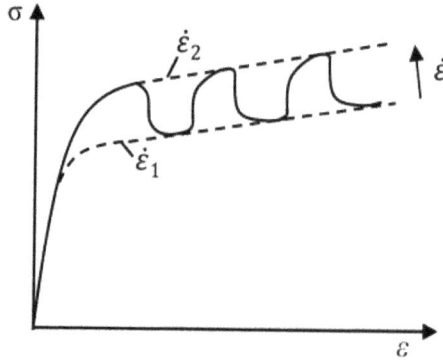

Fig. 3.4 Stress–strain curve for a strain rate jump test for a superplastic material.

Strain rate jump tests (**Fig. 3.4**) are often used to determine the *m*-value for a material at high temperatures:

$$m = \frac{\log \sigma_2 - \log \sigma_1}{\log \varepsilon_2 - \log \varepsilon_1}. \tag{3.9}$$

The *m*-value of a material is related to temperature, deformation rate and microstructure of the material, such as grain size, which could change during forming processes. For superplastic materials, the *m*-value is normally above 0.5. A high *m*-value of a material enables more complex-shaped components to be formed.

3.1.2 Calculation of plastic strains

As discussed earlier (reference Eq. (3.3)), the key challenge to calculate the stress components for solving a system problem is to calculate plastic strains. Various laws can be used for calculating plastic strain rate. Here the two most commonly used laws, (i) power-law and (ii) sinh-law, are introduced.

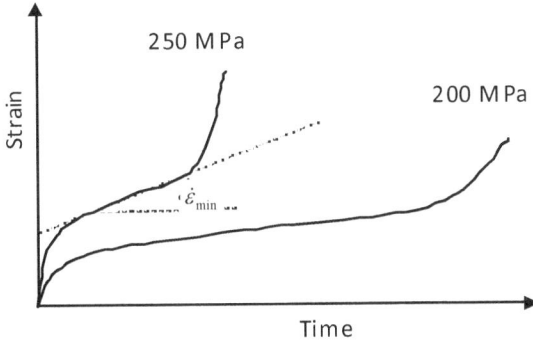

Fig. 3.5 Creep curves for different stresses.

a. Power-law

The power-law is commonly used to represent creep deformation of materials. **Fig. 3.5** shows creep curves with *minimum creep rate*, $\dot{\varepsilon}_{min}$. A linear relationship between the stress and minimum creep strain rate $\dot{\varepsilon}_{min}$ is observed in log-log coordinate system, as shown in **Fig. 3.6**. The slope of the straight line, n, gives the stress dependent creep rate function. The equation can be given as

$$\dot{\varepsilon}_{min} = \frac{d\varepsilon_{min}}{dt} = \dot{\varepsilon}_0 \sigma^n, \tag{3.10}$$

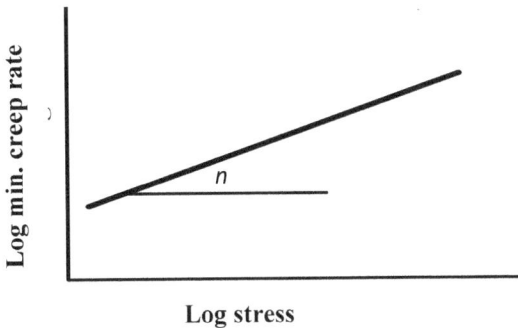

Fig. 3.6 Relationship of stress and minimum creep rate.

where $\dot{\varepsilon}_0$ is a parameter related to the reference creep rate. By the introduction of temperature effect, the above equation can be written as:

$$\dot{e}_{min} = \dot{\varepsilon}_0 \sigma^n \exp\left(-\frac{Q_c}{RT}\right), \tag{3.11}$$

where Q_c is related to the activation energy of the material, R ($R = 8.31$ J/(mol·K)) is the gas constant and T the temperature in Kelvin. The temperature-dependent variables will be discussed further in later chapters.

Equation (3.10) can be readily extended into general viscoplastic problems through various assessments. For example, **Fig. 3.7** shows stress–strain curves at a fixed temperature with different strain rates obtained from uniaxial tensile tests. We can observe that:

(i)   The flow stress increases as the deformation rate increases.

(ii) The ductility increases as the strain rate decreases.

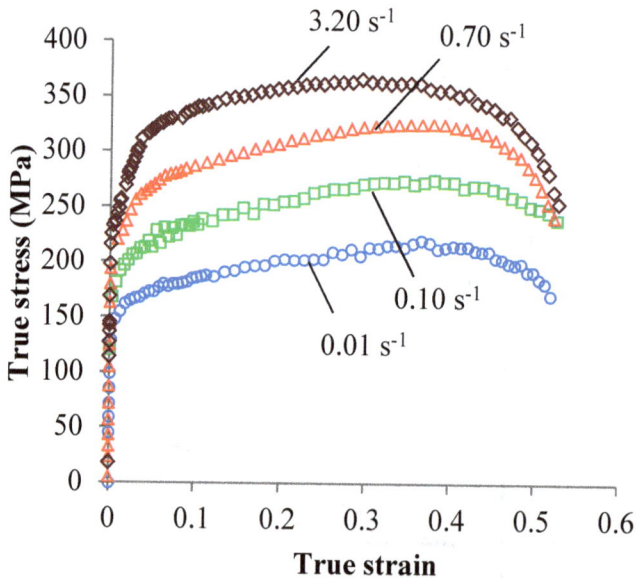

Fig. 3.7 Experimental true stress–strain curves for a boron steel deforming at 923 K with different strain rates (Li, 2013).

If the flow stresses at a strain of 0.005 s^{-1}, 0.05 s^{-1} or 0.1 s^{-1} for the different strain rate curves are chosen, we will have relationships shown in **Fig. 3.8**, where the relationship for other temperatures are also given. It can be seen from **Fig. 3.8** that for a given temperature, the relationship between the flow stress and strain rate is approximately on a straight line in log-log scales. In the figure, for any strain rate, the stress increments with increasing strain level reveal the strain hardening characteristic: with respect to any strain level, the increase of stress with increasing strain rate exhibits the strain rate hardening characteristic. The strain rate hardening exponent is represented by the slope of trend lines, n, and corresponds to the power exponent in the equations. For example, n is 0.062 at the strain of 0.005 s^{-1} and increases to 0.107 at the strain of 0.1 s^{-1}.

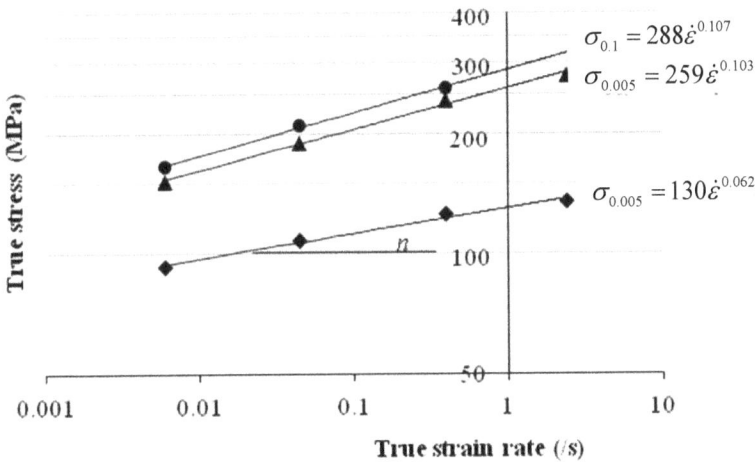

$$\sigma_{0.1} = 288\dot{\varepsilon}^{0.107}$$
$$\sigma_{0.005} = 259\dot{\varepsilon}^{0.103}$$
$$\sigma_{0.005} = 130\dot{\varepsilon}^{0.062}$$

Fig. 3.8 Effect of strain rate on flow stress at strain rates of 0.005 s^{-1}, 0.05 s^{-1} and 0.1 s^{-1} ($\sigma_{0.005}, \sigma_{0.05}, \sigma_{0.1}$) for a boron steel tested at 973 K (Li, 2013).

The features observed in **Fig. 3.8** provide the evidence that the viscoplastic flow of the material can be expressed using the power-law relationship, which is often written as

$$\dot{\varepsilon}^p = \frac{d\varepsilon^p}{dt} = \dot{\varepsilon}_0 \left(\frac{\sigma}{K}\right)^n, \tag{3.12}$$

where $\dot{\varepsilon}_0$ is the reference strain rate and gives the correct unit for the equation (e.g. s^{-1}). K is a constant in MPa, for example.

In continuum damage mechanics, the failure process of a material can be modelled using a damage variable, ω. If the relationship of lifetime (or ductility) and strain rate for a material is linear in log-log scales, the damage evolution variable can also be expressed using a power-law relationship. This will be detailed in Chapter 6. An example damage equation is given below.

$$\dot{\omega} = \frac{d\omega}{dt} = C \left(\frac{\sigma}{\sigma_0}\right)^\eta, \tag{3.13}$$

where, C, σ_0 and η are material constants to be determined from experimental data.

b. Sinh-laws

For some materials deformed at certain conditions a hyperbolic sinh-law can be used to express the viscoplastic flow, creep strain development and damage evolution. Examples are given in **Fig. 3.9** for copper and an aluminium alloy. The vertical axis (effective stress) for both materials is linear and the horizontal axis (rupture time, t_R) is in log scale. Linear relationships are observed for all the test conditions.

In this case, the hyperbolic sinh-law can be used to express damage evolution. The simplest form for a uniaxial case can be written as

$$\dot{\omega} = \frac{d\omega}{dt} = D \sinh(B\sigma), \tag{3.14}$$

where D (unit: e.g. h^{-1}) and B (unit: e.g. MPa^{-1}) are material constants.

If the values of $B\sigma$ are large enough, the following relationship can be obtained:

$$\dot{\omega} = D \sinh(B\sigma) = D\frac{1}{2}(e^{B\sigma} - e^{-B\sigma}) \approx \omega_0 \exp(B\sigma). \tag{3.15}$$

Taking natural log for the above equation gives

$$\ln \dot{\omega} = \ln \omega_0 + B\sigma. \tag{3.16}$$

In Eq. (3.16), the damage evolution is directly related to the rupture time represented in **Fig. 3.9**. Thus in the axes of linear σ and $\log t_R$ the relationship is linear and their damage evolution can be expressed using a sinh-law. The value B is related to the slope of the strain lines, which in these cases is stress-state-dependent.

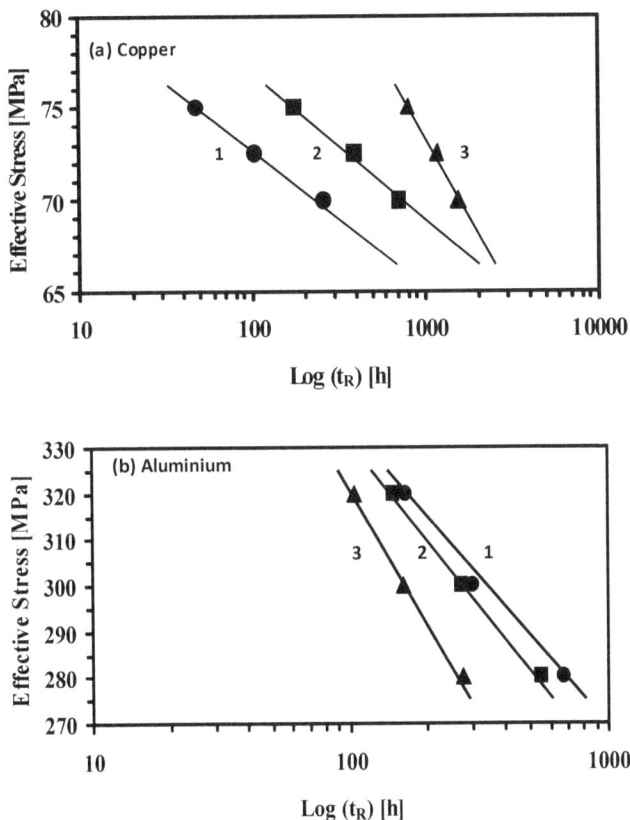

Fig. 3.9 Effective stress versus rupture time for (a) copper and (b) an Al-alloy at various stress-states. The numbers indicate types of stress state: 1 – pure tension; 2 – tension-torsion; 3 – pure torsion (Lin, Kowalewski and Cao, 2005).

Similarly, if the vertical axis (stress) in **Fig. 3.8** is linear (not in log) and the stress (linear)–strain rate (in log) relationship is linear, then the sinh-law can be used to express the viscoplastic flow of the material. That is,

$$\dot{\varepsilon}^p = \frac{d\varepsilon^p}{dt} = A\sinh(B\sigma), \tag{3.17}$$

where A and B are constants.

Based on the above understanding of the mathematics and physics of the power-law and the sinh-law, more appropriate expressions can be selected for the prediction of viscoplastic flow and rupture of materials under different deformation conditions. It should be noted that, for a given material, the dominant deformation mechanism would change with temperature, deformation rate and microstructure, which may change dynamically during a process. This would cause difficulties in choosing appropriate laws. However, for a certain range of deformation rate, temperature and microstructure, a dominant deformation can be defined, which supports materials modelling work.

3.1.3 *Viscoplastic potentials*

a. *Power-law*

Multiaxial flow for viscoplastic materials, represented by a power-law, can be obtained by defining a viscoplastic potential and by assuming von-Mises behaviour for perfect viscoplasticity, i.e. no strain hardening is considered. The potential equation for a power-law can be in the form of

$$\psi = \dot{\varepsilon}_0 \frac{K}{n+1} \left(\frac{\sigma_e}{K} \right)^{n+1},\tag{3.18}$$

where σ_e ($\sigma_e = \left(3S_{ij} \cdot S_{ij}/2 \right)^{1/2}$) is effective stress defined in Chapter 2, and S_{ij} ($S_{ij} = \sigma_{ij} - (\sigma_{11} + \sigma_{22} + \sigma_{33})\delta_{ij}/3$) are the deviatoric stresses. By differentiating the equation in terms of deviatoric stresses, the following relationship can be obtained

$$\dot{\varepsilon}_{ij}^p = \frac{\partial \psi}{\partial S_{ij}} = \frac{3}{2} \frac{\partial \psi}{\partial \sigma_e} \frac{S_{ij}}{\sigma_e} = \frac{3}{2} \frac{S_{ij}}{\sigma_e} \dot{\varepsilon}_0 \left(\frac{\sigma_e}{K} \right)^n = \frac{3}{2} \frac{S_{ij}}{\sigma_e} \dot{\varepsilon}_e^p,\tag{3.19}$$

where

$$\dot{\varepsilon}_e^p = \frac{\partial \psi}{\partial \sigma_e} = \dot{\varepsilon}_0 \left(\frac{\sigma_e}{K} \right)^n\tag{3.20}$$

is the effective plastic strain rate, which is similar to the uniaxial case.

b. *Sinh-laws*

Similar to creep damage constitutive equations (Lin, Hayhurst and Dyson, 1993), sinh-law based viscoplastic constitutive equations for modelling multiaxial deformation of a material can be formulated by assuming an energy dissipation rate potential of the form

$$\psi = \frac{A}{B}\cosh(B\sigma_e) \qquad (3.21)$$

Similarly, by differentiating the equation in terms of deviatoric stresses, there is

$$\dot{\varepsilon}_{ij}^p = \frac{\partial \psi}{\partial S_{ij}} = \frac{3}{2}\frac{\partial \psi}{\partial \sigma_e}\frac{S_{ij}}{\sigma_e} = \frac{3}{2}\frac{S_{ij}}{\sigma_e}A\sinh(B\sigma_e) = \frac{3}{2}\frac{S_{ij}}{\sigma_e}\dot{\varepsilon}_e^p, \qquad (3.22)$$

where

$$\dot{\varepsilon}_e^p = \frac{\partial \psi}{\partial \sigma_e} = A\sinh(B\sigma_e),$$

which is similar to the form of uniaxial case.

The above potentials are formulated based on the assumption of the rigid perfect viscoplasticity condition: the absence of hardening and an elastic region. With the consideration of elastic regions and possible hardening mechanisms, the potential equations will become more complicated. This will be discussed in later chapters.

3.2 Hardening in Plastic Deformation

3.2.1 *Isotropic hardening*

a. *Strain based hardening law*

In this case, the evolution of the loading surface in elastic-plastic deformation is controlled by a scalar variable, R, which can be related to either the dissipated plastic work, the accumulated plastic strain, or the dislocation density. Isotropic hardening is represented in **Fig. 3.10,** which shows:

(i) The initial yield surface with the initial yield stress of σ_Y at the loading condition. It should be noted that both σ_Y and k are used to represent the yield surface in illustrations. For an elastic-plastic problem, the uniaxial yield stress is easy to determine in tests. However, in elastic-viscoplastic tests, the initial yield stress of a material is difficult to determine, since creep may take place at very low stress. Thus k is normally used to represent the threshold stress below which it is assumed that creep does not occur.

J. Lin

(ii) As plastic deformation takes place, hardening occurs due to the accumulation of the dislocations. This enables the yield surface to expand to the surface of $\sigma_Y + R$.

(iii) At the unloading and reloading condition, the new yield point would be $\sigma_Y + R$. The feature of the expansion of the yield surface due to plastic deformation is known as *strain hardening*.

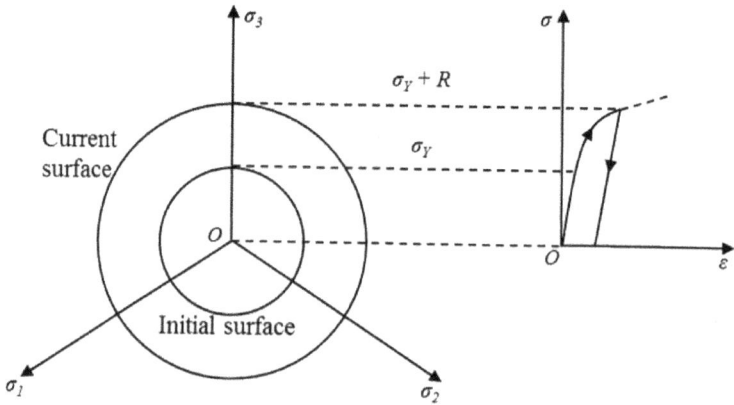

Fig. 3.10 Isotropic hardening representation in the space of principal stresses in loading.

In this particular case, the expansion of the yield surface is the same in all the directions, which is known as *isotropic hardening, R*. This can often be expressed as a function of plastic strain, such as

$$\dot{R} = \frac{dR}{dt} = b(Q - R)|\dot{\varepsilon}^p|, \tag{3.23}$$

where Q and b are material constants. By integrating the equation, the isotropic hardening R can be explicitly expressed as a function of plastic strain.

$$R = Q\left[1 - \exp(-b|\varepsilon^p|)\right]. \tag{3.24}$$

$\dot{R}$ approaches zero when R reaches the constant value, Q (**Fig. 3.11**). b is a scale factor to rationalise the effect of plastic strain on hardening features. If the value of b is high, isotropic hardening develops very quickly and R approaches Q at a lower plastic strain ε^p.

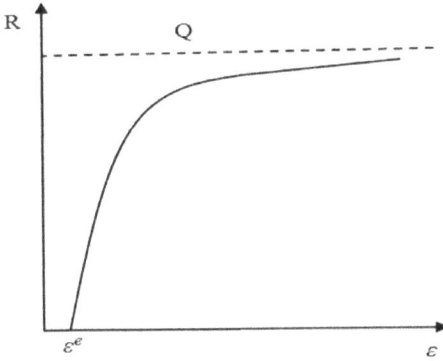

Fig. 3.11 Variation of isotropic hardening.

b. Dislocation based hardening law

Isotropic hardening can also be expressed directly in a function of dislocation density. Dislocations mainly concentrate on sub-grain boundaries, with the average dislocation density being expressed as ρ. Taking the strain hardening and recovery of dislocations into account in the same way as is frequently used in connection with creep type processes, but neglecting annealing and recrystallisation, the dislocation density rate can be described as (Sandstrom and Lagneborg, 1975)

$$\dot{\rho} = \frac{d\rho}{dt} = \frac{1}{bl}|\dot{\varepsilon}^p|, \tag{3.25}$$

where b is the Burgers vector and l is the dislocation mean free path. Equation (3.25) shows the initial increment of dislocations during plastic deformation as being proportional to plastic strain. By including the dynamic recovery of dislocation in plastic deformation, Eq. (3.25) can be written as (Lin and Dean, 2005):

$$\dot{\rho} = \frac{d\rho}{dt} = b(Q - \rho)|\dot{\varepsilon}_p|. \tag{3.26}$$

By integrating the equation, we have

$$\rho = Q\left[1 - \exp(-b|\varepsilon^p|)\right]. \tag{3.27}$$

As can be seen, this is similar to Eq. (3.23). The hardening of the material is directly related to the dislocation density, as discussed by Sandstrom and Lagneborg (1975), where it is argued that

$$R \approx \alpha Gb\sqrt{\rho}, \tag{3.28}$$

where α is a constant of about 0.5–1.0 and G the shear modulus of the material. For engineering applications, this equation can be simplified as (Lin, Liu, Farrugia et al., 2005)

$$R = B\rho^{1/2},$$
(3.29)

and the evolutionary form is

$$\dot{R} = \frac{1}{2}B\rho^{-1/2}\dot{\rho}.$$
(3.30)

Equation (3.30) shows that the evolution of isotropic hardening is directly related to dislocation density rate, which is represented by Eq. (3.26), and is a function of plastic strain rate. These two equations have similar functions to Eq. (3.23) where the isotropic hardening rate is directly related to plastic strain rate. The trend of isotropic hardening development represented by the dislocation concept follows the same shape of the curve shown in **Fig. 3.11**.

It is noted that the above equations only apply to room temperature deformation (cold forming processes) and where only dynamic recovery of dislocation is considered. At high temperature forming conditions annealing takes place and recrystallisation may also occur. These will reduce the dislocation density and thus reduce the isotropic hardening or result in negative hardening (i.e. softening). The isotropic hardening is then not only directly related to plastic strain but will also be time-dependent (strain rate-dependent), temperature-dependent and microstructure-dependent. These will be discussed in later chapters.

c. Effect of isotropic hardening on flow stress

In the power-law elastic-viscoplastic formulation (Eq. (3.12)), we can introduce the initial yield stress (threshold stress), k, and the isotropic hardening, R. Then the equation becomes

$$\dot{\varepsilon}^p = \dot{\varepsilon}_0 \left(\frac{\sigma - R - k}{K}\right)^n.$$
(3.31)

In the multiaxial case the stress σ can be replaced by von-Mises stress σ_e or $J_2' = \left(3S_{ij} \cdot S_{ij}/2\right)^{1/2}$. It should be noted that both the initial yield stress, k, and the isotropic hardening, R, are scalars. By rearranging the equation it becomes

$$\sigma = k + R + K \left(\frac{\dot{\varepsilon}^p}{\dot{\varepsilon}_0}\right)^{1/n}.$$
(3.32)

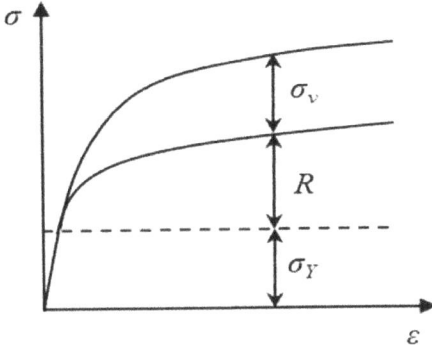

Fig. 3.12 Stress components in an isotropic hardening deformation condition.

Fig. 3.12 shows the stress components in an isotropic hardening deformation condition, where $\sigma_v = K(\dot{\varepsilon}^p/\dot{\varepsilon}_o)^{1/n}$ is the stress due to the viscoplastic strain rate. The flow stress can be divided into three stress parts: the initial yield stress, isotropic hardening and the stress due to the viscoplastic strain rate. For plasticity, the viscoplastic stress part can be ignored.

3.2.2 *Kinematic hardening*

a. Linear kinematic hardening

Kinematic hardening is *anisotropic hardening*, which corresponds to the translation of the yield surface. The kinematic hardening variable, which is represented by **X**, is of a *tensorial nature*; it indicates the present position of the yield surface. By the consideration of the initial yield stress and kinematic hardening and ignoring the isotropic hardening, the current yield surface becomes

$$f = J_2'(\mathbf{S}-\mathbf{X})-k = 0 . \tag{3.33}$$

Figure 3.13 shows, schematically, the movement of the yield surface in a typical loading and unloading condition in the space of principal stresses and the corresponding tension-compression stress–strain relationship. For the convenience of presentation, k is again used as the initial yield stress of the material in the multiaxial case.

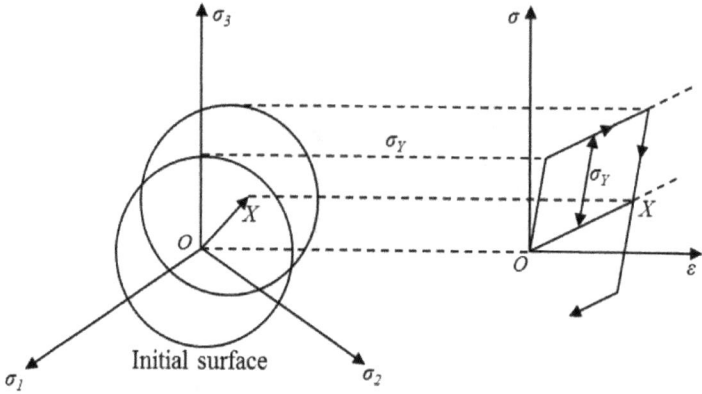

Fig. 3.13 Kinematic hardening in the space of principal stresses in loading and unloading.

This figure shows that:

- At the tension stage, the yield locus is a circle with the radius of k (or σ_Y) in the space of principal stresses, σ_1, σ_2 and σ_3. The centre of the circle is at the origin O.

- Linear hardening is assumed and the flow stress increases proportionally as the plastic deformation continues.

- At the compression stage, the centre of the yield circle moves to the new position of $\mathbf{X}$. The radius of the yield surface stays the same (k or σ_Y). The translation of yield surface is due to the kinematic hardening tensor $\mathbf{X}$.

Generally speaking, for an isotropic material, plastic deformation is supposed to take place at constant volume. Only the deviatoric part of the tensors $\boldsymbol{\sigma}$ and $\mathbf{X}$ appears. That is,

$$J_2'(\mathbf{S}-\mathbf{X}) = \left[\frac{3}{2}\left(S_{ij}-X_{ij}\right):\left(S_{ij}-X_{ij}\right)\right]^{1/2}. \tag{3.34}$$

This has been verified for a number of materials under cyclic loading conditions (Lemaitre and Chaboche, 1990). In the case of linear kinematic hardening, the rate equations can be represented as

$$\dot{\mathbf{X}} = \frac{d\mathbf{X}}{dt} = C_0\dot{\boldsymbol{\varepsilon}}^p \text{ or } \dot{X}_{ij} = \frac{dX_{ij}}{dt} = C_0\dot{\varepsilon}_{ij}^p, \tag{3.35}$$

where C_0 is a constant, which is related to the kinematic hardening rate. This is the simplest kinematic hardening law and is commonly used in modelling work.

b. *Multi-state variables*

Chaboche and Rousselier (1983) introduced two kinematic hardening variables to enable the hardening behaviour of a material under cyclic loading conditions to be accurately modelled. Normally, one state variable is used to model the initial hardening behaviour and the other to model the hardening behaviour at the late stage of the deformation. These can be expressed as (Lin, Dunne and Hayhurst, 1996)

$$\dot{X}_1 = \frac{dX_1}{dt} = C_1 \dot{\varepsilon}^p - \gamma_1 X_1 \dot{\varepsilon}_e^p \tag{3.36}$$

$$\dot{X}_2 = \frac{dX_2}{dt} = C_2 \dot{\varepsilon}^p - \gamma_2 X_2 \dot{\varepsilon}_e^p \tag{3.37}$$

$$\mathbf{X} = \mathbf{X}_1 + \mathbf{X}_2 \tag{3.38}$$

where C_1, γ_1, C_2 and γ_2 are material constants, which are determined through the best fitting of experimental data. $\dot{\varepsilon}_e^p$ is the effective plastic strain rate.

Equations (3.36) and (3.37) have similar features. The first term expresses the linear relationship between the hardening and plastic strain where the hardening rate is defined by the constant C (C_1 and C_2). The second term is related to dynamic recovery of the hardening and characterised by the parameter γ_1 and γ_2 respectively.

To model more complicated kinematic hardening behaviour of materials, multiple kinematic hardening variables can be introduced and the above equations can be generalised in the form of

$$\dot{X}_k = \frac{dX_k}{dt} = C_k \dot{\varepsilon}^p - \gamma_k X_k \dot{\varepsilon}_e^p, \text{ where } k = 1,2,...,N \tag{3.39}$$

$$\mathbf{X} = \sum_{k=1}^{N} \mathbf{X}_k, \tag{3.40}$$

where N is the number of kinematic hardening variables. The multiple kinematic hardening variables can be incorporated into the yield function in the same way as represented in Eq. (3.33).

In cold sheet metal stamping processes, kinematic hardening is important for the investigation of the bending and unbending behaviour, so that residual stress and springback can be accurately predicted in FE process modelling (Brunet, Godereaux and Morestin, 2000). However, obtaining experimental data for fairly high compressive strains from sheet metals at cyclic loading conditions is not easy (Zhao and Lee, 2000) mainly due to the buckling problem in sheet metal compression testing. Significant efforts have been made in this research area with the aim of obtaining improved data to determine kinematic hardening equations for springback prediction.

c. Isotropic and kinematic hardening

In cold metal forming conditions, sheet metal stamping conditions in particular, both isotropic hardening and kinematic hardening are important and should be accurately modelled to predict important features such as localised thinning and springback.

Figure 3.14 shows the stress conditions for the stamping of a box-type panel component. The isotropic hardening in stretching is important to reduce the localised thinning in stamping conditions. However, the material is subjected to bending and unbending conditions. At the bending and unbending regions, the material is subjected to tension first, then compressive loading, which makes the kinematic hardening become important for the prediction of residual stresses and springback at the unloading stage.

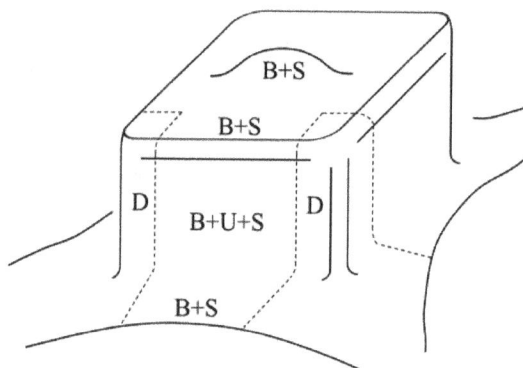

Fig. 3.14 Stress and loading conditions in stamping a box component.
B – bending; S – stretching; U – unbending; D – compressing.

Figure 3.15 shows the yield surface changes with the combined isotropic hardening and kinematic hardening. At the initial loading condition, the initial yield surface is shown with the yield stress of k. At the cyclic loading condition, kinematic hardening $\mathbf{X}$ (note that it is a tensor here) occurs, and the centre of the initial yield surface moves from O to the position of O'. In addition, the initial yield surface has expanded to the new radius of $k+R$, due to isotropic hardening. Therefore isotropic hardening expands the yield surface and the kinematic hardening translates the centre of yield surface.

The function for the combined isotropic and kinematic hardening can be in the form of

$$f = J_2'(\mathbf{S}-\mathbf{X}) - R - k = 0,\tag{3.41}$$

where the isotropic and kinematic hardening can be modelled using the equations given above.

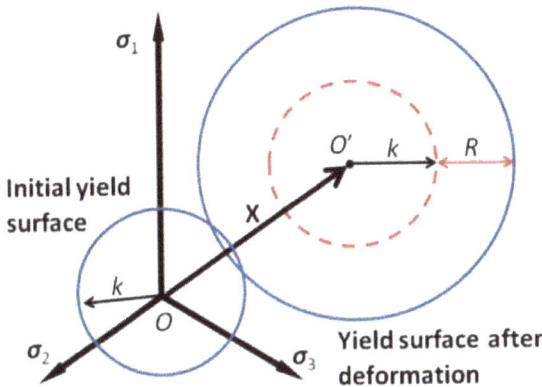

Fig. 3.15 Isotropic and kinematic hardening in the space of principal stresses.

3.3 State Variables and Unified Constitutive Equations

As discussed above, individual physical phenomena of a material at different deformation conditions can be modelled by the introduction of state variables, such as isotropic hardening, kinematic hardening etc. The details will be discussed in later chapters. In this section, a simple set of unified constitutive equations are introduced as an example to illustrate the theories.

3.3.1 *Unified constitutive equations*

For a uniaxial case, using a simple power-law viscoplastic problem with the consideration of isotropic hardening and kinematic hardening, the evolutionary equation set can be written as

$$\dot{\varepsilon}^p = \dot{\varepsilon}_0 \left(\frac{(\sigma - X) - R - k}{K} \right)^n \text{sgn}(\sigma - X) \tag{3.42}$$

$$\dot{R} = \frac{dR}{dt} = b(Q - R)|\dot{\varepsilon}^p| \tag{3.43}$$

$$\dot{X}_1 = \frac{2}{3} C_1 \dot{\varepsilon}^p - \gamma_1 X_1 |\dot{\varepsilon}^p| \tag{3.44}$$

$$\dot{X}_2 = \frac{2}{3} C_2 \dot{\varepsilon}^p \tag{3.45}$$

$$\dot{X} = \dot{X}_1 + \dot{X}_2 \tag{3.46}$$

$$\dot{\sigma} = E\dot{\varepsilon}^e = E(\dot{\varepsilon}^T - \dot{\varepsilon}^p) \tag{3.47}$$

Equations (3.42)–(3.47) are a set of unified constitutive equations, which can be used to model the stress–strain response for cyclic loading conditions. In particular:

- The isotropic hardening, R, occurs between the initial (monotonic) and the stabilised loading cycles.

- Two kinematic hardening variables, X_1, and X_2, are introduced to model the hardening behaviour of the material at the stabilised loading cycles. One is linear hardening and the other is nonlinear hardening.

- The above three hardening variables would affect the viscoplastic flow of the material and given in Eq. (3.42). If $\langle (\sigma - K) - R - k \rangle \leq 0.0$, the material is within the elastic region and hence $\dot{\varepsilon}^p = 0.0$. However, if $\langle (\sigma - K) - R - k \rangle > 0.0$, the plastic strain rate is calculated using Eq. (3.42).

- Once the plastic strain is calculated from Eq. (3.42), the flow stress can be calculated from Eq. (3.47). The total strain rate is represented by $\dot{\varepsilon}^T$. For a uniaxial tensile test, this is a strain rate applied, e.g. $\dot{\varepsilon}^T = 0.1$ s^{-1}. This equation can also be written as $\sigma = E(\varepsilon^T - \varepsilon^p)$, where ε^T is the measured strain from the test.

Within the equation set, E is the Young's modulus, k is the initial yield stress (or threshold stress), and $\dot{\varepsilon}_0$, K, b, Q, C_1, γ_1 and C_2 are material constants to be determined by fitting of experimental data. The detailed technique of the fitting will be introduced in Chapter 7.

3.3.2 *Solution of the unified constitutive equations*

Equations (3.42)–(3.47) are a set of *ordinary differential equations* (ODE), which normally cannot be solved analytically. Thus a numerical integration method should be used to solve these equations. The simplest integration method is the explicit Euler method (Leonhard Euler, Swiss, 1707–1783). For a simple uniaxial loading case with a constant loading rate of $\dot{\varepsilon}^T = 0.1$ s^{-1}, the integration process is outlined below.

- Firstly, the initial conditions, i.e. $t = 0$, should be given for all the variables, which are

$$\varepsilon^p = 0.0; \ R = 0.0; \ X_1 = 0.0 ; \ X_2 = 0.0; \ \sigma = 0.0.$$

- With a given time increment, Δt, the rates of the variables can be calculated from Eqns. (3.42)–(3.47).

- Then the variables can be updated via an integration method. With the use of explicit Euler method, these are:

$$\varepsilon^T_{i+1} = \varepsilon^T_i + \dot{\varepsilon}^T_i \cdot \Delta t \tag{3.48}$$

$$\varepsilon^p_{i+1} = \varepsilon^p_i + \dot{\varepsilon}^p_i \cdot \Delta t \tag{3.49}$$

$$R_{i+1} = R_i + \dot{R}_i \cdot \Delta t \tag{3.50}$$

$$X_{1,i+1} = X_{1,i} + \dot{X}_{1,i} \cdot \Delta t \tag{3.51}$$

$$X_{2,i+1} = X_{2,i} + \dot{X}_{2,i} \cdot \Delta t \tag{3.52}$$

$$X_{i+1} = X_{1,i+1} + X_{2,i+1} \tag{3.53}$$

$$\sigma_{i+1} = E(\varepsilon^T_{i+1} - \varepsilon^p_{i+1}) \tag{3.54}$$

$$t_{i+1} = t_i + \Delta t \tag{3.55}$$

where $i = 0,1,2,...,N$ is an integration iteration index and $i + 1$ indicates the integrated values which are based on the accumulation of the current values.

- The integration stops as a defined value is reached; for example, the predefined maximum strain is reached.

This is known as the solution for initial value problems. Once the initial values (at $t = 0$) of the variables are known, the equation set can be solved and stress–strain curves can be plotted.

Higher order of integration methods can be used for solving the problems, such as the fourth-order *Runge–Kutta* method. Implicit methods can also be used for the problems with automatic step-size control and error assessment.

The main difficulties in the integration of the multiple constitutive equations are the selection of *step size* for time increment and the control of *integration errors*. Step size is related to the integration errors. Furthermore, the different equations have different units and the errors calculated from individual equations cannot be compared directly. This is a difficult problem in integrating the equation set accurately. More details related to the integration of the equations are given in Chapter 7.

3.3.3 *Determination of constants within the equations*

The constants within the unified constitutive equations can be determined by the fitting of experimental data, which are normally in the form of stress–strain curves. For most of the cases, trial-and-error methods are used. Numerical optimisation methods have also been developed (Li, Lin and Yao, 2002). The details are presented in Chapter 7.

Many constants within the unified constitutive equations have significant physical meanings. Even if they could not be determined at the initial stage of the fitting, the ranges of the constants can at least be defined based on their physical significance. This would support the fitting process and enable meaningful results to be obtained. For some constants, the part of the stress–strain curve they control needs to be known so that the values of the constants can be determined by fitting that particular part of the curve.

An example is discussed here to help understand the process. The detailed determination procedures are presented in Chapter 7.

Consider a uniaxial cyclic loading case shown in **Fig. 3.16**. The yield stress is influenced by cyclic loading. For a strain range ($\Delta \varepsilon$ = constant) controlled cyclic load test, the initial yield stress is given by k. As cyclic loading proceeds, the hysteresis loop expands and the maximum stress increases with the number of cycles (**Fig. 3.16 (a)**), until a stabilised state is reached. This is due to the dislocation network structures being stabilised. The expanding of the yield surface at the initial cycles can be modelled using isotropic hardening R, which can be expressed in the evolutionary form:

$$\dot{R} = b(Q - R)\dot{\varepsilon}^p. \tag{3.56}$$

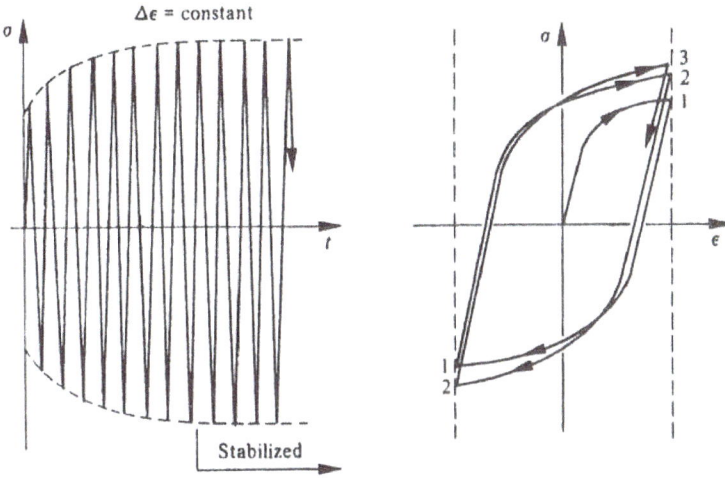

(a) Strain controlled cyclic plasticity (Lemaitre and Chaboche, 1990).

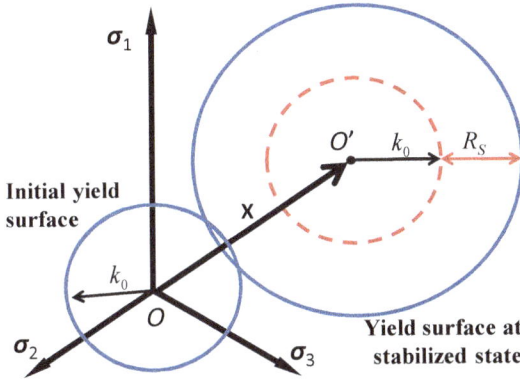

(b) Yield loci with isotropic and kinematic hardening.

Fig. 3.16 illustration of isotropic and kinematic hardenings in cyclic loading (Lin, Zhan and Zhu, 2011).

By integrating the equation and considering the accumulated plastic strain to the stabilised cycles:

$$R_S = Q\left[1 - \exp(-2b\Delta\varepsilon^p \cdot N)\right], \tag{3.57}$$

where Q is a constant to indicate the asymptotic value of R_S, which is the stabilised or saturated state of the isotropic hardening. At the Nth cycle of loading, the stabilised state is achieved. The yield surface has been expanded to

$$k = k_0 + R_S,$$ (3.58)

which is shown in **Fig. 3.16 (b)**, and, the centre of the yield surface has been moved to the new position of O'.

To simplify the modelling problems, the cycles before the stabilised state is reached are ignored (this takes a few cycles for many cases, and compared to the number of cycles to failure it is very small). This then requires only the kinematic hardening for the cyclic loading case to be modelled with the equations, written as:

$$\dot{\varepsilon}^p = \dot{\varepsilon}_0 \left(\frac{(\sigma - X) - k}{K} \right)^n \text{sgn}(\sigma - X)$$ (3.59)

$$\dot{X}_1 = C_1 \dot{\varepsilon}^p - \gamma_1 X_1 |\dot{\varepsilon}^p|$$ (3.60)

$$\dot{X}_2 = C_2 \dot{\varepsilon}^p - \gamma_2 X_2 |\dot{\varepsilon}^p|$$ (3.61)

$$X = X_1 + X_2$$ (3.62)

$$\sigma = E(\varepsilon^T - \varepsilon^p),$$ (3.63)

where $\dot{\varepsilon}_0$ (s^{-1}) is a unit parameter and K, n, C_1, γ_1, C_2, γ_2 and E are temperature-dependent material constants to be determined from experimental data. The methods for the determination of the material parameters arising in the constitutive equations employ multi-dimensional optimisation technique to minimise the errors between experimental and computed data.

Normally, the Arrhenius law is used to express the temperature dependency of the parameters. However to cover a wide range of temperatures, simplified empirical equations are commonly used as well. Here, the temperature-dependent parameters determined by Lin, Dunne and Hayhurst (1996) for copper tested between 20 °C and 500 °C are summarised below, where simplified expressions were used:

$$k = 153.4 \left[1.139 - \exp(-1.319 \Delta \varepsilon) \right] \exp(-2.0429 \times 10^{-3} T),$$

$$C_1 = 1.0863 \times 10^5 \exp(-1.9399 \times 10^{-3} T),$$

$$C_2 = 5950.3 \exp(-2.0733 \times 10^{-3} T),$$

$$\gamma_1 = 723.95 \exp(1.65134 \times 10^{-3} T),$$

$\gamma_2 = 108.12 \exp(1.9798 \times 10^{-3} T)$,

for $T \leq 423$ K

$K = 4.5$

$n = 33.269 \exp(-5.5057 \times 10^{-4} T)$

and, for $T > 423$ K

$K = 8.1881 \times 10^{-7} T^3 - 1.7865 \times 10^{-3} T^2 + 1.2944T - 285.35$

$n = -7.1457 \times 10^{-7} T^3 - 1.5686 \times 10^{-3} T^2 + 1.1435T - 283.48$.

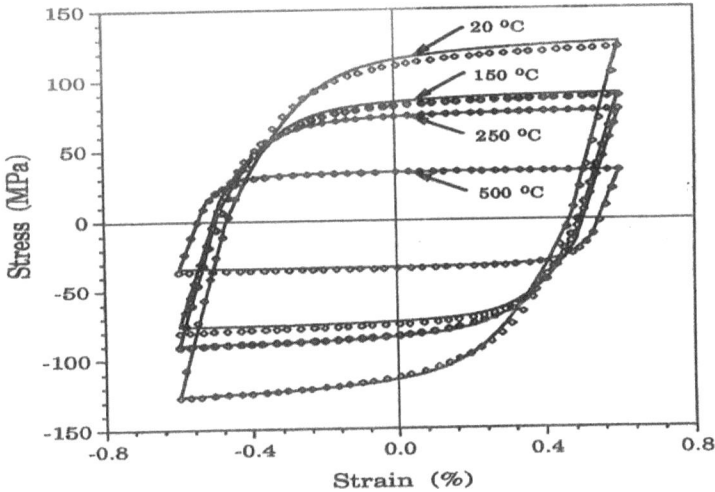

Fig. 3.17 Comparison of computed (solid curves) and experimental (symbols) hysteresis loops at temperatures between 20 °C and 500 °C for a strain range of ±0.6% and strain rate of 0.006% s⁻¹ (Lin, Dunne and Hayhurst, 1996).

In the above equations, T is the absolute temperature in K. The predicted results are shown in **Figs. 3.17, 3.18, 3.19** and **3.20**.

In **Fig. 3.17**, it can be seen that for the same strain range tests, the flow stress increases with the decrease of the temperature, and the hardening decreases with the increase of the temperature. That means, the flow stress curves become more flat as the temperature increases. This is due to more recovery (or annealing) taking place at high temperatures.

J. Lin

At a low temperature of 50 °C, as shown in **Fig. 3.18**, little creep and little recovery take place, thus the deformation is mainly time-independent plasticity dominant. In this case, the power n in Eq. (3.59) is 27.8 (n = 27.8). As the value of the power n increases, Eq. (3.59) becomes more time-independent plasticity dominant.

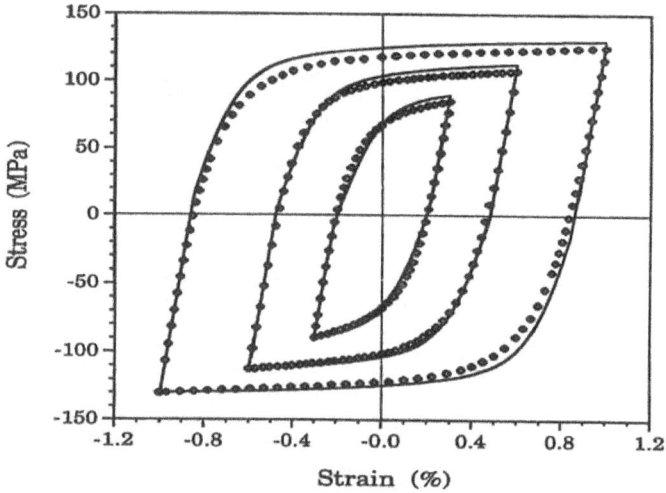

Fig. 3.18 Comparison of computed (solid curves) and experimental (symbols) hysteresis loops for predominantly time-independent plasticity behaviour at the strain ranges: ±0.3%, ±0.6% and ±1.0% and a strain rate of 0.006% s^{-1} and a temperature of 50 °C (Lin, Dunne and Hayhurst, 1996).

Figure 3.19 shows the cyclic viscoplasticity behaviour of the material. Since more creep and static recovery due to annealing takes place at high temperature and low strain rate conditions, these would result in a lower flow stress and relatively low hardening. At high temperature (500 °C) deformation conditions, the value of the power n is 6.8 (n = 6.8). This value is used to model the viscoplasticity behaviour of the material.

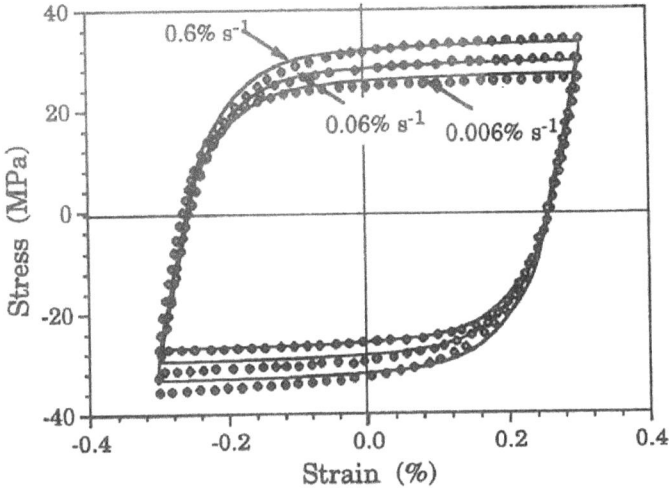

Fig. 3.19 Comparison of computed (solid curves) and experimental (symbols) hysteresis loops for predominantly time-dependent viscoplasticity behaviour for strain rates of 0.006%, 0.06% and 0.6% at a strain range of ±0.3% and a temperature of 500 °C (Lin, Dunne and Hayhurst, 1996).

In general, Eqs. (3.59)–(3.63) are a set of viscoplasticity equations and used to model the time-dependent stress–strain response in deformation. The nature of the equations can be controlled by the value of power n. That means if the n value is low, normally $n = 1 - 7$, Eq. (3.59) could be used to represent the elastic-viscoplasticity behaviour and creep can be modelled. If the n value is high, it approaches the elastic-plastic behaviour of a material. To illustrate the behaviour of the equation, Eq. (3.59) can be simplified by omitting the kinematic hardening parameter X and the unit parameter $\dot{\varepsilon}_0$. The equation then becomes:

$$\dot{\varepsilon}^p = \left(\frac{\sigma - k}{K}\right)^n. \tag{3.64}$$

This equation can be converted to:

$$\sigma = k + K(\dot{\varepsilon}^p)^{1/n}, \tag{3.65}$$

where k is the yield surface of the material at the stabilised state, and the second part of the equation, $\sigma_v = K(\dot{\varepsilon}^p)^{1/n}$, represents the viscoplastic behaviour of the material, as was discussed in the early sections. **Table 3.1** shows the values of σ_v for different n values for two strain rates of

1 s^{-1} and 0.01 s^{-1}. In this case, $k = 200$ MPa is chosen. $\Delta\sigma_v$ is defined as the difference between the values of σ_v calculated using the strain rates of 1.0 s^{-1} and 0.01 s^{-1} respectively, i.e. $\Delta\sigma_v = (\sigma_v)_{1.0} - (\sigma_v)_{0.01}$.

Table 3.1 Comparison of values of σ_v (MPa)$[\sigma_v = 200(\dot{\varepsilon}^p)^{1/n}]$ and $\Delta\sigma_v$ (MPa) at the plastic strain rates of 1.0 s^{-1} and 0.01 s^{-1} for different values of n

n (-)	1	2	5	10	100
1.0 (s^{-1})	200	200	200	200	200
0.01 (s^{-1})	2	20	80	126	191
$\Delta\sigma_v$	198	180	120	74	9

It can be seen that from **Table 3.1**, when $n = 1$, the difference of σ_v for the two strain rates is 198 MPa ($\Delta\sigma_v = 200 - 2 = 198$ MPa). When $n = 100$, the difference is very small, $\Delta\sigma_v = 200 - 191 = 9$ MPa, which minimises the strain rate sensitivity. The viscoplastic equation becomes close to the elastic-plastic formulation. It can be seen that when $n \to \infty$, the time-dependent behaviour of the equations is removed and the equations can be used to model elastic-plastic behaviour of materials. However, when the value of n is too high, Eq. (3.59) becomes very stiff and becomes difficult to integrate.

Another method to reduce the strain rate sensitivity of the general viscoplastic equation (Equation 3.59) is to reduce the value of K. The viscoplastic stress σ_v would then be less strain rate sensitive.

Normally, for the modelling of an elastic viscoplastic problem, the value of K is chosen within a range of 50 MPa to 1000 MPa, and n varies from 1 to 7. If the deformation temperature for a metal is higher, the value of n should be lower, since more viscoplastic behaviour of the material can be observed.

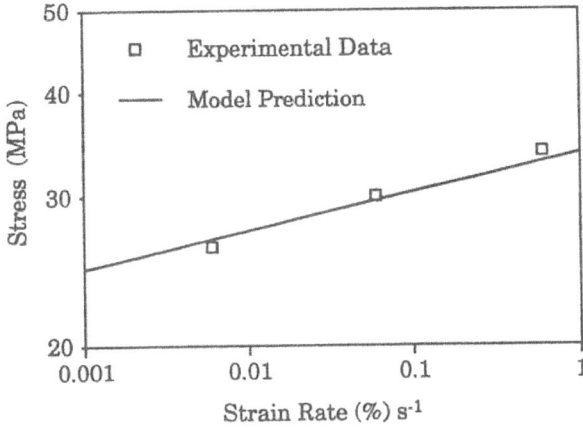

Fig. 3.20 Comparison of computed (solid curves) and experimental (symbols) log stress and log strain rate results for a strain range of ±0.3% and a temperature of 500 °C (Lin, Dunne and Hayhurst, 1996).

Figure 3.20 shows the stress and strain rate relationship in log-log scales. It can be seen that the relationship is linear for both experimental data and the computed results. This indicates that the power-law expression can be used to model the viscoplastic behaviour of a material at the deformation condition.

3.3.4 *Multiaxial constitutive equations*

By the introduction of the tensor representations and the use of von-Mises formulation, the set of unified uniaxial constitutive equations can be expanded into the multiaxial case. They are:

$$\dot{\varepsilon}_e^p = \dot{\varepsilon}_0 \left(\frac{J_2'(S_{ij} - X_{ij}) - R - k}{K} \right)^n \tag{3.66}$$

$$\dot{\varepsilon}_{ij}^p = \frac{3}{2} \frac{S_{ij}}{\sigma_e} \dot{\varepsilon}_e^p \tag{3.67}$$

$$\dot{R} = \frac{dR}{dt} = b(Q - R)\dot{\varepsilon}_e^p \tag{3.68}$$

$$\dot{X}_{ij,k} = \frac{dX_{ij,k}}{dt} = C_k \dot{\varepsilon}_{ij}^p - \gamma_k X_{ij,k} \dot{\varepsilon}_e^p, \text{ where } k = 1,2,\ldots,N \tag{3.69}$$

$$X_{ij} = \sum_{k=1}^{N} X_{ij,k} \tag{3.70}$$

$$\sigma_{ij} = D_{ijkl}\left(\varepsilon_{ij}^{T} - \varepsilon_{ij}^{p}\right).$$ (3.71)

It should be noted that Eq. (3.69) represents multiple equations for the modelling of kinematic hardening phenomena of a material. Similarly, the isotropic hardening equation (Eq. (3.68)) can also be expanded to multiple equations for the modelling of the isotropic hardening feature of a material. However, this is rarely used. In modern modelling techniques, hardening and softening mechanisms will be identified first and the equations are formulated based on the identified mechanisms. Thus not only the stress–strain response, but also the detailed microstructure evolution and hardening softening mechanisms can be modelled individually. These will be detailed in later chapters.

The determined set of unified constitutive equations can be input into a commercial FE solver via a user-defined subroutine for real engineering applications.

In this chapter, both isotropic hardening and kinematic hardening are discussed. This forms the basis for the formulation of unified constitutive equations. In cold metal sheet stamping processes, kinematic hardening is important for the prediction of residual stresses and springback for the formed panels. However, in hot stamping conditions, springback is less important and kinematic hardening is normally ignored, which simplifies the problem.

The application of the unified constitutive equations and hardening variables is introduced in the modelling of cyclic plasticity in this chapter. However, the main purpose of the book is to introduce the unified theories for the modelling of evolution of mechanical and physical properties of metals in metal forming processes, particularly in hot/warm forming applications. Thus cyclic plasticity behaviour of materials will not be considered further.

Chapter 4

Plasticity in Cold Metal Forming

In this chapter, the phenomenological and mathematical modelling of elastic-plastic solids will be discussed in detail. This is based on the early scientific work of the maximum shear stress yield criterion (by Tresca in 1864) and von-Mises yield criterion (by von-Mises in 1913). The theories for anisotropic plastic deformation were developed further by Hill in 1948 (Rodney Hill, British, 1921–2011). Ford (Hugh Ford, British, 1913–2010) developed the slip line field theory and plane strain compression test method in the 1950s. He built metal forming research at Imperial College London in 1948 and, along with his co-workers, successfully developed advanced mechanics theories for metal forming applications, such as cold rolling and extrusion (Ford and Alexander, 1977).

Hardenings due to plastic deformation and anisotropy due to the texture of orientation of materials have been the key focus of research. The concept of isotropic hardening and kinematic hardening and their interactions with general plastic flow equations have been developed e.g. by Lemaitre and Chaboche (1990), and already discussed in Chapter 3. Different yield criteria for isotropic materials have also been introduced in Chapter 2.

Based on the basic modelling techniques and yield criteria introduced in Chapters 2 and 3, this chapter concentrates on the incremental formulation for the plastic deformation of materials. Anisotropic yield criteria are also introduced in this Chapter.

4.1 Application Domains

For metal forming processes, if the forming (or plastic deformation) is carried out at about a temperature of T, where $T/T_m < 1/3$, and T_m is the melting temperature of the material, the plastic deformation is mainly due to dislocation movement. For example, the melting temperature for steel is about 1550 °C (i.e. $T_m = 1550 + 273 = 1823$ K); if it is formed at

J. Lin

a temperature below 334 °C (607K) it is considered as a cold forming and the theory of plasticity can be used without much error introduced.

For an aluminium alloy, the melting temperature is about 660 °C (i.e. $T_m = 660 + 273 = 933$ K). If the material is formed below a temperature of about 38 °C (311K), the forming can be considered as elastic-plastic deformation. In general, if a metal is formed at room temperature (or cold metal forming conditions), time-independent plasticity theory can be used to approximately describe the deformation of the material.

Please note here that "approximately" is used, since the strain rate range for cold metal forming does not vary very much for a particular forming process, for example, in a range of 0.01 s⁻¹ to 10 s⁻¹. Within this strain rate range, the flow stress does not change much. By the use of time-independent plasticity theory, the problem can be simplified. But for some materials, such as lead, their melting temperature is low and viscoplastic behaviour can be observed at room temperature.

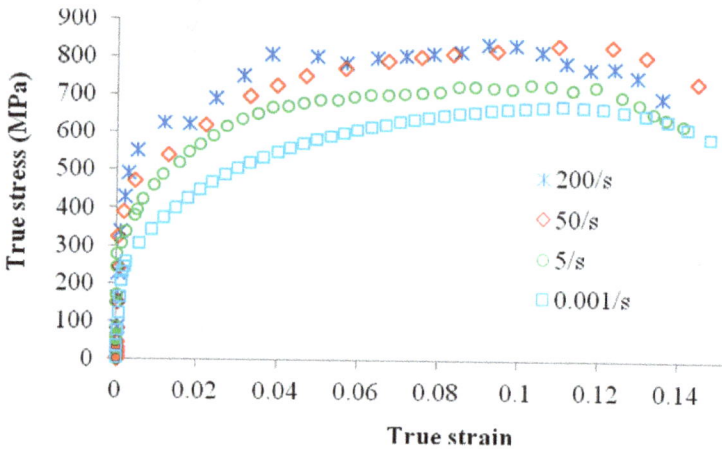

Fig. 4.1 Stress–strain relationships for a boron steel tested at room temperature for a strain rate range of 0.001 s⁻¹ to 200 s⁻¹ (Li, 2013).

For cold impact simulation conditions, material data are normally needed from a large range of strain rates, i.e. 0.001 s⁻¹ to 500 s⁻¹. A big difference of flow stress responses can be observed for the low and high strain rate deformation. This can be seen from the example given in **Fig. 4.1**. In this case, if time-independent elastic-plastic theory is used, computational errors would be big. It would be better to use the time-dependent elastic viscoplasticity theory, which will be discussed in Chapter 5.

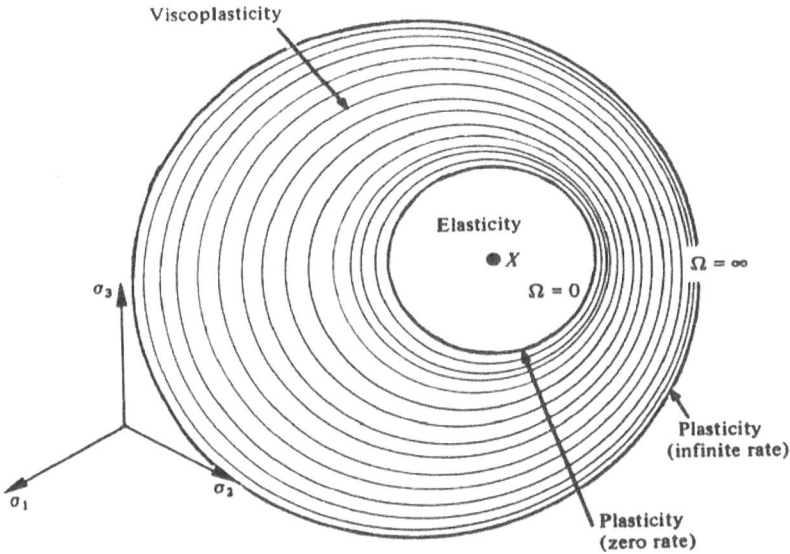

Fig. 4.2 Equipotential flow surfaces illustrating the limitation of time independent plasticity (Lemaitre and Chaboche, 1990).

Time-independent plastic behaviour must be considered as a particular case of the more general viscoplasticity presentations to be introduced in Chapter 5. The application of the time-independent plasticity theory should be based on the understanding of the problems for particular applications. Lemaitre and Chaboche (1990) used the equipotential surfaces Ω to introduce the application domains of plasticity and viscoplasticity. **Figure 4.2** is a schematic illustration of a series of equipotential surfaces (or, surfaces of equal dissipation Ω) in the space of principal stresses. X is the kinematic hardening. On the surface closest to the centre $\Omega = 0$ (or $f = J_2'(S - X) - R - \sigma_Y = 0$), the strain rate approaches zero, whereas the surface farthest to the centre $\Omega = \infty$ is for strain rate of infinity. The elasticity region corresponds to the area of $\Omega < 0$ (or $f = J_2'(S - X) - R - \sigma_Y < 0$). Between the two extremes, i.e. $\Omega = 0$ and $\Omega = \infty$, it is time-dependent viscoplasticity.

Thus time-independent plasticity is applicable to

- infinitely slow loading or to the attainment of asymptotic states under fixed loads;
- extremely high loading rate, where there is no time for recovery or diffusion taking place.

At low temperature deformation conditions, the equipotential surfaces of the two extreme conditions are close and the viscoplastic area is very small. Thus time-independent plasticity theory can be used. However, if a metal is deformed at high temperature, such as in warm and hot forming conditions, the two extreme surfaces are far apart and the viscoplastic area is very large. Thus viscoplasticity theory should be used for warm and hot metal forming.

Plasticity theory can only be used for non-damaging loading conditions. For example, at uniaxial tensile testing conditions, the theory should be used for the strain lower than about 50% of the strain to failure, i.e. approximately $<0.5\varepsilon_f$, where ε_f is the strain to failure at a uniaxial loading condition. That means we only consider the strain hardening region. If the strain is high, the flow stress may drop due to the micro-damage of the material, which will be discussed in Chapter 6.

The above limitations for the use of time-independent plasticity theory are only an indication. It is problem dependent. However, an accurate computational result can only be obtained if a correct theory is used for a particular application. Thus the selection of the right theory for a problem is important.

4.2 Incremental Methods and Hardening Laws

4.2.1 *Uniaxial behaviour*

a. Plastic flow and hardening

In plastic flow, the total strain ε^T can be divided into reversible (or elastic) strain ε^e and the irreversible (or inelastic) strain ε^p. This has been introduced in Chapters 2 and 3. The elastic strain corresponds to the change of *inter-atomic distance* of the material under external loading, while the plastic deformation implies *slip movements* of atoms. We have

$$\varepsilon^T = \varepsilon^e + \varepsilon^p$$

$$(4.1)$$

or

$$\varepsilon^e = \varepsilon^T - \varepsilon^P$$

The total strain ε^T is the measured strain from experiments (measurable state variable). The plastic strain ε^P is calculated from constitutive equations, such as

$$\varepsilon^P = g(\sigma) \quad \text{for} \quad |\sigma| \geq \sigma_Y ;$$

$$\varepsilon^P = 0 \qquad \text{for} \quad |\sigma| < \sigma_Y \text{ within elastic domain.}$$

If the plastic strain can be calculated, the elastic strain can be worked out easily according to Eq. (4.1). Thus the plasticity theory is mainly related to how to calculate plastic strain accurately with the consideration of hardening and other physical properties of the material under elastic-plastic deformation conditions.

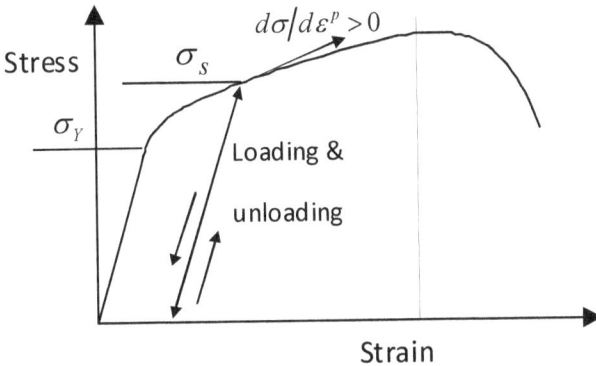

Fig. 4.3 Elastic-plastic deformation of materials.

On the physical level, hardening is due to an increase in the dislocation density: the dislocations have a tendency to interlock and block each other. To a first approximation, the increase in the elastic limit follows the increase in the stress, and it is the approximation which constitutes the theoretical basis of classical plasticity. Thus, for monotonic loading, the current limit of elasticity, also called the plasticity threshold or the yield stress, is equal to the highest values of stress previously attained, which is known as σ_S (**Fig. 4.3**). For a material with positive hardening, $d\sigma/d\varepsilon^P > 0$, the natural elastic limit (the initial yield stress) σ_Y is the smallest value of the yield stress , since the yield stress is a function of the history of plastic deformation.

σ_S is known as the current yield stress, or the flow stress, as indicated in **Fig. 4.3**. σ_S increases with the strain hardening of the material under loading.

Any point on a monotonic hardening curve can therefore be considered as a representative point of the plasticity threshold, and characteristic hardening law may be written as

$$\sigma_S = g^{-1}(\varepsilon^p) . \tag{4.2}$$

Plastic flow occurs only if $\sigma = \sigma_S$, i.e.

$\sigma < \sigma_S$, $d\varepsilon^p/dt = 0$ ($d\varepsilon^p/dt$ is the plastic strain rate and t is the time).

$$\sigma = \sigma_S, \ d\varepsilon^p/dt \neq 0$$

A number of analytic expressions have been proposed to model the hardening function g. We will use the one resulting from a calculation which is based on dislocation theory and which shows that the yield stress is proportional to the square root of the dislocation density ρ,

$$\sigma_S = k\rho^{1/2} ,$$

where k is a constant. In reality, the dislocation density is never zero for a material. If it is assumed that the dislocation density of a material at the initial state (virgin material without plastic deformation) is ρ_0 corresponding to the elastic limit, σ_Y then, there is

$$\sigma_S = \sigma_Y + k(\rho - \rho_0)^{1/2} .$$

In terms of macroscopic strains, in analogy with the above relation for cold deformation conditions, the above equation can be written in a general form of

$$\sigma_S = \sigma_Y + K\left(\varepsilon^p\right)^n . \tag{4.3}$$

This is known as Ramberg–Osgood equation and can be easily inverted to give

$$\varepsilon^p = \left(\frac{\sigma_S - \sigma_Y}{K}\right)^{1/n} , \tag{4.4}$$

where K is the coefficient of plastic resistance and n is the strain hardening exponent. The initial yield stress σ_Y can be determined according to 0.2% of proof stress. The values of K and n can be determined by the best fitting of the plastic part of a stress–strain curve.

If we know two pairs of data points of the plastic part of the stress–strain curve $(\sigma_{S,1}, \varepsilon_1^p)$ and $(\sigma_{S,2}, \varepsilon_2^p)$, then the values of K and n can be determined according to the two pairs of the data points. Taking the logarithm for Eq. (4.3), for the two pairs of data points, the following two equations can be obtained:

$$\log\left(\sigma_{S,1} - \sigma_Y\right) = \log K + n\log\left(\varepsilon_1^p\right)$$

$$\log\left(\sigma_{S,2} - \sigma_Y\right) = \log K + n\log\left(\varepsilon_2^p\right).$$

Taking the difference for the above equations,

$$\log\left(\sigma_{S,2} - \sigma_Y\right) - \log\left(\sigma_{S,1} - \sigma_Y\right) = n\left[\log\left(\varepsilon_2^p\right) - \log\left(\varepsilon_1^p\right)\right]$$

Then,

$$n = \frac{\log\left(\sigma_{S,2} - \sigma_Y\right) - \log\left(\sigma_{S,1} - \sigma_Y\right)}{\log\left(\varepsilon_2^p\right) - \log\left(\varepsilon_1^p\right)}. \tag{4.5}$$

If we know the two data points and the initial yield stress of the material, the value n can be calculated according to Eq. (4.5). Then the K value can be calculated easily according Eq. (4.3) using either pair of the data points.

Example: An aluminium alloy is deformed at room temperature and the experimental stress–strain data is shown in **Fig. 4.4** as symbols. The yield stress is taken as the proof stress, $\sigma_Y = 146$ MPa.

Chose two pairs of data point on the stress–strain curve: A (0.0254, 410) and B (0.1512, 484). According to Eq. (4.5), there is

$$n = \frac{\log\left(\sigma_{S,2} - \sigma_Y\right) - \log\left(\sigma_{S,1} - \sigma_Y\right)}{\log\left(\varepsilon_2^p\right) - \log\left(\varepsilon_1^p\right)} = \frac{\log\left(484 - 146\right) - \log\left(410 - 146\right)}{\log\left(0.1512\right) - \log\left(0.0254\right)}$$

$$n = 0.138.$$

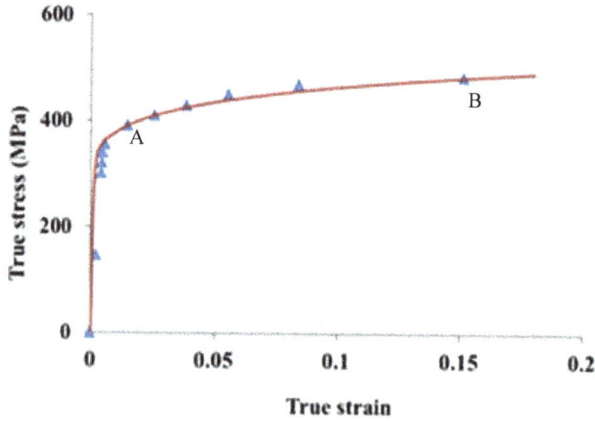

Fig. 4.4 Determination of the values of K and n in Eq. (4.3) according to experimental data (symbols).

According to Eq. (4.3) and using the first pair of the data point, we have,

$$K = \frac{\sigma_{S,1} - \sigma_Y}{\left(\varepsilon_1^p\right)^{1/n}} = \frac{410 - 146}{0.0254^{0.138}} = 438 \text{ (MPa)}.$$

Then, we have $\sigma_Y = 146$ MPa, $K = 438$ MPa and $n = 0.138$ (or $N = 1/n = 7.24$), and the determined constitutive equation for the material is

$$\sigma_S = \sigma_Y + 438\left(\varepsilon^p\right)^{0.138}$$

or

$$\varepsilon^p = \left(\frac{\sigma_S - \sigma_Y}{438}\right)^{7.24}.$$

The solid curve shown in **Fig. 4.4** is the predicted stress–strain relationship according to the determined constitutive equation. It can be seen that the predicted stress–strain curve passes through the chosen two pairs of the experimental data points. Thus the quality of the predicted stress–strain relationship to experimental data is related to the data points chosen for the determination of the values of the constants K and n.

However, the determined constitutive equation can be used to predict the experimental data fairly well.

The values of K and n can also be determined via the best fit to the experimental data using an optimisation method. This is a preferred method and could approximate the overall experimental data better for most cases. For more complicated constitutive equations, it is difficult to determine the values of material constants analytically, and, optimisation techniques should be used. Chapter 7 will introduce the detailed procedures and techniques for the determination of values of constants arising within a set of constitutive equations from experimental data.

b. Constitutive equations for elastic perfectly plastic solids

The elastic perfectly plastic stress–strain relationship for a material is shown in **Fig. 4.5**. In this case, the strain hardening is zero.

If $|\sigma_S| < \sigma_Y$

then

$$\varepsilon^T = \varepsilon^e = \sigma/E$$

This is within the elastic domain.

If $|\sigma_S| = \sigma_Y$,

then

$\varepsilon^T = \varepsilon^e + $ arbitrary ε^p of the same sign as the stress σ, as shown in **Fig. 4.5**.

This is the simplified case. For a material with perfectly plastic

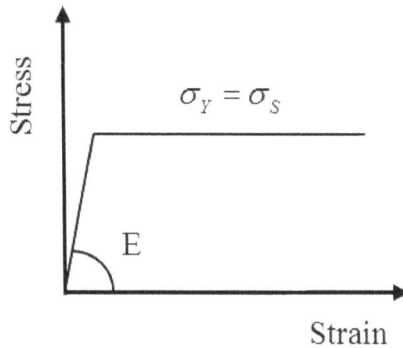

Fig. 4.5 Elastic-plastic deformation of materials.

behaviour, load-controlled tests are very difficult to conduct, since once the stress reaches the yield stress, any small increment of the load would result in infinite deformation of the material.

In tube hydroforming simulation, if the material property is perfectly plastic, then it is very difficult to carry out the process simulation. Within the elastic region, the tube is hardly to expand as the hydraulic pressure is applied. Once the material is yielded, any more pressure applied would cause a high rate of plastic deformation, which is similar to an explosive process, and it is very difficult to get stable simulation results.

c. Constitutive equations for elastic-plastic solids with isotropic hardening

The complete model of an initially isotropic elastic-plastic solid subject to monotonic uniaxial loading is the following:

$$\varepsilon^p = \left(\frac{|\sigma| - \sigma_Y}{K}\right)^N \text{sgn}(\sigma), \text{ with } |\sigma| = \sigma_S$$

$$\sigma = E(\varepsilon^T - \varepsilon^p). \tag{4.6}$$

The above expressions constitute the practical ways to represent the hardening law, but other expressions can be used. In the equation, $\sigma = \sigma_S$, which represents the current yield stress of the material under uniaxial tension.

4.2.2 Plastic flow rules based on von-Mises yield criterion

a. Normality

The von-Mises yield criterion is used to illustrate the normality rule in plastic deformation. The von-Mises yield criterion is expressed as (Chapter 2)

$$\left[(\sigma_1 - \sigma_2)^2 + (\sigma_2 - \sigma_3)^2 + (\sigma_3 - \sigma_1)^2\right]^{1/2} = \sqrt{\frac{2}{3}}\sigma_Y$$

The yield surface is a cylinder in the 3D space of σ_1, σ_2 and σ_3 with the radius (*ab*) of $\sqrt{2/3}\sigma_Y$, which is shown in **Fig. 4.6** and is the stress deviator. As stated before, the stress inside of the cylinder represents elastic behaviour of the material. Yield begins when the stress level reaches the surface of the cylinder. The axis of the cylinder *Oa* makes equal angles to the principal axes and is the hydrostatic component of stress, i.e.

$$\sigma = (\sigma_1 + \sigma_2 + \sigma_3)/3 = \sigma_H.$$

Since plastic deformation of metals is not influenced by hydrostatic stress, the generator of the yield surface is a straight line parallel to the axis of the cylinder. As plastic deformation occurs accompanied by work hardening, experiments show that the yield surface expands outward for isotropic hardening materials, maintaining its same geometric shape. Thus, there is

$$d\varepsilon^P = d\lambda\left(\partial f / \partial \mathbf{S}\right),$$

where $d\lambda$ is a multiplier, which is related to the hardening of plastic flow of a material.

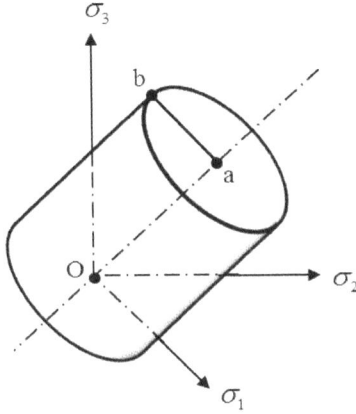

Fig. 4.6   Yield surface for von-Mises criterion in the axes of σ_1, σ_2, σ_3.

If a plane is passed through the yield surface, shown in **Fig. 4.6**, perpendicular to the σ_2 axis, it intersects on the $\sigma_1 O \sigma_3$ plane as an ellipse shown in **Fig. 4.7**. The total plastic strain increment vector $d\varepsilon$ must be *normal to the yield surface*. As a consequence, any acceptable yield surface must be convex about its origin. Because of the normality rule,

- there is no component of the total strain increment vector that acts in the direction of σ_H;
- the hydrostatic stress does not act to expand the yield surface;
- the deviatoric component of stress acts in the same direction as the total strain increment vector;
- the dot product of strain increment and deviatoric stress, $\mathbf{S} \cdot d\varepsilon$, is the magnitude of the plastic work as the yield surface is expanded by plastic deformation.

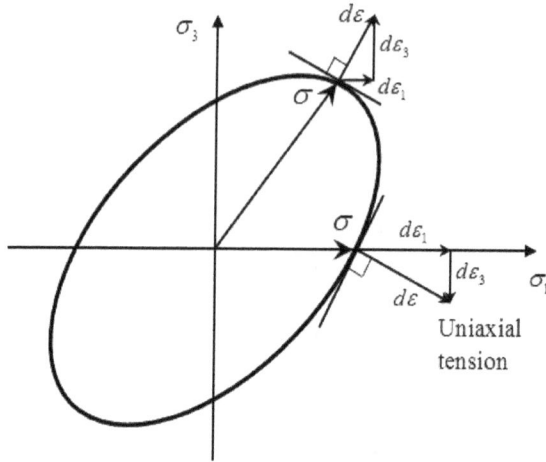

Fig. 4.7 Illustration of normality. $d\varepsilon$ is normal to the yield locus.

The normality rule is useful in constructing experimental yield loci. **Figure 4.7** shows that the total strain vector $d\boldsymbol{\varepsilon}$ is normal to the yield locus. If the yield locus is known, we can establish the ratio $d\varepsilon_1 : d\varepsilon_3$ from the normality rule. For example, for uniaxial tension in the σ_1 direction, we have $d\varepsilon_3 = -0.5d\varepsilon_1$.

b. Incremental plasticity and Levy–Mises Equations

In elastic deformation, the strains are uniquely determined by the stresses through Hooke's law regardless how the stress state was achieved. In plastic deformation, the strains are in general not uniquely determined by the stresses, but depend on the entire history of loading. Thus, in plasticity it is necessary to determine the differentials or *increments of plastic strain* throughout the loading path and then obtain the total strain by integration or accumulation.

For example, a cylindrical specimen with an original length of 10 mm is subjected to a uniaxial cyclic loading. It is extended to 15 mm first, then compressed to 10 mm again. The total strain is

$$\varepsilon = \int_{10}^{15} \frac{dL}{L} + \int_{15}^{10} \frac{dL}{L} = \ln\frac{15}{10} + \ln\frac{10}{15} = 0.405 - 0.405 = 0 \,.$$

If an incremental method is used

$$\varepsilon = \int_{10}^{15} \frac{dL}{L} + \int_{15}^{10} -\frac{dL}{L} = \ln\frac{15}{10} - \ln\frac{10}{15} = 0.405 + 0.405 = 0.81 \,.$$

There are two types of plastic stress–strain relationships:

(i) Incremental theories relate the stresses to the plastic strain increments.

(ii) Total strain theories relate the stresses to the total plastic strain.

Total strain theories simplify the solution of plasticity problems, but in general the plastic strain cannot be considered independent of loading histories. The *effective plastic strain increment* can be defined as

$$d\varepsilon_e^p = \sqrt{\frac{2}{9}\left[\left(d\varepsilon_1^p - d\varepsilon_2^p\right)^2 + \left(d\varepsilon_2^p - d\varepsilon_3^p\right)^2 + \left(d\varepsilon_3^p - d\varepsilon_1^p\right)^2\right]}, \qquad (4.7)$$

where $d\varepsilon_1^p$, $d\varepsilon_2^p$ and $d\varepsilon_3^p$ are plastic strain increments in the principal directions. The above equation can be simplified as

$$d\varepsilon_e^p = \sqrt{\frac{2}{3}\left[\left(d\varepsilon_1^p\right)^2 + \left(d\varepsilon_2^p\right)^2 + \left(d\varepsilon_3^p\right)^2\right]}. \qquad (4.8)$$

The stress–strain relationship for an ideal plastic solid, where the elastic strains are negligible, are known as the Levy–Mises equations. Under uniaxial tension conditions, there are

$$\sigma_1 = \sigma_{11} \neq 0; \sigma_2 = \sigma_3 = 0;$$

and

$$\sigma_H = \sigma_{ii}/3 = \sigma_1/3$$

The deviatoric stresses, which cause the plastic deformation of materials, are

$$S_{11} = \sigma_1 - \sigma_m = \frac{2}{3}\sigma_1; \; S_{22} = S_{33} = -\frac{1}{3}\sigma_1; \; S_{ij} = 0 \; (i \neq j).$$

Then we have

$$S_{11} = -2S_{22} = -2S_{33}.$$

Plastic deformation is incompressible (the volume does not change). For isotropic materials, there is

$$d\varepsilon_1^p = -2d\varepsilon_2^p = -2d\varepsilon_3^p,$$

so that

$$\frac{d\varepsilon_1^p}{d\varepsilon_2^p} = -2 = \frac{S_{11}}{S_{22}},$$

or

$$\frac{d\varepsilon_1^p}{S_{11}} = \frac{d\varepsilon_2^p}{S_{22}}.$$

This can be generalised to the *Levy–Mises equation*:

$$\frac{d\varepsilon_{11}^p}{S_{11}} = \frac{d\varepsilon_{22}^p}{S_{22}} = \frac{d\varepsilon_{33}^p}{S_{33}} = d\lambda. \tag{4.9}$$

This indicates that at any instant of perfect plastic deformation (i.e. no strain hardening), the ratio of the plastic strain increment to the current deviatoric stresses is constant. According to the expressions of deviatoric stresses and stress tensor components, the above equation can be written as

$$d\varepsilon_{11}^p = d\lambda S_{11} = \frac{2}{3}d\lambda\left[\sigma_{11} - \frac{1}{2}(\sigma_{22} + \sigma_{33})\right], \text{ etc.} \tag{4.10}$$

According to the definitions of effective stress and strains, we have

$$d\varepsilon_e^p = \frac{2}{3}d\lambda\sigma_e$$

or

$$d\lambda = \frac{3}{2}\frac{d\varepsilon_e^p}{\sigma_e}.$$

Then the Levy–Mises equations become

$$d\varepsilon_{11}^p = \frac{d\varepsilon_e^p}{\sigma_e}\left[\sigma_{11} - \frac{1}{2}(\sigma_{22} + \sigma_{33})\right]$$

$$d\varepsilon_{22}^p = \frac{d\varepsilon_e^p}{\sigma_e}\left[\sigma_{22} - \frac{1}{2}(\sigma_{33} + \sigma_{11})\right] \tag{4.11}$$

$$d\varepsilon_{33}^p = \frac{d\varepsilon_e^p}{\sigma_e}\left[\sigma_{33} - \frac{1}{2}(\sigma_{11} + \sigma_{22})\right].$$

This is similar to the generated elastic strain–stress relationships, if we change

$$\frac{d\varepsilon_e^p}{\sigma_e} = \frac{1}{E} \text{ (1/Young's modulus), and, (Poisson's ratio) } \nu = \frac{1}{2}.$$

This indicates that in perfect plastic deformation, Young's modulus is related to the proportionality constant $d\varepsilon_e^p/\sigma_e$ and Poisson's ratio is 0.5 (i.e. material deforms at constant volume; yielding is independent of σ_H).

c. Prandtl–Reuss expression

The Levy–Mises equations can only be applied to the problems of large plastic deformation, since elastic strains are neglected. For most cases, the elastic region cannot be ignored and it is necessary to consider both elastic and plastic strains. Prandtl (1925) and Reuss (1930) have proposed equations to deal with elastic-plastic problems.

For elastic-plastic deformation with isotropic hardening, it is based on the following assumptions:

- Plastic incompressibility. Plastic strain occurs at constant volume, $\varepsilon_H = 0$ $(\varepsilon_H = (\varepsilon_{11} + \varepsilon_{22} + \varepsilon_{33})/3 = 0)$ and the flow does not depend on the hydrostatic stress σ_H $(\sigma_H = (\sigma_{11} + \sigma_{22} + \sigma_{33})/3 = Tr(\boldsymbol{\sigma})/3)$. The loading function depends only on the deviatoric stresses and the internal variables, i.e.

$$\partial f / \partial \sigma_H = 0,$$

where f is the yield surface of the material.

- Initial isotropy and isotropic hardening. The loading function depends only on the invariants J_2 and J_3 of deviatoric stress tensor **S**:

$$J_2'(\mathbf{S}) = \sigma_e = \left(\frac{3}{2}\mathbf{S}:\mathbf{S}\right)^{1/2}$$

$$J_3(\mathbf{S}) = \left(\frac{9}{2}\mathbf{S}\cdot\mathbf{S}:\mathbf{S}\right)^{1/3}.$$

- Associated plasticity and normality hypotheses.

$$d\varepsilon^p = d\lambda\left(\partial f / \partial \mathbf{S}\right)$$

$$d\rho = -d\lambda\left(\partial f / \partial R\right).$$

- The choice of von-Mises loading function, independent of the 3rd invariant, in the form of

$$f = J_2'(\mathbf{S}) - R - \sigma_Y = 0,$$

where σ_Y is the initial yield stress of the material in tension.
The hardening curve can be expressed in the form of

$$R = \kappa(\varepsilon_e) = \rho \frac{\partial \Psi}{\partial \varepsilon_e},$$

with $R(0) = \kappa(0) = 0$, where Ψ is the potential and ε_e is the effective strain. Thus the general plasticity flow can be expressed as

$$d\varepsilon^P = d\lambda \frac{\partial f}{\partial S_{ij}} = \frac{3}{2} d\lambda \frac{S_{ij}}{\sigma_e} \tag{4.12}$$

$$d\varepsilon_e = -d\lambda \frac{\partial f}{\partial R} = d\lambda = \left(\frac{2}{3} d\varepsilon^P : d\varepsilon^P\right)^{1/2}, \tag{4.13}$$

where $d\lambda$ equals the effective strain increment $d\varepsilon_e$ and is known as the plasticity *multiplier*. The consistency condition in the presence of plastic flow ($f = 0$ and $df = 0$, where $f = \sigma_e - R - \sigma_Y = 0$) gives

$$df = d\sigma_e - dR = d\sigma_e - \kappa'(\varepsilon_e)d\varepsilon_e = 0,$$

which can be used to express the plasticity multiplier

$$d\lambda = d\varepsilon_e = H(f) \frac{d\sigma_e}{\kappa'(\varepsilon_e)},$$

where $H(f)$ represents the *Heaviside step function*:

$$H(f) = 1, \text{ if } f = \sigma_e - R - \sigma_Y > 0$$

$$H(f) = 0, \text{ if } f = \sigma_e - R - \sigma_Y < 0, \text{ within the elastic domain.}$$

For a material with positive hardening, $R = \kappa(\varepsilon_e) > 0$, there is no plastic flow unless $d\sigma_e$ is positive, i.e. $d\sigma_e > 0$, with the increment of effective stress. For a material with negative hardening (strain softening), $d\sigma_e < 0$, the plasticity multiplier is zero, i.e., $d\lambda = d\varepsilon_e = 0$. We use the symbol $\langle\rangle_+$ to represent this, that is

$$d\lambda = d\varepsilon_e = H(f) \frac{\langle d\sigma_e \rangle_+}{\kappa'(\varepsilon_e)}. \tag{4.14}$$

The elastic-plastic flow equation can be written as

$$d\varepsilon_{ij}^P = \frac{3}{2} d\lambda \frac{S_{ij}}{\sigma_e} = \frac{3}{2} H(f) \frac{\langle d\sigma_e \rangle_+}{\kappa'(\varepsilon_e)} \frac{S_{ij}}{\sigma_e}. \tag{4.15}$$

Taking into account the total and elastic strains and linear isotropic law, the elastic-plastic equation can be expressed as

$$d\boldsymbol{\varepsilon}^T = d\boldsymbol{\varepsilon}^e + d\boldsymbol{\varepsilon}^p , \qquad (4.16)$$

$$d\boldsymbol{\varepsilon}^e = \frac{1+v}{E}d\boldsymbol{\sigma} - \frac{v}{E}d[Tr(\boldsymbol{\sigma})]\mathbf{I} , \qquad (4.17)$$

$$d\boldsymbol{\varepsilon}^p = \frac{3}{2}H(f)g'(\sigma_e)\langle d\sigma_e \rangle_+ \frac{\mathbf{S}}{\sigma_e} , \qquad (4.18)$$

or

$$d\varepsilon_{ij}^p = \frac{3}{2}H(f)g'(\sigma_e)\langle d\sigma_e \rangle_+ \frac{S_{ij}}{\sigma_e} ,$$

where $g'(\sigma_e) = 1/\kappa'(\sigma_e)$ is the isotropic hardening parameter and $\mathbf{I}$ is the identity tensor. The above equation is for the condition of $f = J_2'(\mathbf{S}) - R - \sigma_Y = 0$. Equations (4.16)–(4.18) are known as the *Prandtl–Reuss flow rule*. Here, the multiplier can be expressed as $d\lambda = H(f)g'(\sigma_e)\langle d\sigma_e \rangle_+$.

In a *simple uniaxial tension* test, the non-zero component is $\sigma = \sigma_{11} = \sigma_1$ (the axial stress σ_{11} is the same as the principal stress σ_1), and the non-zero components of deviatoric stress tensor are

$$S_{11} = \frac{2}{3}\sigma , \; S_{22} = S_{33} = -\frac{1}{3}\sigma .$$

We also have $J_2 = \sigma_e = \sigma$, and then the flow law becomes

$$d\varepsilon_{ij}^p = \frac{3}{2}g'(\sigma_e)\langle d\sigma_e \rangle_+ \frac{S_{ij}}{\sigma_e} = \frac{3}{2}g'(\sigma)d\sigma\frac{1}{\sigma}(S_{ij}) .$$

For the tension direction, $S_{11} = 2\sigma/3$, we have

$$d\varepsilon_{11}^p = d\varepsilon^p = g'(\sigma)d\sigma .$$

For the other directions,

$$d\varepsilon_{22}^p = d\varepsilon_{33}^p = -\frac{1}{2}g'(\sigma)d\sigma .$$

Thus, we have

$$d\varepsilon_{22}^p = d\varepsilon_{33}^p = -d\varepsilon_{11}^p / 2 ,$$

or

$$d\varepsilon_{11}^p = -2d\varepsilon_{22}^p = -2d\varepsilon_{33}^p = d\varepsilon_e^p ,$$

since for uniaxial tension, the effective strain is

$$d\varepsilon_e^p = \left\{ \frac{2}{3}\left[(d\varepsilon_{11}^p)^2 + (d\varepsilon_{22}^p)^2 + (d\varepsilon_{33}^p)^2 \right] \right\}^{1/2}$$

$$d\varepsilon_e^p = \left\{ \frac{2}{3}\left[(d\varepsilon_{11}^p)^2 + \frac{(d\varepsilon_{11}^p)^2}{4} + \frac{(d\varepsilon_{11}^p)^2}{4} \right] \right\}^{1/2}$$

$$d\varepsilon_e^p = \left\{ \frac{2}{3}\frac{6}{4}(d\varepsilon_{11}^p)^2 \right\}^{1/2} = d\varepsilon_{11}^p$$

.

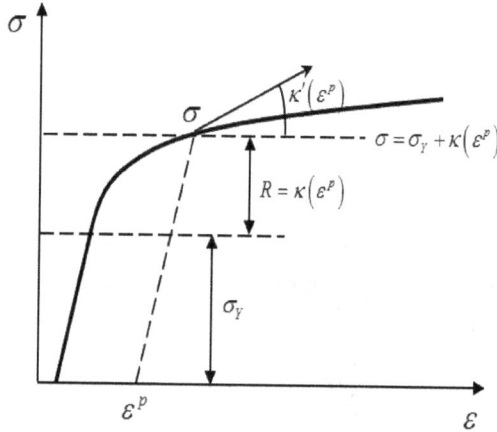

Fig. 4.8 Tension and isotropic hardening.

Similar to the relationship between the tensile stress and effective stress, the effective plastic strain is the same as the axial strain at uniaxial tension tests.

In **Fig. 4.8**, it can be seen that when plastic strain develops, the flow stress, σ ($\sigma = \sigma_Y + \kappa(\varepsilon^p)$, otherwise known as the current yield stress σ_S), increases from the initial yield stress σ_Y to the isotropic hardening (strain hardening) R ($R = \kappa(\varepsilon^p)$). The stress increment rate is the same as the isotropic hardening rate,

$$\frac{d\sigma}{d\varepsilon} = \frac{dR}{d\varepsilon^p} = \kappa'(\varepsilon^p),$$

if we assume

$$\frac{dR}{d\varepsilon^p} = \kappa'(\varepsilon^p) = b(Q - R),$$

where b and Q are material constants. According to, $g'(\sigma_e) = 1/\kappa'(\sigma_e)$, we have

$$d\varepsilon_{11}^p = d\varepsilon^p = g'(\sigma)d\sigma = \frac{1}{\kappa'(\varepsilon^p)}d\sigma = \frac{1}{b(Q-R)}d\sigma.$$

The other strain components can be defined in the similar way.

4.3 Other Plastic Flow Rules

4.3.1 *Hill's anisotropic yield criterion and flow rule*

For metal plasticity, incompressible assumption is used and the anisotropic criteria involve a rotation in stress space, and the generalization of von-Mises criterion is the form of

$$(\mathbf{C}:\mathbf{S}):\mathbf{S}=1, \tag{4.19}$$

where $\mathbf{C}$ is a fourth-order tensor and has symmetric features

$$C_{ijkl} = C_{klij} = C_{jikl} = C_{ijlk}.$$

Based on the stress tensor, it becomes

$$(\mathbf{C}':\boldsymbol{\sigma}):\boldsymbol{\sigma}=1, \tag{4.20}$$

with

$$C'_{ijkk} = C'_{iikl} = 0.$$

In the *Hill criterion*, three symmetric planes are observed during plastic deformation for strain hardening materials. The intersection of the three symmetric planes is the principal axes of anisotropy. The Hill criterion is formulated taking these axes as the reference axes (x_1, x_2, x_3). The yield criterion can be expressed as

$$C'_{1111} = F + H \; ; \; C'_{2222} = F + G \; ; \; C'_{3333} = G + H$$

$$C'_{1122} = -F \; ; \; C'_{2233} = -G \; ; \; C'_{3311} = -H$$

$$C'_{1212} = \frac{1}{2}L \; ; \; C'_{2323} = \frac{1}{2}M \; ; \; C'_{3131} = \frac{1}{2}N.$$

Based on Eq. (4.20) and the symmetric features of C_{ijkl}, we have

$$F(\sigma_{11} - \sigma_{22})^2 + G(\sigma_{22} - \sigma_{33})^2 + H(\sigma_{33} - \sigma_{11})^2 +$$
$$2L\sigma_{12}^2 + 2M\sigma_{23}^2 + 2N\sigma_{31}^2 = 1 \quad , \tag{4.21}$$

where F, G, H, L, M and N are constants defining the degree of anisotropy of a material. For principal axes of orthotropic symmetry, we have

$$F(\sigma_1 - \sigma_2)^2 + G(\sigma_2 - \sigma_3)^2 + H(\sigma_3 - \sigma_1)^2 = 1 ,$$

These constants can be determined from experiments, i.e. three simple tension tests and three simple shear tests:

tension in direction x_1 and yield stress $\sigma_{S,1}$: $F + H = 1/\sigma_{S,1}^2$

tension in direction x_2 and yield stress $\sigma_{S,2}$: $F + G = 1/\sigma_{S,2}^2$

tension in direction x_3 and yield stress $\sigma_{S,3}$: $G + H = 1/\sigma_{S,3}^2$

shear yield stress $\sigma_{S,12}$ in the plane $(0, x_1, x_2)$: $L = \dfrac{1}{2}\sigma_{S,12}^2$

shear yield stress $\sigma_{S,23}$ in the plane $(0, x_2, x_3)$: $M = \dfrac{1}{2}\sigma_{S,23}^2$

shear yield stress $\sigma_{S,31}$ in the plane $(0, x_1, x_3)$: $N = \dfrac{1}{2}\sigma_{S,31}^2$

Normally, for cold rolled metal sheets, they are anisotropic materials and the Hill's criterion can be used for the modelling work. However, simple shear tests are not easy to carry out. If the cold rolled sheet is heated up for hot stamping, the initial anisotropic behaviour of the material may disappear due to phase transformation and annealing.

For sheet metal forming processes, the assumption of plane stress is often used, in which case, the stress in thickness direction is zero, i.e. $\sigma_3 = 0$

$$F(\sigma_1 - \sigma_2)^2 + G(\sigma_2)^2 + H(\sigma_1)^2 = 1 .$$

Then we have

$$(F + H)(\sigma_1)^2 + (F + G)(\sigma_2)^2 - 2F\sigma_1\sigma_2 = 1 . \tag{4.22}$$

Let $\sigma_{S,1}$, $\sigma_{S,2}$ and $\sigma_{S,3}$ be the yield stresses in directions 1, 2 and 3 respectively. Replacing them into Eq. (4.22) then yields

$$(F+H)(\sigma_{S,1})^2 = 1, \text{ or } F+H = 1/\sigma_{S,1}^2$$

$$(F+G)(\sigma_{S,2})^2 = 1, \text{ or } F+G = 1/\sigma_{S,2}^2.$$

Then Eq. (4.22) becomes

$$\left(\frac{\sigma_1}{\sigma_{S,1}}\right)^2 + \left(\frac{\sigma_2}{\sigma_{S,2}}\right)^2 - 2F\sigma_{S,1}\sigma_{S,2}\frac{\sigma_1}{\sigma_{S,1}}\frac{\sigma_2}{\sigma_{S,2}} = 1.$$

For simplicity, we assume that the yield stresses in the plane of the sheet are equal, $\sigma_{S,1} = \sigma_{S,2} = \sigma_Y$. Thus

$$(\sigma_1)^2 + (\sigma_2)^2 - 2F\sigma_Y^2\sigma_1\sigma_2 = \sigma_Y^2, \qquad (4.23)$$

and we have

$$G = H = \frac{1}{2\sigma_{S,3}^2}$$

and

$$F = \frac{1}{\sigma_{S,1}^2} - \frac{1}{2\sigma_{S,3}^2}.$$

For sheet metal applications, if we know the yield stress in the thickness direction, the constants F, G and H can be calculated according to the above relationships. However, the yield stress in the thickness direction of the sheet, $\sigma_{S,3}$, is a difficult property to measure. Thus an anisotropic parameter R is introduced, which is the ratio of the width strain to the thickness strain under tension in one direction of the plane.

$$R = \frac{\varepsilon_2}{\varepsilon_3} = \frac{\ln(w_0/w)}{\ln(t_o/t)}. \qquad (4.24)$$

According to

$$\left(\frac{\sigma_{S,3}}{\sigma_{S,2}}\right)^2 = \frac{1}{2}(1+R),$$

Eq. (4.23), or the yield locus, can be written as

$$(\sigma_1)^2 + (\sigma_2)^2 - \frac{2R}{1+R}\sigma_1\sigma_2 = \sigma_Y^2. \qquad (4.25)$$

If the yield stress in the thickness direction is high, the strain in the thickness direction is low. This is due to texture evolution in cold rolling.

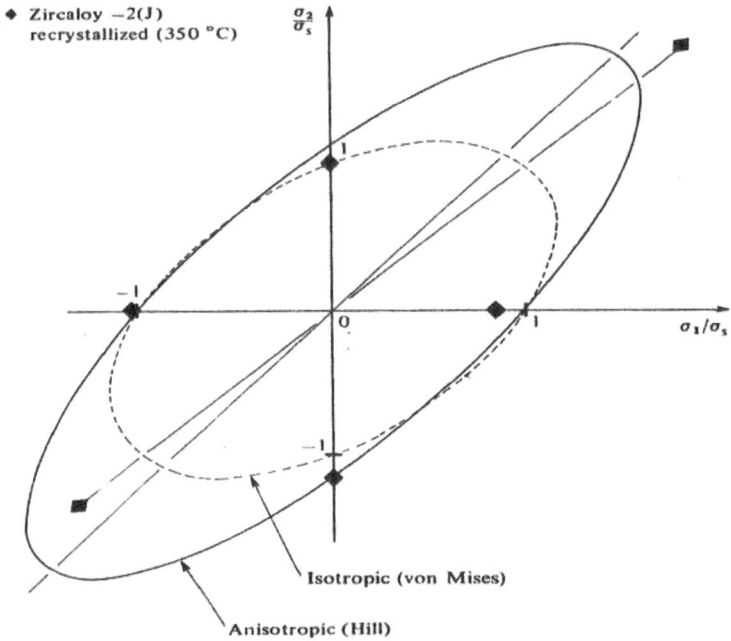

Fig. 4.9 Comparison of the predictions using isotropic von-Mises criterion and Hill's yield criterion (Lemaitre and Chaboche, 1990).

Figure 4.9 shows the comparison of the yield surfaces generated using isotropic von-Mises criterion and Hill's yield criterion. The scales are normalised by the yield stress of the material.

The isotropic flow rule can be extended by the use of the anisotropic yield criterion. With the consideration of isotropic hardening and the initial yield stress of the material, this can be expressed as

$$f = (\mathbf{C}:\mathbf{S}):\mathbf{S} - R - \sigma_Y = 0 \, . \tag{4.26}$$

Then we can get the plastic flow equation,

$$d\boldsymbol{\varepsilon}^p = H(f)g'(\sigma_e)\left\langle (\mathbf{C}:\mathbf{S}):d\mathbf{S} \right\rangle_+ \frac{\mathbf{C}:\mathbf{S}}{R + \sigma_Y} \, . \tag{4.27}$$

This type of hardening rule can be used to describe materials that are initially anisotropic, such as cold rolled metal sheets.

4.3.2 *Isotropic flow rule with Tresca criterion*

For the principal stresses, if $\sigma_1 > \sigma_2 > \sigma_3$, we have, according to the Tresca yield criterion,

$$f = \sigma_1 - \sigma_3 - \tau_S = 0 ,$$

where τ_S is the shear strength of the material. According to the normality rule $d\varepsilon^p = d\lambda(\partial f / \partial \mathbf{S})$, we have

$$\frac{\partial f}{\partial \sigma_1} = 1$$

$$\frac{\partial f}{\partial \sigma_2} = 0$$

$$\frac{\partial f}{\partial \sigma_3} = -1 ,$$

from which

$$d\varepsilon_1^p = d\lambda$$

$$d\varepsilon_2^p = 0$$

$$d\varepsilon_3^p = -d\lambda .$$

The multiplier $d\lambda$ is determined from the hardening curve in simple tension. If the power-law hardening curve is used, we can have

$$d\lambda = d\varepsilon_1^p = g'(\sigma_1 - \sigma_3)d(\sigma_1 - \sigma_3) = \frac{n}{K}\left(\frac{\sigma_1 - \sigma_3}{K}\right)^{n-1} d(\sigma_1 - \sigma_3) .$$

Thus the strain increments for *Tresca yield criterion* can be expressed as

$$d\varepsilon_1^p = g'(\sigma_1 - \sigma_3)d(\sigma_1 - \sigma_3)$$

$$d\varepsilon_2^p = 0 \tag{4.28}$$

$$d\varepsilon_3^p = -g'(\sigma_1 - \sigma_3)d(\sigma_1 - \sigma_3) .$$

4.4 Kinematic Hardening Law

The general kinematic hardening laws have been introduced in Chapter 3. This is commonly used cyclic plasticity modelling (fatigue analysis). In cold sheet metal forming conditions, the material undergoes bending and unbending operations. The kinematic hardening becomes important. However, for sheet metals, it is difficult to determine the material parameters in the kinematic hardening equations, since tension and compressions tests are difficult due to buckling problems for thin sheets.

Considering both the isotropic hardening and kinematic hardening phenomena of a material, the yield function can be expressed as

$$f = J_2(\boldsymbol{\sigma} - \mathbf{X}) - R - \sigma_Y = 0.$$

Then a set of multiaxial constitutive equations can be written as

$$d\varepsilon_{ij}^p = \frac{3}{2} d\lambda \frac{S_{ij} - X_{ij}}{J_2(S_{ij} - X_{ij})} \tag{4.29}$$

$$dR = b(Q - R) d\varepsilon_e^p \tag{4.30}$$

$$dX_{ij} = \frac{2}{3} C d\varepsilon_{ij}^p - \gamma X_{ij} d\varepsilon_e^p \tag{4.31}$$

$$d\lambda = \frac{1}{h} H(f) \left\langle \frac{3}{2} \frac{(S_{ij} - X_{ij}) : d\sigma_{ij}}{J_2(\sigma_{ij} - X_{ij})} \right\rangle, \tag{4.32}$$

where $H(f)$ represent the Heaviside step function: $H(f) = 1$ if $f = \sigma_e - R - \sigma_Y > 0$ and $H(f) = 0$, if $f = \sigma_e - R - \sigma_Y < 0$. The hardening modulus h depends on the kinematic expression (Eq. (4.32)):

$$h = \frac{2}{3} C \frac{\partial f}{\partial \sigma_{ij}} : \frac{\partial f}{\partial \sigma_{ij}} - \gamma X_{ij} : \frac{\partial f}{\partial \sigma_{ij}} \left(\frac{2}{3} \frac{\partial f}{\partial \sigma_{ij}} : \frac{\partial f}{\partial \sigma_{ij}} \right)^{1/2}.$$

With the von-Mises yield criterion, we have $d\lambda = d\varepsilon_e^p$. With the consideration of both kinematic hardening and isotropic hardening, the hardening modulus h becomes

$$h = C - \frac{3}{2} \gamma \frac{X_{ij} : (S_{ij} - X_{ij})}{J_2(\sigma_{ij} - X_{ij})} + b(Q - R). \tag{4.33}$$

The material constants, b, Q, C and γ, characterise the hardening features of the material under plastic deformation, which can be determined from experimental data. The values of the constants are normally determined from uniaxial tension-compression tests.

For uniaxial tension-compression conditions, for the loading direction the above equations can be written as according to the yield function of $f = (\sigma - X) - R - \sigma_Y = 0$:

$$d\varepsilon^p = \frac{1}{h}d\sigma$$

$$dR = b(Q - R)\left|d\varepsilon^p\right| \tag{4.34}$$

$$dX = Cd\varepsilon^p - \gamma X\left|d\varepsilon^p\right|$$

$$h = C - \lambda X \cdot \mathrm{sgn}(\sigma - X) + b(Q - R).$$

Different hardening laws can be used for both kinematic hardening and isotropic hardening. As discussed before, more equations can be used to model both hardening features for materials. These can be superimposed based on certain rules. The simplest method is to add one to another directly.

Kinematic hardening can be treated in a similar manner to a stress tensor and has all the properties that a stress tensor has. For example, the deviatoric components of the kinematic hardening tensor for the uniaxial case are

$$X_{ij} = \begin{bmatrix} \dfrac{2}{3}X & 0 & 0 \\ 0 & -\dfrac{1}{3}X & 0 \\ 0 & 0 & -\dfrac{1}{3}X \end{bmatrix},$$

where $X = X_{11}$ is the kinematic hardening component in the uniaxial tension-compression direction.

The French scientists, Lemaitre and Chaboche (1990), have made significant contributions in the development of kinematic hardening laws. These constitutive equations have been successfully used in cyclic plasticity modelling, such as, by Dunne and Hayhurst (1992) and Lin, Dunne and Hayhurst (1996). However, in metal forming simulations, limited progress has been made in the use of kinematic hardening laws in problems such as the prediction of springback in sheet metal forming. The main problem is that cyclic plasticity tests are very difficult to carry out for thin sheet metals, since buckling could readily take place in large compressive plastic deformation.

Chapter 5

Viscoplasticity and Microstructural Evolution in Warm/Hot Metal Forming

Viscoplasticity is a theory in continuum mechanics that describes the rate-dependent inelastic behaviour of solids. The inelastic behaviour that is the subject of viscoplasticity is plastic deformation. Rate-dependent plasticity is important for transient plasticity calculations. The main difference between rate-independent plasticity and viscoplasticity is that the latter exhibits not only permanent deformation as a load is applied but also continues to undergo a creep flow with time under the applied load. Viscoplasticity theories are often used for metal forming at elevated temperatures. Due to the effects of temperature and deformation, microstructure evolution, such as recrystallisation, grain growth, annealing etc., may take place. This will change the deformation mechanisms and thus affect the viscoplastic flow of the material.

It should be noted that visco-elasticity can occur in some materials. In these cases the time-dependent deformation can be recovered (e.g. Williams, 1973). Such behaviour is most often encountered in polymers but is uncommon in metals.

The first mathematical model for primary creep is represented by Andrade's law (Andrade, 1914). A one-dimensional dashpot model linking the rate of secondary creep to the stress was developed by Norton in 1929. Norton's law was generalised to the multiaxial case by Odqvist in 1934. Although there were significant theoretical developments over the 50-year period, the practical application of the theories was little until 1950s.

Since the first International Union of Theoretical and Applied Mechanics (IUTAM) Symposium "Creep in Structures" organised by Hoff in 1960, there has been major development in viscoplasticity and its applications. The IUTAM symposium of "Creep in Structures" has been held every 10 years since 1960 and has significantly contributed to the development of viscoplasticity theories and applications. The initial focus of the research was on the modelling of secondary creep and later tertiary creep by the introduction of damage parameters. The major contributions include the work of:

- Kachanov (1958) – the introduction of an internal variable representing damage, for the modelling of rupture of metals under high temperature creep;
- Hayhurst (1972) – the development of equations to model stress state effects on damage evolution of metals under high temperature creep;
- Dyson (1987) – the introduction of two internal variables to model different damage mechanisms for superalloys.

The introduction of internal variables representing damage evolution enables tertiary creep development and the lifetime of materials under service conditions at high temperatures to be predicted. In the meantime, significant development has been made in the general viscoplasticity. These contributions include:

- Chaboche (1977) – the introduction of internal variables representing isotropic hardening and kinematic hardening into viscoplastic flow equations for cyclic plasticity applications;
- Dunne and Hayhurst (1991), and Lin, Dunne and Hayhurst (1998) – the introduction of internal variables to model the interaction of creep damage and cyclic plasticity damage with kinematic hardening.

The above work was mainly concentrated on the use of internal variables to model the mechanical behaviour and damage evolution of metals under service conditions.

The dislocation density, represented using an internal variable, has been used to describe the evolution of the microstructural state of a material during viscoplastic deformation. The contribution of the work includes:

- Knocks (1976), and Mecking and Knocks (1981) – for the prediction of creep behaviour of materials;
- Estrin (1991) – turned dislocation models into a versatile tool to describe the mechanical behaviour of a wide range of metals and alloys;
- Lin, Liu, Farrugia *et al.* (2005) – introduced normalised dislocation theory and a model describing the interactive effects of

hardening, recrystallisation and grain growth during viscoplastic flow of materials, which was applied in the hot rolling of steels.

Furthermore, the theories of using internal variables representing physical phenomena of materials in viscoplastic deformation have been generalised and used in a wide range of metal forming applications. For example:

- Zhou and Dunne (1996) – introduced internal variables to model hardening and grain growth in superplastic forming. This model has been further developed by Lin and Dunne (2001) and used for superplastic forming process modelling;
- Lin, Ho and Dean (2006) – introduced internal variables to model precipitate growth and dislocation hardening in creep age forming. This model has been further developed by Zhan, Lin, Dean *et al.* (2011) for modelling of spherical type precipitates nucleated in creep age forming conditions.

In this chapter, the phenomenological and mathematical modelling of elastic viscoplastic solids will be discussed in detail. The hardening laws introduced in Chapters 3 and 4 have similar features to the phenomena observed in viscoplastic deformation, apart from the consideration of static recovery. In particular, strain-induced microstructural evolution will be discussed in detail.

5.1 Application Domains

As discussed in Chapter 4, plasticity is a special case of viscoplasticity. For metals and alloys deformed at high temperatures, the plastic deformation mechanisms could be: (i) the dislocation movement in grains – climb, gliding, deviation, and (ii) grain boundary sliding and grain rotation. These may occur simultaneously in metal forming processes if the work-piece is deformed at a temperature $T/T_m > 1/3$, where T_m is the melting temperature of the material in Kelvin (absolute temperature K). Below that temperature, plasticity theories are usually used, as discussed in Chapter 4.

At low temperature deformation conditions, the equipotential surfaces of the two extreme conditions, as shown in **Fig. 4.2**, are close and the viscoplastic area is very small. Thus, time-independent plasticity theory can be used. However, if a metal is deformed at high temperatures, such as in warm and hot forming conditions, the two extreme surfaces are far and the viscoplastic area is very large. Thus, it justifies the use of viscoplasticity theories for warm and hot metal forming.

Viscoplasticity theories can only be used for non-damaging loading conditions. For example, at uniaxial tensile testing conditions, the theory should be used for strains lower than about 70% of the strain to failure, i.e. approximately $< 0.7\varepsilon_f$, where ε_f is the strain to failure at uniaxial loading conditions. This means that only the strain hardening region is considered. If strain is high, the flow stress may drop due to micro-damage of the material, which will be discussed in Chapter 6.

5.2 Viscoplastic Deformation of Metals

5.2.1 *High temperature creep*

Viscoplasticity is associated with the creep of metals. Thus, it is important to first introduce the basic concept of creep deformation, which often occurs in metals at elevated temperature under stress.

In materials science, *creep* is the time-dependent and permanent deformation of a solid material under constant loading conditions. This kind of *plastic strain* is known as *creep strain*. It can occur as a result of long-term exposure to high levels of stress that are still below the *yield stress* of the material. Creep is more severe in metals that are subjected to high temperatures for long periods and generally increases as they near their melting points. Creep rate normally increases with temperature.

Creep rate is a function of the material properties, applied stress levels, exposure time and temperatures. Depending on the magnitude of the applied stress and its duration the deformation may become so large that a component can no longer perform its function, i.e. it cannot sustain loads. For example, creep of a turbine blade could cause the contact of the blade to the casing, resulting in the failure of the blade and damage to the engine.

a. *Phenomena of creep*

Under a constant load or stress (**Fig. 5.1(a)**), the initial strain is ε_0 as the load is just applied, i.e. at creep time $t = 0$ (**Fig. 5.1(b)**). In the initial stage, known as *primary creep*, the strain rate is relatively high. However, it slows down with increasing time due to strain hardening. Eventually the strain rate reaches a minimum, i.e. $\dot{\varepsilon}_c = \dot{\varepsilon}_{min}$, and becomes near constant. This is due to the balance between work hardening and annealing (thermal softening). This stage is known as *secondary* or *steady-state creep*, which is the most studied and understood. The *creep strain rate* typically refers to the rate in the secondary stage. Stress dependence of this rate depends on the creep mechanism. In the last stage, known as *tertiary creep*, the strain rate increases sharply with time to failure occurring. The time at failure, t_f, is known as the creep lifetime of a structure.

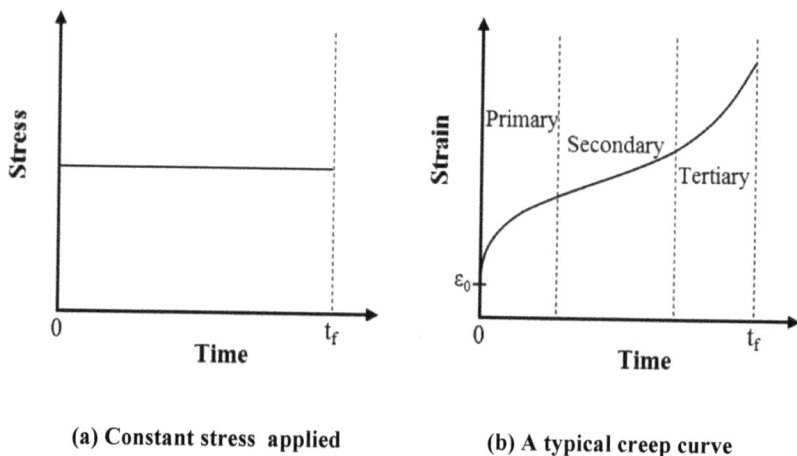

(a) Constant stress applied   (b) A typical creep curve

Fig. 5.1 Creep deformation of a solid under a constant stress (similar to constant load at small deformation conditions).

b. Deformation mechanisms and creep equations

The *secondary creep* can be general modelled using a power-law:

$$\dot{\varepsilon}_c = \frac{d\varepsilon_c}{dt} = C\frac{\sigma^n}{d^\gamma}\exp\left(-\frac{Q}{RT}\right), \tag{5.1}$$

where ε_c is the creep strain. C, n and γ are constants depending on materials and creep deformation mechanisms. Q is the activation energy of the creep mechanism, σ is the applied stress, d is the grain size of the material, R is the gas constant, and T is the absolute temperature in Kelvin. Large grain size would result in less creep (lower creep rate). Thus, single crystal structural components have the highest creep resistance. Castings are normally better than forged components in terms of the high temperature creep resistance since the former normally have larger grain sizes.

At high stress levels (relative to yield stress of a material), creep is controlled by the movement of dislocations. In the case of *dislocation creep*, the creep rate is highly dependent on the stress applied rather than the grain size. By introducing a *threshold stress, k,* below which creep cannot be measured, the modified power-law equation becomes:

$$\dot{\varepsilon}_c = C(\sigma - k)^n \exp\left(-\frac{Q}{RT}\right),$$ (5.2)

where the stress exponent, n, is normally within a range of 3–10. This equation is especially useful for high stress and high creep rate conditions.

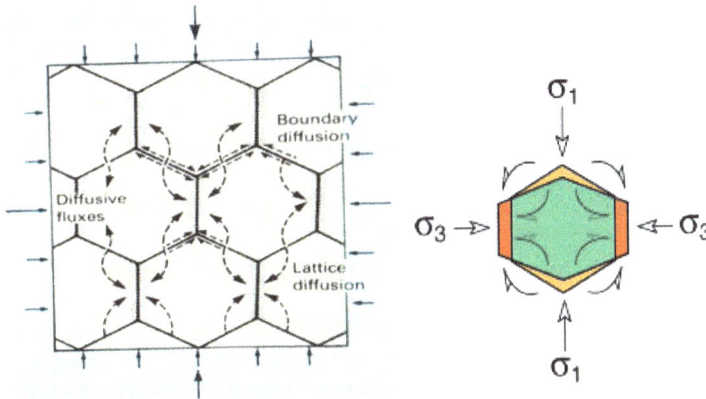

Fig. 5.2 Diffusion along grain boundaries – Coble creep, and the shape of grain changes (E Report by Matthew De Paoli & Matthew Bennett).

Other deformation mechanisms of creep include:

- *Diffusion creep* (or bulk diffusion), also known as *Nabarro–Herring creep*, is where atoms diffuse through the lattice causing grains to elongate along the stress axis. Diffusion creep has weak stress dependence and moderate grain size dependence, with the creep rate decreasing as grain size increases. Nabarro–Herring creep is *strongly temperature-dependent*. For lattice diffusion of atoms to occur in a material, neighbouring lattice sites or interstitial sites in the crystal structure must be free. A given atom must also overcome the energy barrier to move from its current site to the nearby position.

- *Coble creep* is a second form of diffusion-controlled creep. In Coble creep the atoms diffuse along *grain boundaries* to elongate the grains along the stress axis, as shown in **Fig. 5.2**. This causes Coble creep to have stronger grain size dependence than Nabarro–Herring creep. Coble creep is also temperature-dependent as the temperature increase enhances grain boundary diffusion.

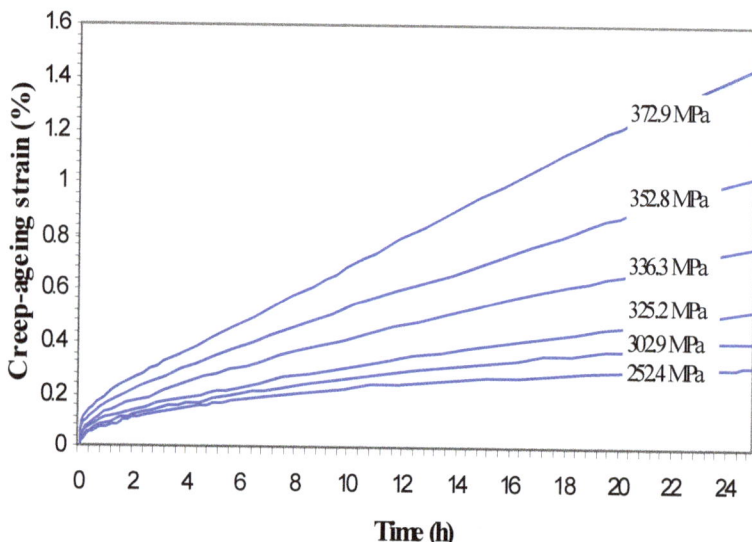

Fig. 5.3 Primary and secondary creep under creep age forming (CAF) conditions for AA7010 at a temperature of 150 °C (Ho, 2004).

Primary creep can be modelled through the empirical equation by the introduction of creep time t and a parameter m:

$$\dot{\varepsilon}_c = C \frac{\sigma^n}{d^\gamma} \exp\left(-\frac{Q}{RT}\right) t^m, \tag{5.3}$$

where m is a material constant between 0 and 1. In many cases, it is about 0.3. **Figure 5.3** shows the primary and secondary creep curves obtained from *creep age forming* conditions at a temperature of 150 °C for AA7010 (Ho, 2004). The material is tested at T4 condition. The details of creep age forming will be introduced in Chapter 8.

Tertiary creep is due to the damage of the material under high temperature creep. This can be modelled by the introduction of damage constitutive equations. More details regarding high temperature creep damage will be introduced in Chapter 6.

5.2.2 *Stress relaxation*

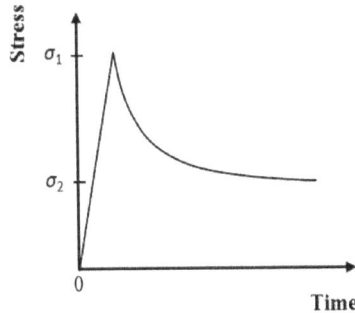

Stress relaxation tests demonstrate the decrease in the stress at high temperatures by maintaining the strain constant at uniaxial loading conditions. The major mechanisms of stress relaxation include:

- Creep, which increases the plastic strain ε^p and thus reduces the elastic strain ε^e, since the total strain ε^T remains constant (according to $\varepsilon^e = \varepsilon^T - \varepsilon^p$, and $\sigma = E\varepsilon^e$);
- Recovery, which occurs due to annealing of metals at high temperature. This reduces the dislocation hardening or work hardening. The static and dynamic recovery of dislocations at hot/warm deformation conditions will be discussed later in the chapter.

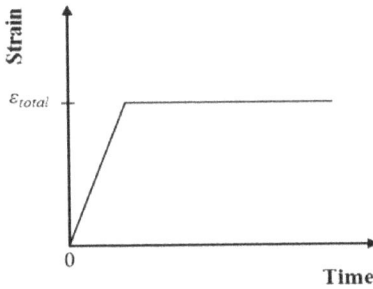

(a) Constant strain is kept

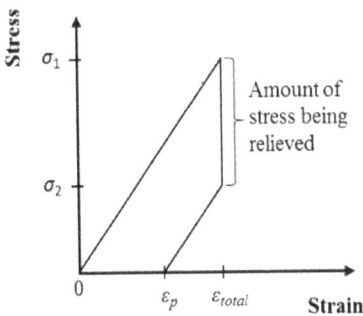

(b) stress relaxation **(c) stress relaxation with time**

Fig. 5.4 Schematic representation of loading and stress relaxation.

Figure 5.4 shows the principle of a stress relaxation test. The specimen is loaded to a certain stress level and the total strain remains constant. The change of load (stress) is measured with time. The test characterises the viscosity and can be used to determine the relationship that exists between the stress and the rate of viscoplastic strain. For the case of stress relaxation we have $\dot{\varepsilon}^T = 0$. According to the rate relationship $\dot{\varepsilon}^T = \dot{\varepsilon}^e + \dot{\varepsilon}^p$ and the linear elasticity relationship $\sigma' = E\dot{\varepsilon}^e$, or $\dot{\varepsilon}^e = \dot{\sigma}/E$, then

$$0 = \frac{\dot{\sigma}}{E} + \dot{\varepsilon}^p, \text{ or } \dot{\varepsilon}^p = -\frac{\dot{\sigma}}{E}. \tag{5.4}$$

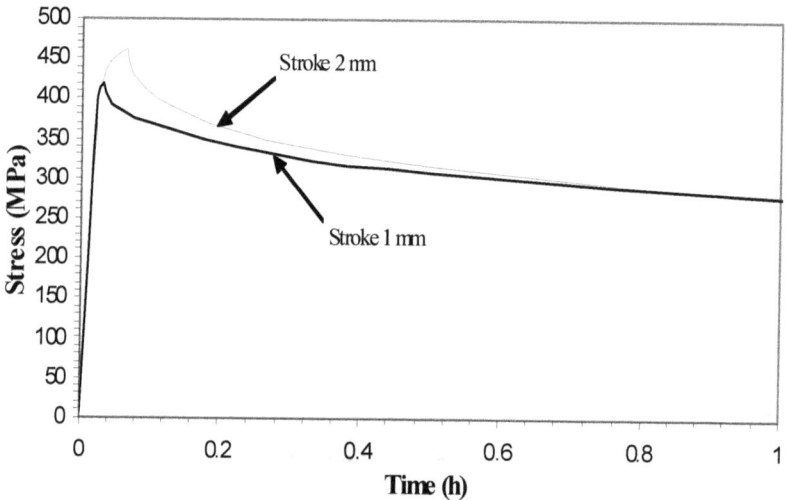

Fig. 5.5 Stress relaxation test results for AA7010 at a temperature of 150 °C. The specimens are held as they are deformed by 1 mm and 2 mm, respectively (Ho, 2004).

Thus, the viscoplastic strain rate can be calculated directly from the rate of stress relaxation. This relationship is based on the thermally activated stress relaxation due to dislocation recovery being ignored. If the temperature is low and stress level is high, then Eq. (5.4) can be used to calculate the plastic strain (or creep strain) or plastic strain rate.

Figure 5.5 shows the stress relaxation test results for AA7010 at a constant temperature of 150 °C. The two test results show that the stress relaxation takes place very quickly at the beginning and slows down with time.

5.2.3 *Basic features of viscoplasticity*

a. Flow stress

Figure 5.6 schematically shows the flow stress response of a material deformed at room temperature and at hot forming temperature. If the material is deformed at room temperature, then the flow stress does not change much as the strain rate increases. This can be approximately modelled using elastic-plastic theories. If the material is deformed at high temperatures, then the flow stress increases significantly with the increase of strain rate. This is due to the mixed effects of high temperature creep, recovery (annealing) and other mechanisms. In this case the viscoplasticity theories must be used to minimise the errors of the modelling.

In plastic deformation, the strain hardening feature is observed and is beneficial for sheet metal stamping processes as it reduces the tendency of localised necking and so increases the formability and drawability of the material. However, the ductility of materials, especially for aluminium, magnesium and titanium alloys, is much lower at room temperature, which is normally not sufficient for stamping complex shaped panel components. Strain hardening is disadvantageous for bulk forming, such as extrusion and forging, as this would increase the forming load significantly and may cause cracks.

In viscoplastic deformation, strain rate behaviour is clearly seen. At high strain rate the dislocation recovery time and creep time are short, which results in high flow stress and high strain hardening (since annealing time is short). These high strain rate work hardening features of the material are extremely important to formability of the sheet metal stamping process. If strain rate hardening, strain hardening and high ductility of materials deformed at high temperatures can be used effectively, more complex-shaped panel components can be formed using warm/hot stamping processes.

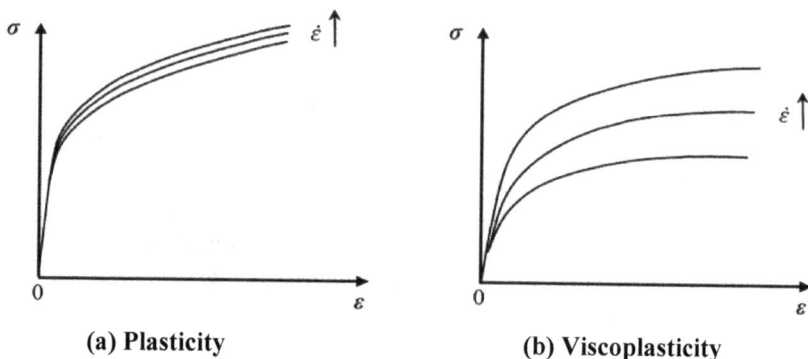

(a) Plasticity   (b) Viscoplasticity

Fig. 5.6 Schematic representation of the stress–strain relationships of a material deformed at the same three strain rates (a) at room temperature (plasticity) and (b) at a hot forming temperature (viscoplasticity).

At a lower deformation rate more time is given for dislocation recovery and this allows creep to take place, which would result in low flow stresses and very low strain hardening (the stress–strain curve becomes horizontal in large deformation). In sheet metal forming processes, strain hardening can be used to reduce necking. However, this low strain hardening feature is favoured in bulk forming, such as forging, rolling and extrusion, where low forces can be applied to deform the material easily.

b. Young's modulus and yield stress

If the material is deformed at room temperature (plastic deformation), little creep takes place. The slope in the elastic region of deformation is the Young's modulus, as shown in **Fig. 5.7**, and the yield stress can be defined using a standard method, e.g. 0.2% proof stress.

If the material were deformed at hot forming conditions, creep would take place and also annealing would occur in the material. Even if the linear stress–strain relationship can be observed from experiments at the initial deformation stage, creep occurs that reduces the slope of the linear part and gives a false value of Young's modulus. Yield stress is even more difficult to measure since creep occurs almost over the whole deformation period. In most modelling processes a threshold stress is assumed below which we "believe" no creep takes place.

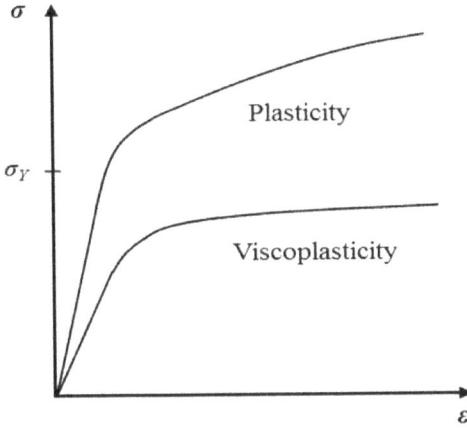

Fig. 5.7 Schematic representation of stress–strain relationships of a material deforming at room temperature (plasticity) and at high temperature (viscoplasticity).

In summary, Young's modulus and yield stress of a material are functions of temperature but they are not obvious at viscoplastic deformation conditions. Moreover, they could not be easily measured from uniaxial tensile stress–strain curves.

5.3 Superplasticity and Deformation Mechanisms

Dominant deformation mechanisms of a material are dependent on deformation temperature, deformation rate, and the initial microstructure. These may change during a forming process since the deformation and material conditions could be changed. In this section we use superplastic forming (SPF) as an example to illustrate viscoplastic deformation mechanisms and microstructural evolution in a practical forming application.

5.3.1 *The process and features*

Superplastic forming (SPF), also known as *gas-blow forming*, is a method of die forming in which a metal sheet is heated to a pliable state, 0.5–0.6 of its melting temperature, then pressurised by a gas to deform the sheet metal into the die cavity, as shown in **Fig. 5.8**. The high temperatures allow the metal to be elongated, or stretched, to much greater degrees without rupture than would be possible in previously utilised cold and warm forming methods. In addition, the metal can be formed into finer details, which requires less overall forming force than traditional methods.

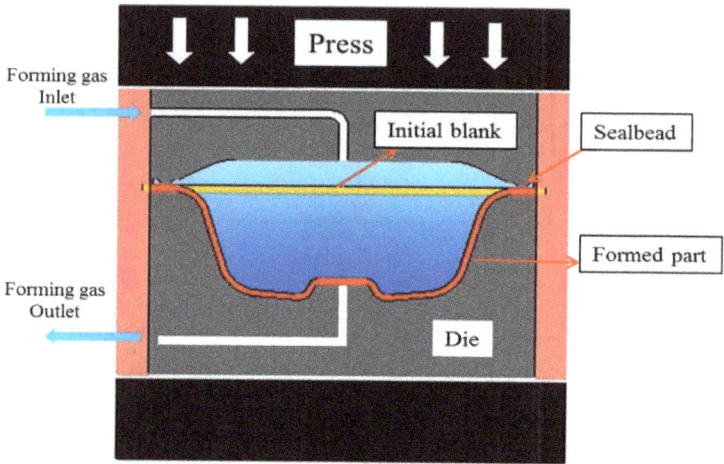

Fig. 5.8 A typical superplastic forming process.

The work-piece is sealed and firmly clamped at the edges, and thus there is no material drawing taking place in SPF. The surface area of the work-piece is increased significantly in a superplastic forming process. This can result in localised thinning. Thus, significant process modelling activities have been carried out to understand and minimise the localised thinning of the material.

Typical features of SPF:

- It is an *isothermal forming* process. The die and the metal sheet are heated together and have the same temperature.

- Extremely large plastic deformation could be observed (**Fig. 5.9**). This is mainly suitable for forming complex-shaped panel components (Pearson, 1934).

- The metal sheet must have *fine grains* and the grain growth rate must be low during forming. The dominant deformation mechanism for SPF is *grain rotation and grain boundary sliding*. This enables large plasticity to be observed. Fine grains are easy to rotate under straining conditions. The grain size is normally below 10 μm.

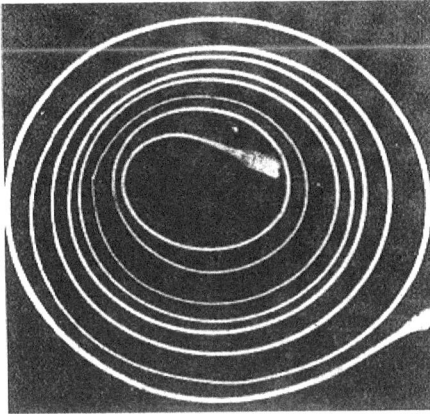

Fig. 5.9 Pearson's well-known photograph of a Bi-Sn alloy with 1950% of elongation (Pearson, 1934).

- SPF only occurs for *fine grained materials* at certain *strain rate* and *temperature* ranges. For the grain size of about 5 μm to 10 μm, the strain rate is within a range of 10^{-5} s^{-1} to 10^{-1} s^{-1}. The strain rate could be higher for smaller grain size materials. The slow deformation rate enables the *grain boundary diffusion* to take place so that grain boundary sliding and grain rotation could occur. The temperature should not be too high because at high temperatures the grain growth rate would be high, making it difficult for grain rotation to take place. For titanium it is around 900 °C (1,650 °F) and for aluminium it is between 450 °C and 520 °C.

- Productivity is low. This is mainly due to: (i) it is an isothermal forming process – die and materials are normally heated together, which takes time, and (ii) the forming rate is low to enable the expected deformation mechanism to take place. Normally the forming cycle varies between 2 minutes and 2 hours. Hence, it is mainly suitable for low volume production, for example forming complex-shaped aerospace panel components.

The hardening mechanisms for SPF alloys would mainly include:

- *Strain hardening.* This is weak as the material is deformed slowly at high temperature and there is sufficient time for annealing to take place. Plastic deformations are associated with two mixed mechanisms; one is grain boundary sliding and grain rotation, as discussed earlier, and the other one is due to dislocations. The amount of dislocations is normally low and could be recovered at low deformation rates. Thus, the strain hardening is relatively low for most of the cases.

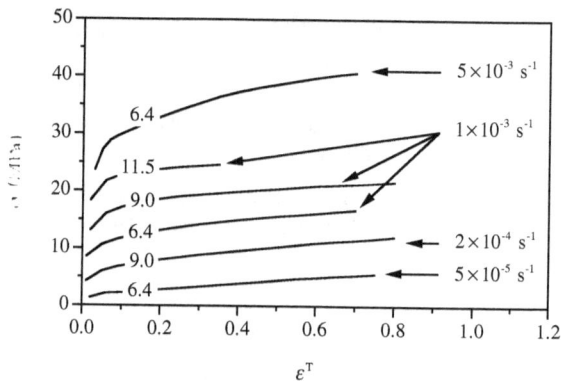

Fig. 5.10 Experimental stress–strain relationships of Ti-6Al-4V at 927 °C for $\dot{\varepsilon}^T = 5 \times 10^{-5}$ s^{-1}, 2×10^{-4} s^{-1}, 1×10^{-3} s^{-1}, and 5×10^{-3} s^{-1}; and $d^0 = 6.4$ μm, 9.0 μm, and 11.5 μm (Ghosh and Hamilton,1979).

- *Grain size.* This has a significant influence on flow stresses, similar to the case in Coble creep. Smaller grain size makes it easier for grain rotation and grain boundary sliding to take place. Thus, the grain growth would increase the flow stress and cause the hardening of the material in superplastic forming processes. This is known as *grain growth hardening.* As shown in **Fig. 5.10**, the flow stress increases with the material of increasing initial grain size (6.4 μm, 9.0 μm and 11.5 μm) at the same strain rate deformation (1×10^{-3} s^{-1}). The grain growth features are shown in **Fig. 5.11**. The grains grow more quickly in the material of a smaller initial grain size. In modelling, we have the relationship of $\dot{\varepsilon}^p = f(d^{-\gamma})$ for the prediction of grain growth in SPF.

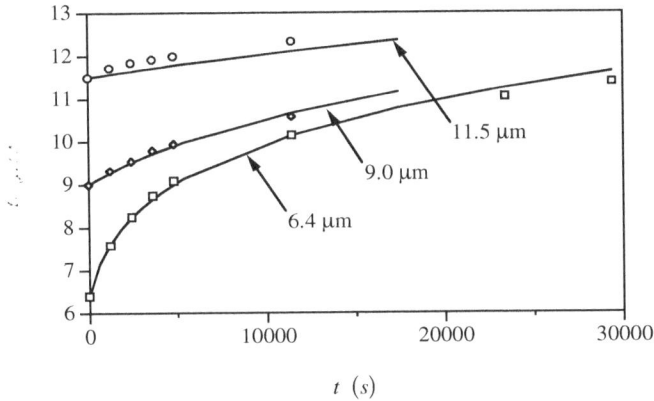

Fig. 5.11 Grain growth of Ti-6Al-4V at 927 °C for the initial grain size of d^0 = 6.4 μm, 9.0 μm, and 11.5 μm (Cheong, Lin and Ball, 2000; Ghosh and Hamilton, 1979).

- *Strain rate hardening.* This is the major hardening mechanism in superplastic deformation. It can be seen from **Fig. 5.10** that as the strain rate increases from 5×10^{-5} s^{-1} to 5×10^{-3} s^{-1} the flow stress increases approximately tenfold (from about 4 MPa to about 40 MPa) for the same initial grain size of 6.4 μm. Since the strain hardening and grain growth hardening in SPF are very limited, the strain rate hardening plays the major role in minimising localised thinning in superplastic forming of complex-shaped components.

5.3.2 *Strain rate hardening, sensitivity and ductility*

In superplasticity, the flow stress σ is particularly sensitive to the rate of deformation. The material behaviour is normally observed at strain rates within the range of 1×10^{-5} s^{-1} to 1×10^{-1} s^{-1}. **Figure 5.12** shows a typical stress–strain-rate relationship for superplastic materials, which is generally described as (Pilling and Ridley, 1989):

$$\sigma = K\dot{\varepsilon}^m, \tag{5.5}$$

where $\dot{\varepsilon}$ is the strain rate which is identical to plastic strain rate $\dot{\varepsilon}^p$, and K a material constant that depends upon the test temperature, microstructure and defect structure of the specimen.

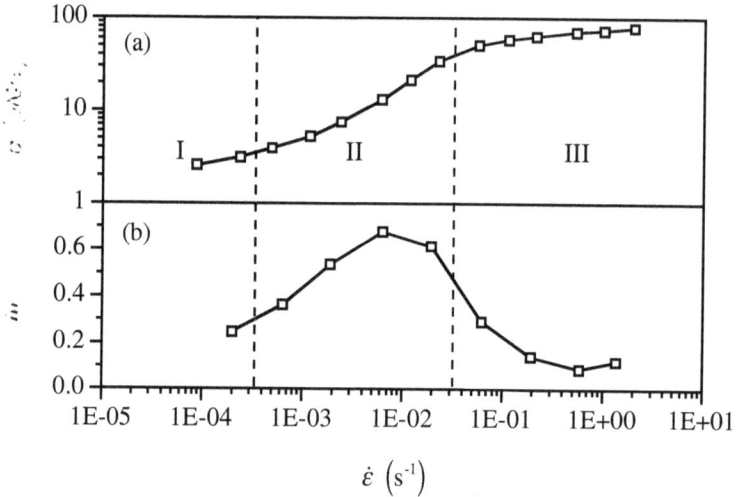

Fig. 5.12 (a) Logarithmic flow stress and (b) strain-rate sensitivity against logarithmic strain rate for Al-Mg eutectic alloy with initial grain size of 10.6 μm deformed at 350 °C (Lee, 1969).

The parameter m is known as the *strain rate sensitivity* or *strain rate hardening* parameter that is defined as:

$$m = \frac{\Delta \ln \sigma}{\Delta \ln \dot{\varepsilon}}. \tag{5.6}$$

If superplasticity is regarded as a type of creep behaviour, Equation (5.5) is often written as (Edington, Melton and Cutler, 1976)

$$\dot{\varepsilon} = \frac{\sigma^n}{k}, \tag{5.7}$$

where n is the *stress exponent* and $n = 1/m$. The value of m is normally greater than or equal to 0.3, with figures in the range of 0.4 to 0.8 being the most common (Pilling and Ridley, 1989). The higher the m-value is, the greater the plasticity of material that could be observed before failure. Normally m varies from 0 to 1.0.

Figure 5.12(b) shows the variation of strain-rate sensitivity as a function of strain rate; the values are high at intermediate strain rates but decrease smoothly to much lower ones at both high and low strain rates. For convenience such variation and the sigmoidal stress–strain-rate profile in **Fig. 5.12(a)** can be divided into three regions: I, II, and III, where Regions I and III correspond to m-values of less than 0.3 at low and high strain rates respectively, and Region II is the superplastic regime that corresponds to *m*-values of equal to or greater than 0.3.

Region III of the high strain rates is generally believed to correspond to recovery-controlled dislocation creep, also termed power-law creep (Pilling and Ridley, 1989). Deformation within this region leads to observations of slip lines and development of high dislocation densities within grains. Also, crystallographic texture within the material increases and significant grain elongations occur as deformation progresses. Increasing strain rate in this region leads to decreasing strain-rate sensitivity.

In the superplastic Region II, high uniform strains are generally observed. It is evident that grain-boundary sliding and grain rotations make a substantial contribution to the total strain. It is generally agreed that three mechanisms operate during superplastic deformation to a greater or lesser extent:

(i)   grain-boundary sliding,

(ii) diffusional creep, and,

(iii) dislocation creep/dynamic recovery.

Generally grain-boundary sliding dominates over the others (Pilling and Ridley, 1989). The fundamental mechanism underlying grain-boundary sliding and grain rotations was not addressed until the analytic explanation developed by Lagos (2000). Currently Lagos's analytical work seems to be the most satisfying account of the *origin* of superplasticity. It shows that grain-boundary sliding is the consequence of a cyclic transport of atoms induced by a closed-loop motion of vacancies between adjacent crystals. The vacancies result from the corrugation of grain boundaries by shear stresses exceeding a critical value.

Other characteristic findings in Region II include limited dislocation activity within the grains; the grains remain equiaxed under straining. Materials showing initial microstructural banding develop a more uniform equiaxed microstructure as deformation progresses. Also, the flow stress decreases and the value of strain-rate sensitivity increases with increasing temperature and decreasing grain size. In addition, the elongation-to-failure in this region tends to increase with increasing strain-rate sensitivity.

Region I is related to very low strain rate and the *m*-value is low. This is the gap between traditional high temperature creep (the strain rate considered is much lower, e.g. 10^{-20} s^{-1} to 10^{-10} s^{-1}) and superplasticity

(superplasticity is also known as high rate creep). The research on the deformation mechanisms is limited and the reasons why the m-value is low at the very low strain rate are vague. The research trend on SPF is to develop smaller grain size materials so that the deformation rate could be high and productivity increased.

5.3.3 Materials modelling for SPF

According to the deformation features of SPF, a set of unified superplastic (creep or viscoplastic) constitutive equations should include the following.

a. Viscoplastic flow

$$\dot{\varepsilon}^p = f(\sigma, R, d)$$

is a function of stress, isotropic hardening R and grain size d. Normally temperature is not considered since SPF is an isothermal forming process.

b. Grain growth

$$\dot{d} = f(\varepsilon^p, \dot{\varepsilon}^p, d)$$

is dependent on plastic strain ε^p, plastic strain rate $\dot{\varepsilon}^p$, and grain size itself d. Time has been considered but it is not explicitly expressed here since they are a set of evolutionary equations and everything is time-related. Again, temperature is not considered in an isothermal forming process for simplicity.

c. Strain hardening (isotropic hardening)

$$\dot{R} = f(\rho, \dot{\rho}, d)$$

is dependent on the accumulation of dislocation density ρ and its rate which are related to plastic deformation rate of the material $\dot{\varepsilon}^p$. At the same time grain size, d, plays an important role in strain hardening. Finer grain size would facilitate grain boundary sliding and grain rotation, which generates fewer dislocations for the same extent of plastic deformation. Thus, strain (or dislocation) hardening would be low. In addition, if the deformation rate is lower, then more time is given for recovery (annealing) and grain boundary diffusion, and the strain hardening is lower.

d. Formability

The formability of material in SPF can be modelled using damage evolution equations, which will be discussed in Chapter 6.

Based on the studies of viscoplasticity in superplastic forming, it is imperative to have a basic knowledge of viscoplastic deformation mechanisms and the relevant physical phenomena to accurately model the behaviour. More details in the development of unified viscoplastic constitutive equations will be given later in this chapter. A set of unified superplastic constitutive equations and a case on materials and process modelling will be given at the end of this chapter as an example.

5.4 Modelling of Viscoplasticity and Hardening

5.4.1 *Dissipation potential and normality law*

The concept of yielding criteria in plasticity is no longer necessary. It has been replaced by a family of equipotential surfaces, ψ, for example

$$\psi = \psi(\sigma, R, y_k),$$

where R is isotropic hardening and y_k ($k = 1,2,3,...$) are internal variables representing individual physical parameters in applications. The equipotential surfaces, ψ, are the surfaces in stress space, the magnitude of which are the same as the strain rate (i.e. the dissipation is the same). The surface of zero potential is the surface without the elastic region.

According to normality laws (i.e. the strain increment is perpendicular to the equipotential surface), the general expression of plastic strain rate tensor is as follows:

$$\dot{\boldsymbol{\varepsilon}}^p = \frac{\partial \psi}{d\mathbf{S}} = \frac{\partial \psi}{\partial \sigma_e} \frac{\partial \sigma_e}{\partial \mathbf{S}} = \frac{3}{2} \frac{\partial \psi}{\partial \sigma_e} \frac{\mathbf{S}}{\sigma_e} \tag{5.8}$$

or

$$\varepsilon_{ij}^p = \frac{3}{2} \frac{\partial \psi}{\partial \sigma_e} \frac{S_{ij}}{\sigma_e},$$

where $\mathbf{S}$ is the deviatoric tensor and σ_e is the effective stress (von-Mises stress), which is defined as

$$\sigma_e = \left(\frac{3}{2} \mathbf{S} : \mathbf{S} \right)^{1/2} = \left(\frac{3}{2} S_{ij} : S_{ij} \right)^{1/2}$$

$$\sigma_e = \left[\frac{3}{2}\left(S_{11}S_{22} + S_{22}S_{33} + S_{33}S_{11} - S_{12}^2 - S_{23}^2 - S_{31}^2\right)\right]^{\frac{1}{2}}$$

or

$$\sigma_e = \left\{\frac{1}{2}\left[(\sigma_{11} - \sigma_{22})^2 + (\sigma_{22} - \sigma_{33})^2 + (\sigma_{33} - \sigma_{11})^2\right] + 3(\sigma_{12}^2 + \sigma_{23}^2 + \sigma_{31}^2)\right\}^{1/2}.$$

Recall the relationships between the natural stress tensor and deviatoric stress tensor:

$$S_{11} = \sigma_{11} - (\sigma_{11} + \sigma_{22} + \sigma_{33})/3 = (2\sigma_{11} - \sigma_{22} - \sigma_{33})/3$$

$$S_{22} = \sigma_{22} - (\sigma_{11} + \sigma_{22} + \sigma_{33})/3 = (2\sigma_{22} - \sigma_{11} - \sigma_{33})/3$$

$$S_{33} = \sigma_{33} - (\sigma_{11} + \sigma_{22} + \sigma_{33})/3 = (2\sigma_{33} - \sigma_{11} - \sigma_{22})/3$$

and

$$S_{12} = \sigma_{12}, \quad S_{23} = \sigma_{23}, \quad S_{31} = \sigma_{31}.$$

To obtain $\dot{\varepsilon}_{11}^p$, we have

$$\dot{\varepsilon}_{11}^p = \frac{\partial \psi}{\partial \sigma_e}\frac{\partial \sigma_e}{\partial S_{11}},$$

where

$$\frac{\partial \sigma_e}{\partial S_{11}} = \frac{\partial\left[\frac{3}{2}\left(S_{11}^2 + S_{22}^2 + S_{33}^2\right) + S_{12}^2 + S_{23}^2 + S_{31}^2\right]^{\frac{1}{2}}}{\partial S_{11}}$$

$$\frac{\partial \sigma_e}{\partial S_{11}} = \frac{1}{2}\sigma_e^{-1/2}\frac{3}{2}(2S_{11}) = \frac{3}{2}\frac{S_{11}}{\sigma_e}.$$

Thus

$$\dot{\varepsilon}_{11}^p = \frac{\partial \psi}{\partial \sigma_e}\frac{\partial \sigma_e}{\partial S_{11}} = \frac{3}{2}\frac{\partial \psi}{\partial \sigma_e}\frac{S_{11}}{\sigma_e}.$$

Similarly, we could have

$$\dot{\varepsilon}_{22}^p = \frac{3}{2}\frac{\partial \psi}{\partial \sigma_e}\frac{S_{22}}{\sigma_e}, \quad \text{etc.}$$

The general rate equations can also be written as

$$\dot{\varepsilon}_{ij}^p = \frac{3}{2}\frac{S_{ij}}{\sigma_e}\dot{\varepsilon}_e^p, \tag{5.9}$$

where $\dot{\varepsilon}_e^p$ is the effective plastic strain rate and given by

$$\dot{\varepsilon}_e^p = \frac{\partial \psi}{\partial \sigma_e}.$$

The other state variables follow the same normality principal.

5.4.2 Viscoplastic flow of materials

a. Odqvist's law

This is a power-law and is similar to Norton's law in that it ignores the elastic domain, i.e. represents *rigid perfect viscoplasticity*.

$$\psi = \frac{K}{n+1}\left(\frac{\sigma_e}{K}\right)^{n+1} \tag{5.10}$$

$$\dot{\varepsilon}_{ij}^p = \frac{\partial \psi}{\partial S_{ij}} = \frac{\partial \psi}{\partial \sigma_e}\frac{\partial \sigma_e}{\partial S_{ij}} = \frac{3}{2}\left(\frac{\sigma_e}{K}\right)^n \frac{S_{ij}}{\sigma_e}, \tag{5.11}$$

where

$$\frac{\partial \psi}{\partial \sigma_e} = \left(\frac{\sigma_e}{K}\right)^n \dot{\varepsilon}_e^p, \text{ and } \frac{\partial \sigma_e}{\partial S_{ij}} = \frac{3}{2}\frac{S_{ij}}{\sigma_e}.$$

For the uniaxial case there is $\sigma_{11} = \sigma_1 = \sigma_e$, i.e. maximum principal stress, effective stress and uniaxial stress are identical.

$$\psi = \frac{K}{n+1}\left(\frac{\sigma_{11}}{K}\right)^{n+1}, \quad \sigma_{11} = \sigma_e$$

and deviatoric stresses

$$S_{11} = \frac{2}{3}\sigma_{11} \text{ (according to } S_{11} = \sigma_{11} - \frac{\sigma_{11}}{3}),$$

$$S_{22} = S_{33} = -\frac{1}{3}\sigma_{11};$$

other stress components are zero. Then

$$\dot{\varepsilon}_{11}^p = \frac{\partial \psi}{\partial S_{11}} = \left(\frac{\sigma_{11}}{K}\right)^n$$

$$\dot{\varepsilon}_{22}^p = \dot{\varepsilon}_{33}^p = -\frac{1}{2}\left(\frac{\sigma_{11}}{K}\right)^n = -\frac{1}{2}\dot{\varepsilon}_{11}^p.$$

According to the above equations, the flow stress for the uniaxial case (ignoring the subscript "11") can be expressed as

$$\sigma_{11} = \sigma = K(\dot{\varepsilon}^p)^{1/n}.$$

Thus, the flow stress is only related to the viscoplastic strain rate and scaled by a coefficient of K. **Figure 5.13** shows the relationship of rigid perfect viscoplasticity of materials. This is equivalent to steady state creep conditions. K and n are material constants (which could be temperature-dependent), to be determined from experimental data.

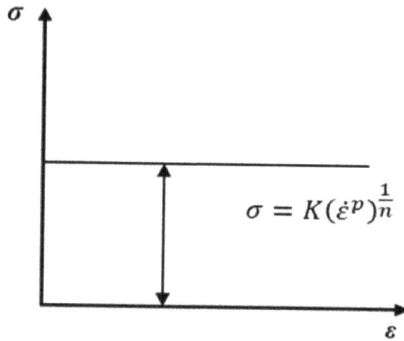

$$\sigma = K(\dot{\varepsilon}^p)^{\frac{1}{n}}$$

Fig. 5.13 Flow stress of rigid perfect viscoplasticity.

b. *With elastic region considered*

By the introduction of the threshold stress, k, the viscoplastic potential Eq. (5.10) can be written in the form of

$$\psi = \frac{K}{n+1}\left(\frac{\sigma_e - k}{K}\right)^{n+1} \tag{5.12}$$

and then we have

$$\dot{\varepsilon}^p_{ij} = \frac{\partial \psi}{\partial s_{ij}} = \frac{\partial \psi}{\partial \sigma_e}\frac{\partial \sigma_e}{\partial s_{ij}} = \frac{3}{2}\left(\frac{\sigma_e - k}{K}\right)^n \frac{s_{ij}}{\sigma_e}. \tag{5.13}$$

In this representation it is assumed that below the threshold stress, k, no plastic flow would occur. That means if $\sigma_e - k \leq 0$, then $\dot{\varepsilon}^p_{ij} = 0$. This is particularly true for high speed forming processes. If the deformation rate is high, then the time for creep occurrence is short. Thus, below a certain stress level, k, no plastic deformation (creep) takes place. As discussed earlier, the yield stress of a material at hot forming conditions is difficult to determine. The value, k, is determined based on certain assumptions.

For a uniaxial case, the equation can be simplified to

$$\dot{\varepsilon}_{11}^p = \dot{\varepsilon}^p = \left(\frac{\sigma - k}{K}\right)^n.$$

By re-arranging the equation the flow stress can be written as:

$$\sigma = k + K(\dot{\varepsilon}^p)^{1/n}.$$

This is known as *elastic perfect viscoplasticity*. **Figure 5.14** shows the stress components of this equation. The total flow stress consists of the elastic region, k, and the stress due to viscoplastic deformation, σ_v. Within a short time the plastic deformation takes place and the stress due to viscoplasticity σ_v would be saturated quickly. This is due to the viscoplastic strain rate approaching the total strain rate, i.e. $\dot{\varepsilon}^p \approx \dot{\varepsilon}^T$. This would result in $\sigma = k + K(\dot{\varepsilon}^p)^{1/n} = k + K(\dot{\varepsilon}^T)^{1/n} =$ constant. Thus, the stress–strain curve becomes horizontal quickly as the plastic deformation occurs.

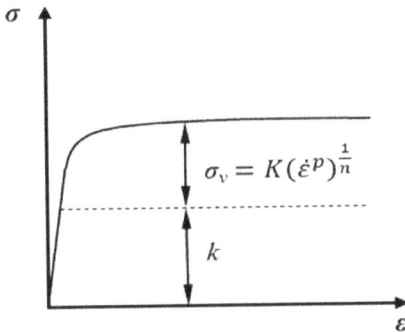

Fig. 5.14 Flow stress for elastic perfect viscoplasticity.

c. With isotropic hardening considered

Hardening often takes place in viscoplastic deformation (hot forming conditions) even if the hardening in viscoplasticity is much lower for the same material deformed at plasticity deformation conditions (cold forming). This is mainly due to the diffusion and recovering mechanisms at high deformation of metals. An internal variable R is introduced to represent isotropic hardening. Based on the previous potentials it can be written in the power-law function

$$\psi = \frac{K}{n+1}\left(\frac{\sigma_e - R - k}{K}\right)^{n+1},$$
(5.14)

which yields:

$$\dot{\varepsilon}_{ij}^p = \frac{S_{ij}}{\sigma_e}\dot{\varepsilon}_e^p$$
(5.15)

and

$$\dot{\varepsilon}_e^p = \left(\frac{\sigma_e - R - k}{K}\right)^n$$

is the effective plastic strain rate. Again, for the uniaxial case it can be written as:

$$\dot{\varepsilon}^p = \left(\frac{\sigma_e - R - k}{K}\right)^n$$

or in the form of:

$$\sigma = k + R + K(\dot{\varepsilon}^p)^{1/n}.$$

Figure 5.15 shows the decomposition of stress components for the flow stress at the uniaxial case. The hardening is due to dislocations, which are generated in plastic deformation. Thus, the hardening rate, $\dot{R}$, is a function of plastic strain or plastic strain rate, which will be discussed later. The strain hardening can also be expressed directly using a power-law relationship as discussed earlier. Again, the stress due to viscoplasticity σ_v would be saturated quickly after plastic deformation and the flow stress is an offset of the hardening curve.

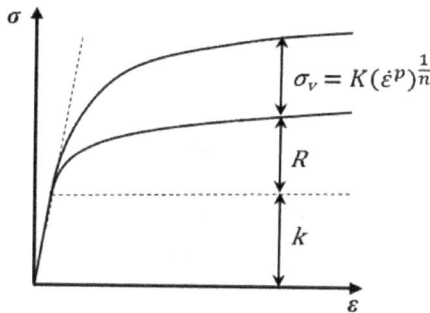

Fig. 5.15 Flow stress for elastic-viscoplasticity with isotropic hardening.

5.4.3 *Isotropic hardening equations*

Hardening is an important phenomenon of metals deformed at room temperature such as under cold forming conditions. That is why it is sometimes called *cold work hardening*. It is well known that work hardening is directly related to the dislocation density which increases monotonically to a saturated state. If recrystallisation and annealing is not considered in a deformation process, then the hardening is directly related to the increment of plastic strain; that is why it is known as strain hardening. At the early stage of development of the hardening equations for viscoplastic applications, the hardening equations for cold deformation were used; these are valid for many applications in viscoplastic deformation.

a. *Power-law (empirical) hardening equation*

Considering strain rate hardening in viscoplasticity, Hollomon's equation is a power-law relationship between the stress and plastic strain and can be written as

$$\sigma = K(\varepsilon^p)^n (\dot{\varepsilon}^p)^m, \tag{5.16}$$

where K is the strength coefficient and n is the strain hardening coefficient. Constant m is the strain rate hardening coefficient in viscoplasticity. Ludwik's equation is similar but includes the threshold stress k:

$$\sigma = k + K(\varepsilon^p)^n (\dot{\varepsilon}^p)^m. \tag{5.17}$$

The strain rate hardening normally occurs at the beginning of the viscoplastic deformation. At hot forming conditions the ductility (strain to failure at uniaxial tensile tests) of a material is high. Once the deformation is stabilised the plastic strain rate $\dot{\varepsilon}^p$ is approximately equal to the total strain rate $\dot{\varepsilon}^T$, i.e. $\dot{\varepsilon}^p \approx \dot{\varepsilon}^T$. Thus, the value of strain rate hardening is constant for a constant strain rate test, which results in $(\dot{\varepsilon}^p)^m = $ constant. Thus Eq. (5.17) becomes

$$\sigma = k + C\left(\varepsilon^p\right)^n \tag{5.18}$$

where $C = K(\dot{\varepsilon}^p)^m$. Differentiating the above equation gives:

$$d\sigma = nC\left(\varepsilon^p\right)^{n-1} d\varepsilon^p \text{ , or, } d\sigma = nC\left(\varepsilon^p\right)^n \left(\varepsilon^p\right)^{-1} d\varepsilon^p .$$

Assuming $\varepsilon^p \approx \varepsilon^T = \varepsilon$ for large plastic deformation, the strain hardening index can be approximated by

$$n = \frac{\varepsilon}{\sigma}\frac{d\sigma}{d\varepsilon}, \text{ or, } \frac{d\sigma}{d\varepsilon} = n\frac{\sigma}{\varepsilon}.$$

It can be seen that the stress increases with strain according to Eq. (5.18). The increasing rate of stress in terms of strain is directly related to the ratio of stress and strain according to the above equation (n is a constant at isothermal deformation conditions). **Figure 5.16** shows the flow stress feature; the hardening rate of flow stress in the stress–strain coordinate system is the same as the isotropic hardening (R) rate. Two curves are in parallel and the viscoplastic stress component σ_v is constant.

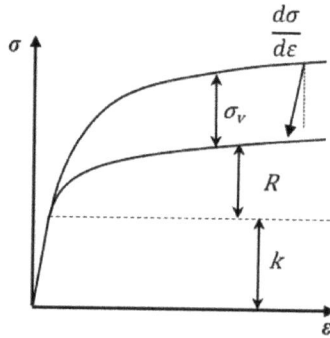

Fig. 5.16 Isotropic hardening with Ludwik's equation.

b. *Phenomenological hardening equations*

By the consideration of primary creep hardening and the viscoplastic deformation directly, a relation can be used to describe the isotropic hardening rate (Lemaitre and Chaboche, 1990),

$$\dot{R} = b(Q_1|\varepsilon^p| + Q_2 - R)|\dot{\varepsilon}^p| + Q_1|\dot{\varepsilon}^p|, \quad (5.19)$$

where b, Q_1 and Q_2 are temperature-dependent parameters. This type of equation is rarely used in viscoplastic applications. A type of simplified rate equation is more commonly used:

$$\dot{R} = \frac{dR}{dt} = b(Q - R)|\dot{\varepsilon}^p|. \quad (5.20)$$

It can be seen that R approaches zero when R reaches the value of Q. b is a factor to model the effect of plastic strain on hardening features. If the value of b is high, isotropic hardening develops very quickly and R could approach to Q at lower plastic strain ε^p.

Equations (5.18), (5.19) and (5.20) can be used to model the strain hardening at both cold and warm metal forming simulations where recrystallisation does not take place. If microstructure evolution occurs, particularly when recrystallisation takes place, dislocation-free grains are generated, which results in the reduction of the dislocation density and material softening. This will be discussed in the following section.

5.5 Dislocation Hardening, Recovery and Recrystallisation

5.5.1 *Evolution of dislocation density*

Taylor (1934) (Geoffrey Ingram Taylor, British, 1886–1975) referred to dislocation as a defect on the atomic scale and defined the mechanism of plastic deformation of metals. The presence of dislocations strongly impacts the mechanical and physical properties of materials. The increase of dislocation density in plastic deformation would improve the strength of materials. This is known as work hardening, strain hardening, or dislocation hardening.

According to Kocks (1976), the plastic strain rate can be expressed as a power-law which takes into account isotropic hardening or dislocation hardening,

$$\dot{\varepsilon}^p = \dot{\varepsilon}_0 \left(\frac{\sigma}{K(\rho)} \right)^n, \tag{5.21}$$

where $K(\rho)$ is an internal variable representing the state of materials. This equation is used to represent the thermally activated plastic flow by dislocation glide. The internal variable $K(\rho)$ is related to the total dislocation density ρ (Kocks, 1976):

$$K(\rho) = M\alpha Gb\rho^{1/2}, \tag{5.22}$$

where G is the shear modulus, b is the magnitude of dislocation Burgers vector, α is a numerical constant, and M is the average Taylor factor defined as

$$M = \frac{\sigma}{\tau} = \frac{\gamma^p}{\varepsilon^p} , \text{ or, } \varepsilon^p M = \gamma^p ,$$

which is directly related to shear stress τ and shear plastic strain γ^p to the axial stress σ and plastic strain ε^p. The average Taylor factor is used to account for polycrystallinity of a material. The dislocation density evolves with plastic deformation. It is well known that the work hardening is directly related to the dislocation density increment in plastic deformation. The dislocation glide resistance is normally related to Gb/L, where L is the characteristic obstacles in the glide plane. Its value is the inverse of the square root of dislocation density, i.e. $1/\sqrt{\rho}$.

There are two concurrent effects which determine the variation of dislocation density ρ: *storage* and *recovery*. The storage of dislocations can be expressed by the shear strain increment $d\gamma^p = Md\varepsilon^p$, in terms of the dislocation density increment $d\rho$ associated with immobilisation of mobile dislocations at impenetrable obstacles after they have travelled a distance L, which is known as the mean free path. From the equations above we have

$$Md\varepsilon^p = bLd\rho \text{ , or, } \frac{d\rho}{d\varepsilon^p} = \frac{M}{bL} \text{ ,}$$

or,

$$\dot{\rho} = \frac{M}{bL}|\dot{\varepsilon}^p| \text{ (\textit{dislocation storage}).} \tag{5.23}$$

The dislocation density accumulates linearly with plastic strain rate as a factor of M/bL. This is particularly true at the initial stage of plastic deformation. The *dynamic recovery* of dislocations occurs due to the annihilation of stored dislocations. This is a process that involves the dislocations leaving the glide planes on which they are stored at impenetrable obstacles. This is furnished by cross-slip of screw dislocations or climb of edge dislocations. The former process is prevalent at cold forming conditions and the latter at warm and hot forming conditions. By including the dynamic recovery, Eq. (5.23) can be expanded to

$$\dot{\rho} = M\left(\frac{1}{bL} - k_2\rho\right)|\dot{\varepsilon}^p| \text{ (include \textit{dynamic recovery}),} \tag{5.24}$$

where k_2, the recovery coefficient, can be written as (Estrin, 1996):

$$k_2 = k_{20}\left(\frac{\dot{\varepsilon}^p}{\dot{\varepsilon}_0^*}\right)^{-1/n} \text{,}$$

which is strain rate and temperature-dependent. k_{20} is a constant.

- At cold forming conditions (room temperature), n and $\dot{\varepsilon}_0^*$ are constants.
- At warm and hot forming conditions, n is a constant and $\dot{\varepsilon}_0^* = \dot{\varepsilon}_0 \exp[-Q/(RT)]$ is expressed using the Arrhenius law with the

activation energy Q associated with dislocation climb. R and T are the gas constant and absolute temperature respectively.

It should be noted that at cold forming conditions, the parameter n is related to stacking fault energy – a quantity strongly influencing the cross-slip probability.

At warm and hot forming conditions annealing takes place which reduces the dislocation density. This is known as *static recovery*. Equation (5.24) can then be expanded to

$$\dot{\rho} = M\left(\frac{1}{bL} - k_2\rho\right)|\dot{\varepsilon}^p| - r \text{ (include \textit{static recovery})},\qquad(5.25)$$

where the static recovery term r can be expressed as (Estrin, 1996)

$$r = r_0 \exp\left(-\frac{U_0}{RT}\right)\sinh\left(\frac{\beta\sqrt{\rho}}{RT}\right).$$

The parameters U_0, β and r_0 could be constants. Static recovery is important under warm and hot forming conditions. For example, it occurs during the interval in multi-pass hot rolling processes, in which the material recovers and changes the conditions for the next pass of rolling.

a. Coarse-grain single-phase material

In an ideal coarse-grain (fewer grain boundaries) and single-phase (no other obstacles) material, the only kind of impenetrable obstacle to dislocation movement is the dislocation structure itself. In this case dislocations are distributed randomly in a cell or sub-grain boundary structure. The mean free path L, which is assumed to be much smaller than the grain size of the material, is proportional to $1/\sqrt{\rho}$, i.e. $L \propto 1/\sqrt{\rho}$. Just considering the dislocation storage and dynamic recovery, Eq. (5.24) can be written as (Kocks, 1976):

$$\dot{\rho} = M\left(k_1\sqrt{\rho} - k_2\rho\right)|\dot{\varepsilon}^p|,\qquad(5.26)$$

where k_1 is a constant. For this representation an initial value of dislocation density should be given. If the initial material is dislocation free, i.e. $\rho = 0$ (no obstacles initially), the dislocation density does not increase during plastic deformation, which is not the case in reality.

b. High density of geometric obstacles

For a material of geometric obstacles being much larger than that of the obstacles caused by the dislocation network, the mean free path L can be replaced by the mean spacing d of the geometric obstacles ($d \ll L$). When a stabilised obstacle structure of a material is considered it assumes that the mean spacing d of obstacles does not change in plastic deformation and the recovery coefficient k_2 is not affected by the obstacles. The dislocation evolution equation can be written as

$$\dot{\rho} = M(k - k_2\rho)|\dot{\varepsilon}^p|, \tag{5.27}$$

where $k = (bd)^{-1}$ (referencing $k = (bL)^{-1}$ in Eq. (5.24)) is a constant. This is suitable for a grain boundary hardening material. In this case the mean spacing d of the geometric obstacles is the smallest characteristic length in the structure and satisfies $d < 10/\sqrt{\rho}$, which is the typical cell or sub-grain size.

Equations (5.26) and (5.27) can be combined to form a "hybrid" model, which can be expressed as

$$\dot{\rho} = M(k + k_1\sqrt{\rho} - k_2\rho)|\dot{\varepsilon}^p|. \tag{5.28}$$

This equation can be used to predict the evolution of dislocation density during plastic deformation for a wide range of materials. However, more material constants are involved which increases the difficulty in determining the equation from experimental data.

c. Particle effects

Precipitates and/or second-phase particles are geometric barriers to dislocation movement. The equations discussed above can be adapted to modelling the material with precipitates and/or second phase. The particles normally increase the strength of materials through effects such as precipitation hardening and strain hardening, which are increased by enhancing the dislocation storage. In this case, the recovery rate may be affected by dislocation–particles interaction which would reduce the recovery rate, e.g. by inhibiting detachment of dislocations from particles (Rossler and Arzt, 1990). For non-shearable particles the dislocation evolution equation can be written as

$$\dot{\rho} = M(k_D + k_1\sqrt{\rho} - fk_2\rho)\dot{\varepsilon}^p, \tag{5.29}$$

where $k_D = (bD)^{-1}$, with D as the mean distance of particles. This equation has the same form of the "hybrid" equation discussed above with the difference being that the parameters, k and k_D, have different physical meaning. In addition, the factor, f, is a function of stress and temperature, $f = f(\sigma, T)$, which is less than unity and reduces the dynamic recovery rate.

d. Normalised dislocation concept

In a real modelling process the dislocation density for a virgin material (before plastic deformation occurs) is very difficult to define as it varies with materials, chemical combinations and processing routes. Thus, it is difficult to define the initial value of the material property accurately in terms of solving an initial value problem. To overcome the difficulties, Lin *et al.* (2005) introduced the normalisation concept for dislocation density. The normalised dislocation density can be defined as

$$\bar{\rho} = \frac{\rho - \rho_i}{\rho} = 1 - \frac{\rho_i}{\rho}, \tag{5.30}$$

where ρ_i is the initial dislocation density and ρ the dislocation density in deformed materials. At the beginning of the deformation ($t = 0$), the dislocation density equals to that of the virgin material, i.e. $\rho = \rho_i$. Thus the normalised dislocation density is zero, i.e. $\bar{\rho} = 0$, as $t = 0$. As the material is plastically deformed the dislocation density would increase according to a certain law and finally reach to its saturated state of dislocation network. At this state we have $\rho \gg \rho_i$ and thus the normalised dislocation density is unity, i.e. $\bar{\rho} = 1$. Thus, the normalised dislocation density varies from 0 (the initial state) to 1 (the saturated state of a dislocation network). Mathematically this is a simple and easy case to deal with.

e. Normalised dislocation evolution laws

By the use of the normalised dislocation concept, Lin, Liu, Farrugia *et al.* (2005) introduced a rate equation for normalised dislocations:

$$\dot{\bar{\rho}} = A(1 - \bar{\rho}^{\gamma_1})|\dot{\varepsilon}^p| - c_1\bar{\rho}^{\gamma_2}, \tag{5.31}$$

where A, c_1, γ_1 and γ_2 are material constants. The equation can be rewritten as

$$\dot{\bar{\rho}} = A|\dot{\varepsilon}^p| - A\bar{\rho}^{\gamma_1}|\dot{\varepsilon}^p| - c_1\bar{\rho}^{\gamma_2}.$$

(i) The first term of equation, $A|\dot{\varepsilon}^p|$, represents the accumulation of dislocations due to plastic deformation. This is similar to Eq. (5.23), and $A \propto M/bL$. The normalised dislocation density increases linearly with plastic strain. The parameter A could be temperature-dependent but it is not sensitive. For most cases, A can be treated as a constant.

(ii) The second term of the equation, $A\bar{\rho}^{\gamma_1}|\dot{\varepsilon}^p|$, represents the dynamic recovery of dislocation density. It limits the maximum value of normalised dislocation density to 1. It can be seen from Eq. (5.31) that when $\bar{\rho} = 1$, the normalised dislocation rate $\dot{\bar{\rho}} = 0$ or negative. γ_1 is a constant and varies from 0.5 to 1. Normally we choose $\gamma_1 = 1$.

(iii) The third term gives the effect of static recovery of dislocation density, i.e. annealing at high temperature. For cold forming conditions this term can be omitted, or $c_1 \approx 0$, since little annealing takes place during a cold forming process. The parameter, c_1, is highly temperature-dependent. At high temperature, annealing takes place very quickly and thus the value of c_1 is high. Normally we have

$$c_1 = c_0 \exp\left(-\frac{Q}{RT}\right),$$

where c_0 is a constant and Q and R are the activation energy and gas constant respectively.

The dislocation density evolution is also related to *recrystallisation* and *grain size*. This will be discussed in the next sub-section.

f. Dislocation hardening

As the hardening is directly related to dislocation density, the hardening can be expressed as a function of the dislocation law:

$$R = B\bar{\rho}^\alpha . \tag{5.32}$$

In general, B can be treated as a temperature-dependent parameter and α as a constant. Normally α varies from 0.3 to 1.0, but for most cases $\alpha = 0.5$, which is directly related to general dislocation hardening laws (Lin, Liu, Farrugia *et al.*, 2005). In this case

$$R = B\bar{\rho}^{1/2} .$$

For consistency, the hardening law can be expressed as an evolutionary equation. By differentiating the above equation,

$$\dot{R} = \frac{1}{2} B \bar{\rho}^{-1/2} \dot{\bar{\rho}}.$$

As discussed in previous chapters, the isotropic hardening increases the yield stress of the material. This can be easily incorporated into the viscoplastic flow laws such as:

$$\dot{\varepsilon}^p = \left(\frac{\sigma - R - k}{K} \right)^n.$$

As mentioned earlier, the isotropic hardening can also be modelled using Eq. (5.21), where K is considered as a function of dislocation density.

5.5.2 *Recrystallisation*

Doherty, Hughes, Humphreys *et al.* (1997) defined recrystallisation as "... the formation of a new grain structure in a deformed metal by formation and migration of high angle grain boundaries driven by the stored energy of deformation". Generally speaking, recrystallisation is a process of destroying strained grains at high temperature. The grains are replaced by strain-free, or "dislocation-free" grains, or new grains. **Figure 5.17** shows the recrystallisation process. New grains are nucleated (b) at grain or sub-grain boundaries of the deformed crystals (a), where the dislocation density is high. As recrystallisation completes (c), grain growth occurs (d). This is one cycle of the recrystallisation process.

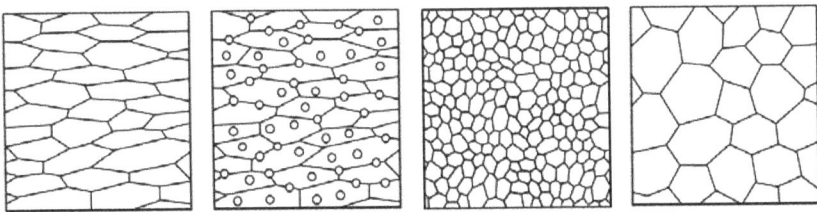

(a) Original material   (b) Nucleation   (c) Recrystallised grains (d) Grain growth

Fig. 5.17 Recrystallisation of metals. (a) to (b): recrystallisation; (c) to (d): grain growth.

Recrystallisation may occur during or after plastic deformation.

- *Dynamic recrystallisation.* This takes place during viscoplastic deformation at high temperature, which reduces the flow stresses because the fine strain-free (dislocation-free) grains are nucleated.

This reduces the dislocation hardening and also facilitates grain boundary sliding and grain rotation, and thus reduces the flow stress. An example is shown in **Fig. 5.18**.

- *Static recrystallisation.* This occurs after plastic deformation such as during two rolling passes of hot rolling of metals and subsequent heat treatment of plastically deformed materials.

In general, it is believed that recrystallisation only occurs when the following three conditions are satisfied.

- *Critical dislocation density.* The prior plastic deformation applied to the material must be adequate to generate sufficient dislocation density and stored energy to drive the nucleation and grain growth. In superplastic forming processes a high amount of plastic deformation could be generated but the dislocation density is low since the plastic deformation is mainly associated with grain rotation and grain boundary sliding. In the meantime, the accumulated dislocations can be recovered due to high temperature annealing during the deformation at a low straining rate. Thus, recrystallisation normally does not occur. This will be further discussed in the rest of the section.

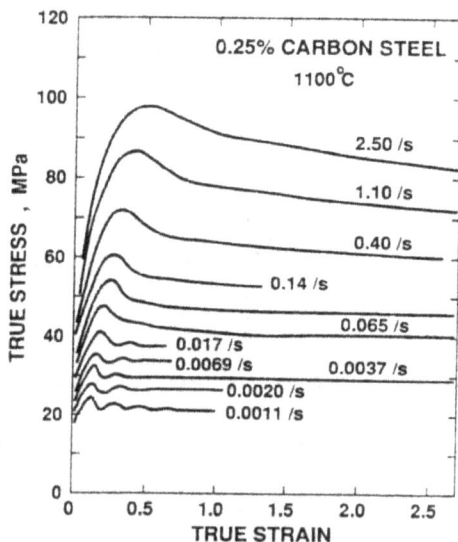

Fig. 5.18 Flow stress curves for 0.25% carbon steel at 1100 °C and strain rates between 1.1×10^{-3} s^{-1} and 2.5 s^{-1} (Rollett, Luton and Srolovitz, 1992).

- *Critical temperature.* Recrystallisation requires a minimum temperature for the occurrence of necessary atomic mechanisms. This recrystallisation temperature decreases with annealing time and with dislocation density.
- *Incubation time.* Sufficient time should be given to enable nucleation to take place for deformed grains. If the temperature is high and/or dislocation density is high, the incubation time is short.

Thus, for a deformed material, if the dislocation density reaches a critical level, recrystallisation would occur given sufficient time and adequate temperature.

a. Driving force for recrystallisation

During plastic deformation, the majority of work is converted to heat with some fraction (about 1%–5%) retained in the material as defects or dislocations. When crystalline materials are deformed at high temperatures, two separate processes destroy the accumulated dislocations. One is dynamic recovery, which leads to the annihilation of pairs of dislocations as well as to the formation of sub-grains. In high stacking fault energy materials such recovery processes completely balance the effects of straining and work hardening, leading to the established steady state flow stress. In materials of moderate-to-low fault energy, dislocation density increases to appreciably high levels; eventually the *local differences in density* are high enough to permit the nucleation of recrystallisation during deformation. Such dynamic recrystallisation eliminates a large number of dislocations and creates dislocation-free grains. The driving force for recrystallisation, i.e. the energy difference ΔE, can be determined by the dislocation density ρ (Doherty, 2005):

$$\Delta E \approx \rho G b^2 , \tag{5.33}$$

where G and b are shear modulus and Burgers vector respectively. The critical value of dislocations for recrystallisation is expressed as (Sandstrom and Lagneborg, 1975)

$$\rho_c = 4\varphi_{\text{surf}} / \left(\tau d^*\right) , \tag{5.34}$$

where φ_{surf} is the grain boundary energy per unit area, d^* is the diameter of the recrystallised nucleus and τ is the average energy per unit length of a dislocation.

b. Empirical equations for modelling of recrystallisation

Figure 5.19 shows the typical recrystallisation kinetics, where t_0 is known as the initial *nucleation period*. At this stage the nuclei form and then grow at almost a constant rate. For spherical grains, the mean radius of recrystallised grain, r, can be expressed as (Humphreys and Hatherly, 2004)

$$\dot{r} = \frac{dr}{dt}(t - t_0).$$

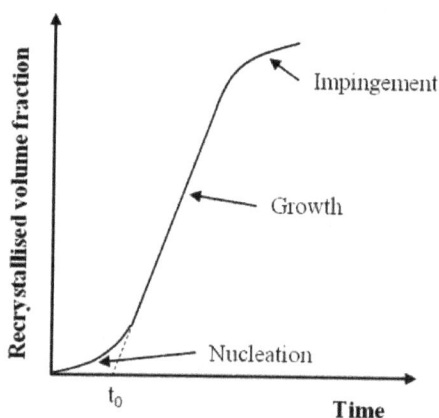

Fig. 5.19 Typical features of nucleation and growth kinetics of recrystallisation.

Based on the spherical grain assumption, the volume fraction, f, of recrystallised material can be expressed as:

$$f = 1 - \exp\left[-\frac{\pi}{3}\dot{N}\dot{r}^3 t^4\right], \tag{5.35}$$

where $\dot{N}$ is the nucleation rate and f varies from 0 to 1. However, in practice few of these are actually valid and alternative models that need to be used.

During recrystallisation a fraction of the grain boundary area is mobile. This fraction varies slightly during the process. It has been shown (Sandstrom and Lagneborg, 1975) that for static recrystallisation this fraction increases with time in the beginning and decreases towards the end of recrystallisation. The velocity $v(\rho)$ of a moving grain boundary is approximately given by the following expression :

$v(\rho) = M\tau\rho.$

During dynamic recrystallisation it is likely that the time variation is smaller since some boundaries may be mobile over several cycles of recrystallisation. Although many models have been proposed for grain boundary movement and the growth of recrystallised grains (Humphreys and Hatherly, 2004), the modelling of the recrystallised volume fraction S normally uses empirical expressions such as (Sakai and Jonas, 1984)

$$f = 1 - \exp\left[-\left(K/D_0\right)t^n\right],$$ (5.36)

where K and n are constants and D_0 is the initial grain size.

c. Dislocation-based evolution equation for recrystallisation

As discussed earlier, recrystallisation is directly related to dislocation density. When the dislocation density reaches a critical value $\bar{\rho}_c$ under a high temperature, recrystallisation takes place if sufficient time is given. The evolution of recrystallised volume fraction S can be expressed as (Lin, Liu, Farrugia *et al.*, 2005):

$$\dot{f} = H[x\bar{\rho} - \bar{\rho}_c(1 - S)](1 - S)^{\gamma_s},$$ (5.37)

where H and γ_s are constants. $\bar{\rho}_c$ is the critical value of normalised dislocation density below which recrystallisation would not occur. It was experimentally observed (Djaic and Jonas, 1972) that there is a need for an incubation time for the onset of recrystallisation and the incubation time varies with the change in values of dislocation density which must exceed the critical value $\bar{\rho}_c$. The phenomenon is described using parameter x which is known as the *incubation factor*. The incubation factor for the onset of recrystallisation is given as:

$$x = A_3(1 - x)\bar{\rho},$$ (5.38)

where A_3 is a temperature-dependent material constant. The equation shows that incubation time decreases with increasing dislocation density. Equation (5.37) ensures that the recrystallised volume fraction f varies from 0 to 1 and its variation is cyclic, depending on the evolution of dislocation density.

Equations (5.37) and (5.38) show that, under hot metal forming conditions, if the normalised dislocation density increases to a critical value $\bar{\rho}_c$, then recrystallisation will occur if sufficient time is given. The formulation is applicable to the modelling of both dynamic and static recrystallisation processes.

5.5.3 *Modelling of grain evolution*

During recrystallisation, new grains are nucleated and the total number increases. Consequently, the average grain diameter d decreases. At the same time, normal grain growth takes place, which works in the opposite direction. Taking only recrystallisation into account, the evolution of the average grain diameter d can be written as (Sandstrom and Lagneborg, 1975)

$$\dot{d} = -d(df_n/dt)\ln N_G,$$

where N_G is the number of new grains per old grain after one cycle of recrystallisation, which may be grain size dependent and f_n is the number of recrystallisation cycles, which can be a non-integer number.

a. *Static and dynamic grain growth*

The static and dynamic grain growth work independently, which is especially important for viscoplastic deformation at a low strain rate such as superplastic deformation. The grain growth rate (Cheong, Lin and Ball, 2000) can be expressed as

$$\dot{d} = M\varphi_{\text{surf}}d^{-r_0} + \alpha\dot{\varepsilon}^p d^{-r_1}, \tag{5.39}$$

where r_0, r_1 and α are constants. The first term of the equation represents the static grain growth, which is directly related to grain boundary mobility M and grain boundary energy density φ_{surf}. The second term describes plastic strain induced grain growth, which was introduced by Cheong, Lin and Ball (2000). Compared with grain refinement due to recrystallisation, grain growth plays a less important role during the deformation process. However, during the interval of hot forming processes static grain growth becomes more important after recrystallisation.

If we consider that the current grain size has the same effect on the static and dynamic grain growth, the above equation can be simplified to (Lin and Dunne, 2001)

$$\dot{d} = (\alpha_1 + \alpha|\dot{\varepsilon}^p|)d^{-r_0}, \tag{5.39}$$

where r_0, α_1 and α are constants to be defined from experimental data.

b. With recrystallisation

In consideration of static grain growth, plastic strain induced grain growth and the grain refinement due to recrystallisation, the average grain size evolution equation may take the following form:

$$\dot{d} = \alpha_1 d^{-\gamma_3} + \alpha_3 \dot{\varepsilon}^p d^{-\gamma_4} - \alpha_2 f^{\gamma_6} d^{\gamma_5}, \tag{5.40}$$

where α_1, γ_3, α_3, γ_4, α_2, γ_6 and γ_5 are material constants. The first term is important to model grain growth during hot forming operations in multipass forming processes following recrystallisation. In superplastic forming processes, recrystallisation does not take place normally, thus the last term can be ignored. Comparing the static grain growth between forming passes and grain refinement due to recrystallisation, the dynamic grain growth, the second term, is less important in multipass metal forming processes under high deformation rate.

5.5.4 Effect of recrystallisation and grain size on dislocation density

a. Grain size effect

According to the deformation mechanisms discussed before, small grain size enables the deformation mechanism of grain boundary sliding and grain rotation to take place, such as in superplastic deformation conditions. Thus, fewer dislocations are generated at the same plastic deformation compared to materials with large grain sizes. To model this effect the normalised dislocation evolution equation, Eq. (5.31), was upgraded by Liu, Lin *et al.* (2006):

$$\dot{\bar{\rho}} = \left(\frac{d}{d_0}\right)^{\gamma_d} (1 - \bar{\rho})|\dot{\varepsilon}^p| - c_1 \bar{\rho}^{\gamma_2}, \tag{5.41}$$

where d_0 is a reference grain size, γ_d, c_1, and γ_2 are material constants and γ_d represents the grain size sensitivity on the dislocation accumulation. A large average grain size could cause a rapid increase of dislocation density under plastic deformation since less grain boundary sliding takes place. As dynamic recrystallisation occurs during plastic deformation, new grains are generated, which reduces the average grain size and thus decreases the dislocation accumulation rate.

b. Recrystallisation effect

Figure 5.20 shows schematically the dislocation density distribution for the old grain and the newly formed grain due to dynamic recrystallisation. When the new grain is nucleated it can be considered as "dislocation-free". However, as the grain grows during the plastic deformation stage, dislocations may increase to facilitate the plastic deformations. The boundary of the new grain moves toward the old grain, where high dislocation density exists. At the new grain boundary the "dislocation-free" can be observed. In general, recrystallisation reduces the average dislocation density of the material.

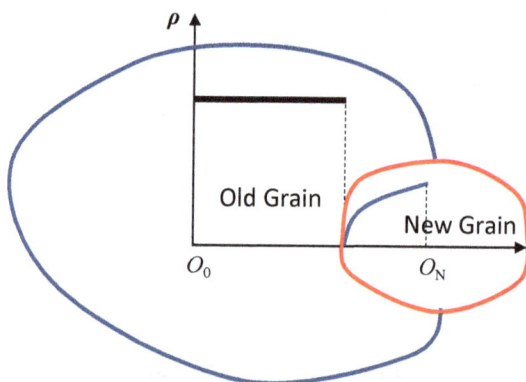

Fig. 5.20 Dislocation density variation along the line O_o–O_N, which are the centre of old grain and new grain.

In the consideration of high temperature deformation mechanisms, namely recrystallisation, static and dynamic recovery, an evolution equation for normalised dislocation density can be of the form (Liu, Lin, Farrugia et al., 2005):

$$\dot{\bar{\rho}} = \left(\frac{d}{d_0}\right)^{\gamma_d} (1 - \bar{\rho})|\dot{\varepsilon}^p| - c_1 \bar{\rho}^{\gamma_2} - \left[c_2 \frac{\bar{\rho}}{1-f}\right]\dot{f}, \qquad (5.42)$$

where c_2 is a material constant. The third term of the equation expresses the effect of recrystallisation on the evolution of dislocation density. Its rate is related to the recrystallised volume fraction of the material and recrystallisation rate. If the normalised dislocation density is zero, then the decreasing rate is zero. Furthermore, the effects of grain size and recrystallisation are taken into account. From the equation above, the following can be deduced:

- Dislocation density increases with plastic deformation. The increasing rate is related to grain size in hot/warm metal forming processes.

- Dislocation density reduces due to the annealing process. This is particularly important for hot/warm forming conditions. In cold forming, this term can be ignored.
- Dislocation density reduces with recrystallisation.

5.5.5 *Effect of grain size on viscoplastic flow of materials*

a. *Power-law*

With the consideration of dislocation hardening and grain size effects, the traditional power-law viscoplastic constitutive equation can be written as

$$\dot{\varepsilon}^p = \left(\frac{|\sigma| - R - k}{K}\right)^n \left(\frac{d}{d_\varepsilon}\right)^{-\gamma_1}. \tag{5.43}$$

If the plastic flow term $|\sigma| - R - k > 0$, then the plastic strain rate can be calculated from the above equation; if $|\sigma| - R - k \leq 0$, then the plastic strain rate is zero. When the material is still within the elastic region, d_ε is a constant or reference grain size. Large grain size d hardens the material when deformed under hot forming conditions. Small grain size facilitates the grain boundary sliding and grain rotation, which makes the plastic deformation easier. The constant γ_1 is used to rationalise the effect of grain size on the viscoplastic flow of materials. It should be noted that the effect of grain size on viscoplastic deformation in Eq. (5.43) is only suitable for hot and warm forming conditions, not cold deformations.

b. *Sinh-law*

Similarly, the grain size effect can be coupled into sinh-law easily by

$$\dot{\varepsilon}^p = \left(\frac{d}{d_\varepsilon}\right)^{-\gamma_1} \sinh[A_2(|\sigma| - R - k)]. \tag{5.44}$$

Again, the term $|\sigma| - R - k > 0$ represents the plastic flow. Otherwise, the material is within the elastic region and the plastic strain rate is zero.

5.6 Examples of Unified Viscoplastic Constitutive Equations

5.6.1 Unified constitutive equations for superplasticity

Depending on material, different internal variables can be selected to form a set of unified constitutive equations for specific applications. The basic rule is to model the dominant mechanical and physical features of the material during plastic/viscoplastic deformation conditions. For simplicity the minimum number of equations should be selected.

In superplastic forming processes, the material is normally deformed at isothermal conditions. The temperature is moderate and the deformation is high. In general cases, recrystallisation does not occur. One of the important features in superplastic forming is grain growth, which affects the viscoplastic flow of the material.

According to Cheng, Lin and Ball (2000), the unified superplastic constitutive equations can be in the forms of:

Viscoplastic flow: $\dot{\varepsilon}^p = \langle \frac{\sigma - R - k}{K} \rangle_+^n d^{-\mu}$

Isotropic hardening: $\dot{R} = b(Q - R)\dot{\varepsilon}^p$

Grain growth: $\dot{d} = \alpha d^{-\gamma_0} + \beta \dot{\varepsilon}^p d^{-\phi}$

Flow stress: $\dot{\sigma} = E(\dot{\varepsilon}^T - \dot{\varepsilon}^p),$

where E is the Young's modulus. The material constants within the constitutive equations are listed in **Table 5.1** and their values are determined from experimental data of Ti-6Al-4V at 927 °C. The detailed determination procedure has been given by Lin and Yang (1999). The determination techniques of the constants will be presented in Chapter 7. The Young's modulus for the material at the deformation condition is $E = 1000$ MPa (Lin and Yang, 1999).

Table 5.1 Material constants determined for Ti-6Al-4V at 927 °C.

α	γ_0	β	ϕ	n	μ	K	Q	b	k
(-)	(-)	(-)	(-)	(-)	(-)	(MPa)	(MPa)	(-)	(MPa)
73.408	5.751	2.188	0.141	1.400	2.282	60.328	3.933	2.854	0.229

Figure 5.21 shows the stress–strain relationships. Symbols represent the experimental data and the solid curves are the computed data from the set of equations given above using the values of the constants listed in **Table 5.1**.

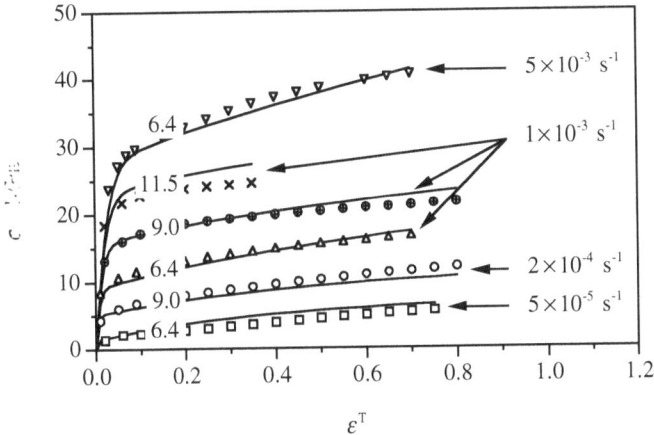

Fig. 5.21 Comparison of fitted (solid curves) and experimental (symbols) (Ghosh and Hamilton, 1979) stress–strain relationships for $\dot{\varepsilon}^T = 5{\times}10^{-5}$, s^{-1}, $2{\times}10^{-4}$ s^{-1}, $1{\times}10^{-3}$ s^{-1}, $5{\times}10^{-3}$ s^{-1}; and $d^0 = 6.4$ μm, 9.0 μm, 11.5 μm.

Figure 5.22 shows the experimental data (symbols) for the relationship of grain size with plastic deformation, and computed data (solid curves) using the equations presented above and the values of constants listed in **Table 5.1**. It can be seen that the average grain size varies with strain rate and viscoplastic deformation.

The above equations can be written in the form of a sinh-law, which has been given by Lin and Dunne (2001). The isotropic hardening equation can also be written as the dislocation-based hardening law, through which the static recovery can be modelled.

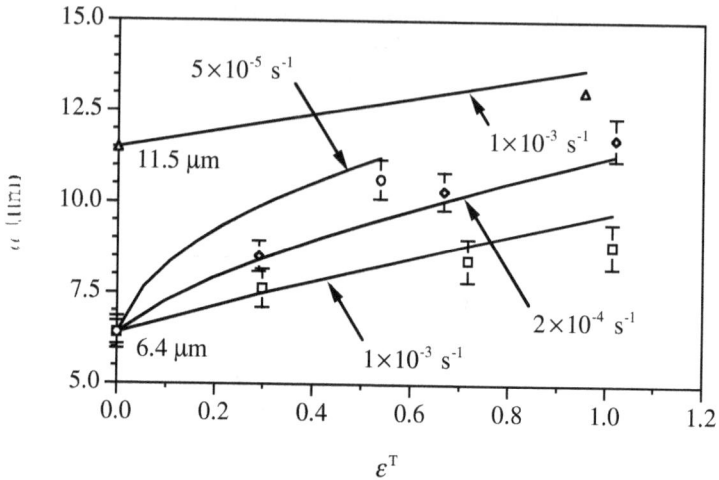

Fig. 5.22 Comparison of fitted (solid curves) (Cheong, 2002) and experimental (symbols) (Ghosh and Hamilton, 1979) grain growth data for $d^0 = 6.4$ μm, 11.5 μm; and $\dot{\varepsilon}^T = 5\times10^{-5}$, s^{-1}, 2×10^{-4} s^{-1}, 1×10^{-3} s^{-1}.

The determined unified superplastic constitutive equations can be generalised to multiaxial constitutive equations using the technique discussed earlier and input into commercial FE codes through user-defined subroutines for industrial applications. This has been performed by Lin (2003), Cheong (2002) and Lin and Dunne (2001).

5.6.2 Unified viscoplastic constitutive equations for hot rolling of steels

In addition to the viscoplastic flow of materials in the hot rolling of steel, recrystallisation is an important physical feature to model. It will change the grain size, dislocation density, work hardening, and thus affect the viscoplastic flow of the material. Lin, Liu, Farrugia, *et al.* (2005) established a set of unified viscoplastic constitutive equations to model the features, which are given below.

Plastic strain: $\quad\quad\quad \dot{\varepsilon}_p = A_1 \sinh[A_2(|\sigma| - R - k)] \cdot d^{-\gamma_4}$

Recrystallisation: $\quad\quad \dot{S} = Q_0 \cdot [x\bar{\rho} - \bar{\rho}_c(1 - S)] \cdot (1 - S)^{N_q}$

Onset of recrystallisation: $\dot{x} = A_0(1 - x)\bar{\rho}$

Dislocation density: $\dot{\bar{\rho}} = \left(\frac{d}{d_0}\right)^{\gamma_d} (1 - \bar{\rho})|\dot{\varepsilon}_p| - c_1\bar{\rho}^{c_2} - \left(\frac{c_3\bar{\rho}}{1-S}\right)\dot{S}$

Isotropic hardening: $\dot{R} = B\dot{\bar{\rho}}$

Grain size: $\dot{d} = \alpha_0 d^{-\gamma_0} - \alpha_2 \cdot \dot{S}^{\gamma_3} \cdot d^{\gamma_2}$

Stress: $\dot{\sigma} = E \cdot \left(\dot{\varepsilon}_T - \dot{\varepsilon}_P\right)$

where $E = 100$ GPa is Young's modulus, A_1, A_2, γ_4, Q_0, $\bar{\rho}_c$, N_q, c_1, c_2, c_3, d_0, γ_d, A_0, B, α_0, γ_0, α_2, γ_2 and γ_3 are material constants. The experimental data reported by Medina and Hernandez (1996) are used here for the optimisation in a C-Mn steel with an average initial grain size of 189 μm for two strain rates $\dot{\varepsilon} = 0.554$ s⁻¹ and 5.224 s⁻¹ at 1100 °C. The material constants within the equations are given in **Table 5.2**.

Table 5.2 Determined constants for the set of unified viscoplastic constitutive equations for a C-Mn steel (Liu, 2004).

A_1 (s⁻¹)	A_2 (MPa⁻¹)	γ_1 (-)	H (-)	$\bar{\rho}_c$ (-)	γ_s (-)
1.81×10^{-6}	3.14×10^{-1}	1.00	30.00	1.84×10^{-1}	1.02

c_1 (-)	γ_2 (-)	c_2 (-)	d_0 (μm)	γ_d (-)	A_3 (-)
16.00	1.44	8.00×10^{-2}	36.38	1.02	40.96

B (MPa)	α_1 (μm)	γ_3 (-)	α_2 (μm)	γ_6 (-)	γ_5 (-)
75.59	1.44	3.07	78.68	1.20×10^{-1}	1.06

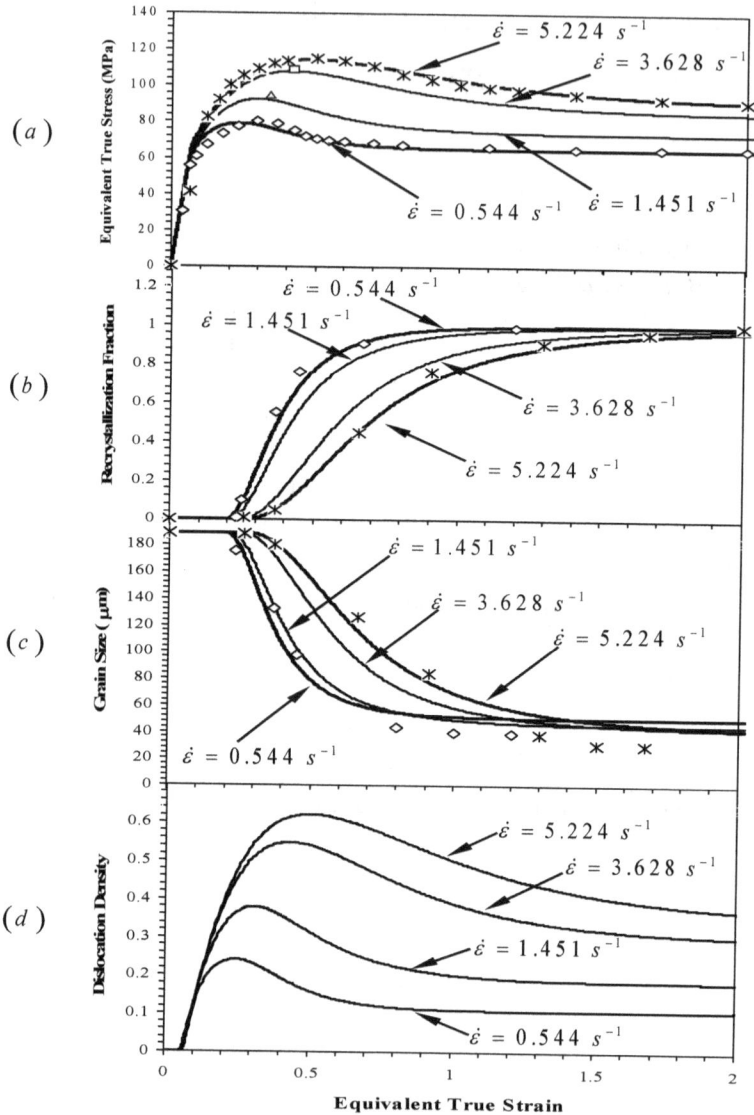

Fig. 5.23 Comparison of experimental data (symbols) with computed (thick curves) and predicted (thin curves) results for different strain rates for the variation of stress, recrystallised fraction, grain size and predicted normalised dislocation density with equivalent true strain, respectively (Liu, 2004).

Figure 5.23 shows the predicted results using the determined constitutive equations. It can be seen that once the physical features of recrystallisation, dislocation density and grain size evolution are well predicted, the softening of the material due to recrystallisation can be accurately captured.

Again, based on the uniaxial constitutive equations, multiaxial constitutive equations can be generated using von-Mises yield criterion. The multiaxial constitutive equations can be input into commercial FE codes in order to predict the mechanical and physical features of materials deformed under hot forming conditions. This has been carried out by Lin, Liu, Farrugia *et al.* (2005), and Liu (2004) for the prediction of rolling processes.

More examples of applying the theories for industrial uses will be given in Chapter 8.

Chapter 6

Continuum Damage Mechanics in Metal Forming

Continuum damage mechanics (CDM) is concerned with the modelling of *degradation/deterioration of materials* under thermal-mechanical deformation and/or ageing. The material deteriorates due to loading and/or ageing to a stage that it could not sustain any load, and then ultimate failure takes place. Damage mechanics is suitable for making engineering predictions about micro-crack initiation, growth, and failure of materials without resorting to a microscopic description that would be too complex for practical engineering analysis. Damage mechanics illustrates the typical engineering approach to model complex phenomena. As said by Krajcinovic (1989) (Dusan Krajcinovic, Croatian/American, 1935–2007), "It is often argued that the ultimate task of engineering research is to provide not so much a better insight into the examined phenomenon but to supply a rational predictive tool applicable in design". Most of the work on damage mechanics uses *state variables* to represent the effects of damage on the structural stiffness, remaining life and failure of the material.

The state variables may be measurable physical phenomena, such as micro-crack density, particle coarsening due to ageing and mobile dislocations, or inferred from the effect that they have on some macroscopic property, such as a decrease in stiffness (Young's modulus), remaining strain to failure etc. A damage activation criterion is needed to predict damage initiation. Damage evolution does not progress spontaneously after initiation, thus requiring a damage evolution model. In plasticity-like formulations, damage evolution is controlled by a hardening function but this requires additional phenomenological parameters that must be found through experimentation, which is expensive, time-consuming, and arduous. On the other hand, micro-mechanics of damage formulations are able to predict both damage initiation and evolution without additional material properties (Gaskell, Dunne *et al.*, 2009; Pan, Balint and Lin, 2011a).

The study of damage mechanics was initiated by Kachanov (1958) in 1958, who introduced a *continuous state variable* to represent damage evolution and predict the creep failure of metals under uniaxial loading. This concept was further developed into ductile fracture and fatigue fracture in the 1970s in the UK (Leckie, Hayhurst and Dyson), France (Lemaitre and Chaboche), Sweden (Hult), Japan (Murakami) and Denmark (Tvergaard) where the use of physically-based state variables for the prediction of material failure was established.

Bulk metal forming is normally considered as compressive forming for ductile metals and failure of materials is not an important issue to study. However, in sheet metal forming processes, tension is the dominant deformation state and *forming limit diagrams/curves* (FLD or FLC) are often used for the prediction of material failure in cold stamping processes (Marciniak, Kuczynski, Pokora, 1973; BSI, 2008).

Although the concept of a damage state-variable was introduced in 1969 (Rice and Tracey, 1969) for the study of material failure in cold forming ($T < 0.3T_m$) and micro-damage has been observed in hot forming ($T > 0.6T_m$) by Cottingham (1966) and Dieter, Mullin and Shapiro (1966), research on the identification of the dominant damage mechanisms and their effect on microstructure evolution during deformation is still rudimentary. Substantial work is necessary for the development of phenomenon-based damage constitutive equations to predict high temperature creep damage and fatigue damage (Lin, Liu and Dean, 2005). Major developments in damage models for metal forming applications in recent years are:

- Gurson (1977) proposed a ductile rupture model which has been used for room temperature metal forming applications.

- Gelin (1995, 1998) introduced predictive techniques modelling isotropic and anisotropic ductile damage in metal forming processes (large straining conditions).

- Cheong, Lin and Ball (2000) introduced a set of constitutive equations with state variables to model damage evolution in superplastic forming.

- Lin, Liu and Dean (2005) introduced continuum damage mechanics theories to model damage evolution in a wide range of metal forming processes, particularly for hot metal forming. This is mainly based on the PhD research work of Cheong (2000) on damage evolution in superplastic forming and Liu (2004).

Research work by Foster (2007), Kaye (2012), Karimpour (2012) and Afshan (2013) further developed the damage evolution models for the prediction of edge cracking in hot rolling.

- Chow (2009) introduced damage models for the prediction of anisotropic sheet metals in cold forming conditions.

- Lin, Mohamed, Dean *et al.* (2014) introduced plane stress damage equations for predicting the forming limit in warm/hot stamping of sheet metals, particularly for boron steel and aluminium hot stamping processes. This is based on the PhD research work of Cai (2010), Mohamed (2011) and Li (2013) on damage evolution and microstructural evolution in hot stamping.

For the applications of creep damage and fatigue damage, the voids are normally considered irreversible. Once the damage has been generated, it cannot be recovered. This is true in cold metal forming as well. However, in hot forming conditions, especially during hot rolling and hot forging, damage may be recovered by annealing and healing processes due to *diffusion bonding* in a compressive stress state. This *damage recovery* process would reduce the defects and consolidate the materials. Thus hot formed materials could be stronger due to the elimination of micro-damages (Afshan, 2013).

6.1 Concept of Damage Mechanics

Metals undergoing continued plastic deformation would suffer from a *degrading microstructure*. Discreet voids or cracks may nucleate and grow within the material and eventually these defects would coalesce to form macro cracks, leading to material failure (Lin, Liu and Dean, 2005). The collective name for these degrading defects is damage.

Damage has previously been concisely described as "The appearance of new surfaces in a material" (Maire, Bordreuil, Babout *et al.*, 2005). This definition reflects the fact that damage is associated with cavities nucleating within the material. However, damage theories may also take into account pre-existing voids such as those deriving from porosity in casting processes. To reflect these facts, damage in this work is defined as "the creation and enlargement of internal cavities within a material".

In addition, particle coarsening due to ageing or some microstructure evolution, which deteriorates or decays the material, can also be considered as damage in a more general concept.

6.1.1 *Definition of damage and damage variables*

a. Definition of damage

Let's consider a solid, which has been deformed and damaged from its original state (virgin material) due to plastic deformation and/or ageing, as shown in **Fig. 6.1**. Let A be the area of a section of the element with its normal $\vec{n}$ (**Fig. 6.1(c)**). On this surface, micro cracks and voids/cavities generated due to plastic deformation and ageing are known as damage. The sum of the areas from the micro voids on this surface forms the total damaged area, A_D. The effective area for sustaining the load is noted as A_E and can be expressed as

$$A_E = A - A_D , \text{ or, } A_D = A - A_E .$$

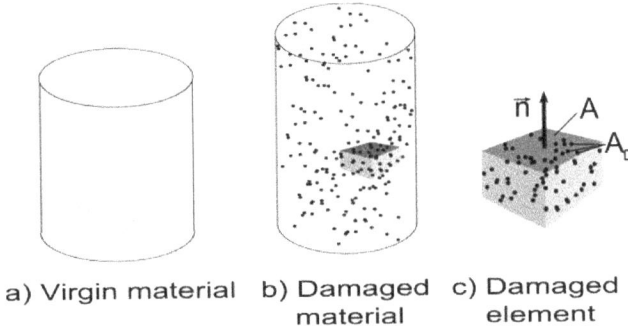

a) Virgin material   b) Damaged   c) Damaged
                       material   element

Fig. 6.1 Definition of damage variable.

Define

$$D_n = \frac{A_D}{A} \tag{6.1}$$

to be the measure of damage, described as the ratio of local damage area relative to the original area in the direction $\vec{n}$. According to this definition, we can see that:

- for the virgin (non-damaged) material, $A_D = 0$, thus $D_n = 0$;
- for the material at failure (break into two parts), $A_D = A$, thus $D_n = 1$;
- the damage variable varies from 0 to 1, i.e. $0 \leq D_n < 1$.

This represents anisotropic material damage, related to material orientation. Isotropic damage consists of micro voids distributed uniformly in all directions. In this case the damage variable is not related to material orientation $\bar{n}$ and is normally represented using a scalar D or ω, known as isotropic damage.

From Eq. (6.1), we have $A_D = A\omega$ for the isotropic material. Then the effective area can be expressed as

$$A_E = A - A_D = A - A\omega = A(1 - \omega) . \tag{6.2}$$

b. Effective stress

Due to damage in a solid, the effective area for sustaining the load has been reduced. This would result in high stresses due to the reduction of the material cross-sectional area. For a uniaxial case, let F be the applied force and A the cross-sectional area of the undamaged material. The traditional stress definition is:

$$\sigma = \frac{F}{A} .$$

But with the consideration of material damage, the actual cross-sectional area for sustaining the load is $A_E = A - A_D$ and the stress definition with the consideration of damage is known as effective stress (please note that this is different from von-Mises stress, which is also known as effective stress). The effective stress can be expressed as

$$\tilde{\sigma} = \frac{F}{A_E} = \frac{F}{A(1-\omega)} = \frac{\sigma}{1-\omega}. \tag{6.3}$$

From this equation, we have

- for the virgin (non-damaged) material, $\omega = 0$, $\tilde{\sigma} = \sigma$;
- for the material at failure (break into two parts), $\omega = 1$, $\tilde{\sigma} = \infty$.

For a damaged material, the elastic strain can be calculated using the effective strain concept, which is

$$\varepsilon^e = \frac{\tilde{\sigma}}{E} = \frac{\sigma}{(1-\omega)E} = \frac{\sigma}{\tilde{E}}, \tag{6.4}$$

where $\tilde{E}$ is the equivalent Young's modulus at the damaged state of the material, i.e. $\tilde{E} = (1 - \omega)E$.

6.1.2 *Specific damage definition methods*

a. *Void/cavity volume fraction*

According to the definition of damage discussed above, the direct measurement for the damage value is the volume fraction of the damaged material and the damage variable can be directly defined as according to Eq. (6.1):

$$\omega = \frac{A_D}{A} = \frac{A - A_E}{A} = 1 - \frac{A_E}{A} , \tag{6.5}$$

where A and A_D are the areas of the virgin and damaged material respectively. A_E is the area of the undamaged material (known as the effective area sustaining the load). At the beginning of the deformation ($t = 0$), $A_E = A$ and $A_D = 0$, then $\omega = 0$; for the material at fracture, $A_D = 0$ ($A_E = 0$), then $\omega = 1$.

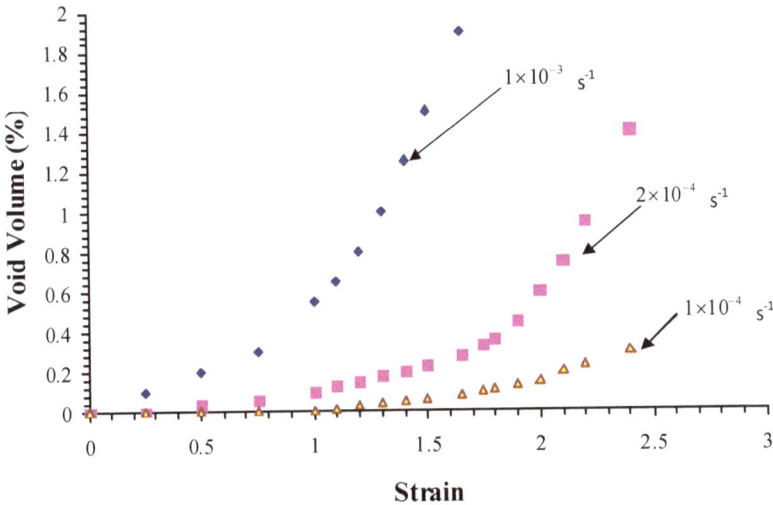

Fig. 6.2 Relations of void volume fraction to strain for different strain rates in superplastically deformed AA7475 alloy (Ridley, 1989).

Figure 6.2 shows the void growth of an aluminium alloy deforming under superplastic conditions. In general, void volume fraction increases with plastic strain. For high strain rate deformation, voids are relatively big and the volume fraction reaches about 2% close to failure for $\dot{\varepsilon} = 10^{-3} \text{ s}^{-1}$. At lower strain rates, the void volume fraction close to failure is significantly lower, at about 0.25% for $\dot{\varepsilon} = 10^{-4}$ s^{-1}. In high temperature creep, the strain rate is much lower than that in superplastic forming, e.g. $\dot{\varepsilon}_c = 10^{-15}$ s^{-1}, and the void volume fraction close to failure should be much lower than that observed at superplastic deformation conditions. This indicates that there would be experimental difficulties when interpreting microstructural examination results of void growth to relate to the assumption of the damage variable defined in Eq. (6.5). Thus determining the damage evolution equations through micro-void examination results proves challenging.

b. Variation of Young's modulus

According to $\tilde{E} = (1 - \omega)E$, the damage variable can be expressed as a function of Young's modulus,

$$\omega = 1 - \frac{\tilde{E}}{E}. \tag{6.6}$$

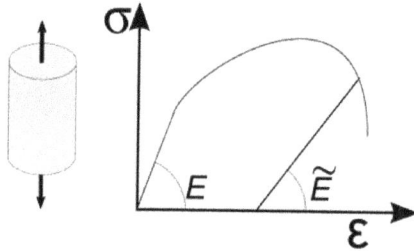

Fig. 6.3 Damage measurement according to Young's modulus.

Figure 6.3 shows the variation of Young's modulus in a loading, unloading and re-loading sequence condition. The values of Young's modulus can be measured according to the loading and unloading curves, and the damage can be calculated from the measured values of Young's modulus. This concept is excellent in theory. However, it is very difficult, in practice, to get the Young's modulus value close to zero, i.e. $\tilde{E} \rightarrow 0$, so that $\omega \rightarrow 1$ close to failure state of the material.

c. Damage definition for cyclic loading

Cyclic plasticity/viscoplasticity tests are normally carried out for the calibration of fatigue damage. These cyclic tests can be either strain range controlled or stress range controlled. **Figure 6.4** shows experimental results for pure copper (Dunne and Hayhurst, 1992), under strain range controlled cyclic viscoplastic loading conditions. It can be seen that, at the initial stage of the cyclic tests, the stress range increases. This is due to isotropic hardening, which takes place only over the first few cycles.

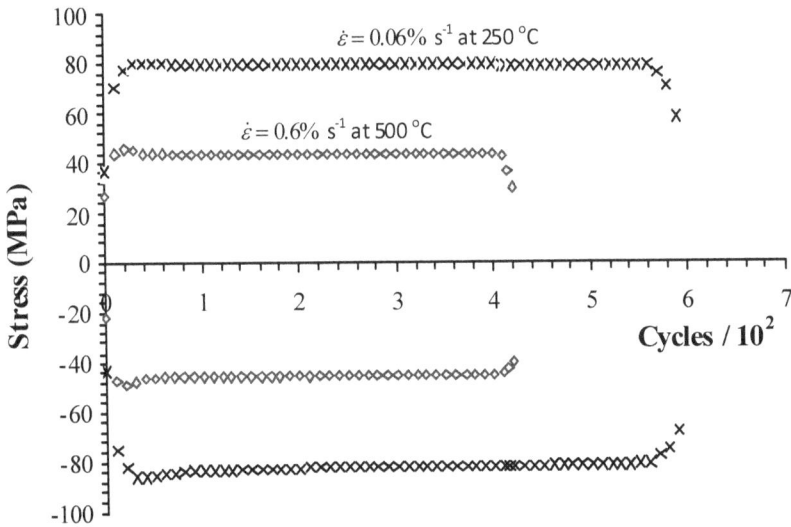

Fig. 6.4 Variation of stresses for strain range controlled cyclic tests for copper (Dunne and Hayhurst, 1992).

The drop of the stress range at the late stage of the cyclic tests is believed to be due to the accumulation of fatigue damage and also creep damage if the values of strain and temperature are sufficiently high. Thus, the damage can be defined according to the reduction of stresses for strain range controlled tests, for example,

$$\omega = \frac{\Delta\sigma^S - \Delta\sigma}{\Delta\sigma^S} = 1 - \frac{\Delta\sigma}{\Delta\sigma^S}, \qquad (6.6a)$$

where $\Delta\sigma^S$ is the stress range during the stabilised cycles, where the isotropic hardening has been saturated. $\Delta\sigma$ is the stress range from the stabilised condition to failure. At the beginning of the stabilised cycle, $\Delta\sigma = \Delta\sigma^S$, then $\omega = 0$; at the time close to failure, $\Delta\sigma \to 0$ and thus $\omega = 1$.

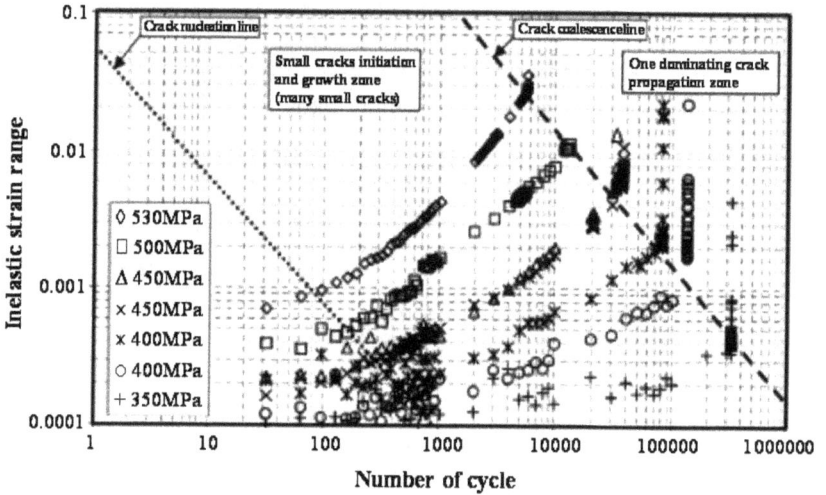

Fig 6.5 Variation of strain range for stress controlled cyclic tests for a steel
at different stress amplitudes (Socha, 2002).

If the cyclic plasticity test is stress range controlled, ratcheting takes
place. Ratcheting results in an increased strain range, as shown in
Fig. 6.5, for a steel tested at different stress ranges (Socha, 2003). It is
believed that the increased strain range under the same cyclic stress
range is due to material softening under cyclic loading, resulting from the
accumulation of micro-damage. Based on this assumption, a damage
variable can be defined for the measure of damage progression,

$$\omega = \left(\frac{\Delta \varepsilon^i - \Delta \varepsilon_0^i}{\Delta \varepsilon_f^i - \Delta \varepsilon_0^i} \right)^{1/C} , \tag{6.7}$$

where C is a constant, $\Delta \varepsilon_0^i$, $\Delta \varepsilon^i$ and $\Delta \varepsilon_f^i$ are initial, current and final
values of the inelastic strain range. The damage variable ranges from 0,
the initial state, to 1, at which failure takes place. The definition in
Eq. (6.7) provides a convenient measure of damage evolution for cyclic
loading conditions. The strain range variation over the cyclic loading is
easier to measure since small changes of effective stress (the effective
cross-sectional area reduces due to material damage) would result in
significant changes in strain magnitudes. This enables damage to be
calibrated more accurately. The parameter C provides flexibility for

representing the damage evolution, so that the effect of damage on the strain range variation can be well predicted.

d. Damage definition from creep curves

During high temperature creep deformation, creep rate increases in the tertiary stage due to the accumulation of damage. Thus the damage variable in creep deformation could be defined according to the creep rate.

$$\omega = \left(\frac{\dot{\varepsilon}-\dot{\varepsilon}_{min}}{\dot{\varepsilon}}\right)^{1/C} = \left(1 - \frac{\dot{\varepsilon}_{min}}{\dot{\varepsilon}}\right)^{1/C}, \tag{6.8}$$

where $\dot{\varepsilon}$ and $\dot{\varepsilon}_{min}$ are the current and minimum creep rates, as shown in **Fig. 6.6**. At the beginning of the secondary creep, $\dot{\varepsilon} = \dot{\varepsilon}_{min}$, thus $\omega = 0$; at the stage close to failure, $\dot{\varepsilon} \gg \dot{\varepsilon}_{min}$ and this results in $\omega \approx 1$. Again the parameter C provides extra flexibility for the definition of the damage variable. The value of C could be unity, i.e. $C = 1$. The horizontal axis t/t_f in **Fig. 6.6** varies from 0 to 1, where t_f is the creep rupture time.

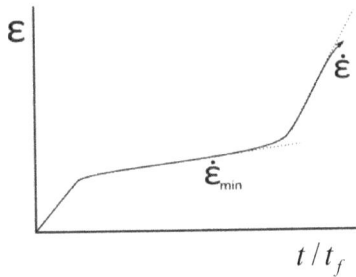

Fig. 6.6 Damage measurement according to creep rate.

6.2 Damage Mechanisms, Variables and Models

As discussed before, damage variables are introduced to represent individual damage phenomena for different applications. The dominant damage mechanisms are related to materials, processing routes, deformation conditions and deformation mechanisms of individual materials. **Figure 6.7** schematically shows typical damage mechanisms observed from different applications and materials, which will be discussed together with their mathematical expressions in the following section.

6.2.1 *Kachanov creep damage equation*

Kachanov (1958) proposed the first empirical model to describe damage evolution in creep deformation conditions. This took the form of

$$\dot{\omega} = \left[\frac{\sigma}{A_0(1-\omega)}\right]^\gamma, \tag{6.9}$$

where A_0 and γ are coefficients for a specific material. This expression yields very low values for the primary and secondary creep. This is the first time to use an internal variable to express damage and material degradation under creep rupture conditions. This simple model has been used extensively for the initial stage of the application of continuum damage mechanics theories applied to engineering problems. However, the model is not mechanism-based.

Significant research has been carried out since then for the identification of damage mechanisms for different materials deforming in different conditions. But the main research was concentrated in the applications of high temperature creep and fatigue.

6.2.2 *Damage due to multiplication of mobile dislocations in high temperature creep*

The micro-damage mechanism responsible for this behaviour is a progressive accumulation of mobile dislocations as metal creep proceeds at high temperatures (Dyson, 1988). It has been termed *mobile dislocation strain-softening* (Othman, Lin, Hayhurst *et al.*, 1993) and is shown schematically in **Fig. 6.7(a)**. Creep in this case is therefore not controlled by dislocation recovery but is best thought of in terms of the kinetics of *dislocation multiplication and subsequent motion* (Ashby and Dyson, 1984). For example, for nickel superalloys, by treating the velocity of dislocations around γ' particles, whose spacing is more compact than that of the dislocations, as one of diffusive drift it is possible to model secondary and tertiary creep in superalloys under various loading conditions by introducing mobile dislocation damage. An example of a typical damage model is given as (Lin, Hayhurst and Dyson, 1993a, 1993b):

$$\dot{\omega}_1 = C(1 - \omega_1)^2 \dot{\varepsilon}_c, \tag{6.10}$$

where ω_1 $(= 1 - \rho_i/\rho)$ is mobile dislocation damage and its evolution increases proportionally to creep rate $\dot{\varepsilon}_c$. The initial dislocation density, ρ_i, is influenced by the material's processing route, whilst ρ is the current dislocation density. The parameter C reflects the propensity of the material for entering tertiary creep and its magnitude is inversely proportional to ρ_i. This type of damage causes material softening, i.e. increasing creep rates, but does not determine the failure of materials (Dyson, 1988).

6.2.3 *Damage due to creep-constrained cavity nucleation and growth*

The damage mechanism, shown in **Fig. 6.7(b)**, is grain boundary cavity nucleation and growth. Its presence or absence is strongly sensitive to alloy composition and processing route; for example, it is clearly absent in single crystals. Its presence reduces the load-bearing cross-sectional area and so accelerates creep and this, in turn, increases the rate at which the damage grows.

At low stresses the damage is void-like; at high stresses, the voids may link to form grain boundary cracks (Dyson and McLean, 1983). Many metals can fracture by this mechanism alone, though it is more usual for other mechanisms to contribute. *Grain boundary cavity formation* is a kinetic phenomenon and its influence on deformation resistance and fracture mode depends critically on cavity nucleation rate and growth rate. When both rates are high, there is potential for a strong coupling between cavitation and creep rate through the mechanism of creep-constrained cavity growth, leading to rapid tertiary creep (Dyson, Verma and Szkopiak, 1981). This damage normally occurs during long-term high temperature creep at low stress levels.

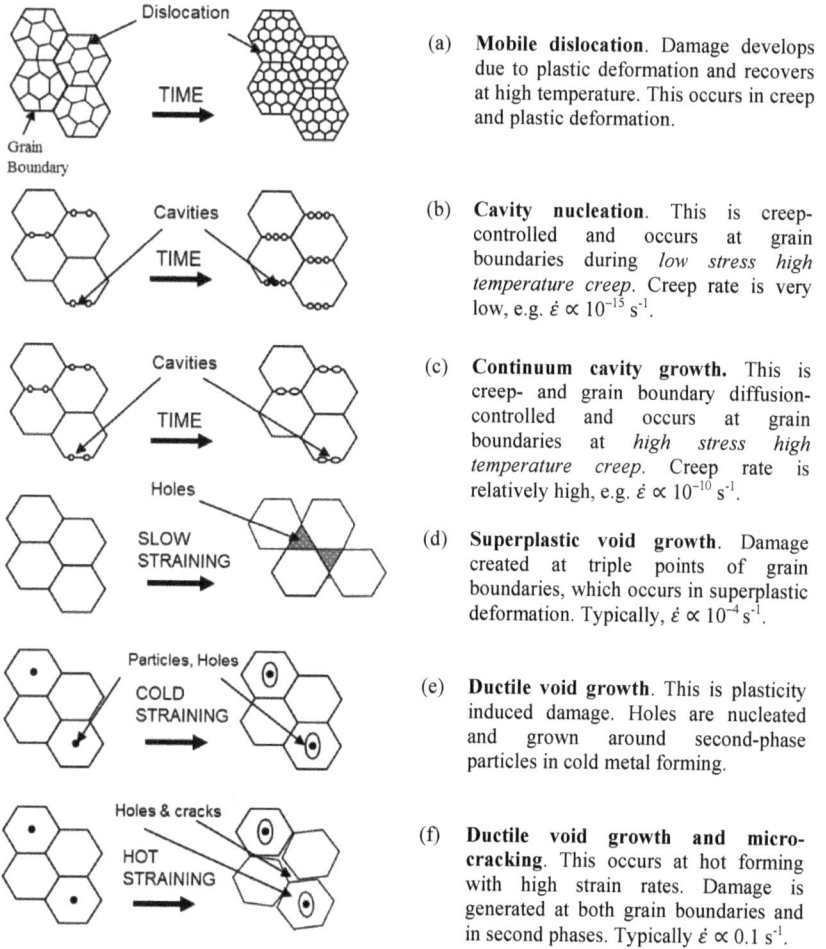

(a) **Mobile dislocation**. Damage develops due to plastic deformation and recovers at high temperature. This occurs in creep and plastic deformation.

(b) **Cavity nucleation**. This is creep-controlled and occurs at grain boundaries during *low stress high temperature creep*. Creep rate is very low, e.g. $\dot{\varepsilon} \propto 10^{-15}\,\text{s}^{-1}$.

(c) **Continuum cavity growth**. This is creep- and grain boundary diffusion-controlled and occurs at grain boundaries at *high stress high temperature creep*. Creep rate is relatively high, e.g. $\dot{\varepsilon} \propto 10^{-10}\,\text{s}^{-1}$.

(d) **Superplastic void growth**. Damage created at triple points of grain boundaries, which occurs in superplastic deformation. Typically, $\dot{\varepsilon} \propto 10^{-4}\,\text{s}^{-1}$.

(e) **Ductile void growth**. This is plasticity induced damage. Holes are nucleated and grown around second-phase particles in cold metal forming.

(f) **Ductile void growth and micro-cracking**. This occurs at hot forming with high strain rates. Damage is generated at both grain boundaries and in second phases. Typically $\dot{\varepsilon} \propto 0.1\,\text{s}^{-1}$.

Fig. 6.7 Schematics showing dominant damage mechanisms for particular deformation conditions (Lin, Liu and Dean, 2005).

Failure by this mechanism implies low ductility for the material. **Figure 6.7 (b)** shows cavities observed at grain boundaries. A typical equation for modelling this type of damage evolution is given by Dyson (1990) and Lin, Hayhurst and Dyson (1993a):

$$\dot{\omega}_2 = D\dot{\varepsilon}_c, \tag{6.11}$$

where D ($D = \varepsilon_f/3$ and ε_f is the strain at failure under uniaxial tension) is a material constant.

The damage variable ω_2 ($= \pi d^2 N/4$) describes grain boundary *creep-constrained cavitation*. N is the number of constrained grain boundary facets per unit area and d the cavity diameter. Creep cavitation can either be nucleation or growth controlled, which is linear to creep strain rate $\dot{\varepsilon}_c$, and the failure criterion is $\omega_2 = 1/3$ (Dyson, 1990).

6.2.4 *Damage due to continuum cavity growth*

This type of damage is schematically shown in **Fig. 6.7 (c)**. It is similar to cavity nucleation and growth, shown in **Fig. 6.7 (b)**, and occurs during high temperature creep, but *at high stress levels*. The difference in fracture mode, resulting from the two damage mechanisms, shown in **Figs. 6.7 (b)** and **6.7 (c)**, was observed by Dyson and Loveday (1981). Again, creep damage occurs as voids or cracks nucleate, often on grain boundaries, as shown in **Fig. 6.8 (a)**. A void can grow during creep by diffusion of atoms away from it, or by the plastic flow of the material, which surrounds it, or by the combination of both (Ashby and Dyson, 1984). If the void growth is controlled by boundary diffusion alone, matter diffuses out of the growing void and deposits onto the grain boundary. If surface diffusion is rapid, matter is distributed quickly within the void allowing its shape to remain near-spherical.

A void can grow by power-law creep of the surrounding matrix. Towards the end of material life, when damage is extensive, this mechanism always, ultimately, takes over. A model has been developed by Cocks and Ashby (1982) to calculate void growth rate based on grain boundary diffusion mechanisms. In simple tension, the zone within the damaged area extends a little faster than the rest of the material, by a factor of $1/(1 - \omega_3)$, where $\omega_3 = r_h^2/l^2$ is the damage due to continuum cavity growth; here l is the spacing of the growing voids and r_h the void radius. Matter is also constrained by its surroundings so that it dilates, causing the void to grow in volume, thereby increasing the damage. The rate of damage growth due to continuum cavity growth has been given by Cocks and Ashby (1982), as

$$\dot{\omega}_3 = [1/(1 - \omega_3)^n - (1 - \omega_3)]\dot{\varepsilon}_{\min}, \qquad (6.12)$$

where $\dot{\varepsilon}_{\min}$ is the minimum creep rate and n a constant.

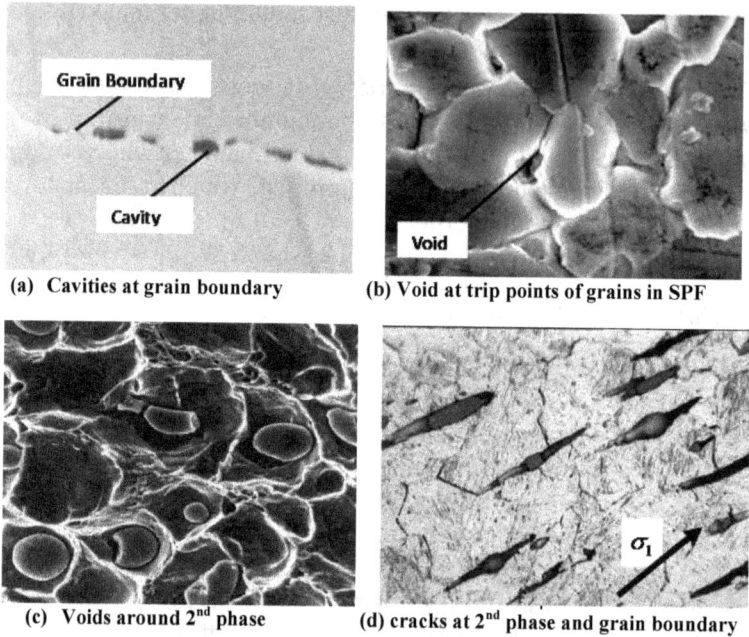

(a) Cavities at grain boundary   (b) Void at trip points of grains in SPF

(c) Voids around 2nd phase   (d) cracks at 2nd phase and grain boundary

Fig. 6.8 Damage in (a) high temperature creep (Evans and Wilshire, 1985),
(b) superplastic forming (Neih, Wadsworth, Sherby, 1997), (c) cold
deformation and forming (Baker and Charles, 1972), and, (d) hot forming.

6.2.5 *Damage and deformation mechanisms in metal forming*

Void growth occurring *at triple points of grains* is often found in
superplastic forming (SPF), especially for aluminium alloys, as shown in
Fig. 6.8 (b) and schematically represented in **Fig. 6.7 (d)**. Superplastic
deformation rates are much higher than creep rates. The fine-grained
microstructure at high temperatures encourages grain boundary sliding
and grain rotation. This dominant deformation mechanism causes the
voids to be created at triple points of grains. Although the two types of
damage equations described above can be used to model the softening of
superplastic materials with good accuracy, there are no physically-based
equations available to model void nucleation and growth at triple points
due to grain rotation and grain boundary sliding.

   In *cold metal forming* processes, multiplication of dislocations is the
dominant deformation mechanism and no grain boundary sliding takes

place (Li, Bilby and Howard 1994). This causes voids to be nucleated around second phases (Zheng *et al.*, 1996), normally within grains, as shown in **Figs. 6.7 (e)** and **6.8 (c)**. Once a micro-void has been nucleated in a plastically deforming matrix, by either the debonding or cracking of a second-phase particle or inclusion, the resulting stress-free surface of the void causes a localised stress and strain concentration in the adjacent plastic field (Tvergaard, 1990). With continuing plastic flow of the matrix the void will therefore undergo volumetric growth and shape changes, which amplifies the distortion imposed on the remote uniform strain-rate field (Staub and Boyer, 1996). The early work on damage model development was based on modelling the rate of change of the void radius, $\dot{R}_k$ ($k = 1, 2, 3$), in the principal directions of strain for the remote strain rate field (Rice and Tracey, 1969). This will be detailed in the following sections.

Damage in *hot forming* conditions has been observed in different forms, since the dominant deformation mechanisms vary with material microstructure, temperature and deformation rate. **Figure 6.7 (f)** schematically shows the damage mechanisms, which can be *"creep type"* damage, i.e. at grain boundaries. This is mainly due to the grain boundary sliding and grain rotation at high temperature deformation conditions. If the strain rate is low, grain size is small and temperature is high, then grain boundary type damage is dominant. We also can see that the damage initiates around second-phase particles. This type of damage is known as *"plasticity type"* damage and is mainly due to dislocations. If the temperature is low, strain rate is high and/or grain size is large, then more plasticity type damage can be observed. **Figure 6.8 (d)** shows the existence of both types of damage (Lin, Liu and Dean, 2005).

Figure 6.9 shows that, at high deformation rates, damage is mainly observed at the interface of second-phase particles. This is due to the accumulation of dislocations, which results in debonding. However, for the same material deforming at the same temperature, but at a low strain rate, a significant amount of cracks was observed at the grain boundaries. Thus, when the deformation rate is changed, the dominant damage mechanisms can be changed in hot metal forming.

Here we introduced the concept or definition of *"creep-type"* damage and *"plasticity-type"* damage. Each type of damage is related to different deformation and damage mechanisms.

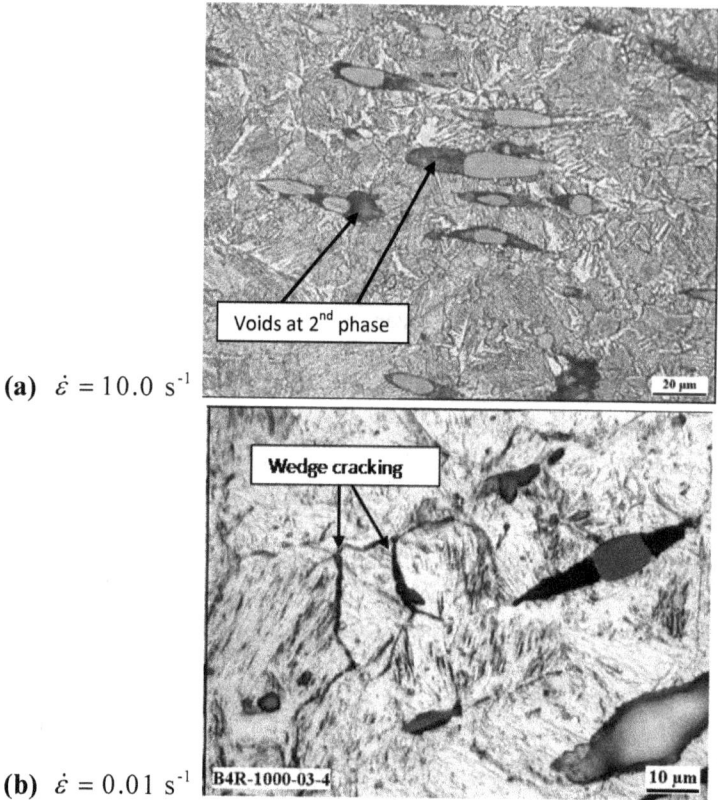

Fig. 6.9 Damage evolution in hot forming ($T = 1000$ °C) of a free cutting steel at (a) high and (b) low strain rates (Lin, Liu and Dean, 2005).

a. Creep-type damage

Creep-type damage is based on grain boundary diffusion processes and is therefore time- and temperature-dependent. The rate of deformation also has a significant effect on the mechanism by which creep occurs. Globular cavities, which occur at stress levels lower than peak stress, lead to grain cracks. Many low-ductility metals can fracture by the mechanism of creep-constrained grain boundary cavity nucleation and growth. Both the rate of the cavity nucleation and growth rate influence deformation resistance and fracture mode.

In some cases the microstructure also alters the supported mechanisms. Superplastic void growth only occurs with a fine grain size of less than 10 μm, temperatures above 0.7 T_m and strain rates of 0.0001 s^{-1} to 0.1 s^{-1} (Cheong, 2002). The dominant deformation mechanisms are grain rotation and grain boundary sliding. Balluffi, Allen *et al.* (2005) suggested that grain boundary diffusion, which is required for accommodation of grain rotation and grain boundary sliding, is 10^6 times greater than bulk diffusion under similar conditions. Vetrano, Simonen *et al.* (1999) have reported that if distribution of matter within the material adjacent to grain boundaries fails to meet imposed requirements by deformation rate, then stresses at grain boundaries are not relaxed sufficiently and subsequently cavities nucleate.

b. *Plasticity-induced damage*

In cold metal forming processes, there is no grain boundary sliding and multiplication of dislocations is the dominant deformation mechanism, causing voids to nucleate around second-phase particles within grains (Brust and Leis, 1992). The inclusions are known to promote damage by acting as nucleation sites and further to this act as stress-raisers to promote damage growth. Nucleation can be strain- or stress-based. Stress-based nucleation will be affected by the distribution of inclusions. In hot metal forming conditions, if the forming rate is very high, meaning there is little time for diffusion to take place, then the dominant deformation mechanism is dislocation-based and thus plasticity-induced damage can be observed.

In summary, *plasticity-type damage* occurs when voids nucleate in a plastically deforming matrix by cracking or debonding between matrix and second-phase particles or inclusions. This causes localised stresses and strains at the void surface, which does not bear any load (Tvergaard and Needleman, 2001). This continues and amplifies the distortion; the micro-cavities or micro-voids grow with increasing tensile strains until macroscopic failure takes place.

6.3　Modelling of Stress State on Damage Evolution

6.3.1　*Stress-state damage models for high temperature creep*

The generalisation of uniaxial creep damage constitutive equations to multiaxial stresses, by Leckie and Hayhurst (1977), has been achieved by making the assumption that the influence of continuum damage on the deformation rate process is of a scalar character, and by introducing a homogeneous stress function which reflects the stress-state effects at the time of rupture. Multiaxial creep damage constitutive equations can then be written as

$$\dot{\omega} = \dot{\omega}_0 \Delta^\nu \frac{1}{(1+\eta)(1-\omega)^\eta}, \tag{6.13}$$

where ν, $\dot{\omega}_0$ and η are constants. The effects of multiaxial stress-state on damage evolution can be modelled by defining

$\Delta = \sigma_1/\sigma_0$ for copper,

and

$\Delta = \sigma_e/\sigma_0$ for aluminium alloys,

where σ_1 and σ_e are the maximum principal stress and effective stress, respectively. This is not very convenient for modelling different materials and is difficult to model the creep rupture behaviour of a material when related to both the maximum principal stress and effective stress. Most engineering materials obey mixed criteria so that the rate of increase of damage can be dependent on a combination of σ_1 and σ_e. A more flexible damage evolution equation, based on this work, was proposed by Hayhurst (1983):

$$\dot{\omega} = \frac{[\alpha(\sigma_1/\sigma_0)+(1-\alpha)(\sigma_e/\sigma_0)]^\nu}{(1-\omega)^\eta}. \tag{6.14}$$

The weighting parameter α, which varies from 0 to 1 defined according to experience, was introduced to rationalise the effects of stress-state on damage evolution (Hayhurst, Dimmer and Morrison, 1984). For example, it was determined that $\alpha = 0.7$ for copper, where damage evolution is mainly controlled by the maximum principal stress, and $\alpha = 0$ for aluminium alloys of which damage evolution is mainly controlled by effective stress. When $\alpha = 0$ or 1, Eq. (6.14) is equivalent to the damage equation in Eq. (6.13). Equation (6.14) enables creep rupture behaviour and lifetime to be well predicted for a range of stress-states and materials deformed under creep deformation.

Another stress-state damage equation was developed by Kowalewski, Lin and Hayhurst (1994) for aluminium alloys in the following form:

$$\dot{\omega} = DN \left(\frac{\sigma_1}{\sigma_e}\right)^n \dot{\varepsilon}_e, \tag{6.15}$$

where D is a material constant and n could be a function of effective stress (Lin, Kowalewski and Cao, 2005).

The damage ω describes grain boundary creep-constrained cavitation, the magnitude of which is strongly sensitive to alloy composition and processing route. The parameter N is used to indicate the state of loading, e.g. for σ_1 tensile $N = 1$ and for σ_1 compressive $N = 0$. In the equation set the damage evolution depends on the maximum principal stress as well as the effective stress.

Equation (6.15) has been further developed by Lin, Kowalewski and Cao (2005) to take the form

$$\dot{\omega} = D \left(\frac{(\sigma_1 + |\sigma_1|)/2}{\sigma_e}\right)^\gamma \dot{\varepsilon}_e, \tag{6.16}$$

where $\gamma = \beta \sigma_1$.

The parameter γ varies linearly with the maximum principal stress. The constant β is used to express the stress-state effects on the damage evolution of materials, and moreover, to model lifetimes and tertiary creep deformation behaviour of materials. The term $(\sigma_1 + |\sigma_1|)/2$ is introduced here to ensure $\dot{\omega} = 0$, when σ_1 is compressive. For a positive maximum stress, the equation is similar to Eq. (6.15), i.e.

$$\dot{\omega} = D \left(\frac{\sigma_1}{\sigma_e}\right)^{\beta \sigma_1} \dot{\varepsilon}_e.$$

- $\beta < 0$ indicates that the damage evolution of the material exceeds the effective stress control (a case typical for aluminium alloys), and the presence of a low value of σ_1 would reduce the creep lifetime.
- $\beta > 0$ indicates that the damage evolution is under the control of the effective stress.
- $\beta = 0$ indicates that the lifetime and tertiary creep of the material is controlled by the effective stress only.

It has been determined by Lin, Kowalewski and Cao (2005) that for pure copper $\beta = 4.245e-2$ (MPa^{-1}) and for aluminium $\beta = -7.1e-3$ (MPa^{-1}). This has been determined via combined tension-torsion creep tests with different stress states (Lin, Kowalewski and Cao, 2005).

6.3.2 *Stress-state damage models for hot forming*

Stress-state behaviour of superplastic damage has been assessed by Pilling and Ridley (1986) for the case of increasingly compressive stress states. In their work, tensile and biaxial deformation was conducted under increasing pressures leading to the formula

$$\dot{\omega} = \dot{\omega}_0 \left(1 - 2\frac{P}{\sigma_e}\right)$$

in which P is the superimposed pressure (whereby a positive number represents a compressive stress-state to the sheet material). The equation has been re-written in terms of the stress triaxiality (Nicolaou and Semiatin, 2003):

$$\dot{\omega} = \dot{\omega}_0 \left(\frac{1}{3} - 2\frac{\sigma_H}{\sigma_e}\right). \tag{6.17}$$

Unlike other formulas, the equation allows damage to accumulate to some degree in compressive environments.

Constitutive models have been developed specifically for hot deformation. Liu, Lin, Dean and Farrugia (2005) have produced a complex but logical model capable of modelling dislocation build up and associated material hardening, dislocation recovery and recrystallisation, grain growth and grain boundary and inclusion damage. The model has phenomenological-based nucleation and growth parts to each damage mechanism, but coalescence is not treated as a separate mechanism, rather as a rapid growth of voids. The equations map the interactions of the various material mechanisms in a phenomenological manner. A simple stress state relationship is introduced of the form

$$\dot{\omega} = \dot{\omega}_0 \left(\frac{J_1}{\sigma_e}\right)^n, \tag{6.18}$$

in which $J_1 = 3\sigma_H$. An n-value of 2.0 is given for their applications. The above equation has been further improved by Foster (2006) with the assumption that inclusion-related damage expands in the form of oval voids parallel to the straining direction, as shown in **Fig. 6.10**. The void does not affect the cross-sectional area over which the flow stress acts. This assumption has been validated via micro-mechanics analysis by Foster, Lin, Farrugia *et al.* (2007).

Based on an averaging concept for ideal elliptical damage features, inclusion related damage growth may be modelled by the equation

$$\dot{D}_i = z_1 \omega \dot{\varepsilon}_p^\gamma \cdot \sinh\left(2\frac{(n_d - 1/2)}{(n_d + 1/2)}\frac{\sigma_H}{\sigma_e}\right) \Big/ \sinh\left(\frac{2}{3}\frac{(n_d - 1/2)}{(n_d + 1/2)}\right), \tag{6.19}$$

in which n_d is a material constant. Exponent γ allows time dependency into the equation, for instance to account for possible rate dependence of strain localisation. Material constant z_1 is formed from material specific growth characteristics and will be affected by the initial damage area (given by the inclusion area fraction). The spatial parameter ω has been defined as (Foster, 2006)

$$\omega = (d_{av}/l_{av})^2.$$

In Foster's work (Foster, 2006), it is assumed that all inclusions are potential void sites, and the level of interaction between voids will be a function of i) the average inclusion diameter (d_{av}), whereby large inclusions will produce larger and further reaching stress discontinuities and ii) the average distance between inclusions (l_{av}), whereby inclusions will have a diminishing effect on each other the further they are apart. This is similar to the definition given by Cocks and Ashby (1982) for damage due to cavity growth: $\omega_3 = r_h^2/l^2$, where r_h is the void radius, and l is the spacing between voids.

The damage coalescence rate can be modelled using (Foster, Lin, Farrugia and Dean, 2011)

$$\dot{D}_c = z_2 \cdot \langle \frac{\sigma_1}{1-D_c} \rangle_+ \cdot \omega \cdot \sinh\left(z_3 \cdot \sqrt{\bar{\rho}} \cdot D_i\right) \cdot \dot{\varepsilon}_p, \tag{6.20}$$

in which $\sigma_1/(1 - D_c)$ is the maximum principal stress acting on the non-damaged area, z_3 is a material constant and z_2 is a material constant dependent on temperature.

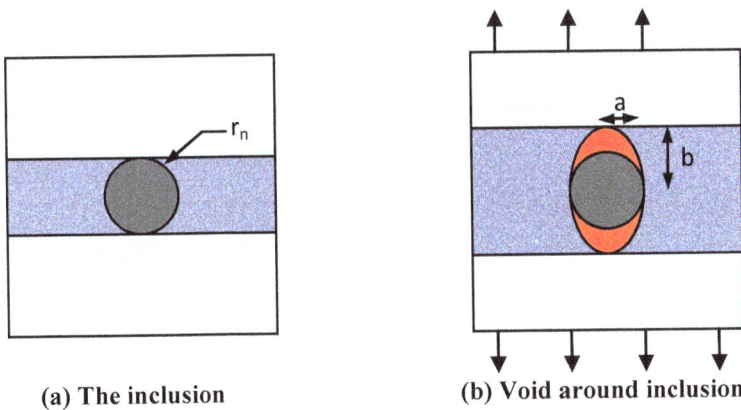

(a) The inclusion   (b) Void around inclusion

Fig. 6.10 Idealised inclusion related damage feature (Foster, 2006).

6.3.3 *2D stress-state model for hot stamping*

In sheet metal forming processes, the formability (failure or starting of necking) is normally modelled using forming limit diagrams/curves (FLD/FLC). FLDs for metal sheets are commonly determined using a stretch forming test either with a hemispherical punch of 100 mm diameter (Nakajima *et al.*, 1968) or a flat punch (Marciniak *et al.*, 1973). Waisted blanks with various major/gauge widths are used, to provide different strain state (major strain/minor strain) conditions.

Figure 6.11(a) shows FLDs for forming three different metals at room temperature (Ali and Balod, 2007) in the coordinate system of principal strains (major strain and minor strain). It can be seen that for the plane strain condition, i.e. the minor strain is zero, the formability (measured as the major strain) for all the materials is lowest. Common features for all three metals are that the formability drops sharply from the uniaxial tensile condition to that of plane strain, then increases slowly to the biaxial stretching condition. A single FLD curve can be used to represent the formability of a sheet metal forming at room temperature.

(a)

(b)

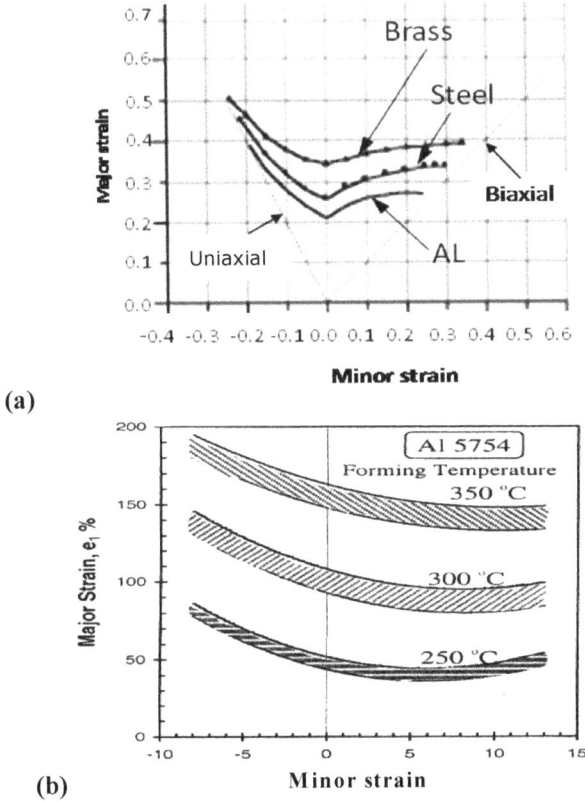

Fig. 6.11 Typical FLDs for (a) an aluminium alloy, mild steel and brass at room temperature (Ali and Balod, 2007), and (b) AA5754 at elevated temperatures at $\dot{\varepsilon} = 1.0$ s^{-1} (Li and Ghosh, 2004).

The FLDs obtained for AA5754 at temperatures of 250 °C, 300 °C and 350 °C and a strain rate of 1 s^{-1}, shown in **Fig. 6.11 (b)** (Li and Ghosh, 2004), have different features from those at room temperature. This shows that in hot stamping conditions, it is not possible to use a single FLD to assess the formability of a sheet metal, since the formability (and ductility) varies with temperature and strain rate, both of which change spatially in a forming process. This is the driving force for the development of 2D based CDM equations for the prediction of the FLC of sheet metals in hot stamping conditions. This theory can also be used for room temperature stamping processes.

In a stamping process, plane stress is normally assumed. The 2D stress-state dependent damage evolutionary equation developed by Lin, Mohamed, Balint, *et al.* (2014) is in the form of

$$\dot{\omega} = \frac{\Delta}{(\alpha_1+\alpha_2+\alpha_3)^\varphi} \langle \frac{\alpha_1\sigma_1+3\alpha_2\sigma_H+\alpha_3\sigma_e}{\sigma_e} \rangle^\varphi \cdot \frac{\eta_1\sigma_e}{(1-\omega)^{\eta_2}} \left(\dot{\varepsilon}_e^p\right)^{\eta_3}, \tag{6.21}$$

where

- α_1, α_2 and α_3 are weighting parameters that represent the damage evolution and are introduced to rationalise the effects of the stress state and turn on the shape of the FLD. These parameters are mainly controlled by the maximum principal stress, effective stress and/or hydrostatic stress and are defined according to experience. It has been determined that α_1 and α_3 vary from 0 to 1. However, α_2 ranges from −1 to 1 since the effect of hydrostatic stress could be in a negative direction. If any of α_1, α_2 or α_3 is zero, then the associated stress has no contribution to the damage process of the material for sheet metal forming conditions.

- φ is a parameter which controls the effect of multiaxial stress values and their combination on damage evolution, thus controlling the formability.

- Δ is a correction factor representing data obtained from uniaxial tensile tests and Marciniak or Nakazima formability tests, since different strain measurement methods are normally used.

For the uniaxial case, the equation can be reduced to

$$\dot{\omega} = \frac{\eta_1}{(1-\omega)^{\eta_2}} \sigma \dot{\varepsilon}_p^{\eta_3}$$

where η_1, η_2 and η_3 are temperature-dependent material constants which can be determined by fitting to experimentally obtained uniaxial stress–strain data (Lin, Mohamed, Balint, *et al.*, 2014).

The detailed solution for such problems will be given via case studies in Chapter 8. The damage equations discussed here should not be used independently but need to be incorporated into a set of viscoplastic constitutive equations for an overall solution.

6.4 Modelling of Damage in Cold Metal Forming

6.4.1 *Rice and Tracey model*

The early work on the development of damage models was based on modelling the rate of change of the void radius, $\dot{R}_k$ ($k = 1, 2, 3$), in the principal directions of strain rate for the remote strain rate field (Rice and Tracey, 1969). Based on an average concept, the average rate of void growth in a unit cell is given (Li, Bilby and Howard, 1994) by

$$\dot{R} = R \cdot N \exp\left(\frac{3}{2}\frac{\sigma_H}{\sigma_e}\right)\dot{\varepsilon}_e^p, \tag{6.22}$$

where R is the radius of an initially spherical isolated void and N is a constant. In this case, voids are assumed to be spherical.

Once the void is no longer spherical, the rate of change for the three radii corresponding to the three directions of principal plastic strain rate is different. This is due to the ellipsoidal void having a different stiffness in each of these directions. Rice and Tracey's formula with a power-law of the radii ratio is introduced by Boyer, Vidal-Sallé and Staub (2002), where the spacing of the voids is sufficiently great. The growth of the void can then take the form (Boyer, Vidal-Sallé and Staub, 2002)

$$\dot{R}_k = \left[(1 + E)\dot{\varepsilon}_k + D\left(\frac{2}{3}\dot{\varepsilon}_L\dot{\varepsilon}_L\right)^{1/2}\right]R, \tag{6.23}$$

where *(k, L)* = 1, 2, 3. The values of $(1 + E)$ and D are related to the strain hardening of materials.

6.4.2 *Strain energy model*

This model was originally developed by Cockcroft and Latham (1968), and involved the integration of the maximum principle stress σ_1 with respect to the equivalent plastic strain ε_e:

$$D = \int_0^{\varepsilon_f} \frac{\sigma_1}{\sigma_e} d\varepsilon_e. \tag{6.24}$$

This is the accumulation of plastic strain energy represented in a simple form. Brozzo, Deluca *et al.* (1972) further developed the Cockcroft and Latham model and explicitly included the hydrostatic (mean) stress, as it had been found to decrease the growth of voids and permitted analysis of multiaxial states of stress on creep-constrained cavity growth. A similar approach was used by Leroy (1981), where the difference in principal and hydrostatic stress was taken:

$$D = \int_0^{\varepsilon_f} \frac{2\sigma_1}{3(\sigma_1 - \sigma_H)} d\varepsilon_e .$$

(6.25)

This model assumes that once the accumulated plastic strain energy (normalised) reaches a certain pre-defined value, failure will take place. This has been developed for assessing the formability of materials.

6.4.3 Gurson's model

The first quantitative analysis of the growth mechanics of an isolated void in a nonlinear material under triaxial loading conditions was given by McClintock (1968) for cylindrical voids. The initiation of fracture in McClintock's model is based on the condition of voids touching the adjacent voids and despite its simplifications exhibits some fundamental features of ductile fracture. A later approach is the Rice and Tracey (1969) growth model, which is based on dilatational growth analysis of a single spherical void in a material under a uniform stress state. Needleman (1972) applied Rice and Tracey's approach to a doubly periodic square array of circular cylindrical voids under plane strain conditions. Needleman's work inspired Gurson (1977) who proposed an approach for obtaining an approximate yield surface for a material containing long cylindrical or spherical voids.

Gurson's model generates a yield function to accommodate for the effect of void growth on softening (Gurson, 1977). The model is based on spherical voids dilating inside an incompressible matrix. An important aspect of the Gurson model is that it accounts for pre-damaged material. It requires two material constants in order to be implemented. The original Gurson model was further developed by Tvergaard and Needleman (2001) to include coalescence (Viggo, 1985).

The derivation of the Gurson yield function for spherical voids is based on the mechanics of void growth under an axisymmetric stress state. The yield function of the Gurson model depends on the von-Mises or equivalent stress, σ_e, the hydrostatic stress, σ_H, the flow stress of the matrix material $\bar{\sigma}$ and void volume fraction, f:

$$\Phi(\sigma_e, \bar{\sigma}, f, \sigma_H) = \left(\frac{\sigma_e}{\bar{\sigma}}\right)^2 + 2f \cosh\left(\frac{3\sigma_H}{2\bar{\sigma}}\right) - (1 + f^2) = 0 .$$

(6.26)

Figure 6.12 illustrates the Gurson model's yield surfaces for different values of void volume fraction f. The yield function is reduced to the classical von-Mises yield criterion if $f = 0$.

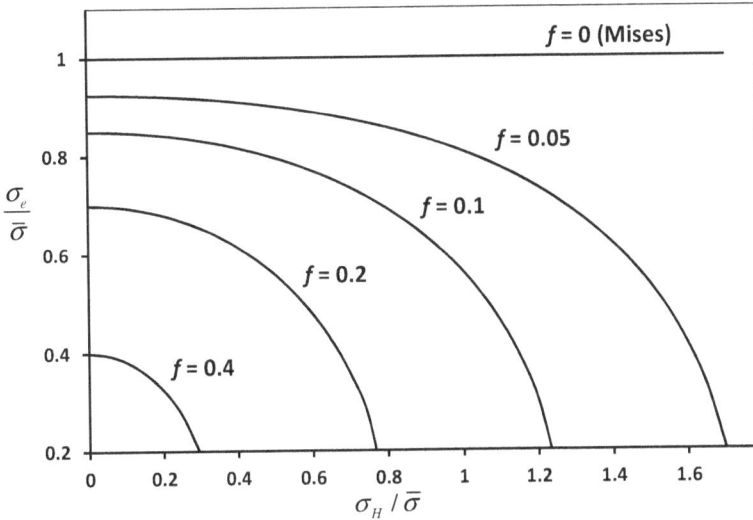

Fig. 6.12 **Schematic of the yield surface in the $\sigma_H/\bar{\sigma}$ - $\sigma_{eq}/\bar{\sigma}$ plane (Afshan, 2013).**

The increase in void volume fraction could be due to the growth of existing voids and also due to the nucleation of new voids. The Gurson model treats the void volume change due to both cases as that of a single void using homogenisation.

The rate of increase of void volume fraction is based on the rate of void nucleation and growth, which is given by

$$\dot{f} = \dot{f}_n + \dot{f}_g. \tag{6.27}$$

The voids that are already nucleated will grow purely due to plastic deformation (Gurson, 1977). The rate of increase in void fraction is given by

$$\dot{f}_g = (1 - f)\dot{\varepsilon}_{ij}^p I_{ij}, \tag{6.28a}$$

where $\dot{\varepsilon}_{ij}^p$ is a plastic strain rate tensor and I_{ij} is the second-order unity tensor. The void nucleation rate $\dot{f}_n$ is defined as

$$\dot{f}_n = A_\varepsilon \left(\frac{EE_t}{E-E_t}\right) \dot{\varepsilon}_e^p + \frac{1}{3} B_\sigma(\sigma_H), \tag{6.28b}$$

where A_ε and B_σ are the void nucleation intensities depending on the plastic strain increment and increase of hydrostatic stress σ_H respectively; E is the Young's modulus and E_t is the current tangent modulus of the material matrix.

The parameters A_ε and B_σ are chosen so that void nucleation follows a normal distribution (Chu and Needleman, 1980). Thus, for nucleation controlled by the plastic strain, the parameters are specified as

$$A_\varepsilon = \left(\frac{1}{E_t} - \frac{1}{E}\right) \frac{f_n}{s_n \sqrt{2\pi}} \exp\left[-\frac{1}{2}\left(\frac{\varepsilon_m^p - \varepsilon_n}{\varepsilon_n}\right)^2\right]$$

$$B_\sigma = 0 \, ,$$

while for the stress controlled nucleation as

$$A_\varepsilon = B_\sigma = \left(\frac{1}{E_t} - \frac{1}{E}\right) \frac{f_n}{\sigma_Y s_n \sqrt{2\pi}} \exp\left[-\frac{1}{2}\left(\frac{\bar{\sigma} + \sigma_H/3 - \sigma_n}{\sigma_Y \varepsilon_n}\right)^2\right],$$

where σ_Y is the yield stress under uniaxial tension, ε_m^p is the effective plastic strain of the material matrix, f_n is the volume fraction of void nucleating particles, ε_n and σ_n are the mean strain and stress for nucleation respectively and s_n is the corresponding standard deviation.

Tvergaard (1981, 1982) studied the bifurcation prediction in the original Gurson model with his numerical analysis and found that the original Gurson model resulted in bifurcation when loads were too small, with strains taking twice the values he found numerically. Tvergaard therefore modified the Gurson model by introducing three fitting parameters, q_1, q_2 and q_3, in Gurson's yield function.

$$\Phi\left(\sigma_e, \bar{\sigma}, f^*, \sigma_H\right) = \left(\frac{\sigma_e}{\bar{\sigma}}\right)^2 + 2q_1 f^* \cosh\left(\frac{3q_2 \sigma_H}{2\bar{\sigma}}\right) - \left[1 + q_3 \left(f^*\right)^2\right] = 0 \tag{6.29}$$

It has been suggested that $q_1 = 1.25 - 1.5$, $q_2 = 1.0$, and $q_3 = (q_1)^2$. Although parameters q_1, q_2 and q_3 have been assumed to be material constants, they can be used for modelling the effect of the stress state on the yield surface.

Another modification to the original Gurson model was related to modelling the complete loss of stress carrying capacity. The Gurson model predicts the total loss of load carrying capacity at $f = 100\%$. This critical void volume fraction, f_c, is unrealistically high and means the total disappearance of the material. The original Gurson model therefore can only model the homogeneous deformation phase (nucleation and growth) and is unable to predict localisation and ductile fracture.

The function $f^*(f)$ was introduced by Tvergaard and Needleman (1984) to model the rapid loss of stress-carrying capacity and therefore to account for the coalescence.

$$f^*(f) = \begin{cases} f & \text{if } f \leq f_c \\ f_c - \dfrac{f_u^* - f_c}{f_f - f_c}(f - f_c) & \text{if } f > f \end{cases} \quad , \tag{6.30}$$

where f_f is void volume fraction at final fracture and $f_u^* = 1/q_1$. Based on experimental (Goods and Brown, 1979) and numerical (Anderson, 1977) results, Tvergaard and Needleman (1984) have chosen the values $f_c = 0.15$ and $f_f = 0.25$. This model is called Gurson–Tvergaard–Needleman model and is used frequently in engineering applications. However, the accuracy of the model is subject to debate (Thomason, 1990). For example, one key aspect the model fails to account for is the interaction of voids.

6.5 Modelling of Damage in Warm/Hot Metal Forming

6.5.1 *Damage modelling for superplastic forming*

Superplasticity can be considered as the high strain rate creep. The work of Cocks and Ashby (1980, 1982) on the incorporation of damage mechanics into a steady-state creep constitutive equation is based on a mechanistic framework they suggested. The assumptions associated with the work are categorised and the two essential equations derived from the mechanistic framework are given: a damage-incorporated plastic strain rate equation and a void growth rate equation. These have been extended to the modelling of damage development in superplastic forming by Lin, Cheong and Yao (2002).

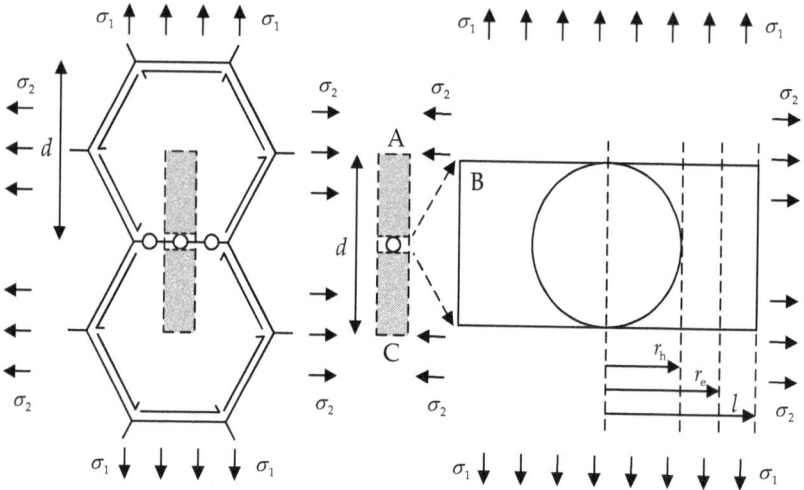

Fig. 6.13 The mechanistic framework of Cocks and Ashby (1982).

a. *Assumptions*

The steady-state creep constitutive equation in the work of Cocks and Ashby (1980, 1982) takes the form

$$\dot{\varepsilon}_{ss}^{p} = \dot{\varepsilon}^{0} \left(\frac{\sigma}{\sigma^{0}} \right)^{n},$$
(6.31)

where $\dot{\varepsilon}_{ss}^{p}$ is the steady-state creep rate, σ the applied stress, and $\dot{\varepsilon}^{0}$, σ^{0}, and n are material parameters. Two classes of assumptions are made. The first is related to the notion underlying the *mechanistic framework*: that the behaviour of a specimen with a homogeneous array of grains and grain-boundary cavities can be characterised by the behaviour of a cavitated cylindrical material isolated from two neighbouring grains among the grain array as shown in **Fig. 6.13**. r_h is the radius of a spherical grain-boundary void, r_e the effective void radius, $2l$ the diameter of the cylinder, which is also the cavity spacing, and d the average grain size. The second is related to the damage mechanisms considered: conventional diffusional and plastic deformation.

For the framework itself it is assumed that:

    i.   The stress field over the cylindrical material is the same as that applied to the specimen as a whole (**Fig. 6.13**).

ii. Grain-boundary sliding occurs and the increase in the volume of the slab containing a spherical grain-boundary void of radius r_h (**Fig. 6.13**) is taken up by a relative rigid-body displacement of the grains on either side.

iii. The slab (**Fig. 6.13**) as a whole, being wide in comparison with its thickness, is constrained by the surrounding material to contract laterally as the cylinder itself does. Therefore,

$$\frac{i}{l} = -\frac{1}{2}\dot{\varepsilon}^p_{ss}. \tag{6.32}$$

For the damage mechanisms considered it is assumed that:

i. Cavity nucleation does not occur. It is mainly concerned with void growth at high stress levels.

ii. Grain-boundary diffusion is driven by the traction acting normal to a grain boundary. Since grain boundaries slide, tensile tractions can appear on certain boundaries, even in samples subjected to compressive stress fields, and can thereby cause voids to grow by diffusion.

iii. The plastic deformation of the material surrounding a void also contributes to void growth.

b. Damage-incorporated plastic strain rate equation

Under the mechanistic framework, the behaviour of a cavitated specimen can be characterised by the behaviour of the cavitated cylindrical material shown in **Fig. 6.13**. Thus, the overall plastic strain rate of the specimen can be described by the plastic strain rate of the cylinder as a whole, which is related to the plastic strain rates of sections A, B, and C in **Fig. 6.13**.

Due to the absence of cavities, the plastic strain rates of sections A and C (**Fig. 6.13**) are equal to that given by Eq. (6.31). To define the plastic strain rate of Section B (**Fig. 6.13**), the concept of an *effective void volume fraction* $f_e = r_e^2/l^2$ is introduced and the *effective* stress experienced by section B is defined as $\sigma/(1 - f_e)$. Thus, the plastic strain rate of section B (**Fig. 6.13**) becomes

$$\dot{\varepsilon}^p_{cc} = \dot{\varepsilon}^0 \left[\frac{\sigma}{\sigma^0(1-f_e)}\right]^n. \tag{6.33}$$

In terms of Eqs. (6.31) and (6.33), the plastic strain rate of the entire cylinder is determined from a geometrical perspective:

$$\dot{\varepsilon}^p = \frac{1}{d}\left[\frac{d-2r_h}{2}\dot{\varepsilon}^p_{ss} + 2r_h\dot{\varepsilon}^p_{cc} + \frac{d-2r_h}{2}\dot{\varepsilon}^p_{ss}\right] = \dot{\varepsilon}^p_{ss}\left[1 + \frac{2r_h}{d}\left(\frac{\dot{\varepsilon}^p_{cc}}{\dot{\varepsilon}^p_{ss}} - 1\right)\right].(6.34)$$

The value of f_e is limited within the range of 0 to 1 and describes the effective damage caused by the damage mechanisms: $f_e = 0$ means no effective damage is incurred while $f_e = 1$ signifies failure. The value of f_e is not required to coincide with the value of void volume fraction $f_h = r_h^2/l^2$ but is always equal to or greater than f_h. The treatment for the description of f_e under coupled diffusional and plastic-strain-controlled void growth is detailed by Cocks and Ashby (1982). The effective damage evolution law to describe the value of f_e under the combined effects of void nucleation, growth, and coalescence is given next.

c. Void growth rate equation

The rate of void growth is also determined from a geometrical perspective. First, consider the volume of the slab $V = \pi l^2(2r_h)$. (**Fig. 6.13**). The rate of change of the volume is expressed as

$$\frac{\dot{V}}{V} = \frac{\dot{r}_h}{r_h} + \frac{2\dot{l}}{l}.$$

By substituting

$$\frac{\dot{r}_h}{r_h} = \dot{\varepsilon}^p_{cc}$$

we have

$$\frac{\dot{V}}{V} = \dot{\varepsilon}^p_{cc} - \dot{\varepsilon}^p_{ss}.$$

Second, differentiate $f_h = r_h^2/l^2$ with respect to time t:

$$\frac{2r_h\dot{r}_h}{l^2} = \dot{f}_h - f_h\dot{\varepsilon}^p_{ss}. \qquad (6.35)$$

Since

$$\frac{2r_h\dot{r}_h}{l^2} = \frac{\dot{v}}{v},$$

where $v = (4/3)\pi r_h^3$ is the volume of the cavity, and

$$\frac{\dot{v}}{v} = \frac{\dot{V}}{V}.$$

Equation (6.25) can be written as

$$\frac{\dot{V}}{V} = \dot{f}_h - f_h\dot{\varepsilon}^p_{ss}. \qquad (6.36)$$

Finally, it can be seen from Eqs. (6.35) and (6.36) that the void growth rate equation takes the simple form

$$\dot{f}_h = \dot{\varepsilon}^p_{cc} - (1 - f_h)\dot{\varepsilon}^p_{ss}. \qquad (6.37)$$

Conceptually, the value of f_h determined from Eq. (6.37) is approximately equal to the void volume fraction obtained experimentally; it is limited within the range of 0 to 1 and is always equal to or less than f_e. In the work of Cocks and Ashby (1980, 1982) that does not consider void nucleation and coalescence, the initial value of f_h is chosen as 0.001 and the final value as 0.25.

d. Determination of effective void volume fraction

Having presented the damage-incorporated plastic strain rate and void growth rate equations, the final piece of building block needed to complete the characterisation of the cylinder (**Fig. 6.13**) is the description for f_e, which is related to the damage mechanisms. For completeness, Cocks and Ashby's (1980, 1982) procedure for determining the f_e due to coupled diffusional and plastic-strain-controlled void growth is given below.

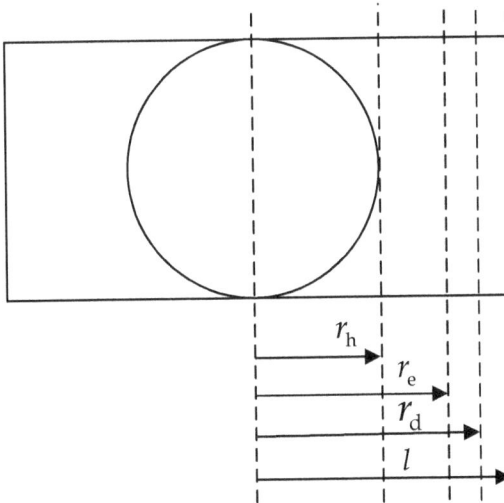

Fig. 6.14. Cocks and Ashby's (1980, 1982) coupled diffusional and plastic-strain-controlled void growth model.

Figure 6.14 shows a schematic representation of the coupled void growth model. As the void expands, matter diffuses from it by boundary diffusion and deposits on the boundary plane over a ring of outer radius r_d. If $r_d = 1$, then void growth is purely by diffusion; if $r_d = r_h$, then the surrounding *power-law zone* totally controls the rate of void growth; and for $l < r_d < r_h$, the void grows by a coupling of the two mechanisms. Within the diffusion zone the stress is relaxed partly by the material that deposits on the boundary plane. This makes the surrounding power-law zone behave as a slab containing a cavity of effective size r_e, where $r_h \leq r_e \leq r_d$. To satisfy equilibrium, the diffusion zone carries a mean stress σ^* that is distributed over the surface area πr_d^2:

$$\sigma^* = \frac{\sigma}{1-f_e} \frac{r_d^2 - r_e^2}{r_d^2} = \frac{\sigma}{1-f_e} \frac{f_d - f_e}{f_d}.$$

(6.38)

σ^* causes a volume of material to flow out of the void and the volume flow rate is expressed as

$$\frac{dV_d}{dt} = \frac{2\pi\Omega D_B \delta\sigma^*}{kT} \frac{\left[\sigma^* - \frac{2\gamma}{r_h}\left(1-\frac{r_h^2}{r_d^2}\right)\right]\left(1-\frac{r_h^2}{r_d^2}\right)}{\ln\frac{r_d}{r_h} - \frac{3}{4} + \frac{r_h^2}{r_d^2}\left(1-\frac{r_h^2}{4r_d^2}\right)}$$

$$\approx \frac{2\pi\Omega D_B \delta\sigma^*}{kT} \frac{1-\frac{r_h^2}{r_d^2}}{\ln\frac{r_d}{r_h}},$$

(6.39)

where Ω is the atomic volume, D_B the grain-boundary diffusion coefficient, δ the grain-boundary thickness, and γ the surface energy. This volume is deposited on to the boundary over an area $\pi(r_d^2 - r_h^2)$, causing the two grains to move apart and results in a local strain rate of

$$\dot{\varepsilon}_{cc}^p = \frac{1}{2r_h\pi(r_d^2 - r_h^2)} \frac{dV_d}{dt}.$$

(6.40)

Equating Eqs. (6.40) and (6.33) and substituting Eqs. (6.28) and (6.39) gives

$$\frac{\phi(1-f_e)^{n-1}}{f_h^{1/2}} \frac{f_d - f_e}{f_d} = f_d \ln\frac{f_d}{f_h},$$

(6.41)

where $\phi = \dfrac{2\Omega D_B \delta}{kTl^3} \dfrac{\sigma}{\dot{\varepsilon}_{ss}^p}$. For the optimum value of f_e, $df_e/df_d = 0$ and hence, Eq. (6.41) reduces to

$$\phi = \frac{f_d f_h^{1/2} \left(1 + 2\ln\dfrac{f_d}{f_h}\right)^n}{\left[(1-f_d)\left(1+\ln\dfrac{f_d}{f_h}\right) + \ln\dfrac{f_d}{f_h}\right]^{n-1}}.$$

Given the initial values for ϕ and f_h, f_e can be determined from the above equation and hence, f_e from Eq. (6.41).

It can be seen that the above procedure for determining f_e is a rigorous one, although without considering the effects of void nucleation and coalescence. However, it possesses a considerable drawback: it imposes a great deal of extra computational effort during the integration of the resulting constitutive equations; each time f_e is to be determined, several numerical interactions are needed. In fact, a Newton–Raphson method was used in the work of Cocks and Ashby (1980, 1982).

e. Effective damage evolution equations

Conceptually, f_e is a scalar value that describes the extend of effective cavitation damage; it intensifies the effective plastic strain rate of a deforming material as cavitation damage accumulates. Its role is analogous to the scalar-valued isotropic hardening R that expands the yield surface of a material as the mobile dislocation density accumulates. It is postulated that the rate of change of f_e can be described as

$$\dot{f}_e = \dot{f}_e^g + \dot{f}_e^c, \tag{6.42}$$

where $\dot{f}_e^g$ is the rate of change of the effective damage due to void nucleation and growth and $\dot{f}_e^c$ the rate of change of the effective damage due to void coalescence.

To establish the description for $\dot{f}_e^g$ it is assumed that void nucleation and growth are only possible during plastic deformation. As such, it is postulated that: $\dot{f}_e^g$ can be expressed in terms of an invariant scalar function of plastic strain components ε_{ij}^p – analogous to the concept of yield criteria – or more directly the effective plastic strain ε_e^p; there exists a mutual relationship between the dissipations of $\dot{f}_e^g$ and ε_e^p, and for every infinitesimal amount of ε_e^p dissipated, it is accompanied by an infinitesimal amount of f_e^g. In other words, $\dot{f}_e^g = \dot{f}_e^g(\dot{\varepsilon}_e^p)$. Thus $\dot{f}_e^g$ can be simply expressed as a power function:

$$\dot{f}_e^g = D(\dot{\varepsilon}_e^p)^{d_1},$$

where D and d_1 are material parameters.

To establish the description for $\dot{f}_e^c$, a *threshold void volume fraction* f_h^* that specifies the onset of void coalescence is introduced. When a void volume fraction f_h surpasses f_h^*, it leads to an abrupt increase in $\dot{f}_e$:

$$\dot{f}_e = C\langle f_h - f_h^* \rangle_+,$$

where C specifies the increase in $\dot{f}_e$ for every unit of $\langle f_h - f_h^* \rangle_+$. Note that a linear relationship between $\dot{f}_e^c$ and $\langle f_h - f_h^* \rangle_+$ is assumed. Then the effective damage evolution equation can be expressed as

$$\dot{f}_e = D(\dot{\varepsilon}_e^p)^{d_1} + C\langle f_h - f_h^* \rangle_+. \tag{6.43}$$

f. Unified superplastic damage constitutive equations

Based on the above analysis for the damage evolution and with the consideration of viscoplastic flow and other physical features discussed in previous Chapters, such as grain growth and strain hardening, a set of multiaxial unified viscoplastic damage constitutive equations can be formulated as (Cheong, 2002):

$$\dot{\varepsilon}_{ij}^p = \frac{3}{2}\frac{S_{ij}}{\sigma_e}\dot{\varepsilon}_e^p$$

$$\dot{\varepsilon}_e^p = \dot{\varepsilon}_{ss}^p\left[1 + \frac{2lf_h^{1/2}}{d}\left(\frac{\dot{\varepsilon}_{cc}^p}{\dot{\varepsilon}_{ss}^p} - 1\right)\right]$$

$$\dot{\varepsilon}_{ss}^p = \langle\frac{\sigma_e - R - k}{K}\rangle_+^n d^{-\mu}$$

$$\dot{\varepsilon}_{cc}^p = \langle\frac{\sigma_e^* - R - k}{K}\rangle_+^n d^{-\mu}$$

$$\dot{R} = b(Q - R)\dot{\varepsilon}_e^p \tag{6.44}$$

$$\dot{d} = \alpha d^{-\gamma_0} + \beta\dot{\varepsilon}_e^p d^{-\phi}$$

$$\dot{f}_h = \dot{\varepsilon}_{cc}^p - (1 - f_h)\dot{\varepsilon}_{ss}^p$$

$$\dot{f}_e = D(\dot{\varepsilon}_e^p)^{d_1} + C\langle f_h - f_h^* \rangle_+$$

$$\dot{l} = -\frac{1}{2}\dot{\varepsilon}^p_{ss}$$

$$\hat{\sigma}_{ij} = 2\nu\dot{\varepsilon}^p_{ij} + \lambda\dot{\varepsilon}^p_{kk},$$

where $\sigma^*_e = \sigma_e/(1 - f_e)$. The equation set can be reduced to the uniaxial form of:

$$\dot{\varepsilon}^p = \dot{\varepsilon}^p_{ss}\left[1 + \frac{2lf_h^{1/2}}{d}\left(\frac{\dot{\varepsilon}^p_{cc}}{\dot{\varepsilon}^p_{ss}} - 1\right)\right]$$

$$\dot{\varepsilon}^p_{ss} = \langle\frac{\sigma - R - k}{K}\rangle^n_+ d^{-\mu}$$

$$\dot{\varepsilon}^p_{cc} = \langle\frac{\sigma^* - R - k}{K}\rangle^n_+ d^{-\mu}$$

$$\dot{R} = b(Q - R)\dot{\varepsilon}^p \qquad\qquad (6.45)$$

$$\dot{d} = \alpha d^{-\gamma_0} + \beta\dot{\varepsilon}^p d^{-\phi}$$

$$\dot{f}_h = \dot{\varepsilon}^p_{cc} - (1 - f_h)\dot{\varepsilon}^p_{ss}$$

$$\dot{f}_e = D(\dot{\varepsilon}^p)^{d_1} + C\langle f_h - f_h^*\rangle_+$$

$$\dot{l} = -\frac{1}{2}\dot{\varepsilon}^p_{ss}$$

$$\dot{\sigma} = E(\dot{\varepsilon}^T - \dot{\varepsilon}^p).$$

For materials without pre-existing cavities, the initial values of f_h and f_e are chosen as 0. If the behaviour of a pre-cavitated material is to be modelled, appropriate initial values need to be given. The initial value of cavity spacing l^0 is regarded as a material parameter that is to be determined together with the others from uniaxial tensile test data.

The material constants within the set of superplastic-damage constitutive equations are: k, K, n, μ, b, Q, α, γ_0, β, ϕ, D, d_1, C, f_h^*, E and l^0. The total strain rate $\dot{\varepsilon}^T$ and initial grain size d^0 are operational and initial microstructural conditions usually quoted with a given experimental stress–strain relationship. However, there are occasions where the value of d^0 is not given. In this case, d^0 is regarded as a material parameter that is to be determined.

The determination of the material constants is commonly carried out by means of optimisation such that the computed behaviour is consistent with uniaxial experimental data. This usually requires numerous iterations. Therefore an efficient and accurate technique for integrating the set of unified, uniaxial constitutive equations is needed. In the meantime, an objective function that provides a robust representation over the agreement between computed and given experimental data is essential. Furthermore, it is important that the optimisation technique itself is suitable for nonlinear problems. These aspects are detailed in Chapter 7.

g. Prediction results

As examples, the material constants within the set of unified superplastic-damage constitutive equations are determined for two aluminium alloys. The determination procedures and numerical methods will be given in Chapter 7. Here the prediction results are given.

(i). Superplastic material of Al-Zn-Mg at 515 °C

To determine the material constants in Eq. (6.45) for Al-Zn-Mg at 515 °C from its experimental (symbols) stress–strain (**Fig. 6.15**) and grain-size (**Fig. 6.16**) data (Pilling, 2001), optimisation methods need to be used to get the best fittings. The material constants determined from the experimental data are listed in **Table 6.1**. The computed (solid curves) stress–strain and grain size evolutions at strain rates $\dot{\varepsilon}^T = 2 \times 10^{-4} \text{ s}^{-1}$, $5 \times 10^{-4} \text{ s}^{-1}$ and $2 \times 10^{-3} \text{ s}^{-1}$ are shown in **Figs. 6.15** and **6.16**, respectively. It can be seen that a good correlation between experimental and computational results is obtained for both flow stress and grain size evolution (Cheong, 2002).

Table 6.1 Material parameters for Al-Zn-Mg at 515 °C (Cheong, 2002).

k (MPa)	K (MPa)	n	μ	b	Q (MPa)	α	γ_0
2.9354e-5	28.7640	1.1299	2.0642	0.1186	5.5769	6.9000e-2	2.4000
β	ϕ	D	d_1	C	f_h^*	E (MPa)	l^0 (μm)
2.6000	5.5000e-5	3.7810e+3	2.3973	32.3739	0.7557	1.000e+3	4.3922

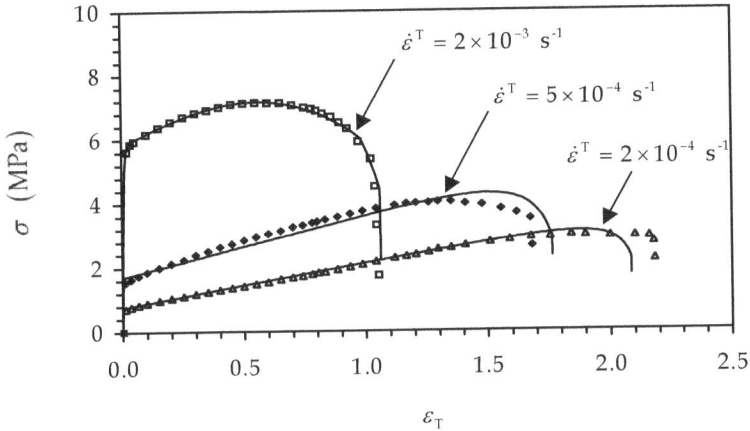

Fig. 6.15. Comparison of computed (solid curves) (Cheong, 2002) and experimental (symbols) (Pilling, 2001) stress–strain relationships for Al-Zn-Mg at various strain rates.

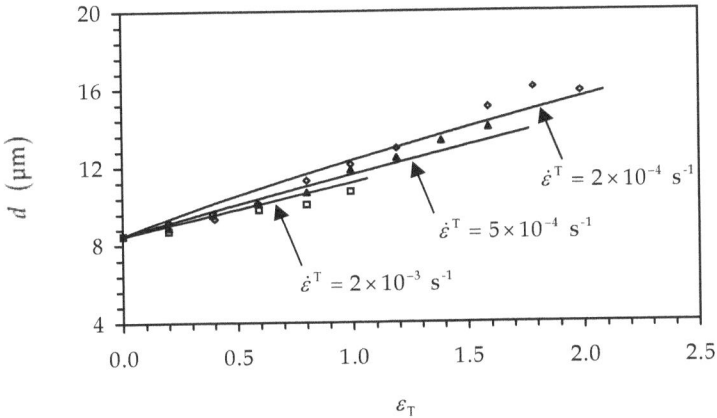

Fig. 6.16. Comparison of computed (solid curves) (Cheong, 2002) and experimental (symbols) (Pilling, 2001) grain-size-strain relationships for Al-Zn-Mg at various strain rates.

(ii). Superplastic material of AA7475 at 515 °C

The determined values for the material constants within the Eq. set (6.45) are listed in **Table 6.2** and the computed results (solid curves) are shown in **Fig. 6.17** (Cheong, 2002). It can be seen that the computed stress–strain relationships show a good agreement with the corresponding experimental data at strain rates $\dot{\varepsilon}^T = 2 \times 10^{-4}$ s^{-1}, 1×10^{-3} s^{-1} and 5×10^{-3} s^{-1}.

Table 6.2 Material parameters for Al 7475 at 515 °C (Cheong, 2002).

k (MPa)	K (MPa)	n	μ	b	Q (MPa)	α	γ_0
1.3602e-2	64.9052	1.5838	1.1789	0.2794	15.0838	12.7187	39.1470
β	ϕ	D	d_1	C	f_h^*	E (MPa)	l^0 (m)
80.6578	0.8831	9.7966e+3	2.9139	1.1977e+3	0.5867	80.3301	16.0781

The unified superplastic damage constitutive equations determined from experimental data can be input into commercial FE codes via user-defined subroutines for the optimisation of processing parameters (Lin, 2003). More details about these applications will be given in Chapter 8.

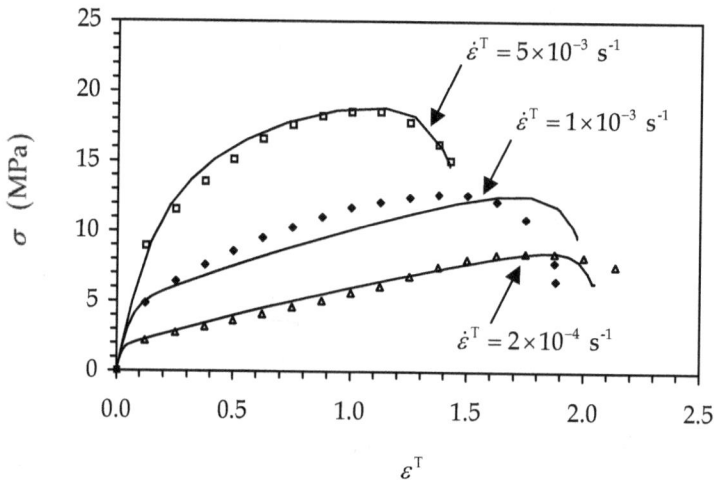

Fig. 6.17 Comparison of computed (solid curves) (Cheong, 2002) and experimental (symbols) (Pilling, 2001) stress–strain relationships for AA7475 at various strain rates.

6.5.2 Grain size and strain rate effects in hot forming

As discussed earlier, depending on deformation rate, temperature and grain size, damage can nucleate either at grain boundaries (creep-type damage) or around second phases due to debonding (plasticity-type damage). Damage models have been developed to model these damage features.

In addition to temperature, the grain boundary damage (creep-type damage) is related to grain size and strain rate. The initial model was developed by Liu (2004) and further improved by Foster (2006) and Kaye (2012) for the modelling of damage evolution in hot rolling processes. This model enables grain boundary type (creep-type) damage, D_{gb}, to be predicted according to the grain size and deformation rate. The rate equation of grain boundary damage can be written as (Liu, 2004)

$$\dot{D}_{gb} = a_4 \cdot \eta \cdot \left(\frac{\sigma_1}{\sigma_e}\right)^{n_2} \cdot \left(\dot{D}_{gb}^N + \dot{D}_{gb}^G\right), \tag{6.46}$$

where $\dot{D}_{gb}^N$ and $\dot{D}_{gb}^G$ are grain boundary damage nucleation and growth respectively and can be expressed as

$$\dot{D}_{gb}^N = a_7 \cdot \left(1 - D_{gb}\right) \cdot \dot{\bar{\rho}}$$

$$\dot{D}_{gb}^G = \left[\frac{1}{\left(1-D_{gb}\right)^{n_3}} - \left(1 - D_{gb}\right)\right] \cdot \dot{\varepsilon}_p.$$

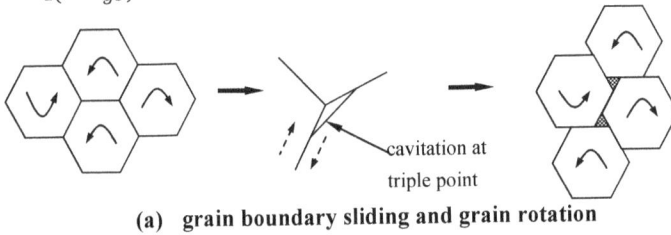

(a) grain boundary sliding and grain rotation

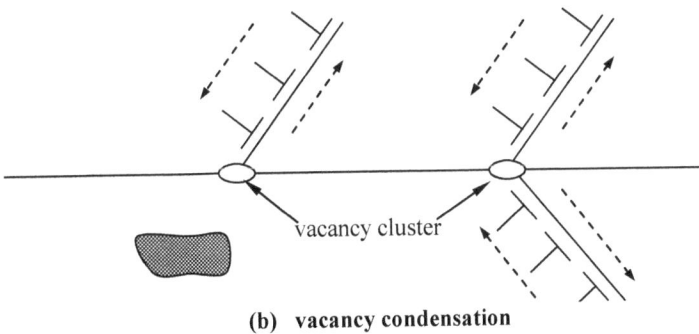

(b) vacancy condensation

Fig. 6.18 Deformation mechanisms responsible for grain boundary damage nucleation (Liu, 2004).

Equation (6.46) describes grain boundary damage, which includes damage nucleation $\dot{D}_{gb}^{N}$, and is related to the normalised dislocation density rate $\dot{\bar{\rho}}$ and growth $\dot{D}_{gb}^{G}$, in a similar way to the ductile void growth model presented earlier (Cocks and Ashby, 1980). The damage nucleation $\dot{D}_{gb}^{N}$ reflects two major possible damage initiation mechanisms: (a) sliding of grains along grain boundaries and grain rotation which could not be fully accommodated by diffusion and (b) vacancy condensation on grain boundaries, as shown in **Fig. 6.18**. Here the damage growth $\dot{D}_{gb}^{G}$ is strain-controlled.

The parameter η in Eq. (6.46) represents the effects of grain size and strain rate on the evolution of grain boundary damage and is given as:

$$\eta = \exp\left(-a_5\left(1 - \frac{d}{d_c}\right)^2\right) \text{ and } d_c = a_6(\dot{\varepsilon}_p)^{-n_1},$$

where a_5, a_6 and n_1 are the material constants. For example, if the grain size is large and deformation rate is high, the value of η is small and thus the grain boundary damage rate $\dot{D}_{gb}$ is low.

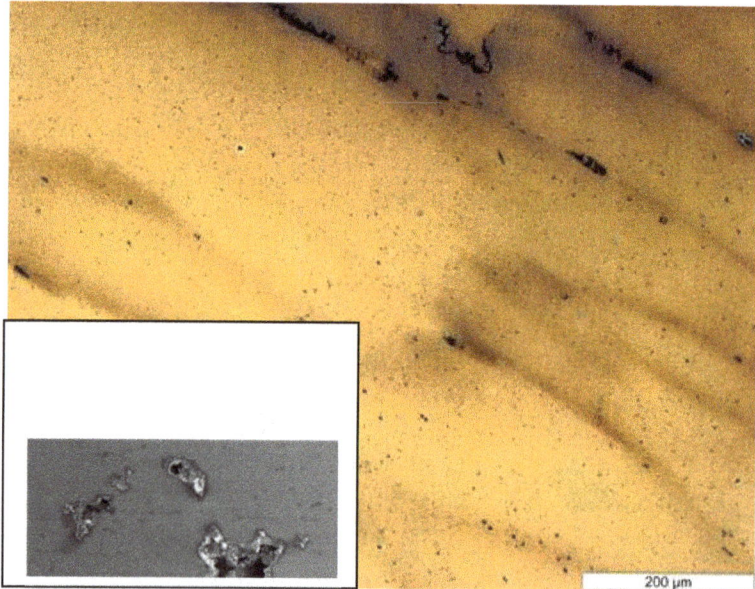

Fig. 6.19 Voids/defects created in cast due to shrinkage during cooling (provided by Dr. R. Qin of Imperial College London).

6.6 Damage Healing in Hot Compressive Forming

Voids and defects can be found in materials, especially from cast of blooms, as shown in **Fig. 6.19**. This kind of porosity can be eliminated or reduced during hot rolling or hot forging processes, i.e. under compressive loading. Then the material would become "stronger" or more "solid". Under hot and compressive loading conditions, voids can be closed and subsequently diffusion-bonded at high temperatures. The bonding mechanisms and bonding models are summarised by Afshan (2013), but will not be detailed in this book.

In addition to the elimination of porosity in hot forging and rolling processes, the diffusion bonding/healing mechanisms have been used for various kinds of metal processing technologies, such as superplastic forming and diffusion bonding for the manufacturing of complex-shaped panel components; hydrostatic isothermal processing (HIPing) and powder forging for the manufacturing of complex-shaped bulk parts using powder materials.

Chapter 7

Numerical Methods For Materials Modelling

In the previous chapters, the basic techniques for formulating phenomenological unified elastic-plastic and elastic-viscoplastic unified constitutive equations have been introduced. Based on these concepts, different types of constitutive equations can be formulated for individual materials and applications. These equations cannot normally be solved analytically and numerical integration methods are required. Additionally, the determination of a set of constitutive equations from experimental data requires optimisation techniques. In this chapter, the relevant numerical methods for solving the constitutive equations, and determining their material constants from experimental data, are introduced.

For isotropic elasticity, the basic material constants are Young's Modulus, E, Poisson's ratio, ν, and the shear modulus, G, which can all be determined directly from the experimental results of simple materials testing, such as, tension/compression and shear tests.

In plasticity or viscoplasticity, if simple relationships are used to model strain hardening and strain rate hardening, such as $\sigma = K\varepsilon^n \dot{\varepsilon}^m$, the material constants K, n and m can be determined from the experimental data using either analytical methods (discussed earlier), or optimisation methods to achieve the best fit to the experimental data. When state variables are used and the unified constitutive equations become more complex, it is very difficult to determine the material constants analytically from experimental data; the use of optimisation methods is recommended for these applications (Lemaitre and Chaboche, 1990).

The numerical techniques for the determination of constitutive equations, particularly for the determination of unified constitutive equations, were only developed in the 1990s and 2000s. This is mainly due to the fast development of computing power in the 1990s, which enabled more complicated unified constitutive equations to be developed and used for the simultaneous modelling of more physical phenomena encountered in real engineering applications, such as high temperature creep rupture, fatigue and materials processing technologies. Due to the complexity of the equations, optimisation methods are recommended for the determination of constants arising in a set of unified constitutive equations from a series of experimental data (Lemaitre and Chaboche, 1990). The following work highlights the development in optimisation techniques for determining constitutive equations over the last two decades:

- Lin and Hayhurst (1993a and 1993b) used a gradient-based optimisation technique and determined the material constants for constitutive equations developed for shoe-upper leathers where stress can be explicitly expressed as a function of strain, and used them for the modelling of a lasting process in leather shoe manufacturing.

- Kowalewski, Hayhurst and Dyson (1994) developed a three-step method to determine the starting values of the constants in a set of creep damage equations for optimisation. Zhou and Dunne (1996) proposed a four-step method to determine material constants for a set of superplastic constitutive equations. The above developments were carried out using a gradient-based optimisation method; the difficulties associated with choosing proper starting values for the constants were highlighted. In addition, extra difficulties were met when the number of constants to be determined in one step was large, e.g. more than five constants (Zhou and Dunne, 1996).

- Lin and Yang (1999) proposed a method of using genetic algorithm (GA)-based optimisation techniques for the determination of unified superplastic constitutive equations, where 13 constants were optimised in one process and the difficulties of choosing proper starting values was avoided.

- Li, Lin and Yao (2002) have used evolutionary programming (EP)-based optimisation techniques to improve the computational efficiency and accuracy of the equation determination. This algorithm has been applied with success to many constitutive equations, and has been found to be particularly useful for determining unified creep damage constitutive equations and general viscoplastic constitutive equations.

- Cao and Lin (2008) formulated an effective unitless objective function for the determination of unified constitutive equations.

The problem of individual differential equations having different unit scales was overcome and weighting factors for the correction were eliminated.

The research continues, and significant progress has been made over the period (Cao, 2006; Lin, Cao and Balint, 2011) for the determination of unified viscoplastic constitutive equations. However, apart from some in-house software codes (Cao, Lin and Dean, 2008b; Pan, Lin and Balint, 2012), the problem has not been thoroughly solved and no expert commercial software systems are available for the determination of the material constants arising in a set of unified constitutive equations. In this chapter, basic concepts for numerical integration, objective functions and optimisation methods for solving/determining unified viscoplastic constitutive equations are introduced and a few case studies are presented.

For the convenience of the discussion in this chapter, we listed a few sets of unified constitutive equations in Appendix A, which cover different applications. The detailed descriptions for the equations are not given in Appendix A. However, the concept and techniques for formulating individual equations within the equation sets have been detailed throughout the book, particularly in Chapters 5 and 6.

7.1 The Numerical Framework in Materials Modelling

7.1.1 *Methods of solving ODE- type constitutive equations*

a. ODE-type unified constitutive equations

The constitutive equations listed in Appendix A are sets of ordinary differential equations (ODE) in terms of time t, and, have the following common features:

$$\frac{d\varepsilon^P}{dt} = f\left(\sigma, y_2, \cdots y_{NE}, T\right)$$

$$\frac{dy_2}{dt} = f_2\left(\varepsilon^P, y_2, \cdots, y_{NE}, T, \frac{d\varepsilon^P}{dt}, \frac{dy_3}{dt}, \cdots, \frac{dy_{NE}}{dt}\right) \tag{7.1}$$

...

$$\frac{dy_{NE}}{dt} = f_{NE}\left(\varepsilon^p, y_2, \cdots, y_{NE}, T, \frac{d\varepsilon^p}{dt}, \frac{dy_1}{dt}, \cdots, \frac{dy_{NE-1}}{dt} \right),$$

where y_i ($i = 2,3,\ldots,NE$) are internal state variables and NE represents the number of ODEs representing the constitutive equations. For example, y_2 could be a hardening variable, y_3 could represent the dislocation density etc. For a given moment in a deformation process, these equations can be solved via numerical integration. In general, these instantaneous values are fed immediately into a "response equation", such as that for the flow stress, σ, which is then calculated at that given moment:

$$\sigma = f\left(y_2, y_3, \varepsilon^p, \varepsilon^T \cdots, T \right),$$

where ε^p, ε^T and T are plastic strain, total strain and temperature respectively.

Notice that the equations are different from the system of ODEs presented in common mathematics books, since the units of the individual equations within a set may be different (the units of individual rate equations are given in Appendix A). This creates significant difficulties in solving the equations using numerical integration methods.

b. Particular features of numerical integration in materials modelling

The ODE-type constitutive equations, Eq. (7.1), can be solved according to the initial values (at time $t = 0$) given for each equation using a numerical integration method. This is also known as the solving of initial value problems.

The multiple ODE-type constitutive equations with different unit scales for individual equations in each set need to be integrated simultaneously using an integration method with sufficient error control algorithms. The existence of different unit scales increases the difficulty of obtaining the solution. Different tolerances are required to control the integration accuracy for individual equations, which are difficult to specify.

When using evolutionary algorithm (EA)-based methods to determine the constitutive equations from experimental data (Lin and Yang, 1999), thousands of generations are needed to get a solution throughout the whole optimisation process, and in each generation hundreds of populations are normally generated. For each population within a generation, the unified constitutive equations need to be accurately integrated; the fitness (error) is assessed by calculating the difference between the integrated and experimental data (Li, Lin and Yao 2002; Cao and Lin, 2008a). Such magnitudes of integrations take a significant amount of computational time, and the error estimations also need to be calculated accurately. To solve the problem efficiently, a minimum number of time increments is required to integrate the equations with controlled accuracy in the least amount of time. In other words, the integration needs to be solved both accurately and efficiently.

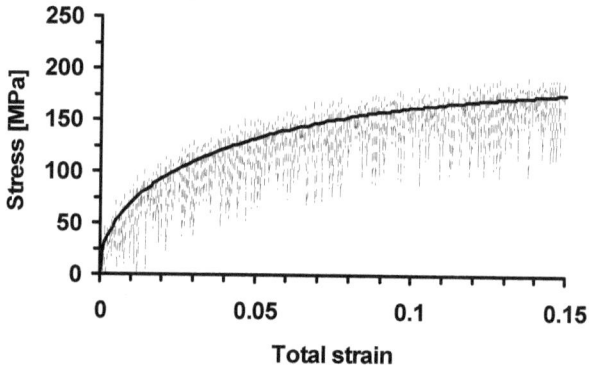

Fig.7.1 Stress–strain curves for equation SET IV in Appendix A with $\dot{\varepsilon} = 0.1$ integrated with step size, $\Delta t = 1.5\times10^{-4}$ (solid line) and 1.5×10^{-3} (dashed line) (Cao, 2006).

Unified viscoplastic constitutive equations are normally highly nonlinear, coupled and mathematically very stiff (Miller and Shih, 1977; Banthia and Mukherjee, 1985; Miller and Tanaka, 1988; Vinod, 1996). If a set of unified viscoplastic constitutive equations (SET I in Appendix A) is to be integrated numerically using a forward Euler method (which will be given later), given a small constant time increment, for example, $\Delta t = 1.5\times10^{-4}$, the stress–strain curve (solid line) is integrated smoothly with 10,000 iterations, as shown by the solid curve in **Fig. 7.1**.

However, by increasing the constant step size, Δt, to 1.5×10^{-3}, oscillation occurs in the stress–strain curve (with 1,000 iterations), as shown in **Fig. 7.1**. The large Δt causes the inaccuracy and instability in the numerical integration, even though the equations themselves have certain good features that are conducive to convergence. The reason for this is as follows.

- If the step size is too big, then the calculated stress is too high in Equation SET I in Appendix A; as it is used in the next calculation step, it will result in a high plastic strain rate, and consequently a high plastic strain increment $\Delta \varepsilon^p$.

- If the total strain increment $\Delta \varepsilon^T$ is constant, then this results in the difference $(\Delta \varepsilon^T - \Delta \varepsilon^p)$ being low, thus the calculated stress is low.

- In the next iteration, the low stress would result in the calculated plastic strain increment $\Delta \varepsilon^p$ being low, which then creates higher stresses. This process repeats and thus generates the oscillatory behaviour in the stress, yielding the poor results.

If the step size for the integration is too large, the convergence cannot be achieved and the computation would not continue. In this book, a number of integration methods are covered. The focus will be on the problem of error control in a system of ODE-type unified constitutive equations with different units.

7.1.2 *Determination of unified constitutive equations from experimental data*

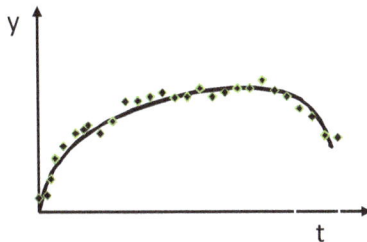

Fig. 7.2 Fitting of experimental data using an optimisation method.

a. The concept of least squares fitting

In materials modelling, it is necessary to determine the values of the constants arising in a set of constitutive equations, such that the best fit to the experimental data of the material can be achieved, as shown in **Fig. 7.2**. A mathematical procedure for finding the best-fit curve for a given set of points (experimental data), is to minimise the sum of the squares of the differences between the points and the curve ("the residuals"). The sum of the *squares* of the residuals is used instead of the absolute values of the residuals because this allows them to be treated as a continuous differentiable quantity.

In general, *vertical residuals* are used to assess the errors of the fitting, as shown in **Fig. 7.3 (a)**. However, *perpendicular residuals* (**Fig. 7.3 (b)**) are also used for many cases. The *linear* least squares fitting technique is the simplest and most commonly applied form of *linear regression,* and provides a solution to the problem of finding the best-fit *straight* line through a set of points.

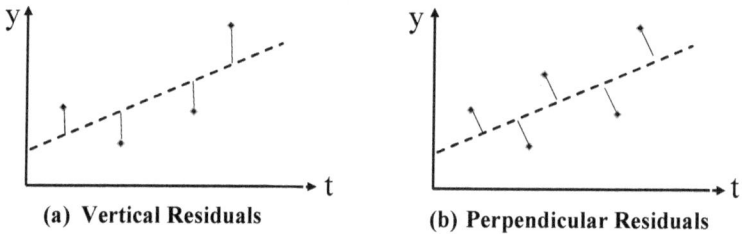

(a) **Vertical Residuals**   (b) **Perpendicular Residuals**

Fig. 7.3 Error assessment methods.

Vertical least squares fitting proceeds by finding the sum of the *squares* of the *vertical* deviations $f(x)$ of a set of M data points

$$f(x) = \sum_{i=1}^{M} \left[g(t_i, x_1, x_2, \ldots x_{NC}) - y_i^e \right]^2. \tag{7.2}$$

From a function

$$y = g(t, x_1, x_2, \ldots x_{NC}), \tag{7.2a}$$

$x = \{x_1, x_2, \ldots, x_{nc}\}$ are the constants to be determined for the best fitting. NC is the number of constants to be determined. Note that this procedure does *not* minimise the actual deviations from the line (which would be measured perpendicular to the given function). In addition, although the *unsquared* sum of distances might seem a more appropriate quantity to minimise, use of the absolute value results in discontinuous derivatives which cannot be treated analytically. The squared deviations from each data point i are therefore summed, and the resulting sum is then minimised to find the best fit line.

The condition for $f(x)$ to be a minimum is that

$$\frac{\partial f(x)}{\partial x_j} = 0 \tag{7.2b}$$

for $j = 1,2,\ldots,NC$. For a linear fit, for example,

$$y = g(t, x_1, x_2) = x_1 + x_2 t .$$

We then have

$$f(x_1, x_2) = \sum_{i=1}^{M} \left[(x_1 + x_2 t_i) - y_i^e \right]^2 ,$$

and the conditions for the best fitting are

$$\frac{\partial f(x_1, x_2)}{\partial x_1} = 2 \sum_{i=1}^{M} \left[(x_1 + x_2 t_i) - y_i^e \right]^2 = 0$$

$$\frac{\partial f(x_1, x_2)}{\partial x_2} = 2 \sum_{i=1}^{M} \left[(x_1 + x_2 t_i) - y_i^e \right]^2 t_i = 0 .$$

These can be solved analytically or numerically. If the number of data points is 2, i.e. $M = 2 = NC$, then the fitted curve will pass all the data points and the error is zero. For a general best-fitting problem, if the number of data points is more than the number of constants to be determined, i.e. $M > NC$, the fitted curve may not pass through all the data points, as shown in **Fig. 7.2**.

b. *Optimisation and the problem description in materials modelling*

In mathematics or computing, to solve an *optimisation* problem is to find the *best solution* from all *feasible solutions*. For many cases, it is very difficult to find the best global solution. In general, an optimisation problem consists of maximising or minimising a function of a real number. Many methods are available for searching for the best solution, such as gradient methods (including the conjugate gradient method, the gradient descent method etc.), and evolutionary algorithms (including genetic algorithms, evolutionary programming etc.). These methods have been introduced in text books and their corresponding computer codes have been developed and embedded in many commercial mathematical systems, such as NAG, MATLAB, MATHEMATICA etc. In this book, optimisation methods will not be introduced in detail.

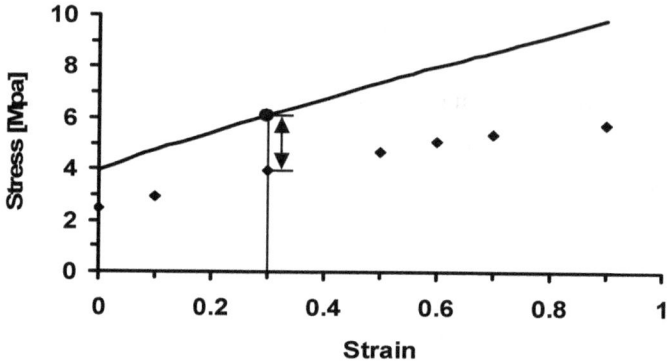

Fig. 7.4 Calculation of error using the conventional least-squares method. The sum of the squared difference between the computed (solid line) and experimental (symbols) data is calculated at the same strain.

The conventional least-squares method has been widely used as an *objective function*, also referred to as a *fitness function*, in which the sum of the squares of the errors between the experimental and computed data is minimised. An objective function for the optimisation problem can be formulated as (**Fig. 7.4**):

$$f(x) = \sum_{i=1}^{M} w_i \left[\sigma_i^c - \sigma_i^e \right]^2 , \tag{7.3}$$

where $f(x)$ is the sum of the squared residuals of stress, which is also known as the *objective function* for optimisation. The parameter x ($x = \{x_1, x_2, ..., x_{nc}\}$) represents the constants x_j ($j = 1, 2, ..., NC$), where NC is the number of constants to be determined (optimised) from the experimental data. σ_i^c and σ_i^e are the computational and experimental stresses for the same strain level i, w_i is a weighting function, which is used to quantify the importance of individual experimental data points in the fitting, and M is the number of experimental stress–strain data points, as shown in **Fig. 7.4**.

The necessary conditions for evaluating the minima of $f(x)$ are given by

$$\frac{\partial f}{\partial x_j} = 0, \quad j = 1, 2, ..., NC . \tag{7.3a}$$

The objective function Eq. (7.2) is a nonlinear function and many solutions exist that satisfy the condition given in Eq. (7.3). A typical nonlinear objective function for a constant x_j is schematically shown in **Fig. 7.5**, where a few local minimum/maximum locations are shown to satisfy the condition (Eq. (7.3a)), such as a_1, a_2 etc.

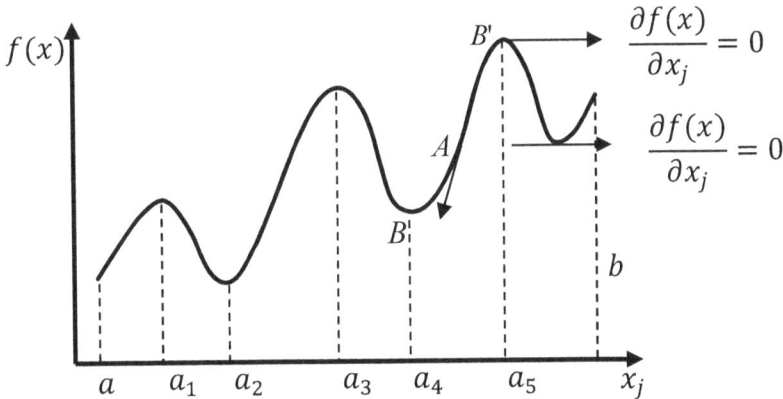

Fig. 7.5 Nonlinear optimisation for a one variable problem. Many local minimum and maximum locations are found.

If the gradient search optimisation method is used, it is very important that correct starting points are chosen. For example, if the starting point is A, the minimum and maximum values obtained from the optimisation would either be at the location of B or B'. This is one of the difficulties of finding the global minimum values in the determination of unified constitutive equations through optimisation techniques.

Various constitutive models have been determined by using the conventional least-squares method (Tong and Vermeulen, 2003; Kajberg and Lindkvist, 2004; Springmann and Kuna, 2005). If a set of creep constitutive equations (SET II: Appendix A) is considered, the lifetime (up to failure of the materials) is characterised by material constants, listed in **Table A.2** in Appendix A, and it becomes difficult to use the conventional least-squares method to solve the problem (Li, Lin and Yao, 2002), since too many constants need to be optimised at the same time to get the best fitting.

c. An example of determining a simple equation

For rate-independent plasticity, the stress–strain relationship of a material can be modelled using the following simple equation as discussed in previous chapters

$$\sigma = k\varepsilon^n,$$

where k and n are material constants to be determined from experimental data. It was mentioned before that given two pairs of experimental stress–strain data, the values of k and n can be determined analytically. For example, using the experimental stress–strain data points (3) and (9) indicated in **Fig. 7.6**, the values of k and n are 179.99 MPa and 0.14 respectively. The stress–strain curve computed using these values (dashed curve) is shown in the figure; it can be seen that the curve only passes the two experimental data points.

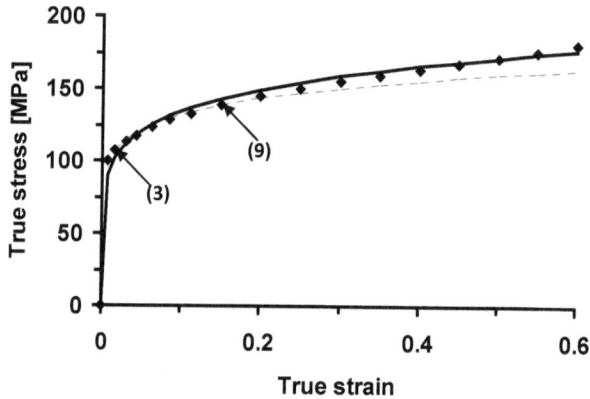

Fig.7.6 Determination of k and n for $\sigma = k\varepsilon^n$ for a pearlitic steel at 20 °C. The stress–strain curves are obtained using analytic (dashed line) and optimisation (solid line) methods (Cao, 2006) from experimental data (symbols) from Zener and Hollomon (1944).

An optimisation curve fitting method can also be used to determine the constants within the equation, by minimising the differences between the experimental and fitted results. The least squares method is a popular choice used for this particular type of problem. The fitting result using optimisation is shown in **Fig.7.6** with the solid line (where $k = 190.45$ MPa, $n = 0.15$). The equation determined using the optimisation method can predict the overall experimental data better, even though it may not pass any of the experimental data points.

Two constants needed to be determined from the experimental data for this case. This is known as a two-dimensional optimisation problem. However, in a set of phenomenological unified constitutive equations, more than 10 or even 30 constants need to be determined from a series of experimental curves. For larger dimensions and multiple objective problems, we need suitable optimisation techniques.

d. Multi-objective optimisation

In this optimisation process, we may have many experimental stress–strain curves for different temperatures and strain rates. Each curve can be considered as an objective, or collectively, a set of stress–strain curves can be considered as one objective, since all the curves have the same units and the same features over all the data points. We may have experimental data for other physical parameters of a material at high temperature deformation conditions, such as the recrystallised volume fraction of the material and the grain size evolution (grain refinement and grain growth). The evolution of these physical phenomena at hot deformation conditions is represented by state variables in a set of unified constitutive equations. The experimental data for each physical phenomenon should be considered as an objective. Thus a multi-objective optimisation problem can be expressed as

$$\min_{x \in X} \sum_{i=1}^{k} w_i f_i(x), \qquad (7.4)$$

where i represents individual objectives, $i = 1,2,...,k$, and k is the number of objectives. $f_i(x)$ is the ith objective function and $x = \{x_1, x_2,...,x_{NC}\}$ represents the constants x_j ($j = 1,2,...,NC$), where NC is the number of constants to be determined (optimised) from the experimental data. w_i is the weight of the individual objectives, which is very difficult to define in the determination of the unified constitutive equations, since different objectives have different units.

In this book, different methods for formulating objective functions for the optimisation problem will be introduced.

e. Numerical procedures

The overall numerical process for the determination of material constants within a set of unified constitutive equations from experimental data is shown in **Fig. 7.7**. It consists of three key stages:

(i) The first stage is to solve the ODE-type constitutive equations, as listed in Appendix A, using a numerical integration method according to the initial values given. This stage is of prime importance. The accuracy of the numerical integration should be controlled with the consideration of using large time increments for individual numerical integration iterations such that the computational efficiency is maximised. Thus, error assessment and automatic step-size control are the key issues in the numerical integration process. The difficulties of solving the integration problem are highlighted by Cao, Lin and Dean (2008a).

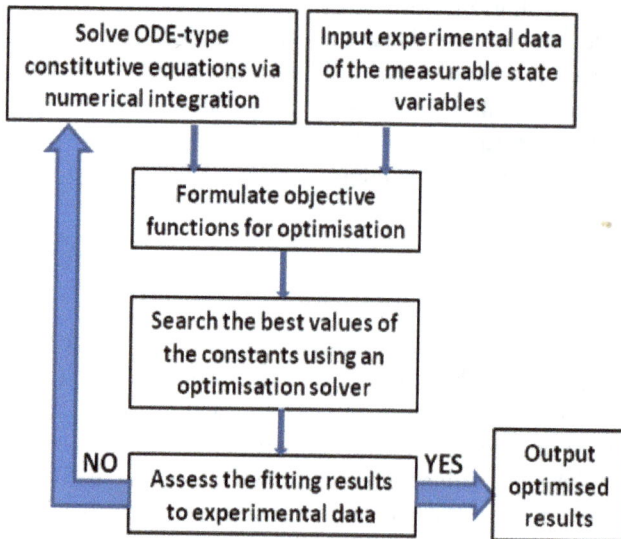

Fig.7.7 Numerical procedures for the determination of material constants through an optimisation process.

(ii) The second stage is to formulate an objective function so that the errors between the computed data and the corresponding experimental data can be effectively assessed. Multiple experimental curves are always considered. This results in multi-objective functions. For example, stress–strain curves for different temperatures and strain rates, grain size evolution during and after a deformation process, recrystallised volume fraction etc. Individual objectives may have different units and the errors are very difficult to assess. This problem has been discussed by Cao and Lin (2008a).

(iii) The third stage is to search for the best values of the constants within the equation set using an optimisation method, or combined optimisation methods, so that the best fit to the corresponding experimental data can be obtained. It is very difficult to obtain the global solution, but evolutionary algorithms such as genetic algorithms (Lin and Yang, 1999) and evolutionary programming (Li, Lin and Yao, 2002), could provide a better chance of solving the optimisation problem and obtaining the global minimum (Cao, 2006; Lin, Cao and Balint, 2011).

In solving a large dimensional problem (more than ten constants within an equation set), a few hundreds of generations are normally required. In one generation, we normally choose 50–100 populations. This kind of selection would result in a large number of numerical integrations for the constitutive equations, which takes a lot of computational time. In the following sections, we will introduce the methods related to each of the key stages in the determination of the material constants in a set of unified constitutive equations.

7.2 Numerical Integration

For the convenience of discussion, the set of ODE-type constitutive equations shown in equation (7.1), can be expressed as

$$\dot{y}_i = \frac{dy_i}{dt} = f_i(y_1, y_2, \dots, y_{NE}, T, \dot{y}_1, \dot{y}_2, \dots, \dot{y}_{NE}), \tag{7.5}$$

with the initial values (at $t = 0$)

$$y_i(t = 0) = y_{0,i} \qquad i = 1, 2, \dots, NE \tag{7.6}$$

where y_i ($i = 1, 2, \dots, NE$) is the variable integrated from the ith ordinary differential equation y_i and NE represents the number of equations representing the state variables. The set of ODEs can be integrated using many numerical methods, which are introduced in this section.

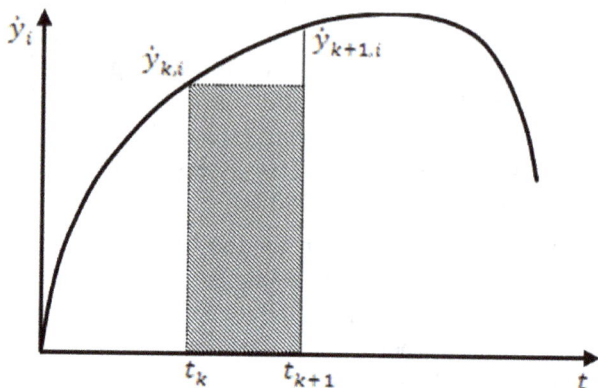

Fig. 7.8 Explicit first-order Euler method for numerical integration.

7.2.1 *Explicit Euler method*

The *Euler method* (Leonhard Euler, Swiss, 1707–1783) is a first-order numerical procedure for solving ordinary differential equations with given initial values for the variables. This is the simplest and the most commonly used method for solving ordinary differential equations; it is also known as the *Euler forward method.*

a. Integration

The geometric representation of the integration is shown in **Fig. 7.8**. For approximating the solution of the initial value problem using the forward Euler method, we have

$$y_{k+1,i} = y_{k,i} + \Delta t_k \dot{y}(t_k, y_{k,i}) = y_{k,i} + \Delta t_k \dot{y}_{k,i} \quad i = 1,2,...,NE \quad (7.7)$$

Equation (7.7) advances a solution (from t_k to t_{k+1}) from $y_{k,i}$ to $y_{k+1,i}$ for the ith variable with the time step size at the kth iteration, where $k = 1,2,3,...,N$, and, N is the number of increments for the integration. It should be noted that the time t (or t_k at the kth iteration) is normally not a variable in a set of constitutive equations (see Eq. (7.5) and the lists of equations in Appendix A). It is implicitly expressed through the evolutionary equations. However, for the generalization purpose, t_k is written within the Eq. (7.7).

$$\Delta t_k = t_{k+1} - t_k , \text{ or } t_{k+1} = t_k + \Delta t_k$$

The step size at the *k*th *iteration* is Δt_k. The step size could also be a constant value throughout the integration, which is the simplest case and is known as constant step-size integration.

b. Local truncation error

The local truncation error of the Euler integration method is the error made in a single iteration. It is the difference between the numerical solution after one step, $y_{k,i}$ (at time t_k), and the exact solution at time $t_{k+1} = t_k + \Delta t_k$. The numerical solution is given by Eq. (7.7). For the exact solution, we use the Taylor expansion

$$y_{k+1,i} = y_{k,i} + \Delta t_k \dot{y}_{k,i} + \frac{1}{2}\Delta t_k^2 \ddot{y}_{k,i} + O(\Delta t_k^3),\tag{7.8}$$

where $\ddot{y}_{k,i}$ is the second differentiation in terms of time, i.e. $\ddot{y}_{k,i} = d\dot{y}_{k,i}/dt$. By ignoring high order terms, the local truncation errors (ΔE_i) for the individual variables introduced by the Euler method are given by the difference between Eqs. (7.8) and (7.7).

$$\Delta E_{k+1,i} = \frac{1}{2}(\Delta t_k)^2 \ddot{y}_{k,i} \qquad i = 1,2,...,NE.\tag{7.9}$$

As a result, the truncation error is of the order Δt_k^2. This shows that for a small Δt_k, the local truncation error is approximately proportional to Δt_k^2. This makes the Euler method less accurate than other higher-order techniques. If a local truncation error is defined, it can be used to estimate the required step size Δt_{k+1} in the numerical integration process.

7.2.2 Midpoint method

The Midpoint method is a one-step method for numerically solving the differential equations listed in Eq. (7.5) with the initial values (at $t = 0$) shown in Eq. (7.6), and is given by the equation

$$y_{k+1,i} = y_{k,i} + \Delta t_k \dot{y}(\bar{t}_k, \bar{y}_{k,i}) \qquad i = 1,2,...,NE.\tag{7.10}$$

The name of the midpoint method comes from the fact that in Eq. (7.10) the differential function $\dot{y}_i$ (Eq. (7.5)) is evaluated at the midpoint of the increment, as shown in **Fig. 7.9**. The difficulty of the midpoint method arises from the need to evaluate the values of the variables at the midpoint of each increment. In the *k*th iteration, we could calculate the midpoints using

$$\bar{t}_k = t_k + \frac{\Delta t_k}{2}$$

$$\bar{y}_{k,i} = y_{k,i} + \frac{\Delta t_k}{2} \dot{y}_{k,i}.$$

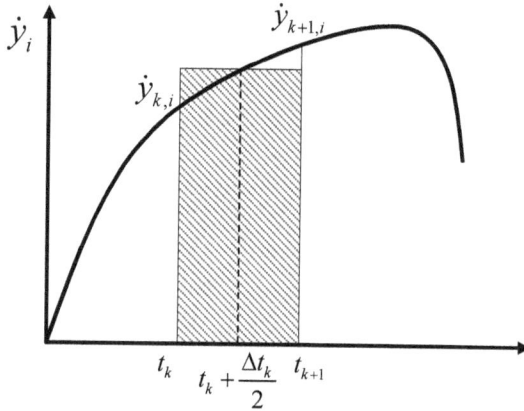

Fig. 7.9 Explicit midpoint Euler method for numerical integration.

The midpoint method is also known as the *modified Euler method*. The local truncation error for the *i*th constitutive equation at each step of the midpoint method is of the order Δt_k^3. Thus, while more computationally intensive than the first-order Euler method, the midpoint method generally gives more accurate results. The method is an example of a class of higher-order methods known as Runge–Kutta methods.

7.2.3 *Runge–Kutta method*

In numerical analysis, Runge–Kutta methods are an important family of iterative methods for the approximation of solutions for a set of ordinary differential equations. The method was developed in 1901 by mathematicians Runge (Carl David Tolmé Runge, German, 1856–1927) and Kutta (Martin Wilhelm Kutta, German, 1867–1944). One of the Runge–Kutta methods is known as the *classical Runge–Kutta method* (RK4), or simply the Runge–Kutta method, and can be used for solving the constitutive equations Eq. (7.5) with the initial values given in Eq. (7.6):

$$y_{k+1,i} = y_{k,i} + \frac{1}{6}\left(K_{1,i} + 2K_{2,i} + 2K_{3,i} + K_{4,i}\right) \quad i = 1, 2, ..., NE, \quad (7.11)$$

where

$$t_{k+1} = t_k + \Delta t_k$$

and

$$K_{1,i} = \Delta t_k \dot{y}_i(t_k, y_{k,i})$$
$$K_{2,i} = \Delta t_k \dot{y}_i(t_k + \Delta t_k/2, y_{k,i} + (\Delta t_k/2)K_{1,i})$$
$$K_{3,i} = \Delta t_k \dot{y}_i(t_k + \Delta t_k/2, y_{k,i} + (\Delta t_k/2)K_{2,i})$$
$$K_{4,i} = \Delta t_k \dot{y}_i(t_k + \Delta t_k, y_{k,i} + \Delta t_k K_{3,i})$$

This is the typical 4th order Runge–Kutta method and the values $y_{k+1,i}$ are approximated from the current values $y_{k,i}$ plus the *weighted average* of four increments. Each increment is the product of the size of the time increment, Δt_k, and an estimated slope specified by the differential function $\dot{y}_i$.

- $K_{1,i}$ is the increment based on the slope at the beginning of the interval for the ith equation, using $\dot{y}_i(t_k, y_{k,i})$;
- $K_{2,i}$ is the increment based on the slope at the midpoint of the interval for the ith equation;
- $K_{3,i}$ is again the increment based on the slope at the midpoint;
- $K_{4,i}$ is the increment based on the slope at the end of the interval.

In averaging the four increments, a greater weight is given to the increments at the midpoint. The weights are chosen such that if $\dot{y}_i$ is independent of y_i, the differential equation becomes a simple integral. The local error for the ith constitutive equation at each step of the Runge–Kutta method is of the order Δt_k^5.

7.2.4 *Implicit Euler method*

The *implicit Euler method*, also known as the *backward Euler method*, is the most basic and commonly used method for accurately solving ODE-type constitutive equations using the least number of increments.

a. Integration

The general form of a first-order implicit method is given by (Press, Flannery, Ateukolsky *et. al.*, 2002):

$$y_{k+1,i} = y_{k,i} + \Delta t_k \dot{y}_{k+1,i}. \tag{7.12}$$

By linearizing the first-order implicit method as in Newton's method we obtain

$$\dot{y}_{k+1,i} = \dot{y}_{k,i} + \frac{\partial \dot{y}_i}{\partial y_j}(y_{k+1,i} - y_{k,i}). \tag{7.13}$$

Incorporating Eq. (7.13) in Eq. (7.12), we have

$$y_{k+1,i} = y_{k,i} + \Delta t_k \left[\dot{y}_{k,i} + \frac{\partial \dot{y}_i}{\partial y_j}(y_{k+1,i} - y_{k,i}) \right]$$

$$\left(I - \Delta t_k \frac{\partial \dot{y}_i}{\partial y_j} \right) y_{k+1,i} = \left(I - \Delta t_k \frac{\partial \dot{y}_i}{\partial y_j} \right) y_{k,i} + \Delta t_k \dot{y}_{k,i}.$$

Then we have:

$$y_{k+1,i} = y_{k,i} + \Delta t_k \left(I - \Delta t_k \frac{\partial \dot{y}_i}{\partial y_j} \right)^{-1} \dot{y}_{k,i}, \tag{7.14}$$

where $y_{k+1,i}$ represents the variable integrated from the ith ordinary differential equation $\dot{y}_i$ using the first-order implicit Euler method. i and j vary from 1 to NE, where NE is the number of differential equations within an equation set. $\partial \dot{y}_i / \partial y_j$ is a matrix with an order of $NE \times NE$ and is known as a *Jacobian matrix* at the current kth iteration in the implicit numerical integration. I is a unit matrix. The equation can be written in matrix form as

$$
\begin{bmatrix} y_{k+1,1} \\ y_{k+1,2} \\ .. \\ y_{k+1,NE} \end{bmatrix} = \begin{bmatrix} y_{k,1} \\ y_{k,2} \\ : \\ y_{k,NE} \end{bmatrix} + \Delta t_k \left(\begin{bmatrix} 1 & 0 & .. & 0 \\ 0 & 1 & .. & 0 \\ : & : & : & : \\ 0 & 0 & .. & 1 \end{bmatrix} - \Delta t_k \begin{bmatrix} \frac{\partial \dot{y}_1}{\partial y_1} & \frac{\partial \dot{y}_1}{\partial y_2} & .. & \frac{\partial \dot{y}_1}{\partial y_{NE}} \\ \frac{\partial \dot{y}_2}{\partial y_1} & \frac{\partial \dot{y}_2}{\partial y_2} & .. & \frac{\partial \dot{y}_2}{\partial y_{NE}} \\ : & : & : & : \\ \frac{\partial \dot{y}_{NE}}{\partial y_1} & \frac{\partial \dot{y}_{NE}}{\partial y_2} & .. & \frac{\partial \dot{y}_{NE}}{\partial y_{NE}} \end{bmatrix} \right)^{-1} \begin{bmatrix} \dot{y}_{k,1} \\ \dot{y}_{k,2} \\ .. \\ \dot{y}_{k,NE} \end{bmatrix}
$$

Based on the implicit Euler scheme, stability can be obtained. The difficulty of the implicit method is in the evaluation of the Jacobian matrix. Calculation of the inverse of the matrix takes significant computational time. However, the step size can be big and convergence is guaranteed.

b. Numerical evaluation of the Jacobian matrix

In attempting to integrate the equation sets implicitly using Eq. (7.14), one of the key problems is developing a method to calculate the Jacobian matrix accurately and efficiently. The partial derivatives $\partial \dot{y}_i / \partial y_j$ are defined as derivatives of a function of multiple variables when all but the variable of interest are held constant during the differentiation (Abramowitz and Stegun, 1972). An analytical Jacobian matrix is difficult to obtain for a set of general unified constitutive equations, which constrains the use of implicit numerical methods in the integration.

The numerical Jacobian matrix extends the ideas of using numerical methods to efficiently generate partial derivatives for NE number of equations at the current kth iteration. For example, the partial derivatives in the Jacobian matrix can be calculated using (Abramowitz and Stegun, 1972)

$$\frac{\partial \dot{y}_i}{\partial y_j} = \lim\nolimits_{\Delta h_i \to 0} \frac{\dot{y}_i\big|_{y_i + \Delta h_i} - \dot{y}_i\big|_{y_i}}{\Delta h_i}, \tag{7.15}$$

where Δh_i is a fraction of the current increment Δy_i for the ith equation and is defined as:

$$\Delta h_i = \alpha \cdot \Delta y_i \tag{7.16}$$

where $\alpha (\approx 0 \sim 1)$ is a factor. Theoretically, $\partial \dot{y}_i / \partial y_j$ can be calculated easily from Eq. (7.15) as $\Delta h_i \to 0$ but in practice, it is difficult to calculate for multiple complex equations, where $\dot{y}_i\big|_{y_i + \Delta h_i}$ cannot be simply calculated from $\dot{y}_i(y_1, \ldots, y_i + \Delta h_i, \ldots, y_{NE})$ since a small increment for variable y_i affects the values of other variables within a complex equation set (Lin, Cao and Balint, 2012).

A flow chart for numerically calculating the Jacobian matrix is shown in **Fig. 7.10**. It describes the overall structure for defining the Jacobian matrix over one increment of the integration. During the calculation process, particular attention needs to be paid to the value of Δh_i. When $\Delta y_i = 0$, $\dot{y}\big|_{y_i + \Delta h_i} = \dot{y}\big|_{y_i}$, so the partial derivatives $\partial \dot{y}_i / \partial y_j = 0$. The key points to calculating $\partial \dot{y}_i / \partial y_j$ are: (i) the determination of the value α and (ii) the calculation of each partial derivative for every iteration within an equation set. The sensitivity studies for the α values are carried out in Section 7.3. It is recommended that $\alpha = 0.1$ for most applications (Cao, Lin and Dean, 2008a).

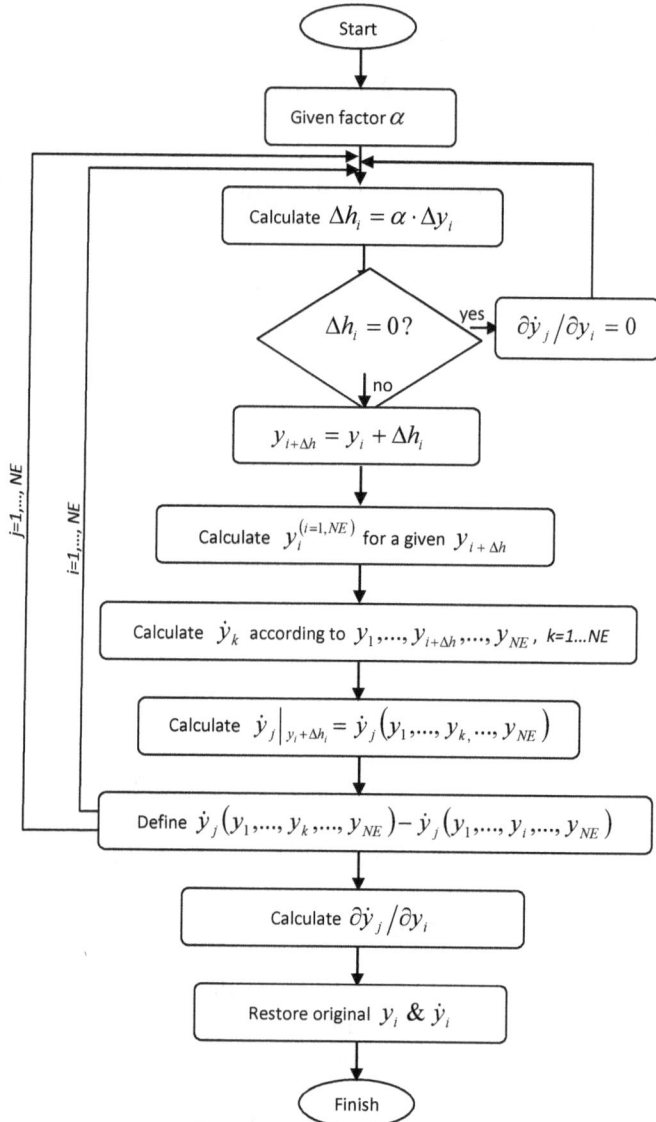

Fig. 7.10 Flowchart showing the algorithm of numerically calculating the Jacobian matrix for an increment of the integration (Cao, Lin and Dean, 2008a).

7.3 Error Analysis and Step-size Control Methods

7.3.1 *Error and step-size control*

Truncation errors defined in numerical integration methods can be used to control the step size. For example, we can define the truncation errors for the first order Euler method for the ith variables as

$$O_i\left(\Delta t_k^2\right) \leq Tol_i.$$ (7.17)

So, the error should be less than the specified tolerance Tol_i for the equation.

a. *Use of truncation error*

If the truncation error is used, there is

$$O_{k,i}\left(\Delta t_k^2\right) = \frac{1}{2}(\Delta t_k)^2 \ddot{y}_{k,i} = Tol_i.$$

The step size of the ith equation (variable) for the next iteration $(k+1)$ can be estimated from the current error and the tolerance specified.

$$\Delta t_{k+1,i} = \sqrt{\frac{2Tol_i}{\ddot{y}_{k,i}}},$$

and the step size for the next iteration can be estimated from

$$\Delta t_{k+1} = Min\left\{\Delta t_{k+1,i}\right\}, \text{ where } i = 1,2,...NE .$$ (7.18)

This method can be expanded to higher order truncation errors. It should be noted that the step size for the next iteration is estimated from the current iteration of the integration. After integration the error should be checked; if it exceeds the specified truncation error, then the integration for this iteration should be rejected and the step size should be reduced for the next trial, until the error is within the specified tolerance for all the equations.

b. *Variation of strain rate*

The step size can also be controlled, such that the difference between two iterations is maintained within a certain tolerance. That is,

$$\left|\dot{y}_{k+1,i} - \dot{y}_{k,i}\right|.$$ (7.19)

This is a very simple and effective method. There are many other methods of controlling integration errors and thus the step size, but these were mainly developed for solving general mathematical problems in numerical integration. The ODE-type unified viscoplastic constitutive equations contain particular features and so a new method has been developed particularly for this type of problem by Cao, Lin and Dean (2008a), which will be introduced later in this section.

7.3.2 Unit problems in unified constitutive equations

When dealing with the unified viscoplastic/creep constitutive equations, the inconsistency in the units of the individual equations needs to be taken into account. **Table 7.1** shows the values of the differential equations $\dot{y}_i$ and truncation errors $O_i(\Delta t_k^2)$ calculated for equation SETs I and II listed in Appendix A. The two SETs of equations are integrated using the first-order implicit method with the same number of iterations ($N = 10,000$). Equation SET I (see Appendix A) contains three differential equations with different units (**Table 7.1**), which are integrated for $\dot{\varepsilon}^T = 0.1$ s^{-1}. The rates for equations (I.1), (I.2) and (I.3) and the truncation errors $O_i(\Delta t_k^2)$ for the equations are given for $t = 1.26 \times 10^{-3}$ s. At this time value, the maximum difference is 1×10^4 for rates $\dot{y}_i$ and 1.3×10^5 for the truncation errors $O_i(\Delta t_k^2)$. It can be seen from **Table 7.1** that the unit of (I.3), MPa s^{-1}, is different from those in (I.1) and (I.2), s^{-1}. Even if the units of (I.1) and (I.2) are the same, the values are significantly different; the maximum value of the dislocation density rate is 10 times that of the plastic strain rate.

Table 7.1 Comparison of the rates and truncation errors between equations within equation SETs I & II listed in Appendix A. L = I and II represents the equation sets (Cao, 2006).

		(L.1)	(L.2)	(L.3)	(L.4)	Max/Min
SET I	$\dot{y}_i\big|_{t=1.26e-3s}$	9.2e-2 (s^{-1})	0.95 (s^{-1})	954.2 (MPa s^{-1})		1.0e4
	$O_i(\Delta t^2)\big|_{t=1.26e-3s}$	4.2e-9	3.2e-8	5.3e-4 (MPa)		1.3e5
SET II	$\dot{y}_i\big|_{t=2.7h}$	2.2e-4 (h^{-1})	1.1e-2 (h^{-1})	6.1e-5 (h^{-1})	6.2e-4 (h^{-1})	188.1
	$O_i(\Delta t^2)\big|_{t=2.7h}$	2.6e-5	2.5e-3	5.2e-9	7.1e-5	4.5e5

Similar investigations are carried out for equation SET II, which is integrated for $\sigma = 241.3$ MPa. The results are listed in **Table 7.1** for comparison. Huge differences between the equations for an equation SET make it impossible to choose a suitable value of tolerance for the investigation. It can be seen from the table that different rate equations within an equation set have different units, hence the truncation errors are also approximated with different units. Therefore, the error values are difficult to compare because the units of the variables have different scales.

Figure 7.11 shows the different unit scales of natural truncation errors for the equations in SET I with $\dot{\varepsilon}^T = 0.1$ s^{-1}. The three equations have been integrated using the first-order implicit method introduced above with the number of iterations set to 10,000 and the natural truncation errors estimated using Eq. (7.9) for the individual equations. Due to the different unit scales, the error in (I.3) is much higher than those in (I.1) and (I.2), and such differences are almost unchanged during the whole integration process. This indicates that the error in the third equation controls the step size in the error estimation for the entire integration process, if a single tolerance is used. To obtain similar errors, the difference in tolerances specified for Eqs. (I.1) and (I.3) should be about 10^6.

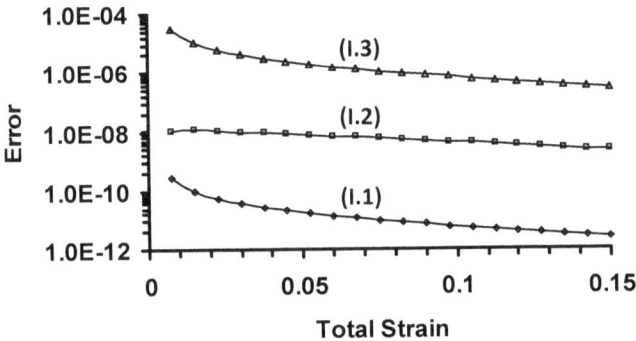

Fig. 7.11 Comparison of natural truncation errors of the integration for equation SET I listed in Appendix A (Cao, Lin and Dean, 2008a).

The inconsistency in units between equations causes problems, as the unified constitutive equations are numerically integrated simultaneously. It is impossible to define a single tolerance that is suitable for all the equations within a set. To obtain similar values of the truncation error requires different tolerances, Tol_i, to be defined for individual equations, which is very difficult. For a simple set of equations, these tolerances can be defined on the basis of experience. For more complex equation sets, such as SET III in Appendix A, however, it is very difficult to define an acceptable tolerance, Tol_i, for every equation that controls the integration accuracy.

7.3.3 Unitless error assessment method

It is due to the problem of inconsistent units that an adaptation of the truncation error needs to be applied such that the unit values associated with the integrated variables can be transferred into unitless values. This can be achieved by introducing a normalisation method. The normalised truncation error $\overline{O}_i(\Delta t_k^2)$, can be defined as

$$\overline{O}_i(\Delta t_k^2) = \frac{O_i(\Delta t_k^2)}{|\Delta y_{k,i}|}, \qquad (7.20)$$

where the truncation error $O_i(\Delta t_k^2)$ for the ith equation of the kth iteration is estimated from the difference between the higher and lower order approximations. $\Delta y_{k,i}$ is the increment of the integrated values for the ith equation. This normalisation method has the advantage of defining a tolerance Tol that is independent of the magnitude and unit of the solution to each equation.

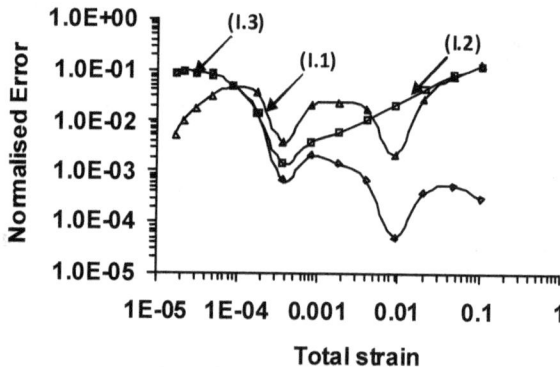

Fig. 7.12 Comparison of normalised truncation errors of the integration for equation SET I listed in Appendix A (Cao, Lin and Dean, 2008a).

Figure 7.12 shows the normalised errors for Eqs. (I.1), (I.2) and (I.3) in equation SET I listed in Appendix A, with the same number of iterations (N = 10,000). During the integration process, Eqs. (I.3) and (I.2) have the highest error for low strains. At higher strains, the Eq. (I.2) error becomes the highest and remains so to the end. This indicates that the step size of the integration could be controlled by different equations within a set if a single normalised tolerance is used.

Figure 7.11 showed that the step size of the integration is always controlled by Eq. (I.3), if the natural error estimation method and a single tolerance are used. By using normalisation, the truncation errors are transferred to unitless values, and the difference between the normalised errors obtained from the individual equations is reduced significantly with good features of the error variables over the whole integration history. This provides a better chance of controlling the step size using a single specified tolerance, making the error assessment much easier.

The step size for numerical integration can therefore be controlled easily by the use of the normalised truncation error, and a single tolerance, *Tol*, can be defined for an integration process.

7.4 A Case Study for Implicit Numerical Integration

An entire process for numerically integrating a set of unified constitutive equations will be demonstrated in this section. Equation SET I listed in Appendix A is used.

7.4.1 *Implicit integration method*

The first-order implicit integration equation is given as

$$y_{k+1,i}^{(1)} = y_{k,i} + \Delta t_{k+1}\left[1 - \Delta t_{k+1}\frac{\partial \dot{y}_i}{\partial y_j}\right]^{-1}\dot{y}_{k,i}, \tag{7.21}$$

where $y_{k+1,i}^{(1)}$ represents the variable integrated from the ith ordinary differential equation $\dot{y}_i$ using the first-order implicit Euler method. Higher accuracy can be achieved by averaging explicit and implicit Euler methods according to the implicit trapezoid rule (Cao, Lin and Dean, 2008a), given by

$$y_{k+1,i}^{(2)} = y_{k,i} + \frac{1}{2}\Delta t_{k+1}\left(\dot{y}_{k,i} + \dot{y}_{k+1,i}\right). \tag{7.22}$$

This method uses the average of the derivatives of $\dot{y}_{k,i}$ and $\dot{y}_{k+1,i}$ to approximate $y_{k,i}$. The derivatives $\dot{y}_{k+1,i}$ are unknown, and can be obtained by linearization using Newton's method from Eq. (7.13). Then the second-order implicit integration, using the trapezoid rule, can be represented by

$$y_{k+1,i}^{(2)} = y_{k,i} + \Delta t_{k+1} \left[1 - \frac{1}{2}\Delta t_{k+1} \frac{\partial \dot{y}_i}{\partial y_j} \right]^{-1} \dot{y}_{k,i}. \tag{7.23}$$

This method has a second-order truncation error through a Taylor series expansion, thus it is more accurate than Eq. (7.21). Both implicit integration methods have good convergence features, but their disadvantage is that in each step the Jacobian matrix $\partial \dot{y}_i/\partial y_j$ must be inverted to find y_{k+1}. The detailed analysis about the approximation of the Jacobian matrix will be discussed later.

7.4.2 *Normalised error estimation and step-size control*

a. Error estimation

The normalised truncation error $\overline{O}_i(\Delta t_k^2)$ can be defined as

$$\overline{O}_i\left(\Delta t_k^2\right) = \frac{O_i\left(\Delta t_k^2\right)}{\left|\Delta y_{k,i}\right|}, \tag{7.24}$$

where the truncation error $O_i(\Delta t_k^2)$ for the ith equation of the kth iteration is estimated from the difference between the higher and lower order approximations defined in Eqs. (7.21) and (7.22):

$$O_i\left(\Delta t_k^2\right) = \left| \frac{1}{2}\Delta t_k \left(\dot{y}_{k-1,i}^{(2)} + \dot{y}_{k,i}^{(2)}\right) - \Delta t_k \dot{y}_{k,i}^{(1)} \right|$$

and

$$\left|\Delta y_{k,i}\right| = y_{k,i}^{(2)} - y_{k-1,i}^{(2)}.$$

$\Delta y_{k,i}$ is the increment in the integrated values for the ith equation using the second-order implicit trapezoid rule (Eq. (7.23). This normalisation method has the advantage of defining a single tolerance *Tol* that is independent of the magnitude of the solution to each equation.

b. Step-size control

The error can be estimated accurately from the difference between Eqs. (7.23) and (7.21) according to the definition

$$\overline{O}_i\left(\Delta t_k^2\right) = \frac{\left|y_{k,i}^{(2)} - y_{k,i}^{(1)}\right|}{\left|y_{k,i}^{(2)} - y_{k-1,i}^{(2)}\right|}.$$

The normalised error can then be extrapolated using

$$\overline{O}_i\left(\Delta t_{k+1}^2\right) = 1 - \left[1 - \Delta t_{k+1}\frac{\partial \dot{y}_j}{\partial y_i}\right]^{-1} \Bigg/ \left[1 - \frac{1}{2}\Delta t_{k+1}\frac{\partial \dot{y}_j}{\partial y_i}\right]^{-1}. \qquad (7.25)$$

Δt cannot be calculated for the next step directly due to the inversion of the Jacobian matrix. Thus the extrapolation method must be used and at each step, the matrix has to be inverted, which requires significant computational time. To solve this problem, the step size is calculated using an approximate method by linearizing the equations (Faruque, Zaman *et al.*, 1996) as in Newton's method, in which all the unknown values $\dot{y}_{k,i}$ are eliminated; the normalised error from Eq. (7.25) for the $(k+1)$th iteration can then be estimated by

$$\overline{O}_i\left(\Delta t_{k+1}^2\right) = \frac{\Delta t_{k+1} \cdot \left|\left(\dot{y}_{k,i}^{(2)} - \dot{y}_{k,i}^{(1)}\right) + \frac{\partial \dot{y}_j}{\partial y_i}\left(\frac{1}{2}\Delta y_{k+1,i}^{(2)} - \Delta y_{k+1,i}^{(1)}\right)\right|}{\left|\Delta y_{k+1,i}^{(2)}\right|}, \qquad (7.26)$$

where $\overline{O}_i\left(\Delta t_{k+1}^2\right)$ is the predicted normalised truncation error for the *i*th equation at the $(k+1)$th iteration. $\Delta y_{k+1,i}$ is the increment of the integrated value, however $y_{k+1,i}$ is unknown at the $(k+1)$th iteration; by assuming the same increment as in the previous integration point, i.e. $\Delta y_{k+1} \approx \Delta y_k$, the error can be approximated. Thus, the normalised error can be approximated by the percentage error in each iteration. Physically, it is an estimate of the relative error of the integrated value. Practically, this relative error is unitless and can be compared directly with the errors calculated from the other rate equations within a set.

By using the normalised truncation error, the different units of error for the individual rate equations have been eliminated. This gives a better opportunity to use a single tolerance to control accuracy. Rearranging Eq. (7.26) and replacing the normalised truncation error $\overline{O}_i\left(\Delta t_{k+1}^2\right)$ with a specified normalised tolerance *Tol*, the following equation can be obtained:

$$\Delta t_{k+1,i} = \frac{Tol \cdot \left| \Delta y_{k,i}^{(2)} \right|}{\left| \left(\Delta \dot{y}_{k-1,i}^{(2)} - \Delta \dot{y}_{k-1,i}^{(1)} \right) + \frac{\partial \dot{y}_j}{\partial y_i} \left(\frac{1}{2} \Delta y_{k,i}^{(2)} - \Delta y_{k,i}^{(1)} \right) \right|}, \quad i = 1, ..., NE \qquad (7.27)$$

$$\Delta t_{k+1} = Min\{\Delta t_{k+1,i}\},$$

where Δt_{k+1} is the estimated step size for the next iteration calculated from the individual equations. The minimum value of $\Delta t_{k+1,i}$ is used to estimate the step size for the next iteration. In practice, if the normalised error is within the defined tolerance, Tol, then Δt_{k+1} is used for the next increment. However, if the normalised error is larger than Tol, the step size is decreased further until it is acceptable. The step size for the next trial can be estimated using (Omerspahic and Mattiasson, 2007)

$$\Delta t_{k+1}^{new} = \beta \cdot \Delta t_{k+1} \cdot \left| \frac{Tol}{\overline{O}\left(\Delta t_{k+1}^2 \right)} \right|^{1/q}, \qquad (7.28)$$

where β is a safety factor which is a few percent smaller than unity in practice, and q is related to the order of the integration method. In this example, $\beta = 0.8$ and $q = 3$ were chosen. The normalisation technique introduced here solves the problem of inconsistent units by enabling the errors calculated from integrating the unified constitutive equations to have unitless values. This scheme aims to keep the error estimation close to the user specified tolerance Tol. The step is rejected when $Tol < \overline{O}_i(\Delta t_{k+1}^2)$ and a new attempt is made with a smaller step size Δt_{k+1}^{new}.

7.4.3 Jacobian matrix and computational efficiency

The numerical procedures for the estimation of the Jacobian matrix $\partial \dot{y}_i / \partial y_j$ have been discussed and detailed in **Fig. 7.10**. These are based on Eqs. (7.15) and (7.16), shown below:

$$\frac{\partial \dot{y}_i}{\partial y_j} = \lim_{\Delta h_i \to 0} \frac{\dot{y}_i|_{y_i + \Delta h_i} - \dot{y}_i|_{y_i}}{\Delta h_i}$$

$$\Delta h_i = \alpha \cdot \Delta y_i,$$

where Δh_i is a small increment in the variable y_i.

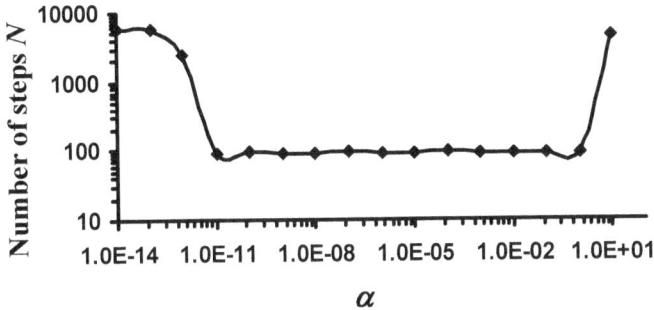

Fig. 7.13 The effects of α values on the number of steps used to numerically integrate equation SET I. (Cao, Lin and Dean, 2008).

The value of α has a significant effect on the accuracy of the partial derivatives $\partial \dot{y}_i / \partial y_j$ numerically calculated using Eq. (7.15), and thus affects the computational results and efficiency. Sensitivity studies have been carried out for a wide range of α values varying from 10^{-14} to 10 (beyond this range, the integration cannot be performed). The effect of the α value on the number of iterations N required to integrate equation SET I is shown in **Fig. 7.13**; the novel step-size control is applied and the tolerance *Tol* is set to 0.1.

From **Fig. 7.13**, it can be seen that the number of iterations required for the numerical integration is acceptable and almost unchanged as the value of α varies between 10^{-12} and 1.0. Such a large range of α values indicates that it is likely that a suitable value can be chosen. Hence the numerical partial derivatives will be acceptable when using a value of α within the above range. But if $\alpha < 10^{-12}$, then the errors increase significantly, and a huge number of iterations are required to achieve the specified tolerance. The reason is that when Δh_i reaches a certain small value, the ratio between the two small quantities in Eq. (7.15) may cause numerical instability.

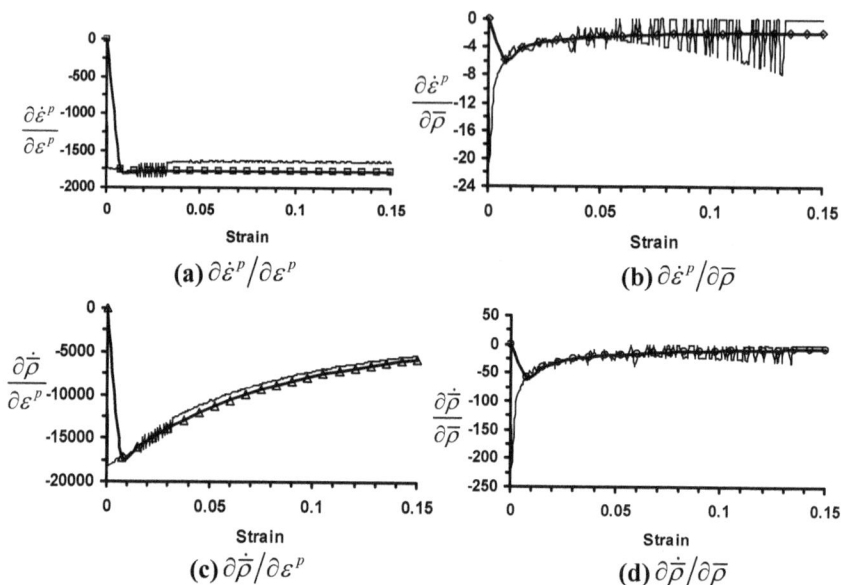

(a) $\partial \dot{\varepsilon}^p / \partial \varepsilon^p$

(b) $\partial \dot{\varepsilon}^p / \partial \bar{\rho}$

(c) $\partial \dot{\bar{\rho}} / \partial \varepsilon^p$

(d) $\partial \dot{\bar{\rho}} / \partial \bar{\rho}$

Fig. 7.14 Comparison of the partial derivatives calculated using analytical (thick curves) and numerical (thin curves $\alpha = 10^{-12}$ and symbols $\alpha = 0.1$) methods for SET I with $\dot{\varepsilon} = 0.1 \text{ s}^{-1}$ (Cao, Lin and Dean, 2008a).

The differences in partial derivatives calculated analytically and numerically using $\alpha = 10^{-12}$ and 0.1 are shown in **Fig. 7.14**. The thick curves represent the values of the partial derivatives, $\partial \dot{\varepsilon}^p / \partial \varepsilon^p$, $\partial \dot{\varepsilon}^p / \partial \bar{\rho}$, $\partial \dot{\bar{\rho}} / \partial \varepsilon^p$ and $\partial \dot{\bar{\rho}} / \partial \bar{\rho}$, in the numerical integration of equation SET I with a strain rate of 0.1, obtained using the analytical Jacobian matrix. The corresponding oscillating thin curves and symbols represent the partial derivatives obtained using the numerically calculated Jacobian matrix, with $\alpha = 10^{-12}$ and 0.1. The figure shows that the partial derivatives become unstable and that the calculation tends to failure as the value of α becomes less than 10^{-12}; thus the accuracy and efficiency of the numerical integrations is affected significantly. If $\alpha > 1.0$, however, then the increment becomes too large and unacceptable, and the error increases dramatically.

(a) $\partial\dot{\varepsilon}^p/\partial\varepsilon^p$

(b) $\partial\dot{\bar{\rho}}/\partial\varepsilon^p$

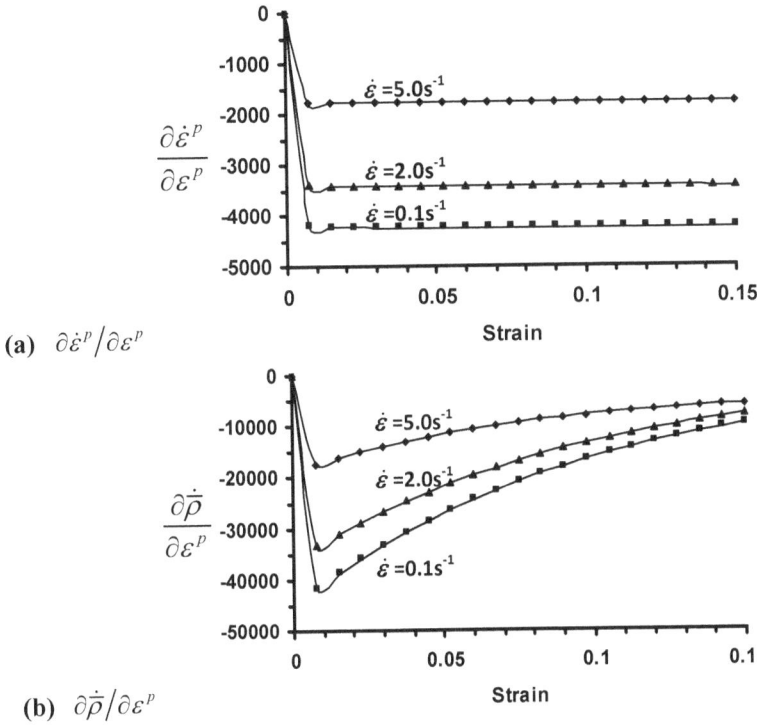

Fig. 7.15 Comparison of the partial derivatives calculated using analytical (solid lines) and numerical (symbols) ($\alpha = 0.1$) methods for SET I equations at different strain rates (Cao, Lin and Dean, 2008).

Investigations have also been carried out using analytical and numerical Jacobian Matrices to solve equation SET I at different stain rates. **Figure 7.15** represents the values of two partial derivatives $\partial\dot{\varepsilon}^p/\partial\varepsilon^p$ and $\partial\dot{\bar{\rho}}/\partial\varepsilon^p$ in the numerical integration of equation SET I with $\dot{\varepsilon} = 0.1\ \mathrm{s}^{-1}$, $2.0\ \mathrm{s}^{-1}$ and $5.0\ \mathrm{s}^{-1}$. A comparison is made between the values of the partial derivatives calculated using the analytical (solid curves) and numerical (symbols) Jacobian matrix. Almost identical results have been obtained for the values of the two partial derivatives throughout the integration history; the same was also achieved for the other two partial derivatives $\partial\dot{\varepsilon}^p/\partial\bar{\rho}$ and $\partial\dot{\bar{\rho}}/\partial\bar{\rho}$ (which are not presented here). This indicates that the Jacobian matrix can be estimated accurately using numerical methods, for this type of constitutive equations.

Figure 7.16 shows a comparison between stress–strain curves obtained with the implicit Euler method, using analytical (solid curves) and numerical (symbols) Jacobian matrices. The equation SET I, which contains three differential equations with different units, is integrated for strain rates of $\dot{\varepsilon} = 0.1\ \text{s}^{-1}$, $2.0\ \text{s}^{-1}$ and $5.0\ \text{s}^{-1}$. For the numerical calculation, the value of $\alpha = 0.1$ is chosen. It can be seen from **Fig. 7.16**, that almost identical results are obtained using both analytical and numerical Jacobian matrices. Sensitivity studies on the α value have been conducted by Cao, Lin and Dean (2008a) for other more complicated unified constitutive equations. From the results of these studies, $\alpha = 0.1$ was recommended for obtaining good results when solving general creep/viscoplastic constitutive equations.

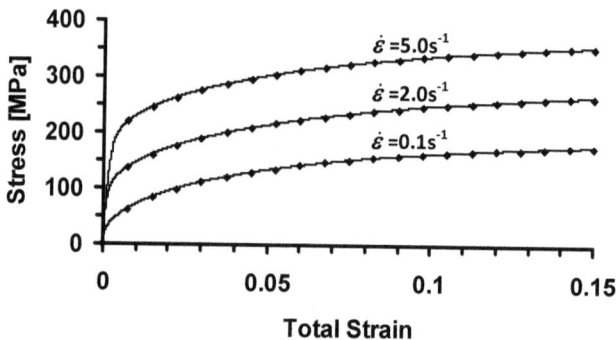

Fig. 7.16 Comparison of Stress–strain curves obtained using analytical (curves) and numerical (symbols) Jacobian matrices for equation SET I with strain rates $\dot{\varepsilon} = 0.1\ \text{s}^{-1}$, $2.0\ \text{s}^{-1}$ and $5.0\ \text{s}^{-1}$ (Cao, Lin and Dean, 2008a).

The natural, O, and normalised, $\overline{O}$ accumulated errors for the integration can be defined as

$$O = \sum_{k=1}^{N} O_{k,i}\left(\Delta t_k^2\right)$$

$$\overline{O} = \sum_{k=1}^{N} \overline{O}_{k,i}\left(\Delta t_k^2\right),$$

where N is the number of iterations for the integration. The natural (O) and normalised ($\overline{O}$) accumulated errors and the number of iterations used to integrate the individual curves are shown in **Table 7.2** for equation SET I. The normalised accumulated error ($\overline{O}$) for the individual differential equations, Eqs. (I.1), (I.2) and (I.3), are of similar magnitudes. Conversely, the accumulated natural errors (O) have hugely different magnitudes. The ratios between the maximum and minimum accumulated errors are significantly lower for the normalised errors than for the natural errors. Using the step-size control techniques outlined previously, the equations can be integrated with 38 iterations.

Similar features are shown for equation SET II, the unified creep damage constitutive equations listed in Appendix A. More iterations are used for equation SET II, since the strain rate is higher, and step size needs to be smaller, hence there are more iterations before the failure of the material.

Table 7.2 Comparison of natural (O) and normalised ($\overline{O}$) accumulated errors when integrating equation SETs I and II. L = I and II represents the equation sets (Cao, 2006)

		(L.1)	(L.2)	(L.3)	(L.4)	Max/Min
SET I	O	3.0e-4	4.9e-2	52.88 (MPa)		1.8e5
$N = 38$	$\overline{O}$	4.5	5.4	5.9		1.3
SET II	O	1.7e-2	2.0e-2	6.3e-8	4.7e-2	7.5e5
$N = 114$	$\overline{O}$	6.4	36.9	3.0	6.4	12.3

7.4.4 *Comments on solving ODE-type unified constitutive equations*

a. Difficulties in integration

The main difficulties encountered when numerically integrating ODE-type constitutive equations are:

(i) The equations are very stiff. A very small step size should be used if explicit (forward) integration methods are used. This is very time-consuming.

(ii) Implicit (backward) integration methods are recommended for solving the ODE-type constitutive equations, which is efficient. The main problem with implicit methods is the evaluation of the

Jacobian matrix. Numerical methods should be used to approximate the components of the Jacobian matrix.

(iii) Different units for different variables within a set of equations are one of the most difficult aspects to deal with in the control of the integration accuracy. This is different from the ODE type of equations discussed in mathematics textbooks, and special methods need to be developed to deal with this.

(iv) The different units cause the difficulties in error assessment (errors calculated from individual equations cannot be compared directly). Thus it is difficult to create an algorithm for time step-size control.

b. Error and step-size control

If the error can be controlled, an automatic step-size control algorithm can be implemented in the integration. To directly compare the errors obtained from the integration of the individual equations, normalisation methods should be used, such that the errors become unitless. In this way, a single tolerance value can be defined for integrating a set of equations, and the step size can be controlled effectively.

c. Efficiency analysis

Low-order implicit (backward) integration methods, such as the first-order implicit Euler method, with normalised error control, are recommend for integrating unified constitutive equations. With automatic step-size control, the equations can be integrated with many fewer increments. This is extremely important for FE simulations, since the equations need to be solved for every integration point of a work-piece mesh, for every increment. Thus, maximising the step size could reduce the number of increments required for solving a process modelling problem, and consequently reduce the computational time significantly.

7.5 Formulation of Objective Functions for Optimisation

As mentioned above, it is important to determine the material constants within a set of constitutive equations from experimental data (Lin *et al.*, 2002). The current technique for determining the constitutive equations is to minimise errors between the experimental and computed data using an optimisation method. The errors can be expressed in an equation $f(x)$ which is known as the *objective function*, or, *fitness function*.

Many optimisation methods, such as gradient-based (Kowalewski *et al.*, 1994) and evolutionary algorithm (EA)-based (Lin and Yang, 1999) methods have been developed for determining constants within constitutive equations from experimental data, and various objective functions have been formulated to assess the errors between the experimental and computed data (Li, Lin and Yao, 2002; Cao and Lin, 2008). The objective functions should be able to efficiently "guide" the optimisation process, to find the best fit to the experimental data. An ideal objective function should have the following properties:

- **Criterion 1** For *a single curve* – all the experimental data points on the curve should be involved in the optimisation, and have an equal opportunity to be optimised, if the experimental data errors have been eliminated previously.

- **Criterion 2** For *multi-curves* – all experimental curves should have an equal opportunity to be optimised. The emphasis of the fitting process should not depend on the number of data points in each experimental curve.

- **Criterion 3** For *multi-sub-objectives* – an objective function should be able to deal with multi-sub-objective problems, in which the units of sub-objectives may be different and all the sub-objectives should have an equal opportunity to be optimised. Different units and/or the number of curves in each sub-objective should not affect the overall performance of the fitting.

- **Criterion 4** *Weighting factors* – the above criteria should be achieved automatically without choosing weighting factors manually, since they are difficult to select in practice.

The basic least squares method for the formulation of objective functions has been given in Eq. (7.2). In this section, specific methods developed for the solution of unified constitutive equations are introduced.

7.5.1 *Specific features of objective functions for materials modelling*

For stress–strain data, objective functions are normally formulated based on minimising the sum of the squares of the differences between computed and experimental data at the same strain level, or residuals.

For example, to determine the material constants in a set of viscoplastic damage constitutive equations (Equation SET III in Appendix A) from stress–strain data, an objective function can be written as

$$f(x) = \sum_{j=1}^{M} \sum_{i=1}^{N_j} w_{ij} \left[\sigma_{ij}^c - \sigma_{ij}^e \right]^2 \qquad (7.29)$$

where $f(x)$ is the sum of the squares of the residuals for stress, which is the objective function for the optimisation. The parameter x ($x = \{x_1, x_2, \ldots, x_j, \ldots, x_{NC},\}$) represents the constants x_j ($j = 1, 2, \ldots, NC$), where NC is the number of constants to be determined (optimised) from the stress–strain experimental data. σ_{ij}^c and σ_{ij}^e are the computational and experimental stresses for the same strain level i of the experimental data of curve j, w_{ij} is a weighting function, and N_j is the number of experimental stress–strain data points for stress–strain curve j, as shown in **Fig. 7.17(a)**. M is the number of experimental curves considered in the optimisation. For this example, only one experimental curve is considered, i.e. $M = 1$, $N_1 = 15$ and $w_{ij} = 1.0$. For a general elastic viscoplastic problem, a group of stress–strain curves will be considered for different temperatures and strain rates.

Figure 7.17(a) shows the experimental curve (symbols) and four curves derived from the equation set (Equation SET III in Appendix A) using different sets of material constants. For the viscoplastic damage material model, strains to failure are implicitly characterised by some of the material parameters, mentioned by Lin, Foster, Liu *et al.* (2007). Four computed stress–strain curves are given in **Fig. 7.17 (a)**, and it can be seen that the computed curve 1 has a poor fit with its low strain to failure. Curve 2 fits best to the experimental data in general. Curve 3 fits quite closely within the experimental data, but beyond that, the curve has a much higher strain to failure. Curve 4 is furthest away from the experimental data with the highest strain to failure.

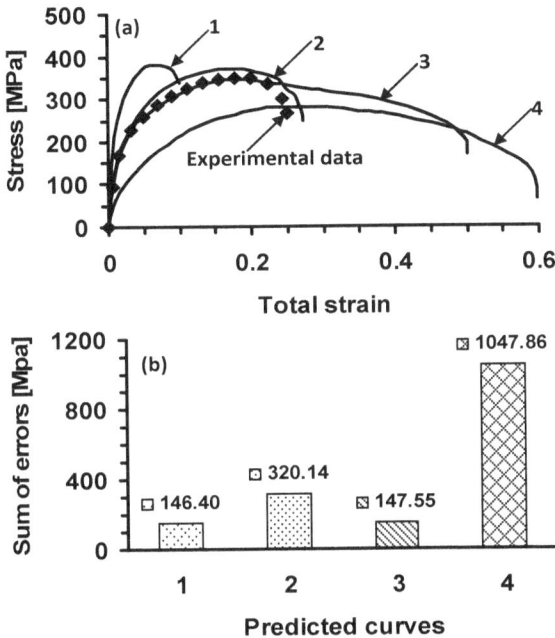

Fig. 7.17 (a) Comparison of experimental (symbols) and computed
(curves 1, 2, 3 and 4) stress–strain curves. (b) Sum of the square of
residuals in stress. (Cao and Lin 2008).

Figure 7.17(b) shows the corresponding sum of the square of the
residuals (stress differences at various strains) using Eq. (7.29) for the
four computed curves in **Fig. 7.17 (a)**. From **Fig. 7.17 (b)**, it can be seen
that the computed curve 4 has the largest sum and curve 1 the smallest. It
is obvious from **Fig. 7.17 (a)** that curve 2 fits best to the experimental
data. However, the sum of the residuals is still high. It is obvious that if
the predicted strain to failure is less than that of the experimental data,
such as for the computed curve 1 in **Fig. 7.17 (a)**, some experimental
data is not included in the error estimation. If the computed strain to
failure is higher than that of the experimental data, such as for computed
curves 3 and 4 in **Fig. 7.17 (a)**, the result of the fitting can also be
misleading.

If the calculation of the sum of the residuals is done in both directions
(strain and stress), then another situation arises: that of different unit
scales, as shown in **Fig. 7.17 (a)**, for the horizontal and vertical axes. The
above example highlights the difficulties encountered in formulating an

objective function in materials modelling. In the following sub-sections, a few methods for formulating objective functions for the determination of constitutive equations are introduced.

7.5.2 *Shortest distance method with correction (OF-I)*

This method was developed by Li, Lin and Yao (2002) and used to determine a set of creep damage constitutive equations (SET II in Appendix A) proposed by Kowalewski, Hayhurst and Dyson (1994). In this method, errors are defined by the shortest distance between the experimental and the corresponding computed creep curves:

$$r^2 = \Delta\varepsilon^2 + \Delta t^2 ,$$

as shown in **Fig. 7.18**, where $\Delta\varepsilon$ is the difference between the experimental and computed data in creep strain (in %), and Δt in creep time (in hours). To compensate for the different scales for strain and time, two weighting parameters α and β were introduced by Li, Lin and Yao (2002). The error for the ith data point of the jth curve is defined by

$$r_{ij}^2 = \alpha\Delta\varepsilon_{ij}^2 + \beta\Delta t_{ij}^2 = \alpha(\varepsilon_{ij}^c - \varepsilon_{ij}^e)^2 + \beta(t_{ij}^c - t_{ij}^e)^2 .$$

By summing r_{ij}^2 for all the data points of each curve with the consideration of weighting factors, the objective function is expressed as

$$f(x) = \sum_{j=1}^{M}\sum_{i=1}^{N_j} w_{ij} r_{ij}^2 + \sum_{j=1}^{M} W_j (t_{N_j j}^c - t_{N_j j}^e)^2 , \qquad (7.30)$$

where w_{ij} is the relative weight for the ith data point of the jth experimental creep curve. W_j is the relative weight for the curve j. The second term in Eq. (7.30) was introduced by Kowalewski, Hayhurst and Dyson (1994) for determining creep damage constitutive equations, to increase the sensitivity of the creep lifetime on error estimation. In the calculation of the values r_{ij}, two scaling factors have been introduced to keep the unit scales compatible and to increase the sensitivity of the objective function. By choosing the weighting parameters properly, reasonably good fitting has been obtained, as presented by Li, Lin and Yao (2002) for a set of creep damage constitutive equations.

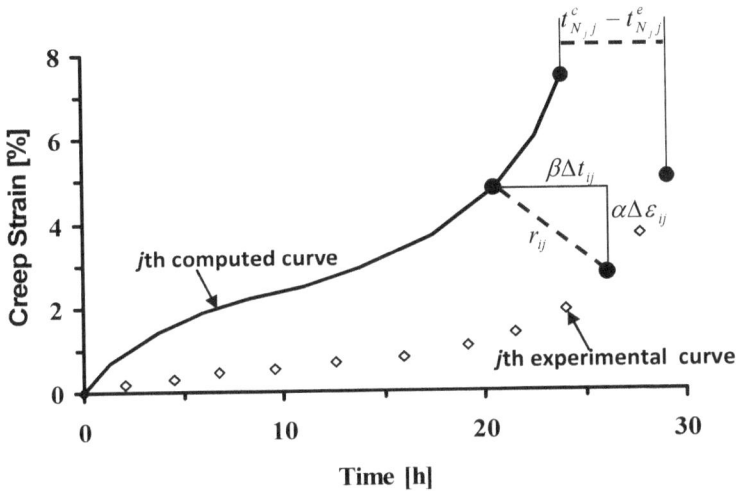

Fig. 7.18 Error definition for shortest distance method with correction term (Li, Lin and Yao, 2002).

The scaling factors α and β can be chosen by the normalisation of the maximum value of each axis, i.e.

$$\alpha = \frac{1}{\varepsilon^e_{N,j}} \quad \text{and} \quad \beta = \frac{1}{t^e_{N,j}},$$

where $\varepsilon^e_{N,j}$ and $t^e_{N,j}$ are the strain and time at failure for the jth creep curve respectively. We can also set them to

$$\alpha = \frac{t^e_{N,j}}{\varepsilon^e_{N,j}} \quad \text{and} \quad \beta = 1.0 \, .$$

Li, Lin and Yao (2002) have chosen $\alpha = 100$ and $\beta = 0.01$ for all the curves, which worked well for the determination of the equations. The introduction of a weighting and various other factors bring a certain flexibility to the objective function, which in turn provides a better chance of obtaining optimised results for the constants within a set of constitutive equations. However, this causes additional difficulty for engineers or other researchers, as they have to choose suitable values for these factors, such that good results can be obtained.

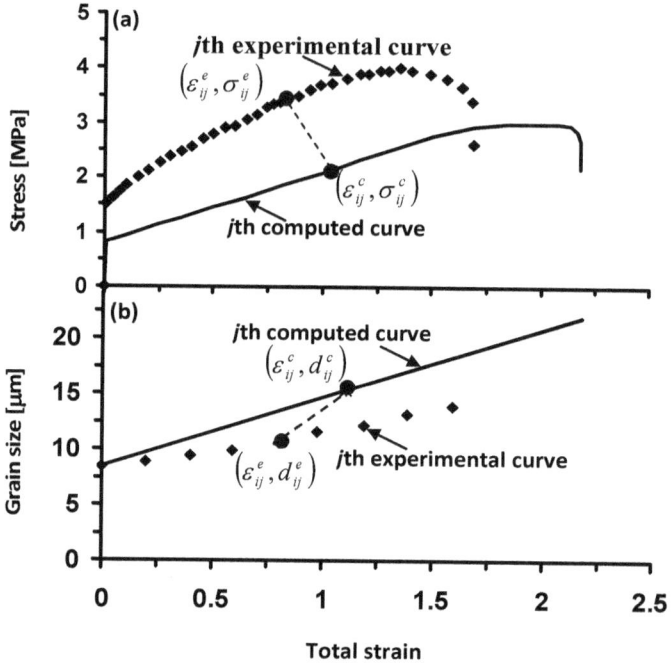

Fig. 7.19 Error definition for (a) stress (MPa) and (b) grain size (μm) (Lin, Cheong and Yao, 2002).

7.5.3 Universal multi-objective function (OF-II)

To reduce the differences between predicted and experimental strains to failure, and the problem of different unit scales, a unitless objective function was introduced by Lin, Cheong and Yao (2002) for the determination of a set of superplastic damage constitutive equations, where multi-objectives are involved. Error definition for the multi-objective problem is shown in **Fig. 7.19** and the objective function takes the form of

$$f(x) = f_\sigma + f_d,$$

$$(7.31)$$

where

$$f_\sigma = \sum_{j=1}^{M} \left\{ \left(\frac{1}{N_j}\right) \sum_{i=1}^{N_j} \left\{ \left[\frac{\sigma_{ij}^e - \sigma^c\left(\varepsilon_{N_j}^c \varepsilon_{ij}^e / \varepsilon_{N_j}^e\right)}{s_{\sigma,ij}} \right]^2 + \left[\frac{\varepsilon_{N_j}^e - \varepsilon_{N_j}^c}{s_j} \right]^2 \right\} \right\}$$

(7.31a)

$$f_d = \sum_{j=1}^{M} \left\{ \left(\frac{1}{N_j}\right) \sum_{i=1}^{N_j} \left\{ \left[\frac{d_{ij}^e - d^c\left(\varepsilon_{N_j}^c \varepsilon_{ij}^e / \varepsilon_{N_j}^e\right)}{s_{d,ij}} \right]^2 + \left[\frac{\varepsilon_{N_j}^e - \varepsilon_{N_j}^c}{s_j} \right]^2 \right\} \right\},$$

(7.31b)

where f_σ is the residual for stress and f_d is the residual for grain size, $s_{\sigma,ij} = 0.1\sigma_{ij}^e$ and $s_{d,ij} = 0.1d_{ij}^e$ for the ith data point of the jth curve and $s_j = 0.1\varepsilon_{N,j}^e$ for the jth curve. In the first terms of Eqs. (7.31a) and (7.31b), computed data is normalised against the corresponding experimental data and errors are transformed into unitless values. This approach enables multiple objectives to be dealt with, in which the residuals can be combined directly to form a multi-objective function without introducing weighting factors.

The second set of terms in Eqs. (7.31a) and (7.31b), which have a similar function to those in Eq. (7.30), are normalised against their largest experimental strain values. This again results in a unitless error count. It ensures that the predicted strains to failure are close to the corresponding experimental ones. The effect is all encompassing, because pushing $\left(\varepsilon_{N,j}^c\right)$ towards $\left(\varepsilon_{N,j}^e\right)$ ultimately means pushing all the predicted points towards their corresponding counterparts in the model. Normalising the sum of errors for all the data points ensures that the two terms have equal importance.

This objective function was successfully used for determining the material parameters arising in a set of superplastic damage constitutive equations, which is expressed by Lin, Cheong and Yao (2002). For all individual experimental data points, it is important to obtain the corresponding computed data accurately. For a pair of experimental and corresponding computed curves, the same function is used to choose the corresponding data σ^c for the ith experimental data point of the jth curve. This ensures that all the experimental data is involved in the optimisation, and that the predicted data will be pushed towards the corresponding experimental data.

7.5.4 *True error definition multi-objective function (OF-III)*

a. *True error definition*

The assessment of residuals between the computed and the corresponding experimental ith data point of the jth curve can be measured by

$$\sigma_{ij}^c / \sigma_{ij}^e , \; \varepsilon_{ij}^c / \varepsilon_{ij}^e , \; t_{ij}^c / t_{ij}^e \; \text{etc.},$$

when dealing with the unified constitutive equations listed in Appendix A, for example. This measure has the advantage of being unitless, which naturally avoids the weighting parameters in OF-I and normalisation factors in OF-II. The ratios of computed data to experimental data provide a measure of relative error. By using a logarithmic scale and squaring the ratios, we could define the error for a pair of data points as

$$E = \left(\ln \frac{\sigma_{ij}^c}{\sigma_{ij}^e} \right)^2 , \; \left(\ln \frac{\varepsilon_{ij}^c}{\varepsilon_{ij}^e} \right)^2 , \; \left(\ln \frac{t_{ij}^c}{t_{ij}^e} \right)^2 \; \text{etc.} \tag{7.32}$$

This is known as the *true error definition*, which is similar to the *true strain definition*. For example, $E = \left[\ln \left(\sigma_{ij}^c / \sigma_{ij}^e \right) \right]^2$ is the "true" error definition for the ith data point of the jth experimental stress–strain curve for stress, where the square is used to increase the sensitivity and to ensure that all errors are positive.

b. *Weighting factors*

An automatic weighting factor for the ith experimental data point of the jth curve is introduced by Cao (2006), and may be expressed as

$$\omega_{ij} = \phi \cdot \varepsilon_{ij}^e \left/ \sum_{j=1}^{M} \sum_{i=1}^{N_j} \varepsilon_{ij}^e \right. , \tag{7.33}$$

where

$$\phi = \sum_{j=1}^{M} N_j$$

is a scaling factor, which is related to the total number of data points, and which increases the sensitivity of the objective functions.

c. True error definition objective function

The same method for finding the corresponding computed and experimental data points defined in OF-II is used here. In consideration of the true error definition and the weighting of the factors defined in Eqs. (7.32) and (7.33), the residuals of the stress–strain curves are defined as

$$
r_{ij}^2 = \omega 1_{ij} \left(\ln \frac{\varepsilon^c \left(\varepsilon_{N_j J}^c, \varepsilon_{ij}^e / \varepsilon_{N_j J}^e \right)}{\varepsilon_{ij}^e} \right)^2 + \omega 2_{ij} \left(\ln \frac{\sigma^c \left(\varepsilon_{N_j J}^c, \varepsilon_{ij}^e / \varepsilon_{N_j J}^e \right)}{\sigma_{ij}^e} \right)^2,
$$

$$(7.34)$$

where $\omega 1_{ij}$ and $\omega 2_{ij}$ are relative weighting factors for ε_{ij}^e and σ_{ij}^e respectively. Based on Eq. (7.33), the weighting factor of the ith experimental data point of the jth curve is calculated using the sum of the strains of each data point. This provides equality in the optimisation of each data point, and the situation of losing features due to the logarithmic scale expression is compensated for. Together with the residual calculation r_{ij}^2, a form of the natural objective function is formulated as follows:

$$
f(x) = \frac{1}{M} \sum_{j=1}^{M} \left\{ \frac{1}{N_j} \sum_{i=1}^{N_j} r_{ij}^2 \right\}. \tag{7.35}
$$

The sum of the residuals is normalised by the number of experimental data points. This ensures that the assessment of residuals is expressed as an average unitless data point. Equation (7.35) essentially provides a natural unitless average error and can be easily used for multi-objective problems. It gives an opportunity to deal with multi-sub-objectives, in which a different number of curves and units may be involved in the optimisation. Together with the automatic weighting factors, it enforces compatibility between each data point, curve and objective, when dealing with multi-objectives. All objectives are equally important and the importance of an objective is not influenced by the amount of data with which it is associated.

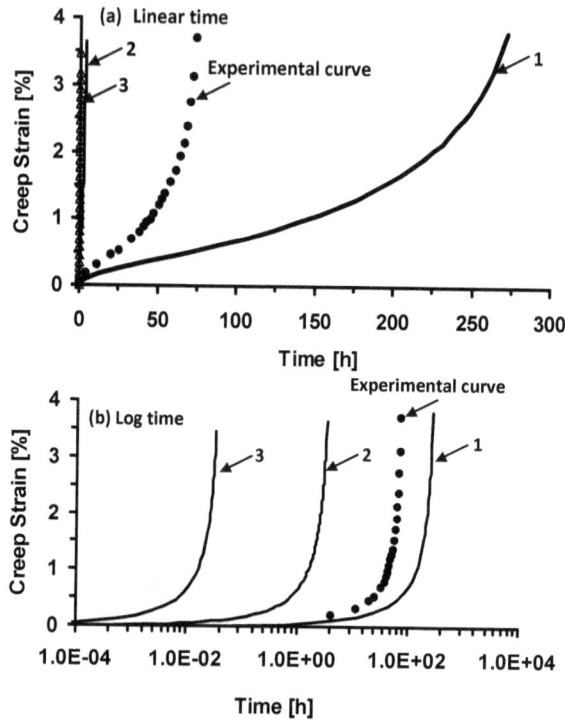

Fig. 7.20 Comparison of experimental and computed (curves 1, 2 and 3) for different values of constant A (10^{-15}, 10^{-13} and 10^{-11}) with linear time (a) and log-time (b) for equation SET-II in Appendix A with stress level of 250 MPa (Cao & Lin, 2008).

7.5.5 *Assessment of the features of the objective function*

The creep damage constitutive equation SET II in Appendix A, for a stress level of 250.0 MPa, is used in the analysis. If five of the material constants listed in **Table A.2** are used and only the constant A is allowed to vary (10^{-15}, 10^{-13} and 10^{-11}), three creep curves are derived as shown in **Fig. 7.20(a)**. It can be seen that computed curve 1 predicts the longest life, which is over three times longer than the experimental one. Computed curves 2 and 3 almost overlap each other.

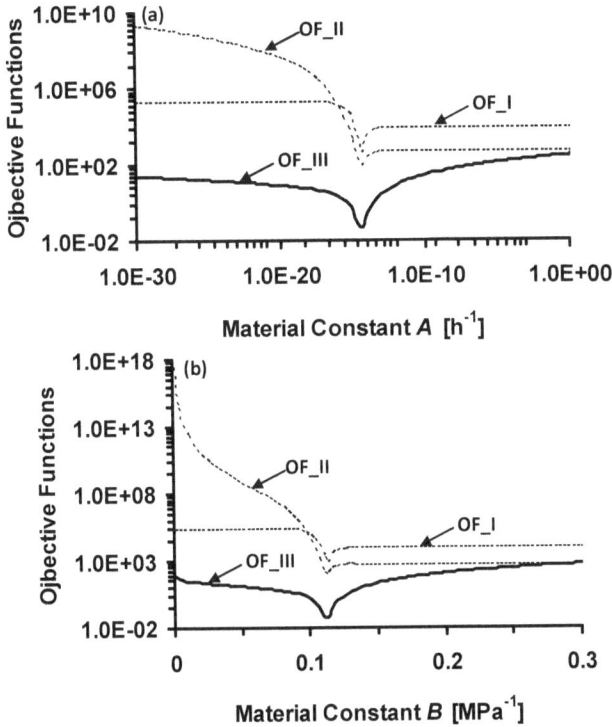

Fig. 7.21 Variation of objective functions (OFs-I, II and III) with material constants (a) *A* and (b) *B* in equation SET-II listed in appendix A (Cao & Lin, 2008).

To view the difference between computed curves 2 and 3, a logarithmic scale in time is used as given in **Fig. 7.20 (b).** The clear difference between computed curves 2 and 3 is shown. However, by using the error definitions for OFs-I and -II, the errors in curves 2 and 3 become similar and less than that of curve 1. Thus, given a bigger value of the constant *A*, the lifetime of the creep curve tends to zero, and the error calculation tends towards t_i^e in both OFs-I and -II. This indicates that none of the error definitions for OFs-I and -II gives a correct error assessment. This also indicates that the logarithmic scale used in **Fig. 7.20 (b)** can enlarge the differences between the shorter computed lifetimes of the creep curve and lessen the longer one. This error assessment phenomenon leads to the true error definition used in OF-III.

The behaviour of the two objective functions (OF-1 and OF-2) with the variation of the material constants A and B is given in **Fig. 7.21**. The characteristics of OF-I have been studied by Li, Lin and Yao (2002) for equation SET II in Appendix A and similar features for OF-II are observed in the figure. Above a certain value of A and B respectively, the sum of the errors remains constant. This indicates that the values of A and B are too big and that the computed lives of the creep curves tend to zero, as do the computed curves 2 and 3 in **Fig. 7.20 (a)**. Thus, the objective functions (OFs-I and -II) for some material constants, such as A and B, contain large plateau regions, where $\partial f(x)/\partial A \approx 0$ or $\partial f(x)/\partial B \approx 0$, which gives a wrong signal in optimisation processes; this causes problems in the determination of the material constants within the constitutive equations.

Compared with OFs-I and -II, large plateau regions are obviated in OF-III due to the logarithmic error definition (or true error definition). This makes the searching process much easier.

In materials modelling, specific objective functions are required to represent the optimisation problem; the features of the objective functions should meet the criteria mentioned at the beginning of the section. Once an objective function is formulated for a specific problem, an optimisation searching algorithm should be used to find out the global minimum of the objective function. The principles of searching techniques are to be discussed in the following section.

7.6 Optimisation Methods for Determining Constitutive Equations from Experimental Data

7.6.1 Concept of optimisation

An optimisation problem can be represented in the following way:

- *Given:* a function $f(x)$ (objective function) where $x = \{x_i\}$, $i = 1, 2, \ldots, NC$ and NC is the number of parameters to be optimised from a domain A of $f(x)$ of real numbers.
- *Sought:* a set of values $\{x_i^*\}$, such that $f(x_i^*) \leq f(x)$ for all $\{x_i\}$ in A, which is known as *minimisation*; or, such that $f(x_i^*) \geq f(x)$ for all $\{x_i\}$ in A, which is known as *maximisation*.

This formulation is known as an *optimisation problem*, and methods for formulating objective functions $f(x)$ have been discussed in the previous section. The domain A of $f(x)$ is called the *search space* or the *choice set*, while the elements of A are called *candidate solutions* or *feasible solutions*. $\{x_i^*\}$ are known as the optimised parameters (or constants).

By convention, the standard form of an optimisation problem is stated in terms of minimisation. Generally, unless both the objective function and the feasible region are convex in a minimisation problem, there may be several local minima, where a *local minimum* $\{x_i^*\}$ is defined as a point for which there exists some $\delta > 0$ for all x such that

$$\left\| x_i - x_i^* \right\| \leq \delta$$

and the expression

$$f(x_i^*) \leq f(x) \tag{7.36}$$

holds; that is to say, in some region around $\{x_i^*\}$ all of the function values are greater than or equal to the value at that point, i.e. $f(x_i^*)$ is the minimum. Numerical methods for optimisation are mathematical algorithms for searching for $\{x_i^*\}$ values that satisfy the condition

$$\left[\frac{\partial f(x)}{\partial x_i} \right]_{x_i = x_i^*} = 0 , \quad i = 1, 2, ..., NC . \tag{7.36a}$$

The determination of material constants in a set of constitutive equations is a non-convex problem in general. Thus such methods are not capable of making a distinction between local optimal solutions and rigorous optimal solutions, and will treat the former as actual solutions to the original problem. Numerical methods that are concerned with the development of deterministic algorithms that are capable of guaranteeing convergence in finite time to the actual optimal solution of a non-convex problem are known as global optimisation methods, as shown in **Fig. 7.5**.

Many optimisation algorithms that can be used for solving the problems encountered in materials modelling have been developed and have been programmed in commercial software systems, such as MATLAB. Typical optimisation methods include gradient-based methods, genetic methods, evolutionary programming methods etc. The detailed theories of these methods can be found in books on numerical methods and numerical recipes. In this section, optimisation search algorithms will only be briefly introduced.

7.6.2 *Gradient-based optimisation method*

One of the commonly used gradient-based optimisation methods is the *conjugate gradient method*, which can be applied to the problem of minimising a nonlinear multidimensional function $f(x)$ by considering its gradient $\nabla f(x)$ within a certain number of iterations, and has the following form (Willima, 2002):

$$x_{k+1,i} = x_{k,i} + \theta_k d_{k,i} \qquad\qquad (7.37a)$$

$$d_{k+1,i} = g_{k+1,i} + \lambda_k d_{k,i}, \qquad\qquad (7.37b)$$

where k is the kth iteration, which is initialised by selecting a starting value of the *i*th material constant $x_{0,i}$ and setting the initial search direction

$$d_{0,i} = g_{0,i} = -\nabla f\left(x_{0,i}\right),$$

where $i \in \{1,2,...,NC\}$ and NC is the number of material constants, $\nabla f(x_{0,i})$ is the gradient of the *i*th material constant for the first iteration and θ_k is a step size obtained by a golden section search in one dimension (Polak, 1971). For calculating $d_{k+1,i}$ for the next iteration in Eq. (7.37b), a search direction and a scalar proposed by Polak (1971) can accomplish the transition to further iterations more gracefully (Dennis and Schnabel, 1983), and these are given by

$$g_{k+1,i} = -\nabla f\left(x_{k+1,i}\right)$$

and

$$\lambda_k = \frac{\left(g_{k+1,i} - g_{k,i}\right) \cdot g_{k+1,i}}{g_{k,i} \cdot g_{k,i}}.$$

This optimisation method is based on conjugate search directions and the steepest descent method. The convergence is faster than the steepest descent method because the previous directions are used as part of a new direction. It is an effective search method for minimising objective functions if the number of parameters to be optimised is less than five (Lin and Hayhurst, 1993; Zhou and Dunne, 1996). However, it is difficult to choose starting values of material constants (Li *et. al.*, 2002) and it is not suitable for solving advanced material models, which contain a large number of material constants.

7.6.3 *Evolutionary programming (EP)-based method*

To overcome the problems of selecting starting values of the constants for optimisation using gradient-based methods, Lin and Yang (1999) proposed genetic algorithms and successfully determined the constants within unified superplastic constitutive equations. Furthermore, Li, Lin and Yao (2002) used evolutionary programming methods and determined a set of creep damage constitutive equations. These works show the power and the suitability of evolutionary algorithms (EA) for the determination of unified constitutive equations in materials modelling. Here, only the basic concept of the evolutionary programming (EP) method is introduced.

EP was firstly proposed as an approach to artificial intelligence (Fogel, 1991). It has been applied with success to determine material constants within unified constitutive equations from experimental data (Cao, 2006). Optimisation by EP can be summarised into two major steps:

1) mutate the solutions in the current population;

2) select the next generation from the mutated and the current solutions.

These two steps can be regarded as a population-based version of the classical generate-and-test methods, where mutation is used to generate new solutions (offspring) and selection is used to test which of the newly generated solutions should survive to the next generation. The classical evolutionary programming method is implemented as follows (Yao, Liu and Lin 1999).

Step 1) Generate the initial population of NC individuals, and set the initial iteration $k = 1$. Each individual is taken as a pair of real-valued vectors, (x_i, η_i), $\forall i \in \{1, \dots, NC\}$, where x_i are material constants and η_i are standard deviations for Gaussian mutations (also known as strategy parameters in self-adaptive evolutionary algorithms) (Li, Lin and Yao, 2002).

Step 2) Evaluate the fitness score for each individual (x_i, η_i), $\forall i \in \{1, \dots, NC\}$, the population based on the objective function, $f(x)$.

Step 3) Each parent (x_i, η_i), $i = 1, \dots, NC$, creates a single offspring (x_i', η_i') for the jth population, for $j = 1, \dots, n$, where n is specified population size, by

$$x_i'(j) = x_i(j) + \eta_i(j)N_j(0,1) \tag{7.38a}$$

$$\eta_i'(j) = \eta_i(j)\exp(\tau'N(0,1) + \tau N_j(0,1)),$$ (7.38b)

where $x_i(j)$, $x_i'(j)$, $\eta_i(j)$ and $\eta_i'(j)$ denote the jth component of the vectors x_i, x_i', η_i and η_i' respectively. $N_j(0,1)$ denotes a normally-distributed one-dimensional random number with mean zero and standard deviation one. $N_j(0,1)$ indicates that the random number is generated anew for each value of j. The factors τ and τ' are commonly set to $\left(\sqrt{2n^{1/2}}\right)^{-1}$ and $\left(\sqrt{2n}\right)^{-1}$. The global factor $\tau' \cdot N(0,1)$ allows for an overall change of the mutability, whereas $\tau \cdot N_j(0,1)$ allows for individual changes of the "mean step sizes" η_i.

Step 4) Calculate the fitness of each offspring (x_i', η_i'), $\forall i \in \{1, ..., NC\}$.

Step 5) Conduct pairwise comparison over the union of parents (x_i, η_i) and offspring (x_i', η_i'), $\forall i \in \{1, ..., NC\}$. For each individual, q opponents are chosen uniformly at random from all the parents and offspring. For each comparison, if the individual's fitness is no smaller than the opponent's, it receives a "win".

Step 6) Select the n individuals out of (x_i, η_i) and (x_i', η_i'), $\forall i \in \{1, ..., NC\}$, that have the most wins to be parents of the next generation.

Step 7) Stop if the halting criterion is satisfied; otherwise, $k = k + 1$ and go to Step 3).

The one-dimensional Gaussian density function centred at the origin is defined by (Bäck and Schwefel, 1993):

$$f_g(x) = \frac{1}{\sqrt{2n}} e^{-\frac{x^2}{2}},$$ (7.39)

where $-\infty < x < \infty$. As a result, the variance of the Cauchy distribution is infinite and Cauchy mutation from Eq. (5.3) is more likely to generate an offspring further away from its parent due to its long flat tails. It is expected to have a higher probability of escaping from a local optimum or moving away from a plateau. This search method has been successfully used for the determination of creep damage constitutive equations (Li, Lin and Yao, 2002).

In the EP method, the range of values of the individual material constants needs to be defined. Many constants have their own physical meanings and should be within a certain range, which is fairly easy to define. However, some material constants have less physical meaning and need to be defined according to experience. In this case, a large range is normally defined, which may have a longer computational time.

The population of individual constants, the number of generations for the optimisation and the tournament size are chosen according to experience and the size of the problem. Normally, the population size in the range of 30–100; the tournament size is half of the population; and the number of generations is between 100 and 500.

7.7 Examples for the Determination of Constitutive Equations

If viscoplastic-type constitutive equations are used, numerical integration should be used for solving the equation set. It is recommended that lower order integration methods are used, such as backward and forward Euler methods with an appropriate step-size control method.

If the number of constants within a set of unified constitutive equations is more than five, EA-based optimisation methods are recommended, which provide a better chance of obtaining a global minimum. An appropriate objective function should be formulated according to the problem. However, it is recommended that OF-III be considered first, which has unique features for solving the problems encountered in materials modelling.

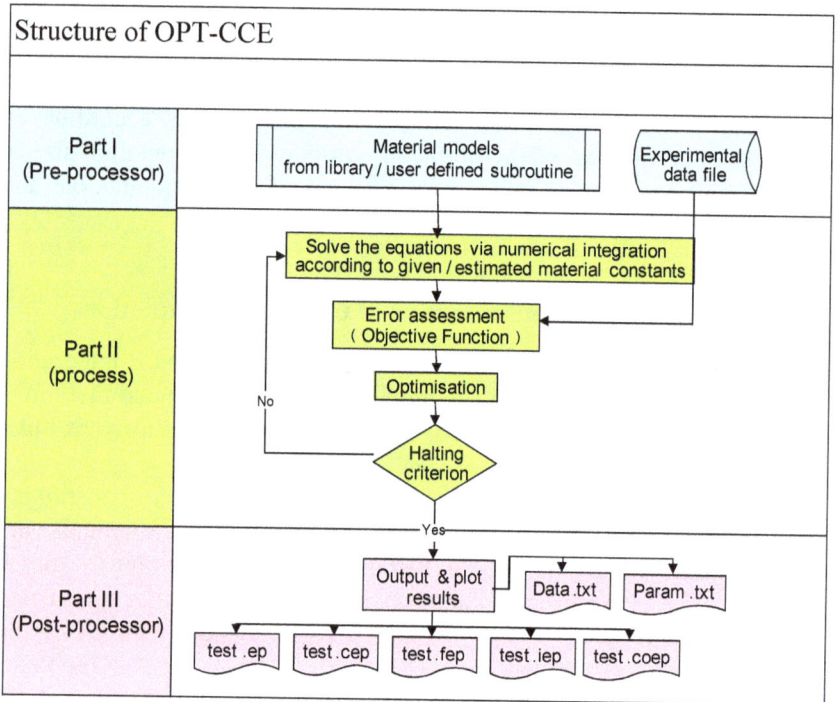

Fig. 7.22 Structure of a system for determining material constants within unified viscoplastic constitutive equations (Cao, Lin and Dean, 2008b).

7.7.1 *System development for materials modelling*

Figure 7.22 shows the structure of a system, known as OPT-CCE (Cao, Lin and Dean, 2008b), developed specifically for the determination of unified viscoplastic constitutive equations from experimental data. The system includes three parts: (i) preprocessing – input necessary information for optimisation; (ii) optimisation – searching for the best values of constants for the best-fitting to experimental data; and (iii) post processing of the optimisation results. The main input window for the preprocessing is shown in **Fig. 7.23**.

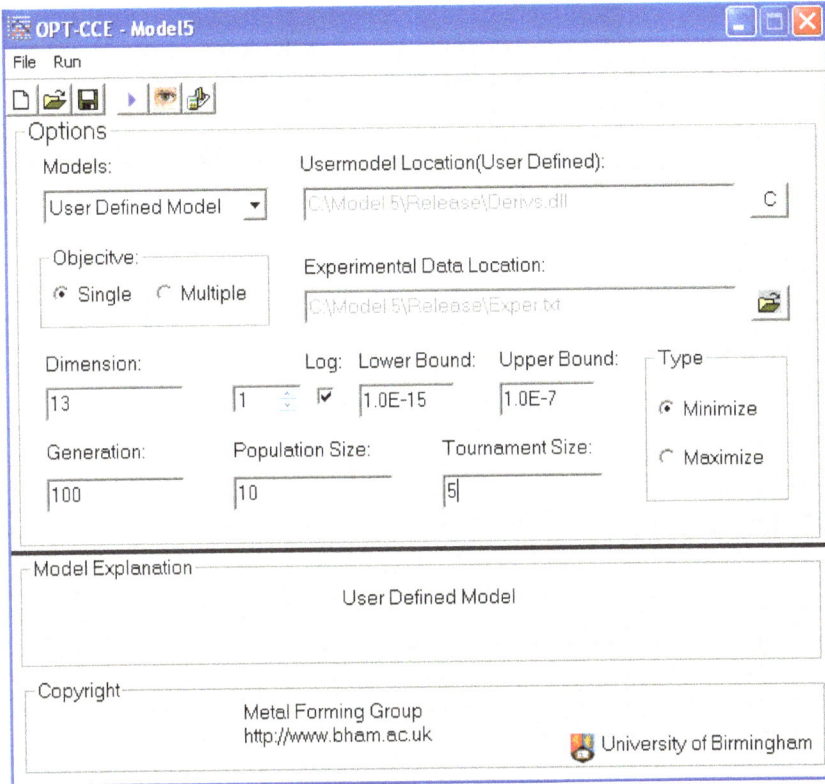

Fig. 7.23 The main input window of OPT-CCE (Cao, Lin and Dean, 2008b).

In the preprocessing part, the input includes:

- Select/input a set of constitutive equations. This allows users to input their own constitutive equations, in addition to selecting a set of embedded constitutive equations.

- Input initial values of individual state variables. By default, they are zero.

- Define the lower and upper bounds of each material constant.

- Input/link experimental data file.

- Parameters for the optimisation, such as population size, tournament size, number of generations etc.

- Select numerical methods: (i) a numerical integration method; (ii) a form of objective function including the number of objectives; and (iii) an optimisation search method.

A few EA-based searching methods are programmed and can be selected for the optimisation in the system (Cao, 2006). Many other searching methods are available from various mathematical solvers, such as MATLAB, which can be linked to this system for performing the searching task.

The fitting results for various state variables and objectives can be viewed graphically at a defined frequency of generations. The final results and selected intermediate results are stored in an output file. More details of the system are given by Cao, Lin and Dean (2008b).

The purpose of this section is not to introduce how to use the system. It is to introduce (i) the general numerical procedures for the determination of unified viscoplastic constitutive equations from experimental data; and (ii) the general terminologies related to the tasks and methodologies.

7.7.2 *Determination of constitutive equations*

The equation SETs used in this sub-section are listed in Appendix A, and will not be repeated here.

a. *Equation SET-II (creep damage constitutive equations)*

For the determination of the constants within the creep damage constitutive equations (Kowalewski, Hayhurst and Dyson, 1994) in SET-II listed in Appendix A, the above introduced methods and software system, OPT-CCE, are used. The experimental data, symbols in **Fig. 7.24**, are given by Kowalewski, Hayhurst and Dyson (1994). This is a fairly simple case, since only six constants are in the equation set and the number of experimental curves is four (for the four stress levels shown in **Fig. 7.24**).

Firstly, the equations are integrated using the implicit Euler method with automatic step-size control. The normalised tolerance for the numerical integration is 0.05, which is used to control the numerical integration accuracy and step size. The initial values for the variables are (at $t = 0$)

$$\varepsilon = 0\,,\ H = 0\,,\ \phi = 0\,,\ \omega = 0\,.$$

The objective function selected for the task is OF-III.

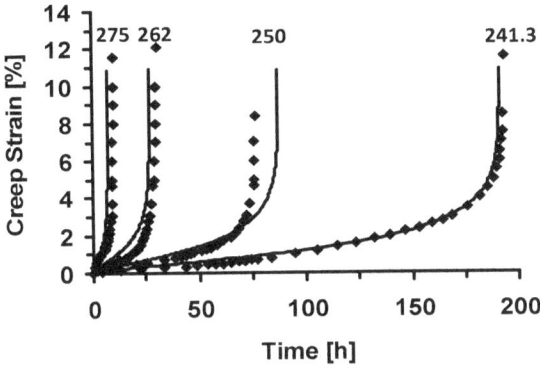

Fig. 7.24 Comparison of experimental (symbols) and computed (solid lines) creep curves for equation SET II in Appendix A for stress levels of 241.3 MPa, 250 MPa, 262 Mpa and 275 MPa (Cao, 2006).

The evolutionary programming method is used and the parameters selected for the optimisation process are: population = 30; tournament = 15; number of generations = 100.

Table 7.3 Definition of the range (lower and upper bounds) of material constants for Equations SET II.

	A (h^{-1})	B (MPa^{-1})	h_0 (MPa)	H^* $(-)$	K_c (h^{-1})	D $(-)$
Lower	1.0e-20	0.05	1.0e+1	0.03	1.0e-10	2.75
Upper	1.0 e-10	0.3	1.0 e+10	0.3	1.0e-1	2.75

The upper and lower bounds of the constants arising in the creep damage constitutive equations (SET-II in Appendix A) for the optimisation process are defined in **Table 7.3**. The value D is related to the uniaxial ductility of the material, which is 2.75 according to the experimental data (Kowalewski, Hayhurst and Dyson, 1994). Thus both the lower and upper bounds are defined as 2.75. For the constants, B and H^*, linear variation is used to select a population of individuals to assess the values of the objective functions. For example, if the population for B is 30, the individuals

B_j, $j = 1,2,3,...,30$, within the range of $(0.05, 0.3)$

can be selected according to the probability theory defined in Eq. (7.39).

The constants A, h_0 and K_c vary significantly, and a huge range is defined. Thus, they are converted to log values. For example, for the constant A, the 30 individuals, A_j, $j = 1,2,3,...,30$, in the range of log-scales $[\ln(1.0 \times 10^{-20}), \ln(1.0 \times 10^{-10})]$ can be selected based on Eq. (7.39). This indicates that the horizontal axis for the Gaussian distribution (Eq. (7.39)) becomes logarithmic, which could increase the sensitivity of the parameters. After 100 generations in the optimisation searching process, the determined values of the material constants are given and listed in **Table 7.4,** where the values are different from those listed in **Table A.2** in Appendix A. However, the fitting results, as shown in **Fig. 7.24**, appear very good; close agreement between the experimental and computed data is obtained for all the creep curves. This indicates that the determined material constants are not unique.

Table 7.4 Determined material constants for Equation SET II.

A (h^{-1})	B (MPa^{-1})	h_0 (MPa)	H^* $(-)$	K_c (h^{-1})	D $(-)$
6.32e-15	0.10	5.28e3	6.19e-2	7.47e-4	2.75

It should be noted that the lower and upper bounds are difficult to define sometimes. They need to be changed according to the optimisation results, especially since some of the constants may have optimised values close to the upper or lower bounds. In this case, an adjustment might be needed.

b. Equation SET-IV (unified viscoplastic constitutive equations)

The equations were originally proposed by Lin, Liu, Farrugia *et al.* (2005) for the modelling of recrystallisation, grain size evolution, hardening and viscoplastic flow of a carbon steel in hot rolling processes. The experimental data available are (i) stress–strain curves; (ii) recrystallised volume fraction, and, (iii) grain size evolution for the material deforming at different strain rates at a constant temperature of 1100 °C.

The optimisation involves multiple objectives, which have different scales. The experimental data (Medina and Hernandez, 1996) for the objectives are shown in **Fig. 7.25** by symbols. Each objective contains multiple curves (two in this case), and the curves have different experimental data points. The formulation of objective function OF-III is used in the work and the problems mentioned above can be overcome automatically.

The implicit Euler integration method is used for solving the unified viscoplastic constitutive equations with the initial values ($t = 0$) of

$$\varepsilon^p = 0, \; S = 0, \; x = 0, \; \bar{\rho} = 0, \; d = 189 \, (\mu m).$$

The integration is carried out at strain rates (simulation of the uniaxial tests) of $\varepsilon = 0.544 \, s^{-1}$ and $5.224 \, s^{-1}$. The integrated values of flow stress, recrystallised volume fraction and grain size are compared with their corresponding experimental data based on the formulation of OF-III (objective function III).

The evolutionary programming method is used and the parameters selected for the optimisation process are: population = 50; tournament = 25; number of generations = 200. The range of constants for the optimisation are defined and listed in **Table 7.5**, where "log" indicates that logarithmic mapping is used for that particular constant; otherwise, a linear system is used for the selection of the individuals for the constant.

Table 7.5 Range of material constants for optimisation for constitutive equation SET IV (Cao, 2006).

$A_1(s^{-1})$ (log)	$A_2(MPa^{-1})$	$k(MPa)$	$\gamma_4(-)$	$Q_0(-)$	$\bar{\rho}_c(-)$	$N_q(-)$
1.0e-8– 1.0e-5	0.01–1.0	1.0–10.0	0.01–1.0	20.0– 50.0	0.1–2.0	1.0– 10.0
$c_1(-)$	$c_2(-)$	$c_3(-)$ (log)	$d_0(\mu m)$	$\gamma_d(-)$	$A_0(-)$	$B(MPa)$
1.0–10.0	0.5–3.0	1.0e-3– 1.0	10.0– 100.0	0.1–3.0	10.0– 100.0	10.0– 100.0
$\alpha_0(\mu m)$	$\gamma_0(-)$	$\alpha_2(\mu m)$	$\gamma_2(-)$	$\gamma_3(-)$		
1.0–5.0	1.0–8.0	30.0– 100.0	0.1–5.0	0.1–5.0		

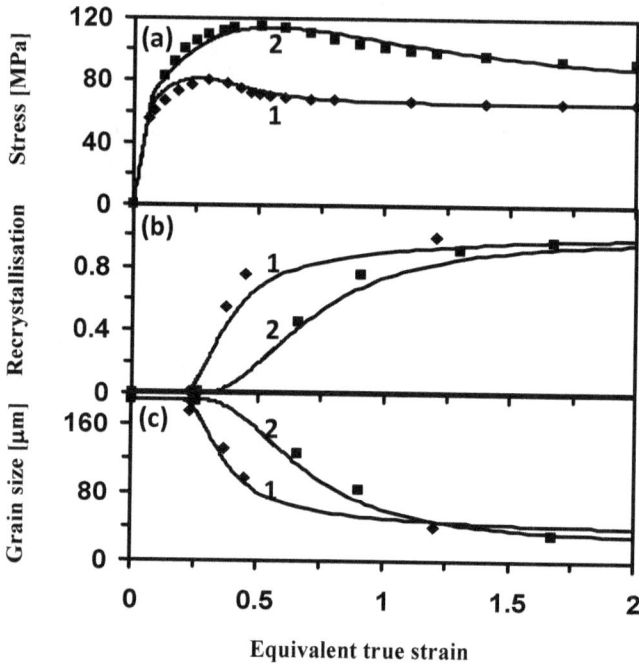

Fig. 7.25 Comparison of experimental (symbols) with computed results (solid curves) for (a) stress; (b) recrystallisation; and (c) grain size with equivalent true strain for equation SET IV in Appendix A. The numbers indicate different strain rates: $\dot{\varepsilon} = 1 - 0.544 \text{ s}^{-1}$ and $2 - 5.224 \text{ s}^{-1}$ (Cao, 2006).

Table 7.6 Material constants for equation SET IV (Cao, 2006).

$A_1(\text{s}^{-1})$	$A_2(\text{MPa}^{-1})$	$k(\text{MPa})$	$\gamma_4(\text{-})$	$Q_0(\text{-})$	$\overline{\rho}_c(\text{-})$	$N_q(\text{-})$
1.29e-7	0.28	4.81	0.12	23.13	0.25	1.00
$c_1(\text{-})$	$c_2(\text{-})$	$c_3(\text{-})$	$d_0(\mu\text{m})$	$\gamma_d(\text{-})$	$A_0(\text{-})$	$B(\text{MPa})$
6.14	1.10	8.05e-2	70.50	1.67	45.97	60.38
$\alpha_0(\mu\text{m})$	$\gamma_0(\text{-})$	$\alpha_2(\mu\text{m})$	$\gamma_2(\text{-})$	$\gamma_3(\text{-})$		
1.74	7.78	91.63	0.11	1.06		

The determined values of the material constants are listed in **Table 7.6** and the computed results are compared with experimental data as shown in **Fig. 7.25**. It can be seen that close agreement between experimental and computed results are obtained. This indicates that the determined unified viscoplastic constitutive equations can be used to predict the viscoplastic flow and microstructure evolution in thermal mechanical processing conditions.

7.7.3 *Discussion*

The key problems in the determination of unified viscoplastic constitutive equations from experimental data are summarised as follows:

(i) The numerical integration and accuracy control of multiple ODE-type equations with different units. Low order implicit integration methods with a normalised error control method are recommended for solving unified viscoplastic constitutive equations.

(ii) The selection of a suitable objective function to satisfy the criteria described in the chapter. The formulation OF-III is recommended. The unit and weighting problems between different objectives can be overcome automatically.

(iii) The selection of a suitable searching method for the optimisation of a large dimensional problem. Evolutionary algorithm methods are recommended, which are available in many mathematical software systems. The key problem is selecting the ranges of individual constants, which might be a difficult job for engineers, who do not know the physical meaning of the equations and constants. Even if significant efforts have been made (Cao, 2006) for solving the problem over the years, this is still an ongoing issue.

Trial and error techniques can be used for the determination of unified constitutive equations from experimental data, but this is very difficult for engineers, who do not have significant experience in materials modelling.

An automatic system is needed for determining unified constitutive equations from experimental data. A multidisciplinary team with significant knowledge of mathematical optimisation, computing engineering and science, and mechanics of materials is required for solving the problem.

Many sets of unified constitutive equations developed in recent years for different applications are listed in Appendix A. The equation sets are determined from the corresponding experimental data using optimisation techniques and the calibrated constants are listed.

Chapter 8

Materials and Process Modelling in Metal Forming Applications

Based on the discussion of basic theories for formulation of unified viscoplastic constitutive equations, which include state variables representing the evolution of physical properties and material damage during thermal mechanical processing, in this chapter a few examples on the use of materials and process modelling theories are introduced for practical metal forming applications. For most cases, unified constitutive equations are given without detailed explanation.

8.1 The Framework for Advanced Metal Forming Process Modelling

Finite element (FE) methods are most commonly used for simulating a forming process. In traditional FE process simulations, the main focus is to investigate material flow and die filling. Thus, stress and strain distribution and velocity fields of material flow can be calculated and displayed for process analysis. Based on simulated distribution of stress, strain and material flow, real forming process parameters can be adjusted to avoid defects and for instance, to obtain an improved material flow route for die filling.

An example of stress distribution in forging a bimetal gear using rigid die set (Politis, 2014) is shown in **Fig. 8.1**. To minimise the computational (CPU) time, the punch, die and counter punch are assumed as rigid. The gear blank is copper ring with lead core, as illustrated in the figure. Modelling and simulation enables material flow and die filling to be observed. In addition to stress and strain distribution, the load-displacement profile during die filling can be obtained and analysed. This allows for blank shape design, tool design and forming process parameters to be optimised so that quality gears can be forged. The details of the analysis have been given in the work of Politis (2014).

In traditional FE process simulation, only the information related to mechanical deformation of a work-piece can be obtained. Usually this is not enough for industrial requirements. Particularly in hot forming processes, an important function is modification of work-piece microstructure, in order to enhance the mechanical properties of formed parts and minimise defects. It is expected that a forged component is stronger and tougher compared to corresponding cast materials. In industry, especially for critical components in aerospace and automotive applications, components should be formed with specified microstructures and mechanical properties, which can be predicted through FE material/process simulation.

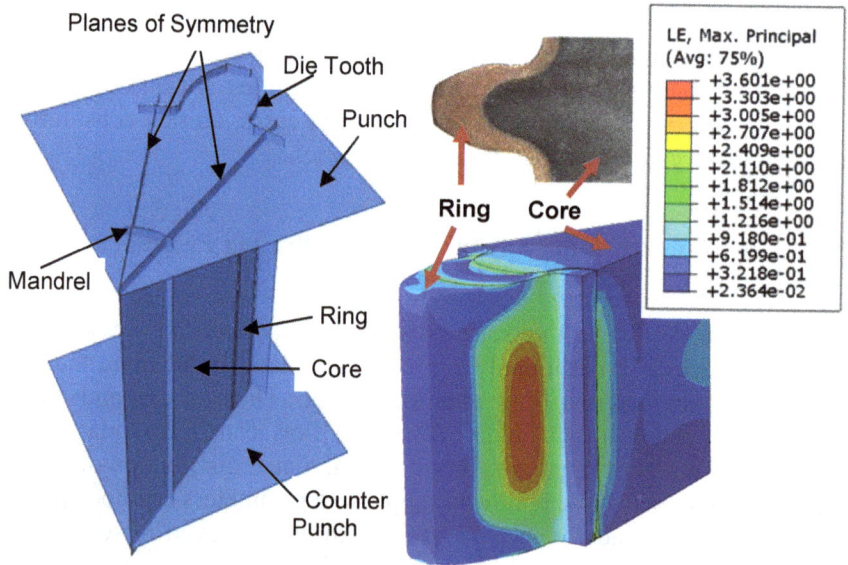

Fig. 8.1 An FE model (left) for the forging of bimetal gears. For an efficient FE analysis only one tooth is used in the FE model. The work-piece blank consists of the outer ring and inner core materials, as shown in the section of a formed part. The colour code (right) represents the maximal principal stress distribution for the formed tooth (Politis, 2014).

The key issue necessary to enable the microstructure and mechanical properties of formed parts to be predicted through process modelling is to have a set of unified constitutive equations, which enable the interactive effects of viscoplastic deformation (for hot/warm forming conditions) and microstructural evolution to be modelled. Once the unified constitutive equations are calibrated from a set of experimental data, embedded in a commercial FE code, forming process modelling can be carried out and the important information related to viscoplastic flow of materials and microstructural evolution and distribution can be obtained.

To establish a good FE model for forming process simulation, definitions of boundary constraints, friction and heat transfer at interfaces of the tool and material blank are very important as well. However, these will not be detailed here.

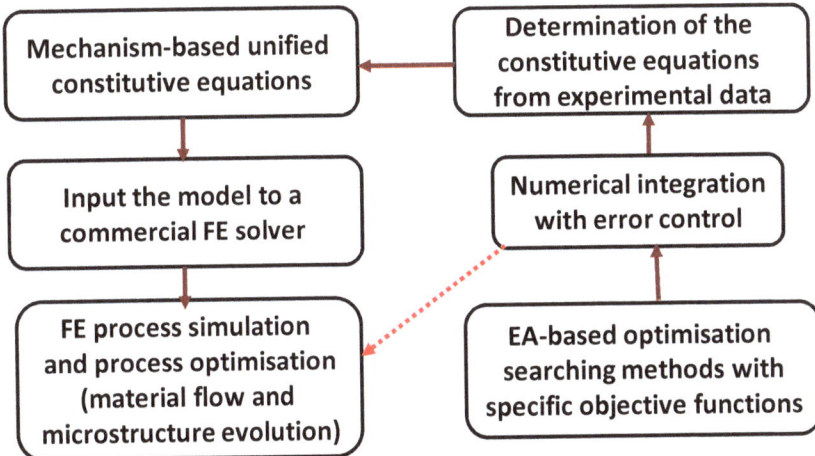

Fig. 8.2 The overall system for materials and metal forming process modelling. This includes the determination of unified constitutive equations from experimental data and the modelling of forming processes using commercial FE codes.

The overall framework for advanced materials and metal forming process modelling is shown in **Fig. 8.2**. This is an integrated process and the theories for individual procedures are discussed in the previous chapters in the book. They are summarised below.

(i)   The first stage is to identify the likely range of temperature, strain and (for elevated temperature) strain rate values likely to be encountered in the process to be modelled. Then constitutive relations for the work-piece alloy are determined by experiment, usually tensile tests for simplicity, within these ranges of process conditions. The outcomes are stress–strain, strain rate, temperature and perhaps microstructural outcomes related to the conditions. Based on this data a set of unified viscoplastic constitutive equations are formulated, which comprise the material model. The techniques for formulating unified viscoplastic constitutive equations are detailed in Chapters 5 and 6.

(ii) An experimental programme is designed to generate data required for the determination of material constants in constitutive equations. Some mechanical and physical data for relevant deformation conditions may be found in the literature. Sometimes, it is very difficult and time-consuming to obtain a full set of experimental data for the calibration of the constitutive equations. Experimental techniques for data generation are not introduced in this book. The quality of the data is related to many aspects, such as specimen design, measurement techniques, facilities used and accuracy control. It is highly recommended that the corresponding testing standards be used and referenced. In addition, interactions between work-piece and forming tool must be determined experimentally. They are named boundary conditions and almost always include friction and, for elevated temperature processes, heat transfer. Mechanical and physical properties of tools are sometimes included in a simulation.

(iii) Determine the material constants arising in the formulated set of constitutive equations, from experimental data, i.e. the equations are calibrated. As discussed in Chapter 7, this consists of three aspects: numerical integration for solving the equations; formulation of objective functions for the problem; and selection of suitable optimisation searching methods, so that the global minimum can be achieved. It is recommended that a low-order implicit integration method with normalised error assessment and control technique be used. The OF-III introduced in Chapter 7 is recommended for assessing the error between the computed and experimental results. Finally, if the number of constants to be determined is large, the use of an evolutionary algorithm is recommended for searching the optimum values of the constants in the equations, thus allowing the experimental data to be well predicted.

(iv) Input the determined unified viscoplastic constitutive equations into a commercial FE code, such as ABAQUS, DEFORM, QFORM, PAMSTAMP, LS-DYNA, MSC-MARC etc., via user-defined subroutines, such as UMAT and VUMAT for ABAQUS. In order to use large step size and control the numerical integration accuracy, the use of an implicit numerical integration method is again recommended. Most advanced FE codes provide suitable subroutines, which enable newly formulated unified constitutive equations to be input into the solver. Moreover, commonly used unified viscoplastic constitutive equations can be embedded into commercial FE codes permanently for industrial users. The physical variables can be represented as state-variables in software systems.

(v) Carry out process simulation. Normally process parameters, tool design and blank-shape design can be optimised via trial-and-error methods according to the simulation results. This optimisation/determination of forming process variables is the main requirement for industry.

(vi) Validate simulation results from experimental results. This is an important step and a benchmark should be designed to assess the reliability and accuracy of the simulation results.

Items (i)–(iv) and (vi) should be carried out by experienced researchers at universities or research institutes. Item (v) is normally performed by process and/or design engineers within companies. Item (vi) could also be carried out by program engineers in software houses. If steps (i)–(iv) and (vi) mentioned above have been completed, then the determined unified viscoplastic constitutive equations can be embedded in commercial FE codes. The process and design engineers at companies can use this as a powerful tool to model the evolution of mechanical and physical properties of alloys in metal forming processes. This enables the mechanical and physical properties of a formed part to be predicted explicitly.

The key issues for formulation and determination of unified viscoplastic constitutive equations are introduced in previous chapters, particularly Chapters 4, 5, 6 and 7. However, a particular interest in microstructural evolution depends on the individual application. In the following sections, a few application examples are given, to illustrate the use of the theories introduced in the book.

8.2 Modelling for Superplastic Forming

Superplastic forming (SPF), also known as *gas-blow forming,* is a method of die forming in which a metal sheet is heated to a pliable state, 0.5–0.6 of its melting temperature, then pressurised by a gas to force it into a die cavity. This forming process has been illustrated in **Fig. 5.8**. **Figure 8.3** shows an example of a SPFed component.

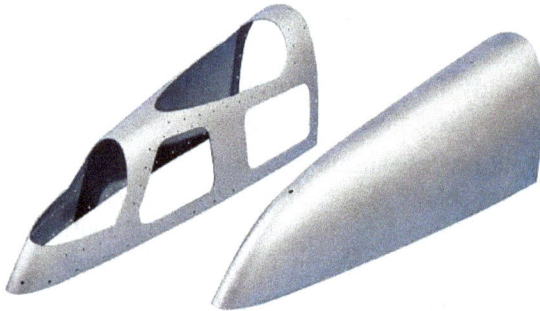

Fig. 8.3 Superplastically formed structure for enclosing the cockpit of aircraft. Supplied by Superform Aluminium (2001).

The material is super-ductile and very complex-shaped panel components can be formed in one operation. The amount of material deformation can easily reach several hundred percent, which causes difficulties in controlling localised thinning for superplastic formed parts. Process modelling is often used for evaluating the effect of process parameters on localised thinning, enabling process optimisation through simulation. In an isothermal SPF process, the localised thinning of a formed part is related to grain size and its distribution, forming rate and the shape of the component to be formed.

8.2.1 *Unified SPF constitutive equations*

It is well known that strain rate hardening is the most important parameter for the control of formability and localised thinning in superplastic forming processes. This is normally modelled using

$$\sigma = K\dot{\varepsilon}^m. \tag{8.1}$$

The strain rate sensitivity or strain rate hardening parameter, *m*, normally varies from 0.3 to 1.0. In this equation, the effect of grain size and plastic deformation on flow stress is ignored. To enhance this expression, Zhou and Dunne (1996) introduced a set of unified constitutive equations, which includes kinematic hardening and grain growth. This equation has been further simplified by inclusion of isotropic hardening only. Theories for the formulation of the unified equations are detailed in Chapters 5 and 6. The methods for solving the equations and determination of the equations from experimental data are presented in Chapter 7.

a. Sinh-law SPF constitutive equations

Sinh-law based elastic-viscoplastic constitutive equations were developed and used by Kim and Dunne (1997) and Lin and Dunne (2001) for a range of materials deforming superplastically. The equations are in the form of

$$
\begin{aligned}
\dot{\varepsilon}^p &= \alpha \sinh[\beta(\sigma - R - k)]\, d^{-\gamma} \\
\dot{R} &= (Q - \gamma_1 R)|\dot{\varepsilon}^p| \\
\dot{d} &= (\alpha_1 + \beta_1 |\dot{\varepsilon}^p|)d^{-y_0} \\
\dot{\omega} &= E(\dot{\varepsilon}^T - \dot{\varepsilon}^p),
\end{aligned}
\tag{8.2}
$$

where ε^T and ε^p are total and plastic strains respectively, *d* is average grain size with an initial value of 6.8 μm. *R* is an isotropic hardening variable, and *E* (= 1000 MPa at 900 °C) is Young's modulus. Material constants within the equation set were determined from uniaxial experimental data for Ti-6AL-4V at 900 °C (Kim and Dunne, 1997) and are listed in **Table 8.1**. The set of differential equations describing superplastic deformation coupled with average grain size can be integrated numerically.

Table 8.1 Values of material constants for Ti-6AL-4V at 900 °C (Kin and Dunne, 1997).

α (s^{-1})	β (MPa^{-1})	γ $(-)$	Q (MPa)	γ_1 $(-)$	α_1 $(-)$	β_1 $(-)$	γ_0 $(-)$
0.242e-6	0.114	1.017	3.974	0.880	0.206e-13	0.690e-9	3.021

b. Power-law SPF constitutive equations

According to Cheng, Lin and Ball (2000), the power-law-based unified superplastic constitutive equations are in the form of

$$\dot{\varepsilon}^p = \langle \frac{\sigma - R - k}{K} \rangle_+^n d^{-\mu}$$
$$\dot{R} = b(Q - R)|\dot{\varepsilon}^p|$$
$$\dot{d} = \alpha d^{-y_0} + \beta|\dot{\varepsilon}^p|d^{-\phi} \qquad (8.3)$$
$$\dot{\sigma} = E(\dot{\varepsilon}^T - \dot{\varepsilon}^p).$$

The values of the material constants within the equation set are determined from experimental data of Ghosh and Hamilton (1979) for Ti-6Al-4V deforming at 927 °C with the initial average grain size of 6.8 μm. These are listed in **Table 8.2** (Cheong, Lin and Ball, 2000). The other equations are similar to the corresponding equations in Eq. (8.2), apart from the grain growth equation, $\dot{d}$. Here the grain size has different effects on static and dynamic grain growth, which has been calibrated using the parameters y_0 and ϕ respectively in Eq. (8.3).

Table 8.2 Material constants determined for Ti-6Al-4V at 927 °C (Cheong, Lin and Ball, 2000).

α $(-)$	γ_0 $(-)$	β $(-)$	ϕ $(-)$	n $(-)$	μ $(-)$	K (MPa)	Q (MPa)	b $(-)$	k (MPa)
73.408	5.751	2.188	0.141	1.400	2.282	60.328	3.933	2.854	0.229

c. SPF damage constitutive equations

With the introduction of damage variables, ductile void growth damage f_h and effective damage f_e, to Eq. (8.3), the unified SPF damage constitutive equations can be written as (Lin, Cheong and Yao, 2002)

$$\dot{\varepsilon}^p = \dot{\varepsilon}_{ss}^p \left[1 + \frac{2lf_h^{1/2}}{d} \left(\frac{\dot{\varepsilon}_{cc}^p}{\dot{\varepsilon}_{ss}^p} - 1 \right) \right]$$

$$\dot{\varepsilon}_{ss}^{p} = \langle \frac{\sigma - R - k}{K} \rangle_{+}^{n} d^{-\mu}$$

$$\dot{\varepsilon}_{cc}^{p} = \langle \frac{\sigma^{*} - R - k}{K} \rangle_{+}^{n} d^{-\mu}$$

$$\dot{R} = b(Q - R)|\dot{\varepsilon}^{p}| \qquad (8.4)$$

$$\dot{d} = \alpha d^{-y_0} + \beta |\dot{\varepsilon}^{p}| d^{-\phi}$$

$$\dot{f}_{h} = |\dot{\varepsilon}_{cc}^{p}| - (1 - f_{h})|\dot{\varepsilon}_{ss}^{p}|$$

$$\dot{f}_{e} = D(|\dot{\varepsilon}^{p}|)^{d_1} + C\langle f_{h} - f_{h}^{*} \rangle_{+}$$

$$\dot{l} = -\frac{1}{2}\dot{\varepsilon}_{ss}^{p}$$

$$\dot{\sigma} = E(\dot{\varepsilon}^{T} - \dot{\varepsilon}^{p}).$$

For materials without pre-existing cavities, the initial values of f_h and f_e are chosen as 0. If the behaviour of a pre-cavitated material is modelled, appropriate initial values need to be given. The formulation of the damage evolutionary equations is given in Chapter 6.

The material parameters associated with the set of superplastic-damage constitutive equations Eq. (8.4) are: k, K, n, μ, b, Q, α, y_0, β, ϕ, D, d_1, C, f_h^*, E and l^0. The total strain rate $\dot{\varepsilon}^T$ and initial grain size d^0 are operational and initial microstructural conditions usually quoted with a given experimental stress–strain relationship. Eq. (8.4) was determined by Choeng (2002) from experimental data of Al-Zn-Mg at 515 °C (Pilling, 2001) and listed in **Table 8.3**. A comparison of the computed and experimental data is shown in **Figs. 8.4** and **8.5**.

Table 8.3 Material parameters for Al-Zn-Mg at 515 °C (Cheong, 2002).

k (MPa)	K (MPa)	n	μ	b	Q (MPa)	α	y_0
2.9354e-5	28.7640	1.1299	2.0642	0.1186	5.5769	6.900e-2	2.4000

β	ϕ	D	d_1	C	f_h^*	E (MPa)	l^0 (μm)
2.6000	5.5000e-5	3.7810e+3	2.3973	32.3739	0.7557	1.000e+3	4.3922

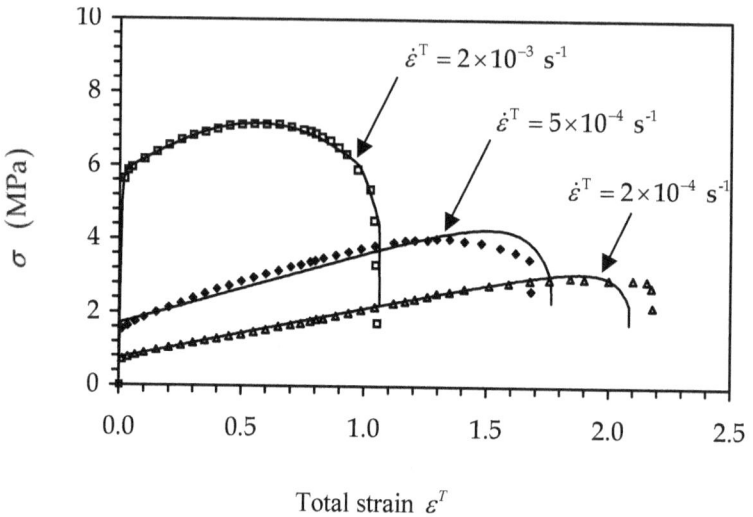

Fig. 8.4 Comparison of computed (solid curves) and experimental
(symbols) (Pilling, 2001) stress–strain relationships for Al-Zn-Mg at
various strain rates (Cheong, 2002).

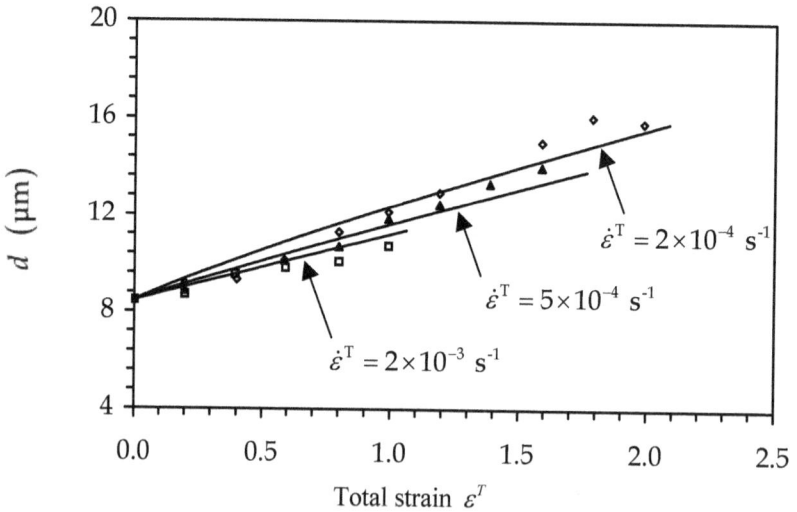

Fig. 8.5 Comparison of computed (solid curves) and experimental
(symbols) (Pilling, 2001) grain size-strain relationships for Al-Zn-Mg at
various strain rates (Cheong, 2002).

d. Multiaxial unified SP constitutive equations

The uniaxial sinh-law model Eq. (8.2) can be generalised by consideration of a dissipation potential function. First consider the strain rate equation without the hardening and grain size variables, which then reduces to

$$\dot{\varepsilon}^p = \alpha \sinh[\beta(\sigma - k)].$$

This equation can be generalised for multiaxial conditions by assuming an energy dissipation potential of the form

$$\psi = \frac{\alpha}{\beta}\cosh(\beta\sigma_e - k).$$

Assuming normality and the associated flow rule, the multiaxial relationship is given by

$$\dot{\varepsilon}_{ij}^p = \frac{\partial\psi}{\partial S_{ij}} = \frac{3}{2}\frac{\partial\psi}{\partial\sigma_e}\frac{\partial\sigma_e}{\partial S_{ij}} = \frac{3}{2}\frac{S_{ij}}{\sigma_e}\alpha\sinh(\beta\sigma_e - k) = \frac{3}{2}\frac{S_{ij}}{\sigma_e}\dot{\varepsilon}_e^p.$$

On re-introduction of the hardening and grain growth variables the effective plastic strain rate $\dot{\varepsilon}_e^p$ for the sinh-law material model can be written as:

$$\dot{\varepsilon}_e^p = \frac{\partial\psi}{\partial\sigma_e} = \alpha\sinh[\beta(\sigma_e - R - k)]\,d^{-\gamma},$$

and then the set of multiaxial viscoplastic constitutive equations, implemented within a large strain formulation, can be written (Lin and Dunne, 2001)

$$D_{ij}^p = \frac{3}{2}\frac{S_{ij}}{\sigma_e}\dot{\varepsilon}_e^p$$

$$\dot{R} = (Q - \gamma_1 R)|\dot{\varepsilon}_e^p|$$

$$\dot{d} = (\alpha_1 + \beta_1|\dot{\varepsilon}_e^p|)d^{-\gamma_0} \tag{8.5}$$

$$\hat{\sigma}_{ij} = GD_{ij}^e + 2\lambda D_{kk}^e$$

in which D_{ij}^p is the rate of plastic deformation, D_{ij}^e the rate of elastic deformation, $\hat{\sigma}_{ij}$ the Jaumann rate of Cauchy stress, and G and λ are the Lame elasticity constants.

In a similar way, the other two sets of unified SP constitutive equations, Eqs. (8.3) and (8.5), can be written in the multiaxial case. The forms for the power-law potentials are detailed in Chapter 3.

8.2.2 *FE model and numerical procedures*

A finite element simulation for superplastic forming of a rectangular-section box was carried out using the commercial solver ABAQUS. The superplastic forming process consists of clamping a flat metal sheet against a die, the surface of which forms a cavity in the shape required. Gas pressure is applied to the opposite side of the sheet, forcing it to acquire the die shape. This is a high-temperature isothermal forming process. The maximum strain rate over the deforming sheet is controlled to be close to the optimum deformation rate of the material, and this is achieved by varying the applied gas pressure. The FE model is shown in **Fig. 8.6**. Due to its symmetry only a quarter of the rectangular box is considered for the FE forming simulation.

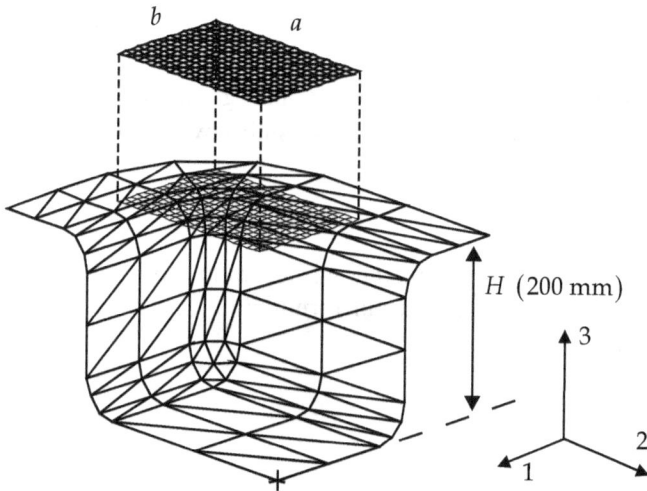

Fig. 8.6 FE model for the modelling of superplastic gas-blow forming of a rectangular box. Dimensions of the material blank are defined by $a = 1.6H$ and $b = 1.1H$ where H is the height of the formed part (Cheong, 2002).

a. *Material blank*

The metal sheet for a quarter of the rectangular box is considered for the analysis and shown in **Fig. 8.6**. The original material blank has a uniform thickness of 1.25 mm. Four-node quadrilateral thin shell elements were used for the analysis.

The viscoplastic flow of the material in gas-blow forming processes is controlled by determined unified superplastic constitutive equations. The multiaxial constitutive equations were embedded into the large strain finite element solver ABAQUS through the user-defined subroutine CREEP and used to simulate the blow-forming of a 3D structural component. To allow for an efficient use of computational power, an adaptive implicit time incrementation scheme is employed. The parameter CETOL, related to the accuracy of the numerical integration is set to 1×10^{-3}; it defines the maximum allowable difference in the plastic strain increments calculated from the plastic strain rates, based on conditions at the beginning and end of an increment.

b. The die

In order to avoid having points "fall off" the rigid die surface during the numerical simulation of the forming process, more than a quarter of the die surface is created. It is always a good idea to extend the rigid surface so that the elements and nodes on the deformable sheet do not slide off it during the simulation.

The die surface is divided into two types of sub-regions, flat and filleted regions, as shown in **Fig. 8.6**. The boundaries of the sub-regions must be located at places where curvature changes suddenly in order to accurately reconstruct the die surface with a number of surface patches small enough to mathematically manipulate with acceptable effort (Lin, Ball and Zheng, 1988).

c. Modelling process

The finite-element analysis is divided into two stages.

- In the first stage, gas pressure is applied quickly and the sheet material deforms in a purely elastic manner. This stage takes only 0.7 s to complete and the applied gas pressure is only 960 Pa.

- Next, the gas pressure is varied such that the maximum strain rate over the deforming material is maintained at a target level $\dot{\varepsilon}^{tgt}$. The material behaviour in this stage is described by a set of unified superplastic constitutive equations presented in Section 8.2.1. The analysis terminates when the deforming material is completely in contact with the die surface. The normal distance from each of the nodes on the sheet to the die surface is monitored by the user-defined subroutine URDFIL. Contact is detected when the normal distance falls below a tolerance of DISTOL $= 1 \times 10^{-3}$ mm.

d. Forming rate control

Figure 8.7 shows superplastic deformation of the box in titanium alloy Ti-6Al-4V, at 900 °C, in which t_f is the time required for the complete forming of the box. The contours represent the magnitude of effective plastic strain rate under a uniformly distributed gas pressure. The forming simulation is carried out using the target strain rate $\dot{\varepsilon}^{tgt} = 1 \times 10^{-5}$ s^{-1}. It can be seen that the effective strain rate is not uniformly distributed over the sheet. Before it is in contact with the bottom of the die surface, the highest strain rate occurs near the centre point of the sheet. The corner is the last part to be fully formed (**Fig. 8.7 (c)**) and the material adjacent to the corner remains stationary due to friction and geometric effects. The thinnest area of the formed part is that near the corner, where tearing is most likely to take place in the superplastic forming process.

The superplastic property of an alloy depends on strain rate; it occurs only within a narrow range of strain rate in which an optimum value exists, which is unique to each material for a given temperature (Ghosh and Hamilton, 1979). However, due to geometric and loading features, it is often not possible to maintain constant and uniform strain rate over the entire component. This is shown in **Fig. 8.7**. FE forming simulations were carried out using two different target strain rates, $\dot{\varepsilon}^{tgt} = 1 \times 10^{-4}$ s^{-1} and $\dot{\varepsilon}^{tgt} = 1 \times 10^{-5}$ s^{-1} (Lin and Dunne, 2001). The comparison of the maximum strain rate histories over the deforming material sheet and the corresponding target strain rates is shown in **Fig. 8.8**. The difference between the achieved maximum and the corresponding target strain rate is within 20% for both cases.

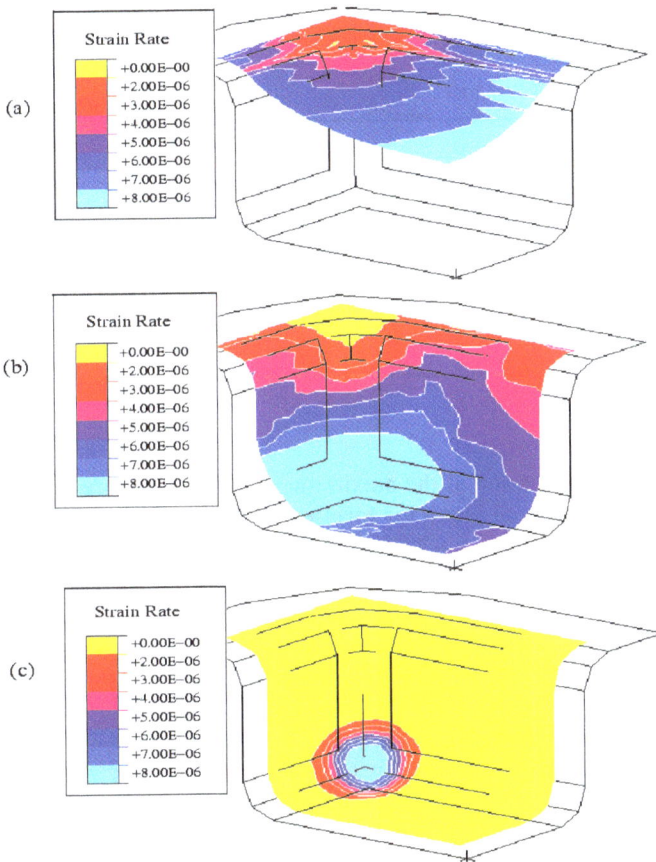

Fig. 8.7 The forming process and effective plastic strain rate field plots at t/t_f = **(a) 0.1, (b) 0.6 and (c) 1.0. The computation was carried out using equation set (8.5) with target forming rate** $\dot{\varepsilon}^{tgt} = 1 \times 10^{-5}$ **s**$^{-1}$ **(Lin and Dunne, 2001).**

The gas pressure histories required to maintain the maximum strain rates over the deforming sheet near the target deformation rates are shown in **Fig. 8.9** for both cases. Higher gas pressure is needed for the higher target strain rate forming because of the higher flow stresses arising at the higher strain rate.

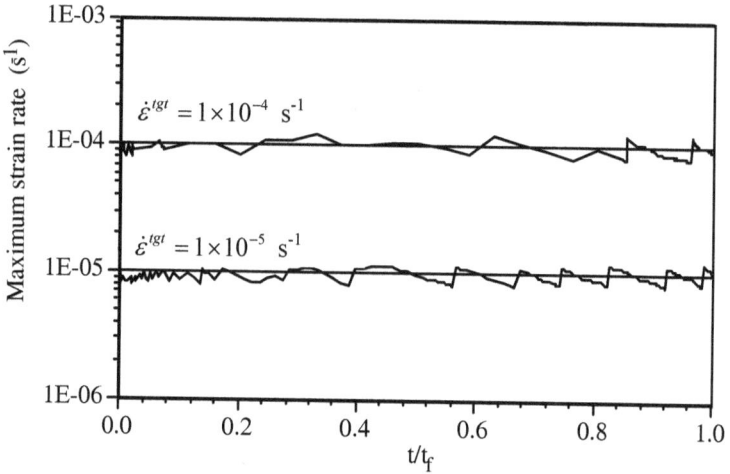

Fig. 8.8   Variation of the maximum strain rate history over the deforming material blank for forming the box with different target strain rates: $\dot{\varepsilon}^{tgt} = 1 \times 10^{-5}$ s^{-1} and $\dot{\varepsilon}^{tgt} = 1 \times 10^{-4}$ s^{-1} (Lin and Dunne, 2001).

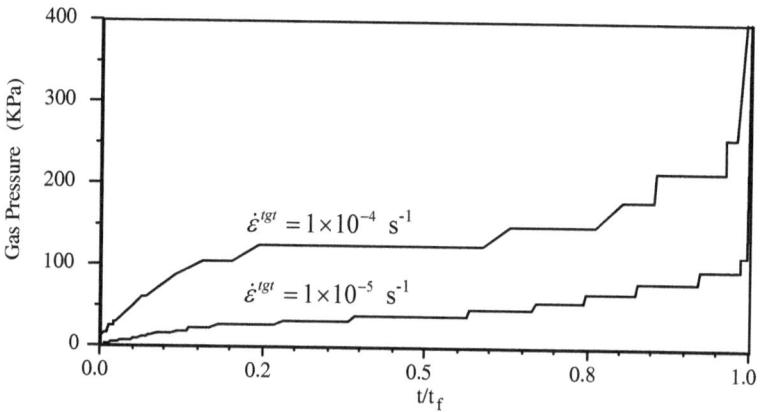

Fig. 8.9   Variation of applied gas pressure history for forming the box with different target strain rates: $\dot{\varepsilon}^{tgt} = 1 \times 10^{-5}$ s^{-1} and $\dot{\varepsilon}^{tgt} = 1 \times 10^{-4}$ s^{-1} (Lin and Dunne, 2001).

e. *Effect of forming rate on localised thinning*

The gas pressure increases continuously with forming time because of increasing geometrical and frictional constraint and, in addition, to overcome material hardening due to grain growth effects. Very high gas pressure is required to fill the corner part of the die, which is the last stage of the forming process. The non-uniform thinning of the formed part was investigated using different target forming rates and is shown in **Fig. 8.10**, where it is seen that higher strain rate forming would result in less localised thinning (**Fig. 8.10(b)**).

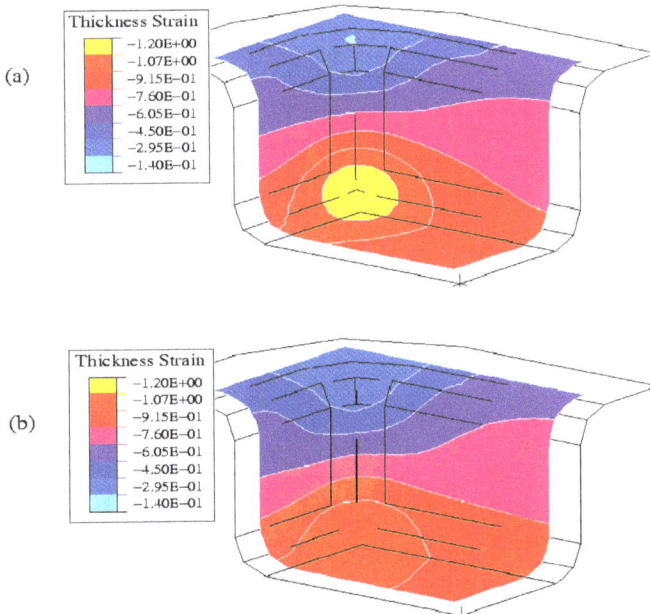

Fig. 8.10 Thickness strain distribution for the formed box with different target strain rates: $\dot{\varepsilon}^{tgt} = 1 \times 10^{-5}$ s^{-1} and $\dot{\varepsilon}^{tgt} = 1 \times 10^{-4}$ s^{-1} (Lin and Dunne, 2001).

f. *Effect of forming rate on grain growth*

By comparison of the field plots (a) $\dot{\varepsilon}^{tgt} = 1 \times 10^{-5}$ s^{-1} and (b) $\dot{\varepsilon}^{tgt} = 1 \times 10^{-4}$ s^{-1} in **Fig. 8.11**, it can be seen that the average grain size over the formed part is larger for the low forming rate (a) than that for the high forming rate (b). This occurs because the time required to form the part is longer with the lower forming rate, which enables further isothermal grain growth to take place. However, the spatial variation in grain size for the formed part is more significant, ($\Delta d = d_{max} - d_{min} \approx$ 1.08 µm) at the higher forming rate (b) than that ($\Delta d \approx 0.5$ µm) at the lower forming rate (a). Both the static and plastic strain-induced grain growth rates decrease with larger grain size. This was experimentally verified by Ghosh and Hamilton (1979) and modelled by Lin and Dunne (2001). The grain size evolution in superplastic forming processes affects viscoplastic flow of the material and thus also the thickness distribution of formed part.

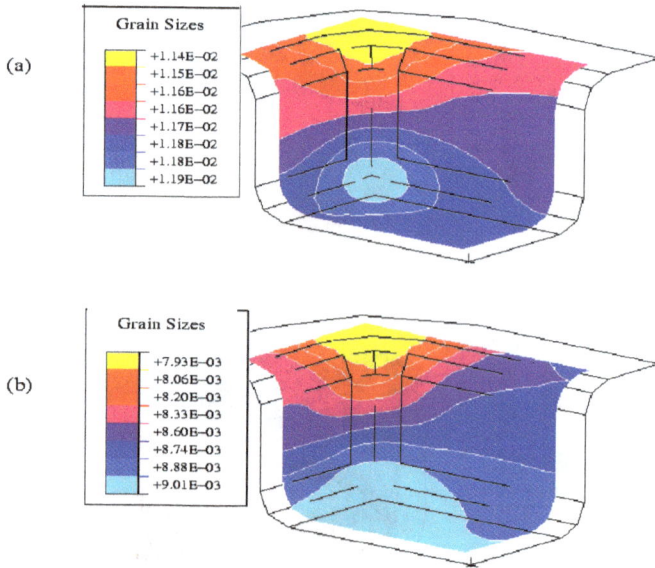

Fig. 8.11 Grain size distribution for the formed box with target strain rates: $\dot{\varepsilon}^{tgt} = 1 \times 10^{-5} \text{ s}^{-1}$ and $\dot{\varepsilon}^{tgt} = 1 \times 10^{-4} \text{ s}^{-1}$ (Lin and Dunne, 2001).

8.2.3 *Grain size effect on forming of the Ti-6Al-4V rectangular box*

To investigate the effects of non-uniform initial grain size, its distribution and target deforming rate on the thinning of Ti-6Al-4V at 927 °C during the superplastic gas-blow forming of the rectangular box, finite-element simulations were carried out using the finite-element model detailed in the previous section (Lin, Cheong and Ball, 2003). The simulations were carried out using Eq. (8.3) with target forming strain rate of 2×10^{-4} s^{-1} and 2×10^{-5} s^{-1}, uniform initial sheet thickness $h^0 = 1.5$ mm and initial grain-size fields as follows:

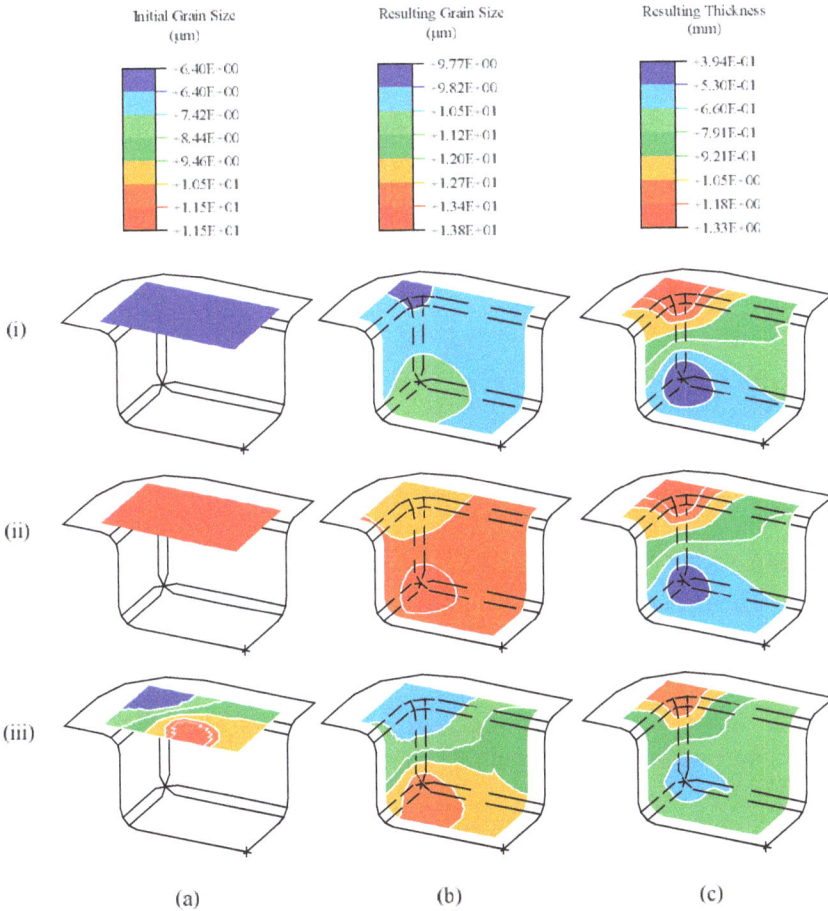

Fig. 8.12 Field plots of (a) initial microstructures; (b) resulting grain size; and (c) resulting thickness for (i) $d^0 = 6.4$ µm; (ii) $d^0 = 11.5$ µm; and (iii) 6.4 µm $\leq d^0 \leq$ 11.5 µm (Lin, Cheong & Ball, 2003).

- $d^0 = 6.4$ µm (uniform) (**Fig. 8.12 (a)(i)**).
- $d^0 = 11.5$ µm (uniform) (**Fig. 8.12 (a)(ii)**).
- 6.4 µm $\leqslant d^0 \leqslant$ 11.5 µm (non-uniform) (**Fig. 8.12 (a)(iii)**).

The non-uniform initial grain-size field 6.4 μm $\leq d^0 \leq 11.5$ μm is obtained by first dividing linearly the sheet material into six regions according to the resulting thickness obtained from the finite-element simulation with $\dot{\varepsilon}^{tgt} = 2 \times 10^{-4}$ s^{-1} and $d^0 = 6.4$ μm (**Fig. 8.12 (c)(i)**), of which the resulting thickness field is 0.394 mm $\leq h^t \leq 1.330$ mm. Then, it is distributed linearly in the range of initial grain size 6.4 μm $\leq d^0 \leq 11.5$ μm over the regions. In particular, the thinnest region corresponds to the coarsest grain size of 11.5 μm and the thickest region the finest grain size of 6.4 μm. The resulting non-uniform initial grain-size field is shown in **Fig. 8.12 (a)(iii)**. The resulting grain-size and sheet-thickness fields at $\dot{\varepsilon}^{tgt} = 2 \times 10^{-4}$ s^{-1}, with the three initial grain-size distributions (**Fig. 8.12 (a)(i–iii)**) are presented in **Fig. 8.12 (b)(i–iii)** and **(c)(i–iii)**, respectively.

a. Effects of initial grain size on gas-pressure history

Fig. 8.13 shows the computed gas-pressure histories for $\dot{\varepsilon}^{tgt} = 2 \times 10^{-4}$ s^{-1} with the three grain-size distributions (**Fig. 8.12 (a)(i–iii)**). It can be seen that the histories can be divided into three distinctive stages:

- For the first approximately 15% of process time, the gas pressures increase with t/t_f where t is time and t_f the time at which the associated forming process terminates.

- Then, they remain at an almost constant level.

- Eventually, they increase with t/t_f again and this stage encompasses approximately 50% of the process time.

The increase in gas pressure seen in the first stage is a result of the rapid increase in stresses mainly due to the hardening effects of grain and dislocation density growth. The sheet with grain sizes $d^0 = 6.4$ μm, $d^0 = 11.5$ μm and 6.4 μm $\leq d^0 \leq 11.5$ μm freely bulge in this stage.

Almost constant gas pressures are observed in the second stage. The hardening due to grain growth and dislocation, modelled by the unified superplastic constitutive equations, may at some point reach a balance with the softening induced by the reduction of sheet thickness; the almost constant gas pressures observed in the second stage (**Fig. 8.13**) are a result of the balance between the material hardening due to grain and dislocation density growth and apparent softening due to material thinning. The sheet with three different grain-size distributions also freely bulges in this stage.

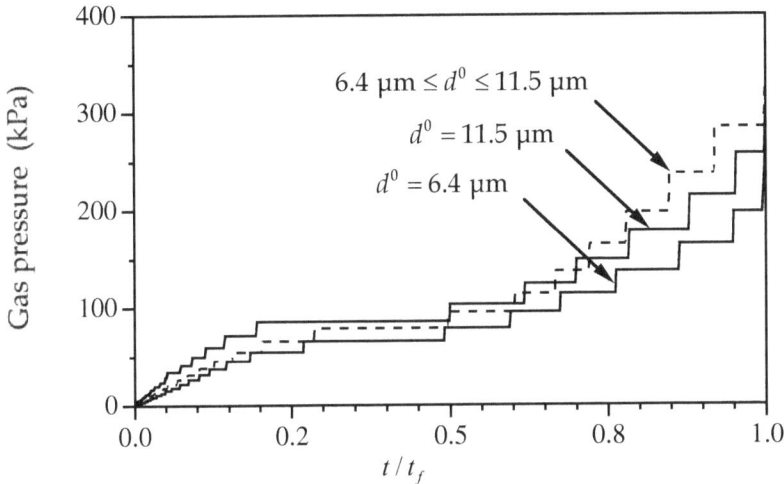

Fig. 8.13 Gas-pressure histories for d^0 = 6.4 µm, d^0 = 11.5 µm and 6.4 µm $\leqslant d^0 \leqslant$ 11.5 µm (Lin, Cheong & Ball, 2002).

In the final stage, gas pressure increases with t/t_f (**Fig. 8.13**). This indicates that part of the sheet has contacted the bottom of the die surface and restrains further deformation of parts yet to fill the cavity. Thus higher pressure is needed to complete the deformation at a rate of $\dot{\varepsilon}^{tgt}$.

The computed gas pressure for d^0 = 11.5 µm is generally higher than that for d^0 = 6.4 µm. This is because of the higher grain-growth hardening in the former sheet material (**Fig. 8.12 (b)(ii)**) due to the larger grain size (**Fig. 8.12 (b)(i)**). Similarly, the computed gas pressure for 6.4 µm $\leqslant d^0 \leqslant$ 11.5 µm lies between that for d^0 = 11.5 µm and 6.4 µm, in the first and second stages (**Fig. 8.13**). However, the gas pressure for 6.4 µm $\leqslant d^0 \leqslant$ 11.5 µm becomes the highest in the final stage. This is because the corner region of the formed part that comes into contact with the die eventually possesses the greatest thickness (**Fig. 8.12 (c)(iii)**), hence the lowest apparent softening due to thinning, and coarsest grain size (**Fig. 8.12 (b)(iii)**) – the highest grain-growth hardening – compared to that in d^0 = 11.5 µm and 6.4 µm.

b. Effects of initial grain size on material thinning

Figure 8.14 shows the histories of the difference between the maximum and minimum resulting sheet thickness $\max\{h\} - \min\{h\}$. Three

J. Lin

computations are carried out at $\dot{\varepsilon}^{tgt} = 2 \times 10^{-4}$ s^{-1} with the three initial grain-size distributions shown in **Fig. 8.12 (a)(i-iii)** while another at $\dot{\varepsilon}^{tgt} = 5 \times 10^{-5}$ s^{-1} with $d^0 = 6.4$ μm.

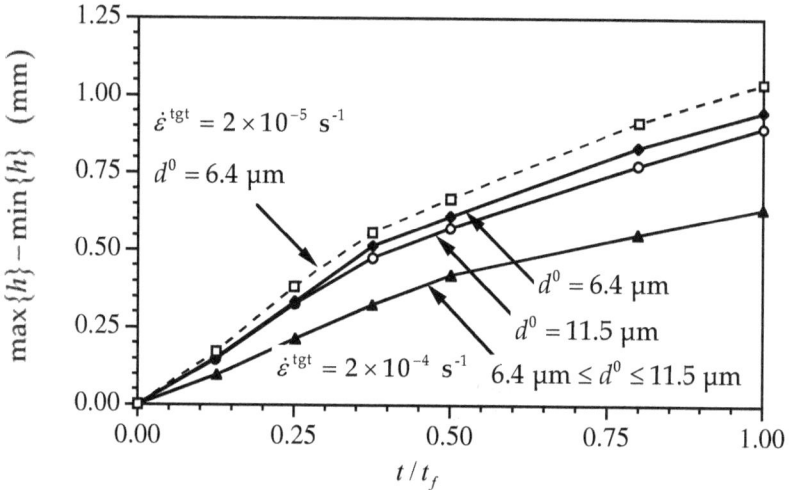

Fig. 8.14 Histories of max{h} – min{h}: three computations were carried out at $\dot{\varepsilon}^{tgt} = 2 \times 10^{-4}$ s^{-1} (solid curves) with $d^0 = 6.4$ μm, 11.5 μm and 6.4 μm $\leqslant d^0 \leqslant$ 11.5 μm and one at $\dot{\varepsilon}^{tgt} = 5 \times 10^{-5}$ s^{-1} (dashed curve) with $d^0 = 6.4$ μm (Lin, Cheong & Ball, 2002).

It can be seen from the figure that there is no significant difference between the profiles of max{h} – min{h} for $d^0 = 11.5$ μm and 6.4 μm at $\dot{\varepsilon}^{tgt} = 2 \times 10^{-4}$ s^{-1}. In particular, the values of max{h} – min{h} for the two simulations at $t/t_f = 1.0$ are 0.949 mm and 0.897 mm, respectively, which give a difference of less than 4% of h^0. The resulting sheet thickness distributions for the two simulations in **Fig. 8.12 (c)(i–ii)** also are not significantly different. This indicates that coarsening uniformly the initial grain size from 6.4 μm to 11.5 μm is not capable of effectively improving the evenness of the resulting sheet thickness. This is because, when the maximum deformation rate is restricted, sheet materials with uniform values of d^0 are very likely to deform in a similar manner; the path of a point on a sheet, from the initial position to the final, is unlikely to be varied significantly.

Using the initial microstructural gradients shown in **Fig. 8.12 (a)(iii)**, it can be seen from **Fig. 8.12 (c)(iii)** that the uniformity over the resulting sheet thickness is greatly improved compared to that in **Fig. 8.12 (c)(i-ii)**. Also, the history of $\max\{h\} - \min\{h\}$ for 6.4 μm $\leq d^0 \leq$ 11.5 μm shown in **Fig. 8.14** is generally lower than the others. In particular, the value of $\max\{h\} - \min\{h\}$ for 6.4 μm $\leq d^0 \leq$ 11.5 μm at $t/t_f = 1.0$ is 0.636 mm, which is lower than that for $d^0 = 6.4$ μm at the same $\dot{\varepsilon}^{tgt}$ by 20% of h^0. The resulting thickness in the corner region for 6.4 μm $\leq d^0 \leq$ 11.5 μm is also the highest among the three (**Fig. 8.12 (c)(i–iii)**). However, its resulting microstructure (**Fig. 8.12 (b)(iii)**) is rather non-uniform compared to the others (**Fig. 8.12 (b)(i–ii)**). This shows that the use of appropriate initial microstructural gradients is feasible for regulating the resulting thickness of a formed part, but with some unevenness over the resulting microstructure.

Fig. 8.14 also shows the history of $\max\{h\} - \min\{h\}$ for the simulation with $\dot{\varepsilon}^{tgt} = 5 \times 10^{-5}$ s^{-1} and $d^0 = 6.4$ μm. It is generally higher than that for $\dot{\varepsilon}^{tgt} = 2 \times 10^{-4}$ s^{-1} and $d^0 = 6.4$ μm. This indicates that for the lower $\dot{\varepsilon}^{tgt}$, the more severe the material thinning.

Tailored grain size distribution sheets can be achieved through localised heat treatment processes, such as laser treatment or friction stir welding type treatment.

8.2.4 *Forming of Al-Zn-Mg box at 515 °C*

The effect of damage on localised thinning in the formed rectangular box, **Fig. 8.6,** is studied using Eq. (8.4). The Al-Zn-Mg blank with size of $a = 1.12H$ and $b = 0.77H$ as shown in the FE model of **Fig. 8.6**, is superplastically formed at a temperature of 515 °C with target strain rates of either $\dot{\varepsilon}^{tgt} = 2 \times 10^{-4}$ s^{-1} or 2×10^{-3} s^{-1}.

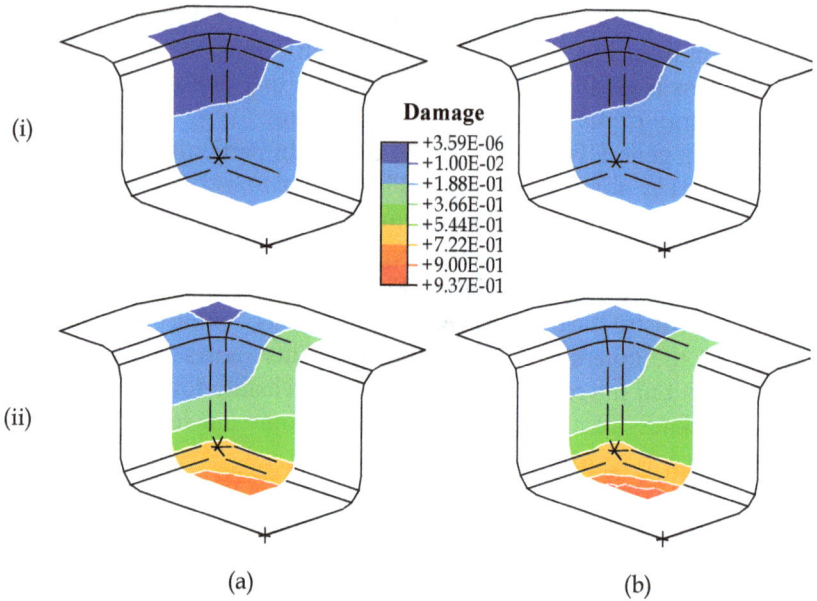

Fig. 8.15 (a) and (b) are field plots of f_h and f_e for Al-Zn-Mg; (i) and (ii) denote $\dot{\varepsilon}^{tgt} = 2 \times 10^{-4}$ s^{-1} and 2×10^{-3} s^{-1}, respectively (Cheong, 2002).

Figure 8.15 shows the distributions of ductile void growth damage f_h **(a)** and effective damage f_e **(b)** over the formed Al-Zn-Mg part with $\dot{\varepsilon}^{tgt} = 2 \times 10^{-4}$ s^{-1} **(i)** and 2×10^{-3} s^{-1} **(ii)**. Making use of the set of unified superplastic-damage constitutive equations, Eq. (8.4), the variations of f_h that are not immediately "visible" to designers and production engineers are modelled and revealed in a quantitative manner. It can be "seen" that the distributions of f_h **(a)(i)** for $\dot{\varepsilon}^{tgt} = 2 \times 10^{-4}$ s^{-1} are generally equal to that of the effective damage f_e **(b)(i)**, approximately. For $\dot{\varepsilon}^{tgt} = 2 \times 10^{-3}$ s^{-1}, the difference is obvious. It also can be seen that damage levels in the formed part are much higher using the higher target forming rate $\dot{\varepsilon}^{tgt} = 2 \times 10^{-3}$ s^{-1}. The reality of such an increase in damage by increase in target forming rate can be justified by the uniaxial stress–strain relationships of the alloy shown in **Fig. 8.4** and is consistent with common experimental observations reported in the literature. For $\dot{\varepsilon}^{tgt} = 2 \times 10^{-3}$ s^{-1} the formed Al-Zn-Mg component records a maximum f_h value of greater than 0.722 and the f_e value is over 0.9 as shown in **Fig. 8.15 (ii) (a)** and **(b)**.

Fig. 8.16 shows thickness distributions of the formed Al-Zn-Mg box with **(i)** $\dot{\varepsilon}^{tgt} = 2 \times 10^{-4}$ s^{-1}, **(ii)** $\dot{\varepsilon}^{tgt} = 5 \times 10^{-4}$ s^{-1} and **(iii)** $\dot{\varepsilon}^{tgt} = 2 \times 10^{-3}$ s^{-1}. Group **(a)** is obtained with consideration of cavitation damage while group **(b)** is without using the equation set Eq. (8.4).

For $\dot{\varepsilon}^{tgt} = 2 \times 10^{-4}$ s^{-1}, the consistent results in **Fig. 8.16 (a)(i)** and **(b)(i)** show that the flow pattern simulated without consideration of cavitation damage is just as reliable as that with consideration of cavitation damage. This indicates that a conical type of spatial material state results at the end of Stage I (as defined in the previous section) for both forming simulation processes. In addition, the development of cavitation damage at the low target forming rate is not important, and hence the material behaviour can be safely simulated without concern regarding the presence of cavitation damage. It can also be seen from **Fig. 8.16 (a)(i)** and **(b)(i)** that the thinnest regions of both the resulting components are the corner regions with similar values.

For the intermediate target forming rate of $\dot{\varepsilon}^{tgt} = 5 \times 10^{-4}$ s^{-1}, the resulting thickness fields shown in **Fig. 8.16 (a)(ii)** and **(b)(ii)** also show that the apparent softening due to cavitation damage is not important. However, the thickness of both the corner regions is now higher and the evenness over the thickness distributions is improved compared to the part formed with $\dot{\varepsilon}^{tgt} = 2 \times 10^{-4}$ s^{-1}. Such improvements are a result of the increased target forming rate that reduces the severity of flow localisation in Stage I.

For the highest target forming rate of $\dot{\varepsilon}^{tgt} = 2 \times 10^{-3}$ s^{-1}, a significant difference is observed between the results in **Fig. 8.16 (a)(iii)** and **(b)(iii)**: the simulation without consideration of cavitation damage (**Fig. 8.16 (a)(iii)**) generates a more even thickness distribution, compared to that in **Fig. 8.16 (a)(ii)**. On the other hand, the simulation with consideration of cavitation damage (**Fig. 8.16 (b)(iii)**) generates a less even distribution, compared to that in **Fig. 8.16 (b)(ii)**. This indicates that the presence of cavitation damage is important at the high target forming rate and should not be ignored.

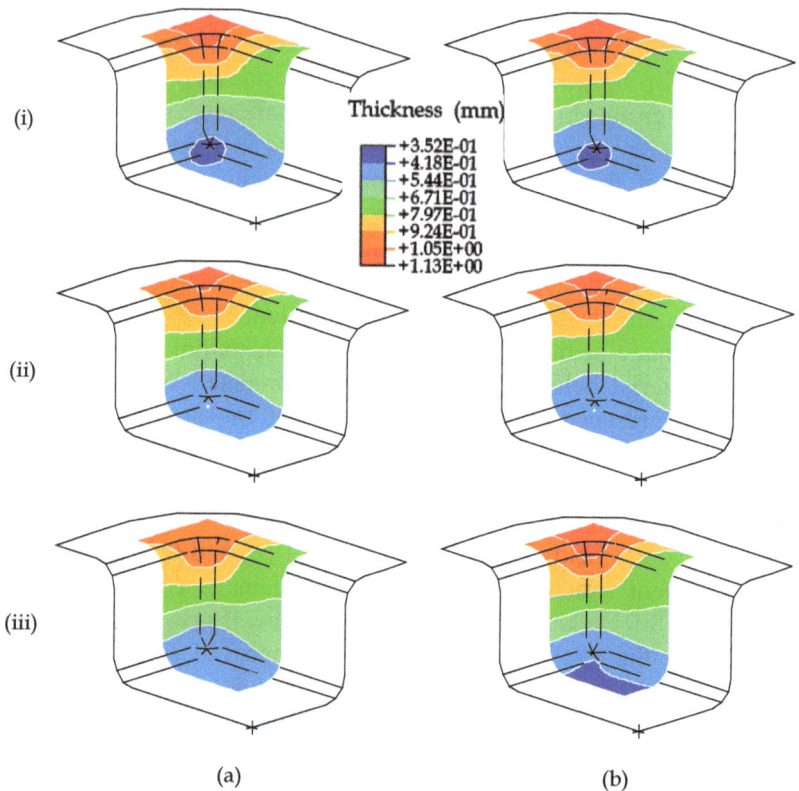

Fig. 8.16 Resulting thickness fields for Al-Zn-Mg; group (a) is obtained without consideration of cavitation damage while group (b) is with consideration of cavitation damage; (i), (ii), and (iii) denote $\dot{\varepsilon}^{tgt} = 2 \times 10^{-4}$ s^{-1}, 5×10^{-4} s^{-1} and 2×10^{-3} s^{-1} respectively (Cheong, 2002).

Without considering the effects of cavitation damage, the severity of flow localisation in Stage I can always be improved by increasing the target forming rate, as well as in Stage II. However, this is not realistic; above a certain level of deformation the effect of increasing the strain rate can result in damage growth. This can be deduced from the uniaxial stress–strain relationships for the material (**Fig. 8.4**). It is to be noted that such behaviour can be simulated only with consideration of cavitation damage, which highlights the robustness and generality of the set of superplastic damage constitutive equations. While the aforementioned results (**Fig. 8.16 (a)(i–ii)**) generated without consideration of cavitation damage do coincide with those in (**Fig. 8.16 (b)(i–ii)**) obtained with consideration of cavitation damage at $\dot{\varepsilon}^{tgt} = 2 \times 10^{-4}$ s^{-1} and 2×10^{-3} s^{-1}, here it is evident that the two modelling methodologies can yield contradictory results at high strain rates. Thus, the modelling of gas-blow forming without consideration of cavitation damage could be dangerous and can sometimes lead to misleading results.

8.3 Creep Age Forming (CAF) of Large Aluminium Panels

8.3.1 *CAF process and deformation mechanisms*

a. Applications

Holman (1989) at Textron Aerostructures conceived a novel forming method and demonstrated its use by applying it to the production of aircraft wingskins. This method, known as *creep age forming* (CAF), has been highly successful for making large, *high-strength aluminium panel* components. It is also called *autoclave age forming*.

Fig. 8.17 A wing panel for Airbus A380 (Watcham, 2004).

CAF takes advantage of the unique material behaviour of certain metal alloys during the process of forming; more specifically, it utilises the stress relaxation and/or creep phenomena that arise when a heat-treatable metal alloy is artificially aged. Examples of parts made by CAF are the wing panel components of the Airbus A380, as shown in **Fig. 8.17**. Potentially, this process can also be used to form the larger outer-skin panels of spacecraft and launch vehicles as shown in **Fig. 8.18**. (Yang, 2013).

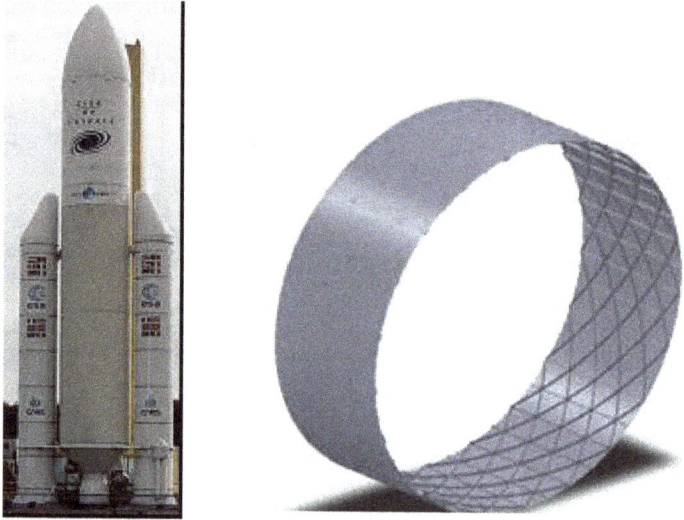

Fig. 8.18 A grid-stiffened out-skin panel for space launch vehicles (Yang, 2013).

b. Forming process and mechanisms

A typical CAF process, as illustrated in **Fig. 8.19(a),** includes the following stages:

- **Stage 1:** A solution heat-treated aluminium plate work-piece, which could be in the T4 condition or partially aged, is placed on a die surface consisting of the profile of the part to be formed.

- **Stage 2:** Utilising vacuum bagging technique, (Zhan, Lin and Dean, 2011) the work-piece is drawn into the die shape by an externally applied pressure. Due to the moderate variation in shape of surfaces required in aerospace panels, deformation of the work-piece is usually mainly elastic. Thus, release of pressure would result in the work-piece returning to its flat shape.

- **Stage 3:** The die and work-piece are placed in an ageing autoclave where a vacuum is used to draw work-piece to conform to the die surface and the alloy is artificially aged for a certain period at elevated temperature. The *ageing temperature and time* are defined according to the heat treatment condition of the material. **Fig. 8.19 (b)** and **(c)** depict the typical stress relaxation and creep behaviour of the work-piece as ageing takes place, which contribute to the plastic deformation in the work-piece.

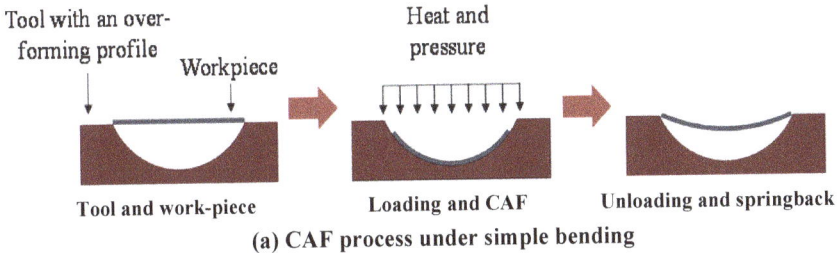

(a) CAF process under simple bending

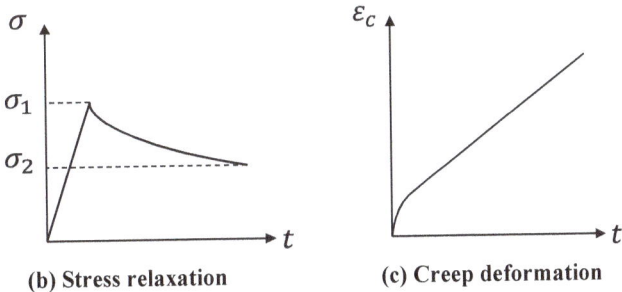

(b) Stress relaxation

(c) Creep deformation

Fig. 8.19 Forming process and mechanisms of CAF under simple bending. Reproduced from Zhan, Lin and Dean (2011).

- **Stage 4:** After ageing, the fixture is cooled down to room temperature, moved out from the autoclave, and loading pressure released. For a 7000 series aluminium alloy (AA7xxx) as an example, the formed part could be in T6 (peak strength) or T7 (over-aged for better toughness) condition. As the accumulated plastic strain during the ageing stage is very small, the resulting residual stress in the work-piece cannot be completely eliminated. Springback is therefore very high, making the control of the shape characteristics of formed components challenging in CAF (Lin, Ho and Dean, 2006).

As a *warm aluminium alloy forming* process, CAF combines the age hardening behaviour (also known as precipitation hardening) of the virgin material with creep deformation together to form high-strength aluminium components. This behaviour is unique to certain heat-treatable metal alloys, with strength that can be varied by ageing. Heat treatable wrought aluminium alloys, such as AA2xxx, AA6xxx and AA7xxx, are very strong and have a wide range of applications in the aerospace industry. In contrast, aluminium alloys in other series, such as AA1xxx, AA3xxx and AA5xxx, do not respond to age hardening and are typically hardened through other mechanisms such as solution hardening and dislocation hardening. It is therefore important to note that aluminium alloy processing in the following text is concerned with only heat-treatable aluminium alloys, unless stated otherwise.

Mechanisms of the deformation that take place in CAF are as follows:

- **Primary and secondary creep under high to medium stress levels.** Creep phenomena that occur at low stress levels are typically disregarded; this is due to (1) the short ageing time (from 8 hours to 30 hours depending on the material, process and required temper condition), and (2) the creep strain that accumulates being insignificant. Creep deformation is not sensitive to grain size in CAF. Primary and secondary creep deformation can therefore be modelled using traditional equations without consideration of creep damage. On the other hand, hardening due to dislocation and ageing are of high importance during primary creep, and this can significantly affect the creep behaviour of the material.

- **Stress relaxation.** This is due to a combination of high stress creep deformation and thermally activated stress relaxation (recovery).

CAF is a slow process due to the slow, lengthy deformation that occurs and the long ageing time that is usually required. Therefore from an economic standpoint, it is more suitable for the production of extra-large panel components in low volume. In fact, its use has already been shown to be highly successful through the forming of various aircraft wing components with predefined microstructure, strength and shape requirements (Zhan, Lin and Dean, 2011).

8.3.2 *Age hardening of aluminium alloys*

As suggested by its name, CAF relies heavily on the age hardening behaviour that is unique to a specific metal alloy. Although its use has been demonstrated only for aluminium alloys, in theory it can also be used for other metal alloys. This is theoretically plausible as long as significant creep and stress relaxation take place at the material's ageing temperature (Jeunechamps, Ho, Lin *et al.*, 2006).

a. Modelling creep ageing

Reliable constitutive equations must first be established in order to enable material behaviour under creep ageing conditions to be modelled. Prior to the development of a set of physically based, *unified creep ageing constitutive equations*, the CAF hardening mechanisms must be understood. In this sub-section, some basic concepts of the heat treatment processes available to aluminium alloys are introduced.

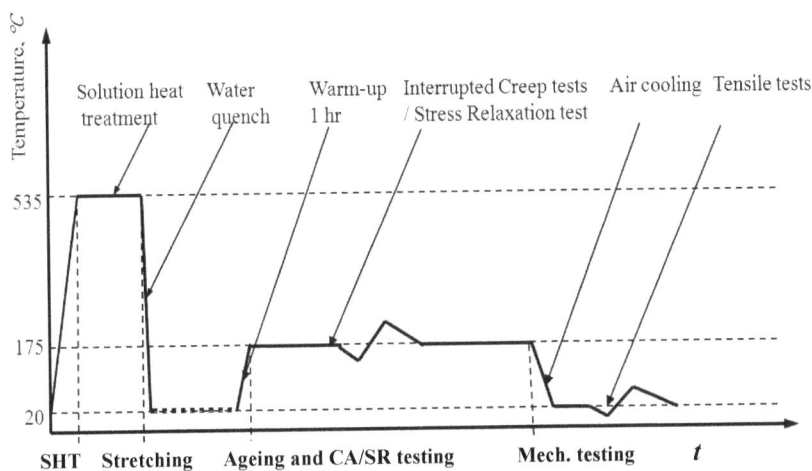

Fig. 8.20 Determination of alloy behaviour through CAF heat treatment, creep ageing (CA), stress relaxation (SR) and mechanical testing.

A typical heat treatment process simulating CAF is shown in **Fig. 8.20**. This must be conducted for the microstructure and mechanical properties of an alloy to be determined.

- *Solution heat treatment (SHT)*. SHT is carried out at a temperature in the range of 470 °C to 570 °C, for approximately 10 minutes to an hour, depending on alloy composition. This enables alloying elements to be distributed uniformly within the aluminium matrix. The material is then rapidly cooled (quenched) from the SHT temperature to room temperature, when a supersaturated solid solution (SSSS) results. The rate of cooling must be sufficiently high to avoid the formation of large precipitates at grain boundaries, which can cause significant reduction in strength. For many alloys, this may also cause stress corrosion cracking (SCC) that is known to reduce fatigue life. The alloy is now in the *T4 condition* and it is relatively soft and ductile.

- *Stretching.* Residual stress, caused by rapid quenching can distort the shape of a work-piece; this is particularly common in alloy sheet. It is often a standard requirement for the component to be stretched by 1–3% strain for residual stress removal.

- *Ageing.* This enables controlled nucleation and growth of precipitates to produce peak strength, (known as the *T6 condition*) with a lower ductility. Alternatively, an alloy can be over-aged (*T7*) for increased ductility accompanied by a slight reduction in strength. Most creep-aged components are therefore deliberately processed to *T7* condition for a good compromise between strength and ductility.

- *Creep ageing/stress relaxation.* Primary and secondary creep behaviour of an alloy is characterised through interrupted creep tests. Stress relaxation tests can also be carried out at specified ageing temperature for different ageing times, if necessary.

- *Mechanical testing.* Tensile (or hardness) testing a creep-aged (or stress relaxed) specimen can reveal mechanical properties after *dynamic ageing. Dynamic ageing* is ageing that occurs under the conditions of stress and plastic deformation (e.g. creep).

Ageing can take place under two conditions:

- *Artificial ageing* is that occurring at a temperature above ambient, for a certain period. As an example, peak strength of AA6082 is reached after ageing for 9 hours at 190 °C. For some alloys, multi-stage artificial ageing is also possible; an AA6xxx alloy may first be aged for 5 minutes at 220 °C (this is known as first stage ageing) and then 180 °C for 30 minutes (second stage ageing). This is a fast ageing process and close-to-peak strength can be achieved. In CAF, however, long term ageing is usually preferred – this is to enable creep or stress relaxation to take place, thus accumulating sufficient plastic strain in the work-piece to ensure permanent shape formation.

- *Natural ageing* is due to formation and growth of precipitates, which cause alloy hardening, at room temperature. Natural ageing can occur readily in certain alloys; apparent age hardening can initiate in some AA7xxx series alloys within only a few days of being left at room temperature. Manufacturers can avoid this by requiring material suppliers to first artificially age the solution heat treated alloy immediately after SHT. This helps microstructural stabilisation at room temperature. Subsequently, manufacturers can perform CAF on the delivered material, which contains second stage ageing.

b. Nucleation and growth of precipitates

Precipitation hardening requires finely dispersed precipitates that occur during the stress ageing process to obstruct the movement of dislocations, and thus increase strength.

The ageing behaviour of some AA7xxx alloys during isothermal heating treatment has been studied (Ho, 2004) and the generally accepted ageing sequence is

$$\alpha_{\text{solid solution}} \rightarrow \alpha + \text{spherical GP zones} \rightarrow \alpha + \eta' \rightarrow \alpha + \eta$$

where α is the aluminium matrix, GP zones are Guinier–Preston zones, η' is a transition phase and η is the equilibrium phase, $MgZn_2$. **Figure 8.21** shows a schematic representation of the microstructural evolution of a 7000 series aluminium alloy during ageing.

As stated above, the ageing process can be divided into several stages. Normally, ageing starts with quenched alloy in which the microstructure is a supersaturated solid solution, (Stage I). This is equivalent to a T4 condition for an AA7xxx alloy. Stage II is initiated by formation of ordered and solute rich clusters, so called GP zones, which are of 1 to 2 atom planes in thickness and have a coherent crystal structure that is similar to the aluminium matrix. Due to this coherency, their interfacial energy is low, easing nucleation of precipitates. Although formation of GP zones promotes the hardness of the alloy, they are small and coherent and can be cut through by dislocations. The number of precipitates increases drastically with time at this stage and the difference in atomic size of the constituents and aluminium, strains the lattice. Hardening is now due to the increased work required to move dislocations through the strained lattice and work required for dislocations to pass through the GP zones. This over-counteracts the decreasing solute strengthening effects

J. Lin

as the concentration of constituent elements decreases. Strength and hardness of the material continues to increase in the initial stages of coarsening as the size of an average particle continues to increase.

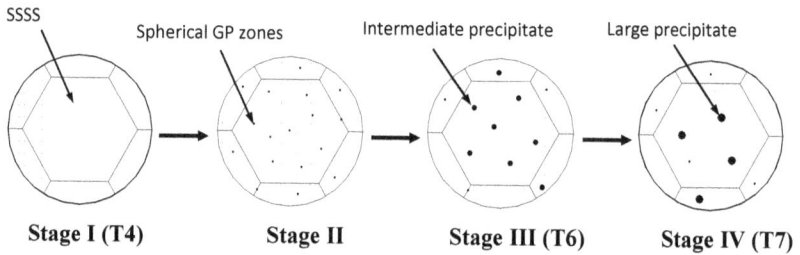

Fig. 8.21 Schematic of an AA7xxx at different stages of ageing:
(I) supersaturated solid solution, (II) nucleation of GP zones (coherent),
(III) precipitate coarsening (semi-coherent), (IV) over-aged condition
(incoherent) (Lin, Ho and Dean, 2006).

Eventually, at Stage III, the GP zones are replaced by the more stable η' phase with a typical size between 1 nm and 10 nm (Ferragut, Somoza and Tolley, 1999), and are semi-coherent with the aluminium matrix. Peak strength of the material is achieved at T6. If ageing continues from here onwards, the size of precipitates continues to increase whilst the number of dissolved constituents decreases steadily until reaching equilibrium. Coherency is lost as the sizes of precipitates increase. Interaction between dislocations and precipitates becomes weaker as the average spacing between precipitates becomes larger. Eventually, at Stage IV, further coarsening gives rise to large incoherent precipitates, known as the η phase. The average spacing between precipitates is now large enough to be bypassed by dislocations by the Orowan bowing process without being cut. Strength is reduced and the alloy is now over-aged (T7). This is illustrated in **Fig. 8.22.**

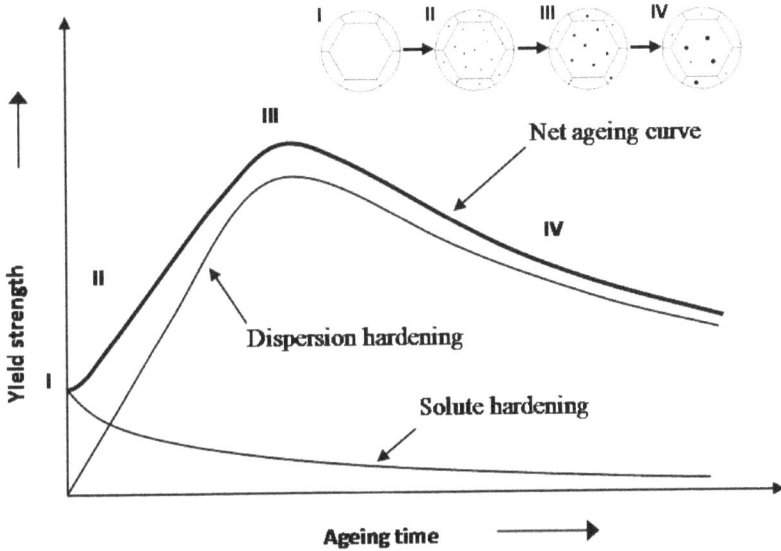

Fig. 8.22 Contributions of solute hardening and dispersion hardening to yield strength of AA7010 as a function of ageing time (Ho, 2004).

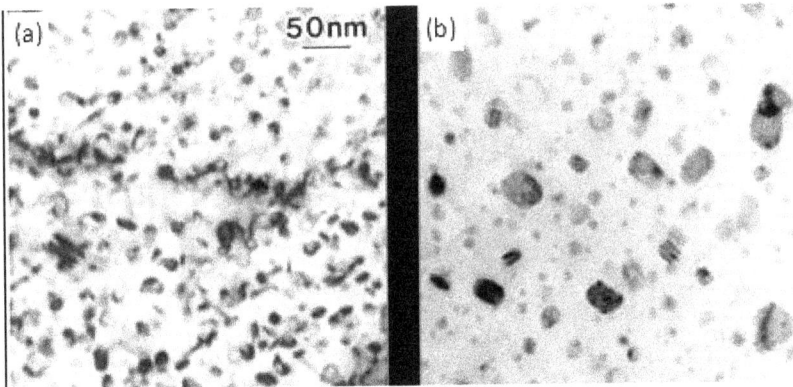

Fig. 8.23 Microstructures of two AA7xxx peak-aged at 150 °C.
(a) Al-1.72Zn-3.4Mg-0.1Ag and (b) Al-1.72Zn-3.4Mg-0.37Cu
(Caraher, Polmear and Ringer, 1998).

Figure 8.23 shows microstructures containing ageing precipitates for two 7000 series aluminium alloys peak-aged (T6) at 150 °C. Precipitates, which are darker in colour and more or less spherical, are evident. The diameter of the largest precipitate is less than 50 nm.

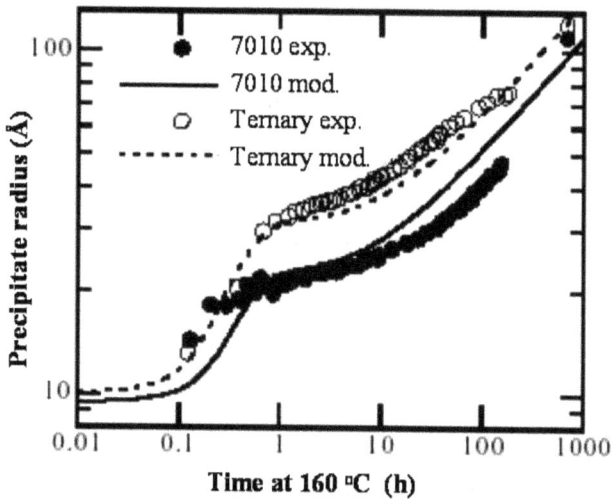

Fig. 8.24 Evolution of radii of precipitates during ageing at 160 °C (Deschamps, Solas and Bréchet, 2005).

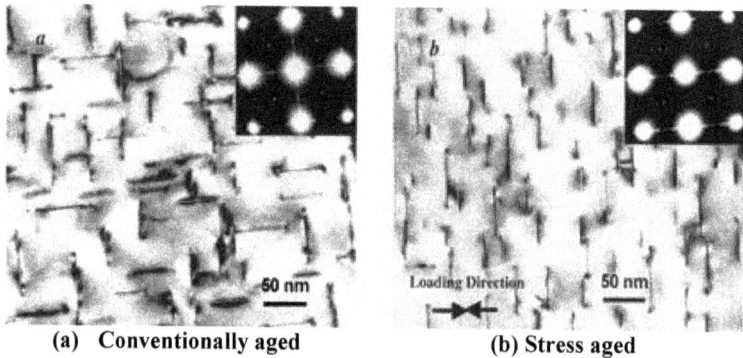

(a) Conventionally aged   (b) Stress aged

Fig. 8.25 TEM-BF images and diffraction patterns of (a) conventionally-aged and (b) stress-aged (40 MPa) Al-4Cu single crystals at 201 °C for 11h (Zhu and Starke, 2001).

Fig. 8.24 illustrates the evolution of the radii of precipitates of two alloys (AA7010 and Ternary, a modified AA7010) during ageing at 160 °C (Deschamps, Solas and Bréchet, 1999).

Distribution and growth of precipitates are also related to the direction of stress applied and this was found to be especially important to 2000 series aluminium alloys (Zhu and Starke, 2001). **Figure 8.25** shows the microstructure of an alloy that has been aged (a) stress-free and (b) under an applied stress of 40 MPa. Under the stress-free condition, the plate-shape precipitates are randomly distributed and mechanical properties are isotropic. On the other hand, the applied stress has changed the orientation of the precipitates, causing material anisotropy.

8.3.3 *Unified creep ageing constitutive equations*

a. Age hardening creep model

This model was developed by Lin, Ho and Dean (2006) and is based on a set of unified creep ageing constitutive equations. Evolution of creep strain, primary creep hardening, precipitate radius and yield strength of components, as well as their inter-relationships during CAF, are considered. Based on the "unified theories" approach, ageing kinetics including assumptions of (1) isothermal ageing, and (2) particle growth at constant volume fraction, the physically-based, unified creep ageing constitutive equations take the form as below:

$$\dot{\varepsilon}_c = A \sinh\left\{ B[(\sigma - \sigma_A)(1 - H)]\left(\frac{C_{ss}}{\sigma_{ss}}\right)^n\right\}$$

$$\dot{H} = \frac{h_c}{\sigma^{0.1}}\left(1 - \frac{H}{H^*}\right)\dot{\varepsilon}_c$$

$$\dot{r}_p = C_0\left(Q - r_p\right)^{m_0}\left(1 + \left|\frac{\dot{\varepsilon}_c}{\gamma}\right|^{m_1}\right) \qquad (8.6)$$

$$\sigma_A = C_A r_p^{m_2}$$

$$\sigma_{ss} = C_{ss}\left(1 + r_p\right)^{-m_3}$$

$$\sigma_Y = \sigma_{ss} + \sigma_A,$$

where A, B, h_c, H^*, n, Q, C_0, C_A, C_{SS}, m_0, m_1, m_2 and m_3 material constants that can be determined using experimental data. Several key features of the evolution of creep strain, ε_c, can be seen:

- Creep rate is not only a function of stress, σ, and dislocation hardening, H, but also age hardening, σ_A, and solute hardening, σ_{SS}.

- The nucleation and growth of precipitates are related to the creep deformation (dynamic ageing).

- C_{SS}, σ_{SS} and n are responsible for softening of the matrix material as precipitates coarsen.

Primary creep is modelled using the variable H, which takes the form of the equation introduced by Kowalewski, Hayhurst and Dyson (1994). It varies from 0 at the beginning of the creep process to H^*, which represents the saturation value at the end of the primary ageing period. Stress sensitivity is controlled through the order of the power of σ.

As precipitates grow monotonically during isothermal ageing, the coarsening kinetic of the ageing mechanism is modelled via the growth of precipitate radius, $\dot{r}_p$. Because CAF is a combination of creep deformation and age hardening, the creep strain rate effect is included in $\dot{r}_p$ to describe this dynamic ageing process.

As dislocation density increases, the effect of age-hardening also increases. The sensitivity of the creep effect is controlled by the constants, γ and m_1. When the creep strain rate is equal to zero, $\dot{r}_p$ models only static ageing behaviour. The form of the equation for $\dot{r}_p$ has four key features:

1. To avoid numerical difficulties, precipitate radius, $\dot{r}_p$ is placed as a numerator instead of a denominator.

2. Q is used to represent the saturation limit for depletion of solute atoms within the aluminium matrix. As the depletion of solute atoms reaches its saturation value, precipitation stops, and precipitates stop growing.

3. m_0 gives flexibility to the equation as different alloys behave differently. For example, $m_0 = 1/3$ for AA7010.

4. $\dot{r}_p$ can model both the static and dynamic ageing behaviour of AA7xxx.

In this model, the use of $\dot{r}_p$ simplifies the ageing mechanism by assuming isothermal ageing (Deschamps *et al.*, 1999). Strengthening contribution from shearable precipitates can arise from a variety of mechanisms, such as chemical hardening or coherency strain hardening, to name but two. The symbols m_2 and C_A describe interactions between dislocations and shearable precipitates whilst assuming coarsening with constant volume fraction.

Table 8.4 Determined constants in equation set (8.6) for AA7010 at 150 °C.

$A(h^{-1})$	B (MPa^{-1})	n (-)	h_c (MPa)	H^* (-)	Q (nm)	γ (h^{-1})
3.97×10^{-7}	0.03	0.03	520	0.2	120	1.83×10^{-4}
C_0 (h^{-1})	C_A (MPa)	C_{SS} (MPa)	m_0 (-)	m_1 (-)	m_2 (-)	m_3 (-)
0.112	31.35	20	$1/3$	0.369	0.5	1.3

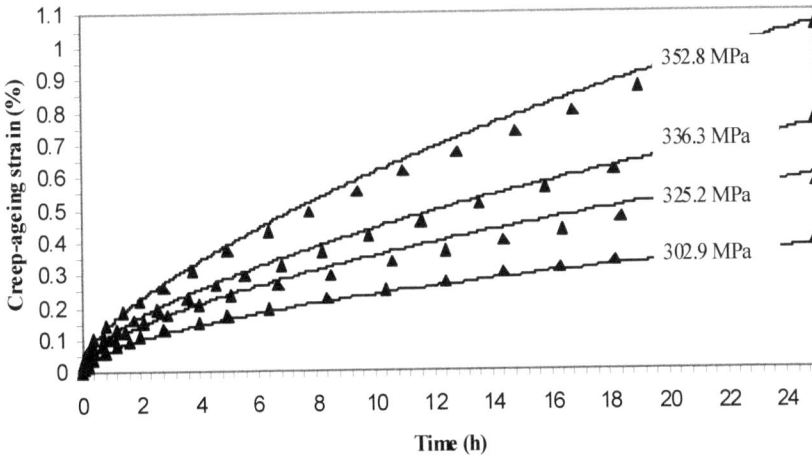

Fig. 8.26 Comparison of predicted (solid line) and experimental (symbols) creep strains for $\sigma = 302.9$ MPa, 325.2 MPa, 336.3 MPa and 352.8 MPa (Ho, 2004).

σ_{SS} approximates the contribution from solute hardening. C_{SS} is not only related to size, but also to the modulus as well as the electronic mismatch of the solute. The symbol m_3 describes the depletion of the solute into precipitates. As the concentration of the solute atoms decreases, the effect of σ_{SS} decreases, becoming a "softening" mechanism. The sum of σ_A and σ_{SS} is the total yield strength of the material, σ_Y. Ho (2004) demonstrated the use of this model using AA7010. Values of the determined constants are listed in **Table 8.4** and the predicted creep strains are shown in **Fig. 8.26**.

b. Modelling of T6 and T7 conditions

Ho's (2004) work is able to model creep ageing conditions only up to T6. Zhan, Lin, Dean *et al.* (2011) introduced the concept of normalised precipitate size to make the modelling of both T6 and T7 possible. Below is the equation set in its uniaxial form.

$$\dot{\varepsilon}_c = A_1 \sinh\{B_1[|\sigma|(1-\bar{p}) - k_0\sigma_y]\} \, sign\{\sigma\}$$
$$\dot{\sigma}_A = C_A \bar{r}^{-m_1}(1-\bar{r})$$
$$\dot{\sigma}_{ss} = C_{ss}\bar{r}^{-m_2}(\bar{r}-1)$$
$$\dot{\sigma}_{dis} = A_2 \cdot n \cdot \bar{p}^{n-1}\dot{\bar{p}} \qquad\qquad (8.7)$$
$$\sigma_y = \sigma_{ss} + \sqrt{\sigma_A^2 + \sigma_{dis}^2}$$
$$\dot{\bar{r}} = C_r(Q-\bar{r})^{m_3}(1+\gamma_0\bar{p}^{-m_4})$$
$$\dot{\bar{p}} = A_3(1-\bar{p})|\dot{\varepsilon}_c| - C_p\bar{p}^{m_5},$$

where A_1, B_1, k_0, C_A, m_1, C_{ss}, m_2, A_2, n, C_r, Q, m_3, γ_0, m_4, A_3, C_p and m_5 are material constants.

In this set of equations, creep rate, $\dot{\varepsilon}_c$, is a function of stress, σ, and normalised dislocation density, $\bar{p}$. Age hardening, σ_A, solute hardening, σ_{ss}, and dislocation hardening, σ_{dis}, altogether contribute to the yield strength, σ_y,, which varies during a CAF process.

Deschamps, Solas and Bréchet (2005) used precipitate radius to simplify modelling of the ageing mechanism. In CAF modelling, this concept was generalised by Zhan, Lin, Dean *et al.* (2011) with the introduction of normalised precipitate size,

$$\bar{r} = \frac{r}{r_c},$$

where r_c is the precipitate size at the peak ageing state, which considers the best match of precipitate size and spacing for the alloy (Ringer and Hono, 2000). When $0 \leq \bar{r} < 1$, under-ageing exists; $\bar{r} = 1$ represents peak-ageing (T6) and $\bar{r} > 1$, over-aged (T7). This approach simplified the modelling process significantly.

The term $(1 - \bar{\rho})$ is used to model primary creep. The normalised dislocation density was introduced by Lin, Liu, Farrugia *et al.* (2005) and is defined by

$$\bar{\rho} = \frac{\rho - \rho_i}{\rho_m},$$

where ρ_i is the dislocation density for the virgin material (the initial state), and ρ_m the maximum (saturated) dislocation density. Thus ρ varies from ρ_i to ρ_m. As a result, $\bar{\rho}$ varies from 0 to 1 with $\rho_i << \rho_m$. The first term of $\dot{\bar{\rho}}$ represents the development of dislocation density due to creep deformation and dynamic recovery. The second term describes the effect of static recovery in the dislocation density at the ageing temperature (Lin, Liu, Farrugia *et al.*, 2005).

$\dot{\sigma}_A$ and $\dot{\sigma}_{SS}$ represent the evolution of age and solute hardening respectively, where C_A describes the interaction between dislocations and shearable precipitates.

$\dot{\sigma}_{SS}$ approximates the contribution from solute hardening, where resistance is caused by solute atoms obstructing dislocation motion. C_{SS} is related to the size, modulus and electronic mismatch of the solute; m_2 describes the depletion of solute into precipitate. As the concentration of the solute atoms decrease, solute strengthening decreases, acting as a "softening" mechanism. $\dot{\sigma}_{dis}$ describes the evolution of dislocation hardening, which is a function of $\bar{\rho}$, as defined by Lin and Liu, Farrugia *et al.* (2005).

Combining $\dot{\sigma}_A$, $\dot{\sigma}_{SS}$ and $\dot{\sigma}_{dis}$ gives the yield strength σ_y – the overall yield strength during the ageing process. σ_y changes dynamically during the creep ageing period due to ageing and dislocation hardening, which affects the creep deformation of the material in CAF conditions.

This equation set was later generalised by Yang (2013) for modelling the nucleation and growth of the precipitates in a 2000 series aluminium alloy.

c. Constitutive relations for AA7055 at 120 °C

AA7055 material parameters for unified creep ageing constitutive equations Eq. (8.7) were successfully determined for a temperature of 120 °C by Zhan, Lin, Dean *et al.* (2011). The determined material constants can be found in **Table 8.5**. It can be seen from the results

(**Figs. 8.27** and **8.28**) that the equation set is able to predict precipitation hardening under static ageing conditions, i.e. the applied stress is 0 MPa (**Fig. 8.27**).

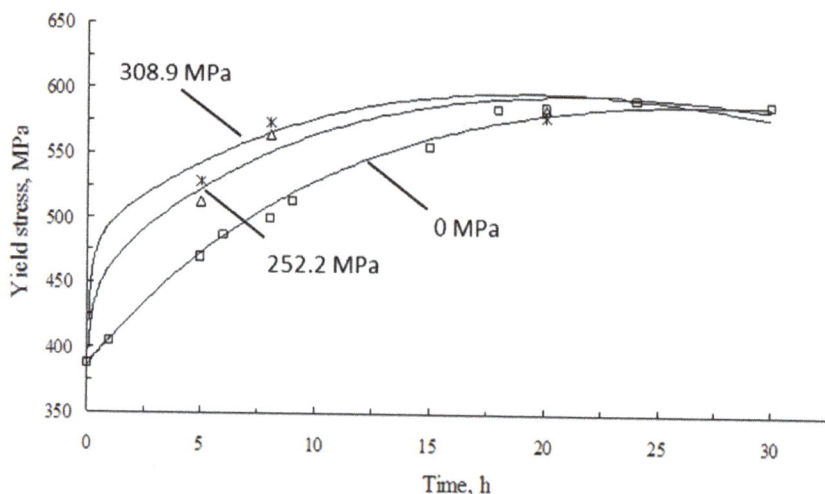

Fig. 8.27 Comparison of experimental (symbols) and computed (solid curves) yield strength variation forAA7055 at 120 °C with different creep stress levels (Zhan and Lin *et al.*, 2011).

Table 8.5 Determined values of the constants for CA constitutive equations Eq. (8.7) forAA7055 at 120 °C (Zhan, Lin, Dean, *et al.*, 2011).

A_1 (h^{-1})	B_1 (MPa)	k_0 (-)	C_A (MPa)	m_1 (-)	C_{ss} (MPa)	m_2 (-)	n (-)	A_2 (-)
5.0E-5	0.0279	0.2	94.3	0.44	20.0	0.4	0.8	291.5

C_r (h^{-1})	Q (-)	γ_0 (-)	m_3 (-)	m_4 (-)	A_3 (-)	C_p (-)	m_5 (-)
0.032	1.69	2.7	1.3	1.98	200.0	0.07	1.3

Figure 8.27 compares the evolution of experimental and predicted yield strength of AA7055 under different creep-ageing conditions ($\sigma = 0$ MPa, 252.2 MPa, 308.9 MPa). It should be noted that at time zero, all the three curves have the same yield stress value of 387 MPa. Close

agreement between experimental and theoretical results exists, showing that the proposed constitutive equations are capable of modelling the creep-ageing behaviour of the alloy at 120 °C up to 30 hours.

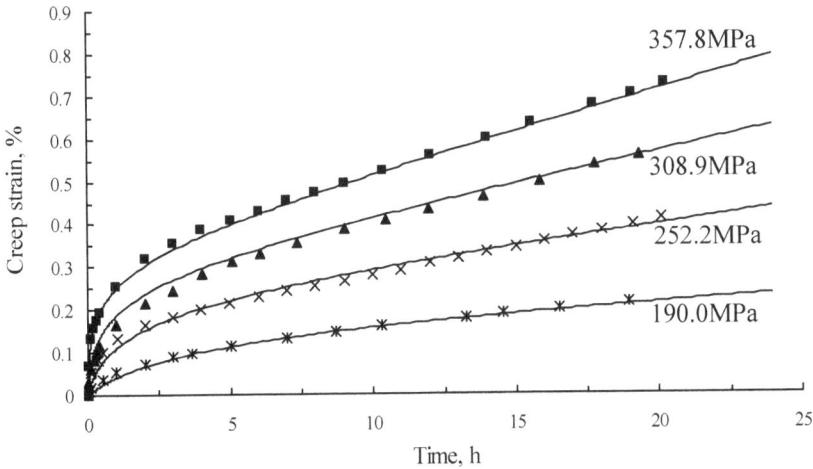

Fig. 8.28 Comparison of experimental (symbols) and computed (solid curves) creep ageing curves for AA7055 at 120 °C with different creep stress levels (Zhan, Lin, Dean, *et al.*, 2011).

Figure 8.28 shows the variation of creep rate for tests at different stress levels (symbols). From the figure, it can be seen that both primary and secondary creep can be predicted well, using the equations.

Creep rate is an important factor in determining creep behaviour. **Figure 8.29** is a comparison of experimental and predicted creep rate at initiation of primary creep (at the beginning of primary creep) and secondary creep (after eight hours of creep testing). Creep rate is very high at the initial stage of primary creep (at 10^{-5} hours of the tests). Due to dislocation and age hardening during creep ageing, the material becomes stronger and thus reduces the creep rate to a steady state creep value – this is known as secondary creep. The minimum creep rate is significantly lower than that of primary creep. Creep rate has an approximate linear relationship with both stress and log creep strain rate. This indicates that a sinh-law-type creep rate constitutive equation is suitable to the material for CAF conditions (see equation set Eq. (8.7)).

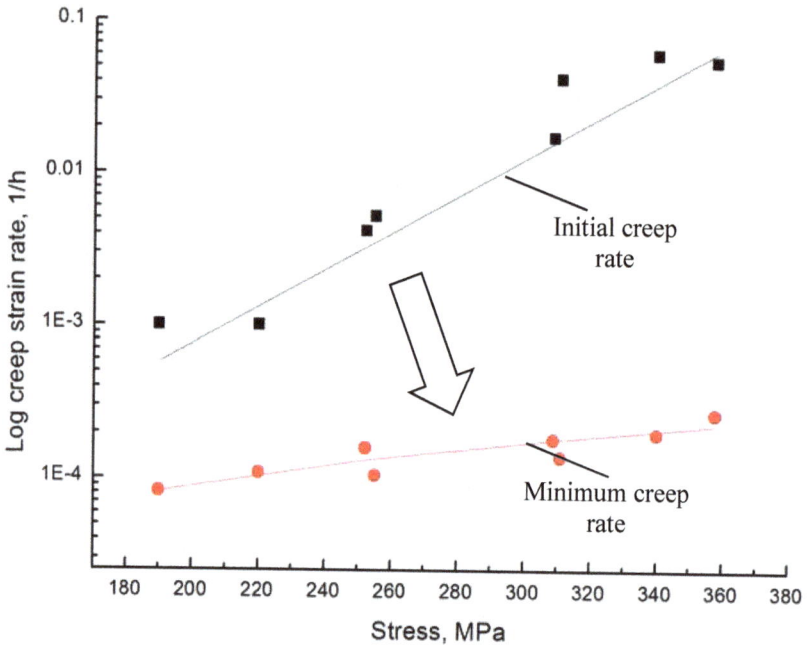

Fig. 8.29 Comparison of experimental (symbols) and predicted (solid lines) creep strain rates at the initial (10^{-5} h) and steady state (8 h) for AA7055 at 120 °C (Zhan, Lin, Dean *et al.*, 2011).

d. *Constitutive relations for AA2219 at 175 °C.*

Yang (2013) generalised the definition of precipitate representation for Eq. (8.7) and extended application of the equation set for AA2219. A range of creep ageing tests was carried out according to the test programme in **Fig. 8.30**, which has similar characteristics to those of **Fig. 8.20**. Interrupted creep ageing and stress relaxation tests were carried out for solution heated specimens at 175 °C. The specimens were then tensile tested at room temperature to determine mechanical properties, as detailed in Section 8.3.2.

Experimental results were then used to construct unified creep-ageing constitutive equations Eq. (8.7) using the optimisation technique detailed in Chapter 7 (Cao and Lin, 2008). Material constants in the unified constitutive equations Eq. (8.7) were determined and listed in **Table 8.6** for tests conducted at a stress level of 210 MPa or below. The constants for tests at stress levels above 200 MPa are given in **Table 8.7**. Material constants from either table for stress levels between 200 MPa and 210 MPa can be used.

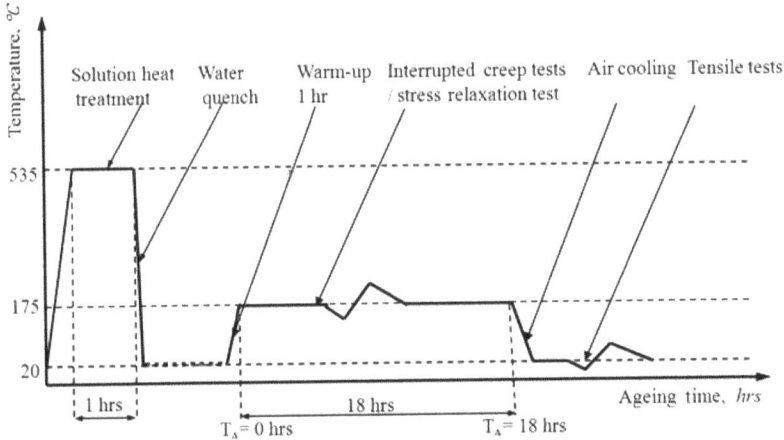

Fig. 8.30 Material heat treatment condition and creep ageing test program for AA2219 at 175 °C (Yang, 2013).

Figure 8.31 shows computed yield stresses. It can be seen that creep deformation increases dislocation density , thus the yield strength of the material aged at a high stress increases, and can reach peak strength within shorter ageing time compared with that for material that has been aged with no stress applied.

Table 8.6 Determined values of constants for CA constitutive equations Eq. (8.7) for AA2219 under a stress of 212.5 MPa at 175 °C (Yang, 2013).

A_1 (h^{-1})	B_1 (MPa)	k_0 (-)	C_A (MPa)	m_1 (-)	C_{ss} (MPa)	m_2 (-)	n (-)	A_2 (-)
3.0e-8	0.0825	0.078	50.1	0.42	15.0	0.85	0.9	520

C_r (h^{-1})	Q (-)	γ_0 (-)	m_3 (-)	m_4 (-)	A_3 (-)	C_p (-)	m_5 (-)
0.032	1.76	2.7	1.6	2.1	65	0.182	1.01

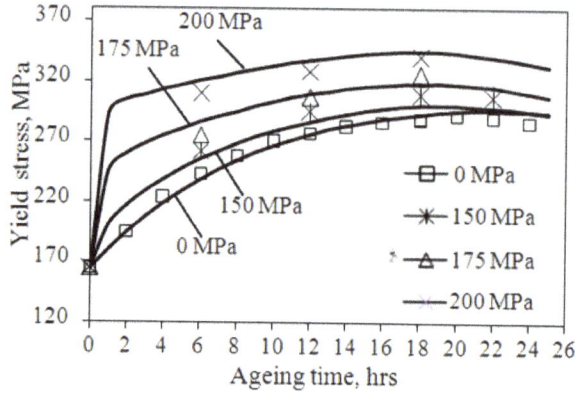

Fig. 8.31 Comparison of experimental (symbols) and computed (solid curves) yield strength aged at different stress levels for AA2219 at 175 °C (Yang, 2013).

Figure 8.32 shows computed primary and secondary creep curves and corresponding experimental data for a large range of stress level, for AA2219 creep aged at 175 °C. It should be noted that these equations can model only primary and secondary creep. When the stress is above 225 MPa, tertiary creep occurs at an early stage of deformation. In general it can be seen that the determined unified creep ageing constitutive equations Eq. (8.7) can be used for predicting well the creep ageing behaviour of AA2219 at 175 °C .

Table 8.7 Determined values of constants for CA constitutive equations (8.7) for AA2219 above a stress of 212.5 MPa at 175 °C (Yang, 2013).

A_1 (h^{-1})	B_1 (MPa)	k_0 (-)	C_A (MPa)	m_1 (-)	C_{ss} (MPa)	m_2 (-)	n (-)	A_2 (-)
1.6e-10	0.101	0.08	50.1	0.42	15.0	0.85	0.9	520

C_r (h^{-1})	Q (-)	γ_0 (-)	m_3 (-)	m_4 (-)	A_3 (-)	C_p (-)	m_5 (-)
0.032	1.76	2.7	1.6	2.1	35	0.318	1.01

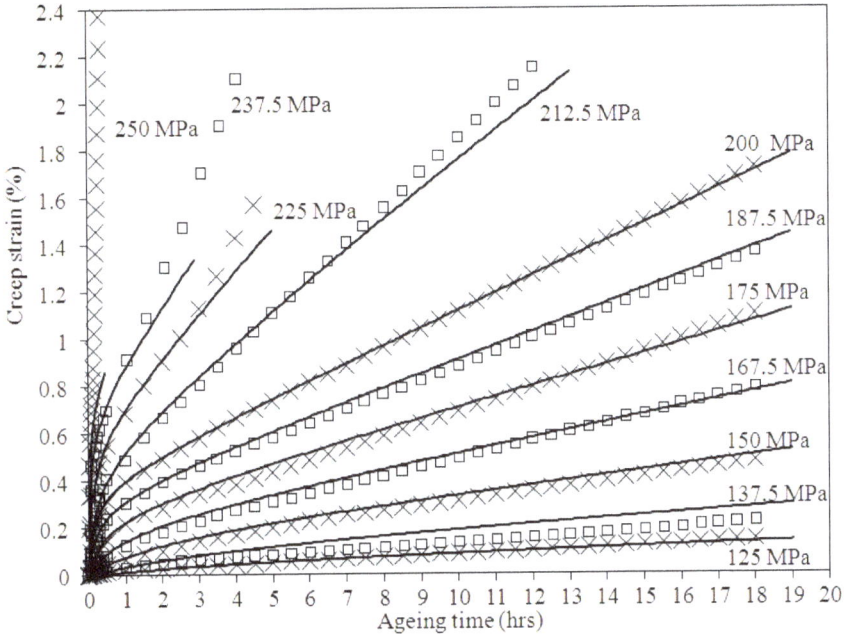

Fig. 8.32 Comparison of experimental (symbols) and computed (solid curves) creep ageing curves for a range of stress levels for AA2219 at 175 °C (Yang, 2013).

Stress relaxation tests have been conducted by the author to validate determined material parameters. **Figure 8.33** shows the stress relaxation test data with different initial stresses applied at 175 °C.

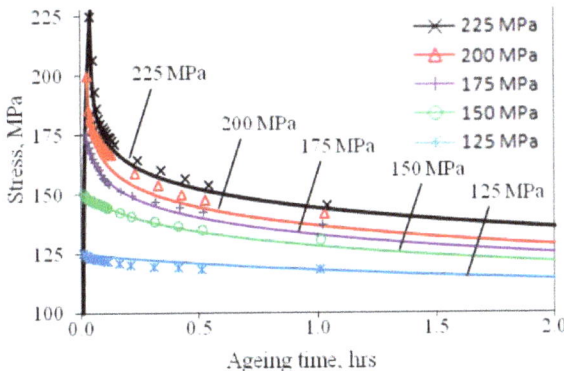

(a)

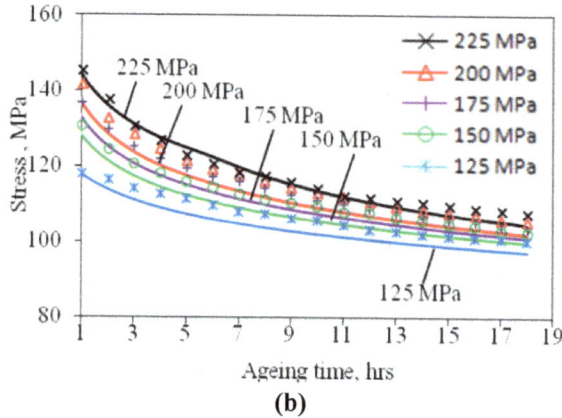

Fig. 8.33 Comparison of experimental (symbols) and predicted (solid curves) stress relaxation for (a) up to 2 hours duration; (b) 1 to 18 hours duration for five initial stress levels of 225 MPa, 200 MPa, 175 MPa, 150 MPa and 125 MPa (Yang, 2013).

Numerically integrating Eq. (8.7) with different initial stress input gives predicted results which are compared with experimental data as shown in **Fig. 8.33 (a)** and **(b)**. Close agreement is achieved with a maximum error of 11.8%. Referring to **Fig. 8.33 (b)**, the predicted stress level is slightly lower than that of experimental data after an hour of ageing. This might be due to the constitutive model being less sensitive to a stress level lower than 137.5 MPa and therefore the time-dependent creep effect is not significant.

Nonetheless, it is seen that the determined parameters can closely predict (i) hardening mechanisms of the material and their effect on yield strength, (ii) primary and secondary creep, and (iii) stress relaxation for AA2219 during CAF.

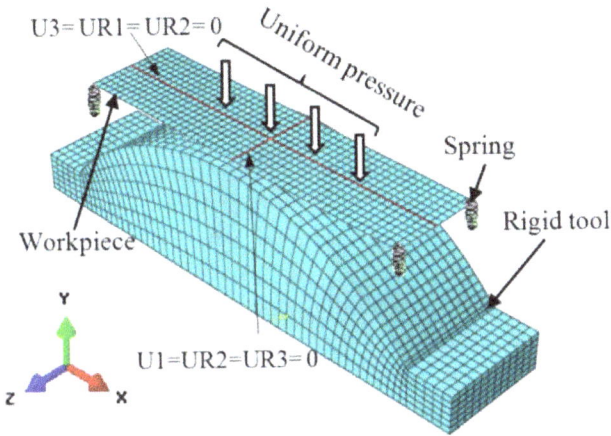

Fig. 8.34 FE model with boundary and loading conditions for forming an AA2219 panel component (Yang, 2013).

8.3.4 *Numerical procedures for CAF process simulations*

a. FE model

An FE model was developed to represent the cylindrical die and work-piece under uniform loading condition, as illustrated in **Fig. 8.34**. The model consisted of a 6 mm thick, initially flat rectangular work-piece (dimensions: $200 \times 50 \times 6$ mm) of aluminium alloy AA2219 held against a die surface, and subsequently subject to CAF at 175 °C for 18 hours. The die surface had a radius of 150 mm and was modelled as a rigid body. Four-node reduced integration shell elements were employed to model the work-piece. A friction coefficient of 0.1 was assigned at the tool/work-piece interface to simulate non-lubricated contact. For the convenience of locating the work-piece on the tool surface and improving computational convergence, four ground springs with a stiffness of 1×10^{-6} N·mm^{-1}, were used to support the weight of the work-piece at its corners (**Fig. 8.34**).

b. Multiaxial constitutive equations

As discussed previously, the sinh-law constitutive equations Eq. (8.7) can be generalised to a multiaxial form by considering a dissipation energy function. The details of this are given in Chapter 3. The multiaxial constitutive equations can be input into a large strain FE solver, such as, MSC.MARC through the user-defined subroutine, CRPLAW, and used to simulate CAF (Yang, 2013).

c. CAF modelling procedures

A three-step analysis was performed, as shown in **Fig. 8.35**. Initially a 10 MPa load was applied incrementally in a static analysis, followed by a visco-step where this load was held over a period of 18 hours, allowing ageing, creep and stress relaxation to take place. Finally, the uniform pressure was removed incrementally, enabling the aluminium work-piece to spring back (Yang, 2013).

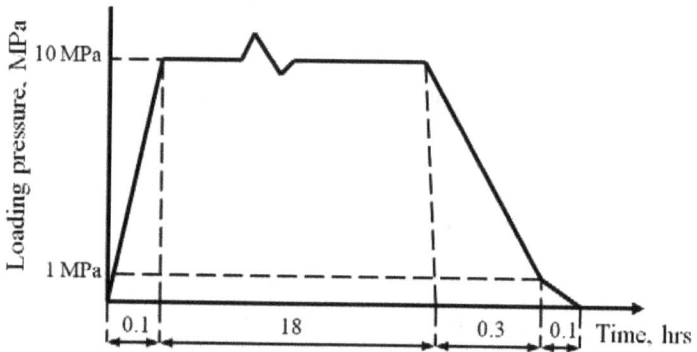

Fig. 8.35 Loading process in the FE simulation (Yang, 2013).

8.3.5 CAF process modelling results

a. Stress relaxation and springback in CAF

Contour plots of the equivalent (Mises) stress distribution are shown in **Fig. 8.36**. **Figure 8.36 (a)** shows the initial fully loaded condition when the ground springs were compressed to overcome their stiffness by the applied pressure and the whole work-piece deformed to fully contact the die surface. At this stage, a relatively uniform stress distribution (190 MPa – 200 MPa) can be observed. The maximum stress (204 MPa) was observed at the centre part of the work-piece. At this stage, the stresses were higher than the yield stress of the material, at about 165 MPa.

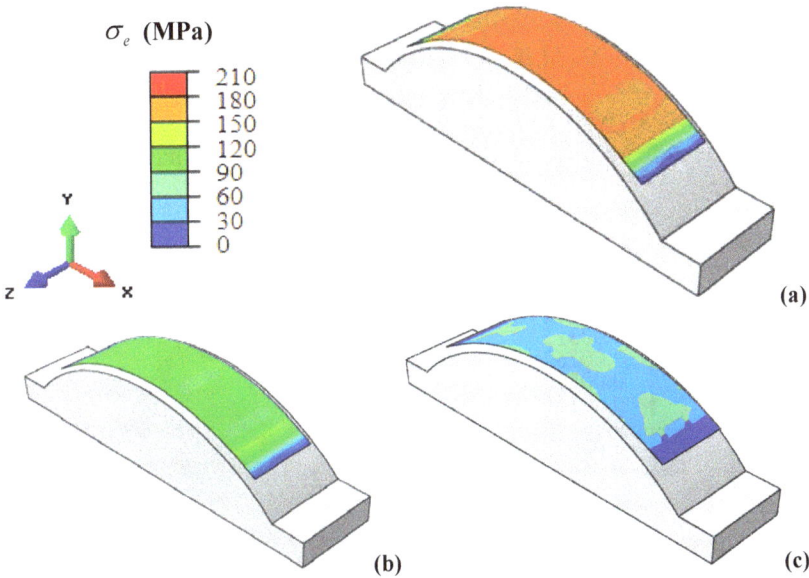

Fig. 8.36 von-Mises Stress distribution at different stages of the CAF process. (a) Deformed to target shape; (b) held for 18 hours; (c) after unloading (Yang, 2013)

At the end of CAF (**Fig. 8.36 (b)**), significant stress relaxation due to creep took place, which reduced the equivalent stress by 80% to a maximum value of 115 MPa. After removal of the applied load, springback occurred and residual stress can be observed in the work-piece. As shown in **Fig. 8.36 (c)**, residual stress was below 76 MPa. Comparison of stress in the three forming stages shows that the equivalent stresses are lower at the edges of the work-piece, but higher at the centre, for all stages. This is because the amount of uniform pressure is insufficient to force the work-piece to fully contact with the die surface (Yang, Davies, Lin *et al.*, 2013).

b. Creep strain and precipitation hardening

Figure 8.37 shows the distribution of surface creep strain during CAF. Maximum creep strain has reached approximately 0.16% after an hour of creep-ageing (**Fig. 8.37 (a)**). The creep strain distribution was uniform except for the ends of the work-piece. After 9 hours of ageing, creep strain in the bulk of the work-piece had increased and high creep strain at the ends had decreased (**Fig. 8.37 (b)**). The majority of the creep strain had a value between 0.36% – 0.38%. At the end of CAF, the most of the work-piece had a creep strain of 0.43% (**Fig. 8.37 (c)**); the creep strain increment is not significant as the creep deformation has reached its steady state.

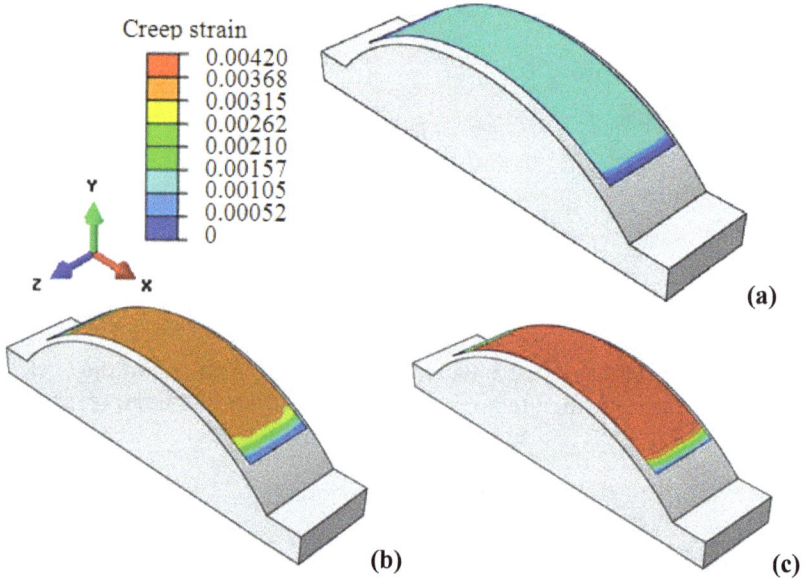

Fig. 8.37 Distribution of creep strain during the ageing period at (a) 1 hour, (b) 9 hours and (c) 18 hours (Yang, 2013).

Apart from modelling creep deformation, the normalised precipitate ratio and yield strength increment of the material during CAF can be predicted. Referring to **Fig. 8.38 (a)**, which relates to $t = 1$ hour, the normalised precipitate size varies from 0.14 in the low deformation area (edges) to 0.32 in the high deformation area (full contact). As ageing time increases from 1 hour to 18 hours, normalised precipitate size grows with creep deformation. At the end of CAF, maximum normalised precipitate size reaches a value of 0.88 while the minimum value is about 0.37. It can be seen that, at the low deformation area (edges), the precipitate size is lower because in this region the generation of precipitates is due to only static ageing. Meanwhile the increment of precipitate size in the high deformation region is due mainly to dynamic ageing (i.e. creep deformation on precipitate growth) (Yang, 2013).

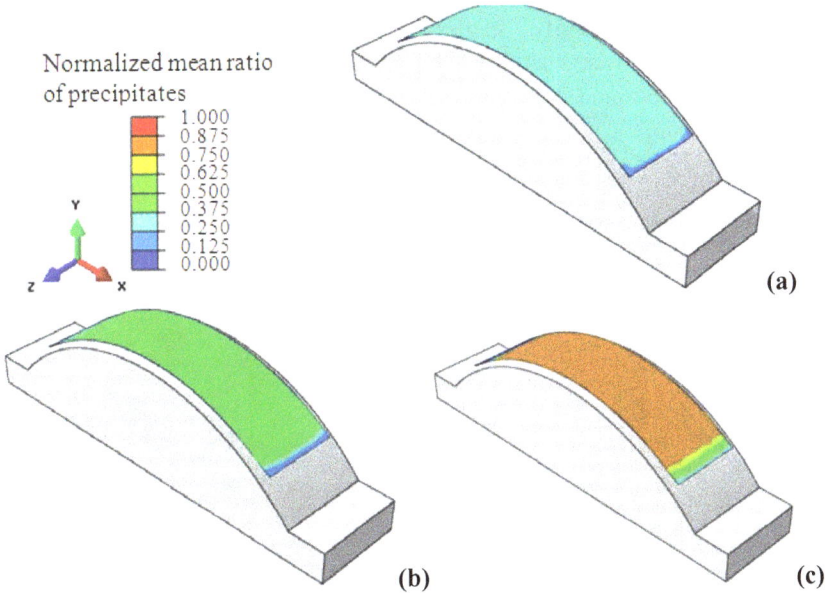

Fig. 8.38 Distribution of normalised mean ratio of precipitates during the ageing period at (a) 1 hour, (b) 9 hours and (c) 18 hours (Yang, 2013).

Evolution of yield strength during CAF is shown in **Fig. 8.39**. As age hardening and solute hardening are functions of precipitate ratio, and dislocation hardening increases the yield strength mainly in the primary stage, the yield strength distribution during CAF simulation is more or less similar to the normalised precipitates distribution. At the end of CAF ($t = 18$ hours), yield strength increments are distributed evenly in the region where full work-piece to die contact was achieved with a value around 318 MPa.

Many CAF FE processing simulations have been carried out using commercial FE codes, such as ABAQUS (Lin, Ho and Dean, 2006; Yang, Davies, Lin *et al.*, 2013), MSC.MARC (Zhan, Lin and Dean *et al.*, 2011) and PAMSTAMP with the use of unified creep age constitutive equations for AA7xxx and AA2xxx alloys. Industrial applications for CAF have been extended also to the forming of high strength, extra-large

panel components. However, greater exploitation of the process is
constrained by low productivity and high springback of formed parts.
Thus the CAF process is mainly suitable for low volume production,
which is a common need for aerospace and aeronautical components.
CAF is already highly successful for forming aluminium parts.
Potentially, the process can also be applied to form any alloy in which
significant creep and/or stress relaxation takes place at their ageing
temperature.

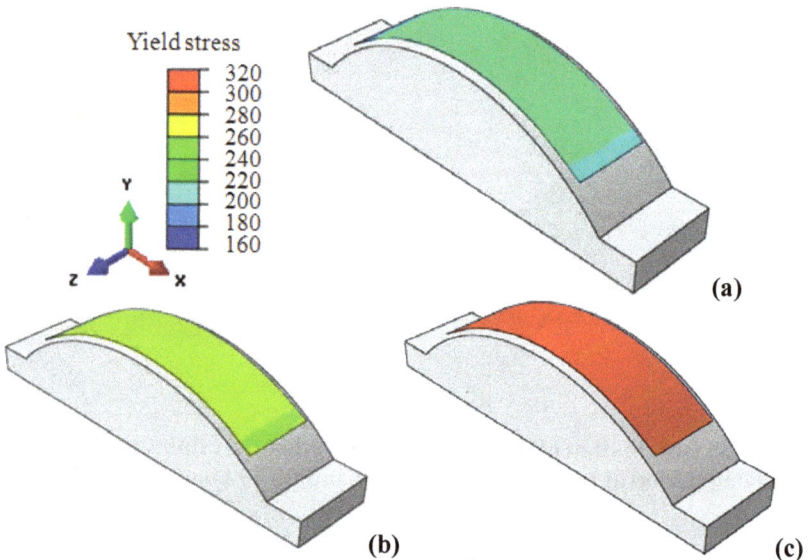

**Fig. 8.39 Distribution of yield stress (MPa) variation during the ageing
period at (a) 1 hour, (b) 9 hours and (c) 18 hours (Yang, 2013).**

8.4 HFQ of Aluminium Panel Components

Solution heat treatment, cold die forming and quenching (HFQ) is an
advanced forming process developed for the stamping of complex-
shaped panel components. This has been divided into one-stage forming
(Foster, Dean and Lin, 2012) and two-stage forming (Lin, Dean, Foster
et al., 2011) depending on the particular work-piece alloy. Currently, the
main success of the process is for the stamping of high strength complex-
shaped aluminium panels for automotive and aerospace applications
(http://www.impression-technologies.com/). Detailed   material   and

process modelling techniques for HFQ of aluminium are introduced in this section.

8.4.1 *Introduction to the HFQ process*

Heat treatment is required for structural aluminium alloys to achieve desired microstructure and thus specified mechanical properties. In CAF, which has been introduced in Section 8.3, the forming is combined with the ageing process. Since the ductility of aluminium alloys does not increase much at their ageing temperature (120 °C – 190 °C), the CAF process is mainly used for forming panel components requiring only small amounts of deformation, such as bending with large bending radii (in the scale of metres), to form aerofoil shapes.

Much higher temperatures (470 °C – 570 °C), are required for the SHT of aluminium alloys and the ductility of the material increases significantly at a temperature of 350 °C or above (Wang, Strangwood, Balint *et al.*, 2011). This alloy characteristic has been utilised for the stamping of complex-shaped aluminium panel components, and has led to formulation of the HFQ process (Foster, Dean and Lin, 2012). HFQ is particularly useful for the stamping of high strength aluminium panels, since the ductility of high strength aluminium is very low at room temperature.

a. *HFQ process*

Figure 8.40 shows the basic stages of the HFQ process (Foster, Dean and Lin, 2012), which consists of:

- *Solution heat treatment (SHT)*. A sheet blank is heated to its SHT temperature for a specified period, until all existing precipitates from previous processes have been "fully" dissolved and the alloying elements are "uniformly" distributed within aluminium matrix. The time required for the material to be fully solution heat treated is related to the previous processes. SHT is normally carried out within a furnace with controlled uniform temperature.

- *Transfer*. After SHT, the blank is transferred to the cold (room temperature or slightly above) die on the press for stamping.

- *Cold die forming and quenching*. The blank is stamped into the die shape and held tightly in the tool to quench it. The cooling rate in the cold die quenching should be sufficiently high to avoid precipitates being formed, especially large precipitates at grain boundaries (which would reduce mechanical properties of the formed parts). The cooling rate is related to the thickness of the material, contact pressure, surface contact conditions etc.

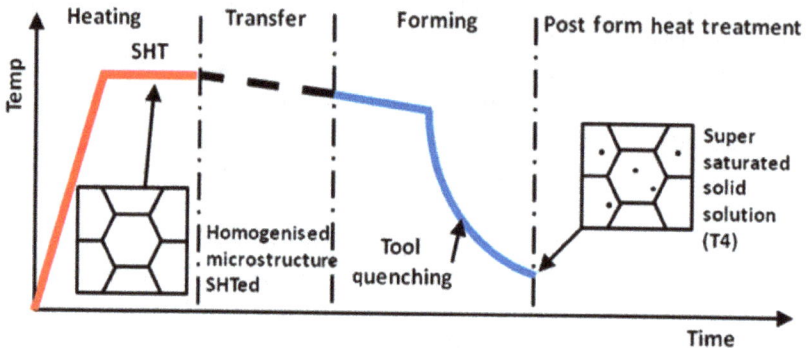

Fig. 8.40 Stages of the HFQ process.

Artificial ageing is needed for age hardening alloys, which is the post heat treatment of the formed parts.

b. Analysis of the process

The HFQ process has been developed with an aim of addressing two main problems encountered in forming complex-shaped aluminium alloy parts by conventional means:

- The formability of aluminium alloys when cold formed (at room temperature) is very low. Therefore forming complex-shaped panel components such as those required for automobile bodies and aerospace structures, is difficult. (Mohamed, 2010).
- Springback is very high in cold formed aluminium sheet due to the low stiffness of aluminium. This makes compensation for springback through tool and process design, as is possible with CAF, very difficult and time-consuming.

The above problems can be effectively solved by the use of HFQ. During HFQ, ductility of the alloy is close to that in superplastic forming (SPF). In addition, in contrast to SPF, the blank material can be drawn into the die and the localised thinning arising in SPF is much reduced in

HFQ. As a result, the productivity of HFQ is much higher than that of SPF.

In addition, quenching in closed dies, as is done in HFQ, effectively solves the problem of thermal distortion during rapid cooling of the formed parts and springback is eliminated (Mohamed, 2010).

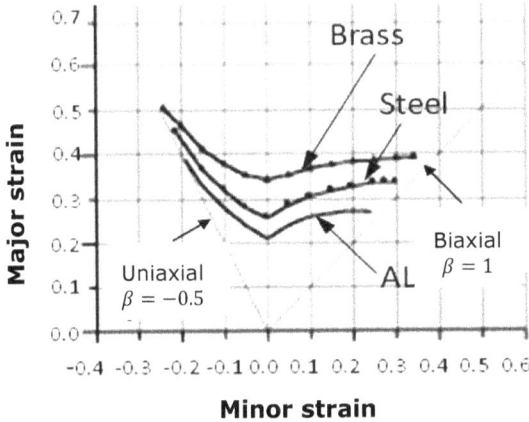

Fig. 8.41 Typical FLCs for an aluminium alloy, mild steel and brass at room temperature (Ali and Balod, 2007). (The ratio of minor strain ε_2, $\beta = \varepsilon_2/\varepsilon_1 = 1$ and major strain ε_1 is equibiaxial straining, and $\beta = -0.5$ is uniaxial straining.)

8.4.2 *Forming limit in hot stamping*

a. Forming limit curve (FLC)

Being able to accurately model the formability of sheet metals at different straining states is of high importance to the success of making parts by HFQ. The cold formability of sheet metals is normally assessed through the use of FLCs (Ali and Balod, 2007). A typical FLC is shown in **Fig. 8.41**.

An FLC shows the critical combination of major and minor surface strains in a metal sheet at the onset of necking failure and provides information for a process engineer to optimise process conditions such as material conditions, tool features and lubrication. The concept of FLCs was introduced by Keeler (1968) and Goodwin (1968), who established a principle for the relationship between the surface principal strains, ε_1 and

ε_2, at the point of necking or fracture. The relationship is presented as a curve: the FLC. If the orthogonal principal strain set at all positions in a deforming sheet lies below the FLC, a sound product will result. On the other hand if it lies above, product quality will be unacceptable.

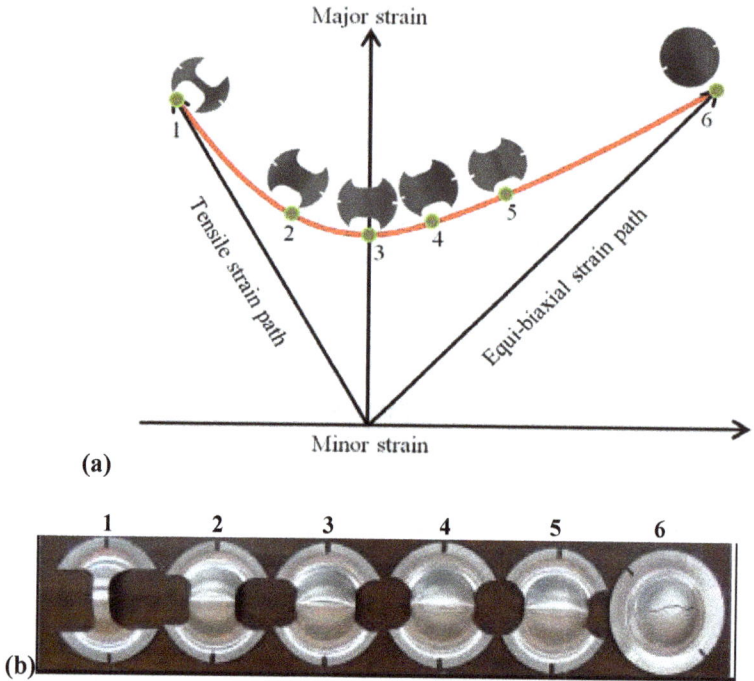

(a)

(b)

Fig. 8.42 (a) Schematic drawing of the test points for an FLC. (b)Test-pieces tested at room temperature at a punch speed of 5 mm·s^{-1} (Shi, Mohamed, Wang *et al.*, 2012).

Standard testing procedures and specimen designs have been developed by Marciniak, Kuczynski and Pokora (1973) for the determination of FLCs for sheet materials. For obtaining the data points shown in **Fig. 8.42 (a)**, the geometry of the test-piece is that of a waisted circular blank with a waist of constant width, as shown in **Fig. 8.42 (b)**. This is based on ISO 12004-2:2008 standard. The method for measuring strains on deformed test-pieces is also described in the standard. The ratio of major to minor strain values in a test-piece can be varied by altering the ratio of waist width to diameter of test-piece. Thus, by using test-pieces with different geometries, a range of values of biaxial

principal strain ratios may be obtained to construct an FLC, as shown in
Fig. 8.42.

In HFQ processes, the blank is heated and the tool is relatively cool.
During deformation, temperature and strain rate vary dynamically in both
time and location in the sheet metal. The ductility of aluminium alloys
increases with temperature and is further enhanced by deforming at low
speed (please notice that it is not necessary to increase drawing depth at
low forming speed, even if the ductility of the material increases). This
will enable the successful formation of complex-shaped aluminium sheet
components (Garrett, Lin and Dean, 2005; Mohamed, Foster, Lin *et al.*,
2012).

Figure 8.43 shows the FLCs of AA5754 for different tool closing
speeds and work-piece temperatures. Forming limit increases as the
forming speed is decreased from 300 mm·s^{-1} to 20 mm·s^{-1}. The forming
limit is also found to have increased the most in the plane strain
condition. A flatter FLC can be observed at 20 mm·s^{-1}. A relatively
larger increase in the forming limit is seen as a result of forming speed
reduction from 75 mm·s^{-1} to 20 mm·s^{-1} than that observed from
300 mm·s^{-1} to 75 mm·s^{-1} (**Fig. 8.43 (a)**). Furthermore, an increase in the
forming limit from 250 °C to 300 °C is about twice that of an increase
from 200 °C to 250 °C under plane strain conditions. These results
indicate that high forming limits can be obtained when the forming
temperature is between 250 °C and 300 °C. The V-shape flattens as
temperature increases and forming speed decreases. From the above
observations, it is concluded that using a combination of low forming
speed and high forming temperature is beneficial for enhancing the
forming limit in an isothermal forming environment, within the limit of
study.

(a) at 250 °C

(b) at 75 mm·s⁻¹

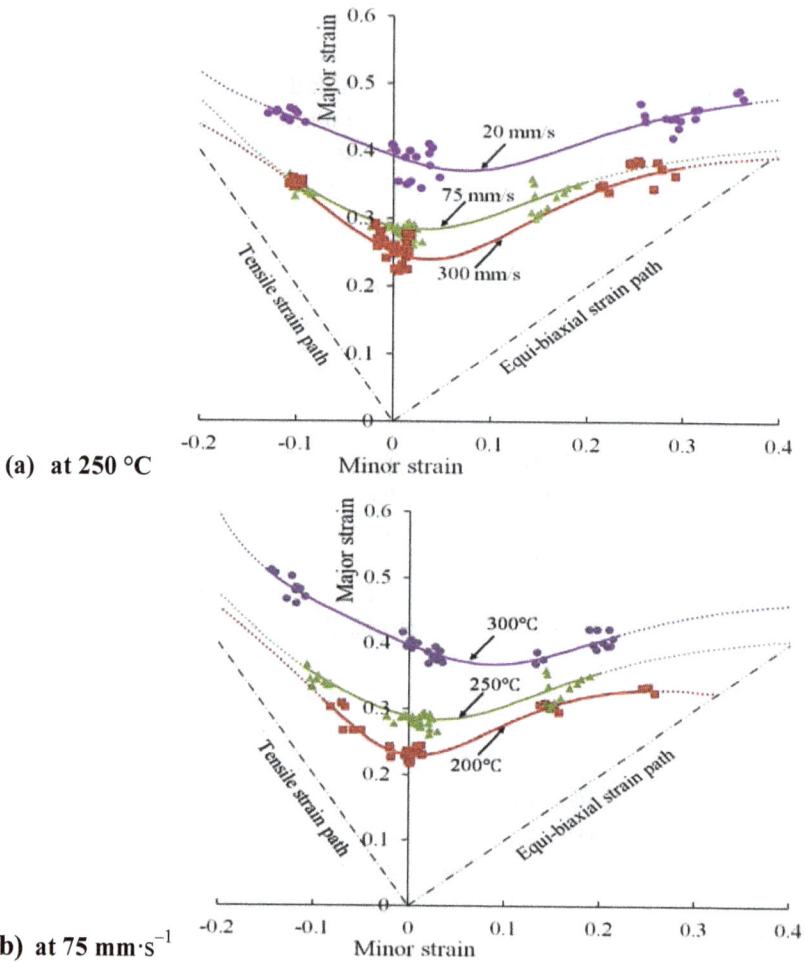

**Fig. 8.43 FLCs obtained by forming tests at different (a) forming speeds
at a temperature of 250 °C and (b) temperatures at a forming speed of
75 mm·s⁻¹ for AA5754 (Shao, Bai, Lin *et al.*, 2014).**

b. Key modelling issues in hot stamping

In a hot stamping process, the strain rate and temperature over the work-piece vary with time and location. As can be seen in **Fig. 8.43**, FLCs are based on constancy of temperature and strain ratio and it is therefore impossible to use a single curve to assess sheet formability. To do so with accuracy requires a mathematical model which is able to take account of varying process conditions within a pressing operation and this is available through a set of viscoplastic damage constitutive equations. The equation set should be able to model:

- viscoplastic flow of the material over a specified range of temperatures and strain rates;
- formability of the material over a specified range of temperatures and strain rates;
- Microstructural evolution in non-isothermal forming conditions. In the case of hot stamping of boron steel parts (Cai, 2010; Li, 2013), phase transformation is important in cold die quenching, which will not be dealt with in this section. In the case of hot stamping of aluminium components, precipitate nucleation and growth in cold die quenching are important. However, the critical cooling rate (that at or below which no precipitation arises) for most of alloys is fairly low and heat is conducted very quickly from aluminium sheet to tools. Thus, in general, precipitate nucleation and growth are not an issue for hot stamping of thin aluminium sheets. To simplify the equation set, this is normally omitted.

8.4.3 *Unified constitutive equations for HFQ aluminium*

Continuum damage mechanics theories introduced in Chapter 6 are used for modelling failure in sheet alloys used for hot stamping. The damage value is accumulated according to plastic deformation, temperature, stress and strain state, during the forming process. Given that the time steps are sufficiently small, temperature and strain rate at a material point are considered as constant over a small increment in time. If the total accumulated damage at a material point reaches a specified value, e.g. 0.7, failure is considered to have occurred.

a. *Unified uniaxial viscoplastic damage constitutive equations*

A set of unified viscoplastic damage constitutive equations presented for hot stamping is based on the framework proposed by Lin and Dean (2005); and Lin, Liu, Farrugia *et al.* (2005). According to Lin, Mohamed, Balint, *et al.* (2014), the equations can be given as

$$\dot{\varepsilon}^p = \left(\frac{\sigma/(1-\omega) - R - k}{K} \right)^n$$

$$\dot{R} = 0.5B\bar{\rho}^{-0.5}\dot{\bar{\rho}}$$
$$\dot{\bar{\rho}} = A(1 - \bar{\rho})|\dot{\varepsilon}^p| - C\bar{\rho}^{n_2}$$
$$\dot{\omega} = \frac{\eta_1}{(1-\omega)^{\eta_3}}(|\dot{\varepsilon}^p|^{\eta_2}) \qquad (8.8)$$
$$\sigma = E(1 - \omega)(\varepsilon^T - \varepsilon^p),$$

where ε^T and ε^p are the total and plastic strains respectively. The evolution of isotropic hardening, R, is a function of the normalised dislocation density, $\bar{\rho}$, defined in the previous sections. Details of the dislocation hardening formulation are given in (Lin, Liu and Farrugia *et al.*, 2005). Damage ω for the uniaxial formulation is a function of strain rate and stress, and affects viscoplastic flow and flow stress (see the first and the last equations in Eq. (8.8)). The damage value $\omega = 0$ represents the initial state of deformation. When the damage level reaches 0.7, it is assumed that failure takes place. According to the character of the damage model and the strain increment in stress–strain curves, as the damage increases from 0.7 to 1.0, is little and could be omitted. This simplification increases computational efficiency significantly in FE process simulation (Mohamed, Foster and Lin *et al.*, 2012). There is no intention here to model a particular damage mechanism of the material; damage is treated as an overall effect in the material during hot/warm forming. This is due to the fact that the dominant damage mechanism may change during a forming process, since deformation rate and temperature vary between locations in a work-piece and with time during deformation.

Parameters K, k, B, C, A, n, η_1, η_2 and E are temperature-dependent material constants, and, η_3 and n_2 are temperature-independent material constants (E is the Young's modulus of the material). The following equations represent the temperature-dependent parameters:

$$K = K_0 \, exp\left(\frac{Q_K}{RT}\right) \qquad\qquad k = k_0 \, exp\left(\frac{Q_k}{RT}\right)$$

$$B = B_0 \, exp\left(\frac{Q_B}{RT}\right) \qquad\qquad C = C_0 \, exp\left(-\frac{Q_C}{RT}\right)$$

$$E = E_0 \, exp\left(\frac{Q_E}{RT}\right) \qquad\qquad \eta_l = \eta_{0_l} \, exp\left(\frac{Q_{\eta_l}}{RT}\right)$$

$$\eta_2 = \eta_{0_2}\, exp\left(-\frac{Q_{\eta_2}}{RT}\right) \qquad A = A_0\, exp\left(-\frac{Q_A}{RT}\right)$$

$$n = n_0\, exp\left(\frac{Q_n}{RT}\right).$$

The equations were numerically integrated using the implicit Euler method with automatic step size control. Optimisation methods and software system introduced in Chapter 7 were used for determining material constants from experimental data.

Table 8.8 Material constants in the viscoplastic damage constitutive equations for AA5754 (Mohamed, Lin, Foster, *et al.*, 2014).

E_0 (MPa)	C_0 (s^{-1})	B_0 (MPa)	k_0 (MPa)	K_0 (MPa)	η_{0_1} (-)	η_{0_2} (-)
8.855	102567	0.7222	2.518	0.702	0.00899	0.8362
Q_E (J/mol)	Q_C (J/mol)	Q_k (J/mol)	Q_K (J/mol)	Q_{η_1} (J/mol)	Q_{η_2} (J/mol)	Q_n (J/mol)
45766	128828	8857	22940	12030	953	14325
η_3 (-)	B_0 (-)	A_0 (-)	Q_A (J/mol)	n_2 (-)	n_0 (-)	
17	0.7222	8.139	6411	1.8	0.6451	

In this case, there are 20 material constants within the temperature-dependent constitutive equations. They were determined from the experimental data of Mohamed, Foster, Lin *et al.* (2012) for AA6082 by Mohamed, Lin, Foster *et al.* (2014) and are listed in **Table 8.8**. The experimental data and the fitting results are shown in **Fig. 8.44** for different temperatures and strain rate (Lin, Mohamed and Balint *et al.*, 2014). A good agreement between experimental and computed data has been obtained.

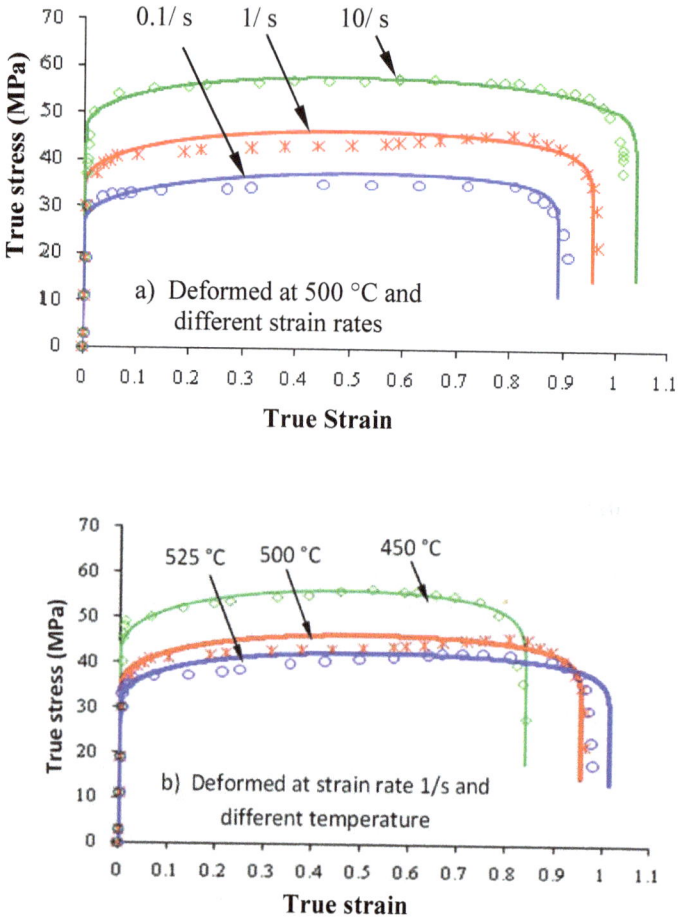

Fig. 8.44 Comparison of computed (solid curves) and experimental (symbols) stress–strain relationships for AA6082 deformed at a) different strain rates at temperatures of 500 °C, and b) different deformation temperatures at strain rates of 1 s⁻¹.

b. 2D viscoplastic damage equations for the modelling of FLCs

In a manner similar to that for creep deformation (Lin, Hayhurst and Dyson, 1993), the general multiaxial power-law viscoplastic equations can be obtained by consideration of a dissipation potential function. With an initial yield stress, k, and ignoring the work hardening and other state variables, an energy dissipation potential can be in the form of

$$\psi = \frac{K}{n+1}\left(\frac{\sigma_e - k}{K}\right)^{n+1},$$

where K and n are material constants. Assuming normality and the associated flow rule, the multiaxial relationship is given by

$$\frac{d\varepsilon_{ij}^p}{dt} = \frac{\partial \psi}{\partial S_{ij}} = \frac{3}{2}\left(\frac{S_{ij}}{\sigma_e}\right)\left(\frac{\sigma_e - k}{K}\right)^n$$

where ε_{ij}^p is the plastic strain tensor. By the introduction of isotropic hardening (R) and damage state variables (ω), the effective plastic strain rate, $\dot{\varepsilon}_e^p$, for the power-law material model can be written as (Lin, Hayhurst and Dyson, 1993)

$$\dot{\varepsilon}_e^p = \left(\frac{\sigma_e/(1-\omega) - R - k}{K}\right)^n.$$

It is worth mentioning again that this equation is tenable only when $\sigma_e/(1-\omega) - R - k > 0$, otherwise, $\dot{\varepsilon}_e^p = 0$. The set of multiaxial viscoplastic constitutive equations, incorporating multiaxial damage evolution, may be written as (Lin, Mohamed, Balint and Dean, 2014)

$$\dot{\varepsilon}_{ij}^p = \frac{3}{2}\frac{S_{ij}}{\sigma_e}\dot{\varepsilon}_e^p$$

$$\dot{R} = 0.5B\overline{\rho}^{-0.5}\dot{\overline{\rho}}$$

$$\dot{\overline{\rho}} = A(1-\overline{\rho})|\dot{\varepsilon}_e^p| - C\overline{\rho}^{n_2} \tag{8.9}$$

$$\sigma_{ij} = (1-\omega)D_{ijkl}\left(\varepsilon_{ij} - \varepsilon_{ij}^p\right)$$

$$\dot{\omega} = \frac{\Delta}{(\alpha_1+\alpha_2+\alpha_3)^\varphi}\left(\frac{\alpha_1\sigma_1 + 3\alpha_2\sigma_H + \alpha_3\sigma_e}{\sigma_e}\right)^\varphi \cdot \frac{\eta_1}{(1-\omega)^{\eta_3}}\left(\dot{\varepsilon}_e^p\right)^{\eta_2},$$

where D_{ijkl} is the elastic matrix of the material. The multiaxial damage equation comes from the uniaxial form with the consideration of the multiaxial stress-state effect (Lin, Mohamed, Balint and Dean 2014). The details of the parameters within the damage equation are given below:

- α_1, α_2 and α_3 are weighting parameters that present damage evolution and are introduced to rationalise the effects of stress state on damage evolution and turn on the shape of FLC. These parameters are mainly controlled by maximum principal stress, effective stress and/or hydrostatic stress and are defined according to experience. In conclusion, it was defined that α_1 and α_3 can vary from 0 to 1. However, if α_2 is changed from 1 to -1, then the effect of hydrostatic stress is negative. If α_1 or α_2 or α_3 is zero, then the particular stress has no contribution to damage in the process of sheet metal forming.

- φ is a multiaxial stress damage exponent, which controls the effect of multiaxial stress values and their combination on damage evolution, and thus controls formability and affects the shape of the FLC.

- Δ is a correction factor representing the tensile data obtained from uniaxial tensile tests and Marciniak or Nakazima formability tests (Marciniak, Kuczynski and Pokora, 1973; Nakazima, Kikuma and Asaku, 1968), since different strain measurement methods are normally used.

The plane stress problem is defined to be a state of stress in a thin sheet in which the normal stress σ_3 (through the thickness of the sheet metal) is small and is assumed to be zero. This must be true on the free surface of a deforming sheet. The numerical procedure for solving the equations for a plane stress situation was detailed by Lin, Mohamed, Balint and Dean (2014) and consists of the following steps:

- Define the deformation case, which is controlled by the total principal strains, ε_1 and ε_2, based on the defined strain rate and strain ratio, $\beta = \varepsilon_2/\varepsilon_1$. $\beta = 1$ represents biaxial stretching, $\beta = 0$ is the plane strain condition and $\beta = -0.5$ is uniaxial tension. In this case, $\varepsilon_1 = \varepsilon_{11}$ and $\varepsilon_2 = \varepsilon_{22}$ in Eq. (8.9). The initial values of all state variables are given as zero when $t = 0$.

- Solve the viscoplastic damage equations Eq. (8.9) according to the initial values and calculate the principal stresses σ_1 and σ_2 using the stress equation in Eq. (8.9), where $\sigma_1 = \sigma_{11}$ and $\sigma_2 = \sigma_{22}$, and calculate the plastic strains, ε_{11}^p, ε_{22}^p etc. according to individual equations in Eq. (8.9).

- Define the values of the ε_1 and ε_2 as the damage value accumulates to 0.7, i.e. $\omega = 0.7$.

ε_1 and ε_2 are a calculated data point on an FLC curve for a particular value of deformation ratio β. As β varies from -0.5 to 1.0, the full FLC curves can be generated, which can be compared with experimental FLCs. The same optimisation procedure can be used to determine the FLC related material constants, α_1, α_2, α_3 and φ in Eq. (8.9).

Due to the lack of experimental FLC data for AA6082 for hot HFQ conditions, the multiaxial part of damage constitutive equations Eq. (8.9) was determined from room temperature FLC data, and was offset to the HFQ condition according to uniaxial tensile data for HFQ conditions.

The symbols, shown in **Fig. 8.45**, are experimental FLC data for room temperature and are used to determine the stress state related material parameters, α_1, α_2, α_3 and φ, in the multiaxial part of Eq. (8.9). The determined values of the constants are

$\alpha_1 = 0.4$, $\alpha_2 = -0.072$, $\alpha_3 = 0.05$ and $\varphi = 4.0$.

A comparison of computed (solid curve) and experimental FLDs (symbols) for AA6082 at 20 °C is shown in the figure. A fairly good agreement between the experimental and computer FLC data has been obtained. In this case, the value of Δ is 0.8. The shape of the FLC is defined by the values of the stress state related material parameters, α_1, α_2, α_3 and φ. It is assumed that as the temperature and forming rate change, the shape of the FLC will remain unchanged. Thus the FLCs for HFQ conditions (different temperatures and strain rates) can be obtained by offsetting the room temperature FLC according to uniaxial data, i.e. $\beta = -0.5$, with the correction parameter $\Delta = 0.8$ (Mohamed, Lin, Foster *et al.*, 2014).

Examples of the offset FLCs are shown in **Fig. 8.45** for different temperatures at a strain rate of 0.1 s^{-1}. The shape of the curves are controlled by the values of parameters, α_1, α_2, α_3 and φ, and, the heights of the curves are dependent on the uniaxial strain ε_1, where $\varepsilon_2 = -0.5 \varepsilon_1$ ($\beta = -0.5$). Significant research needs to be carried out for obtaining experimental data of FLC at hot stamping conditions.

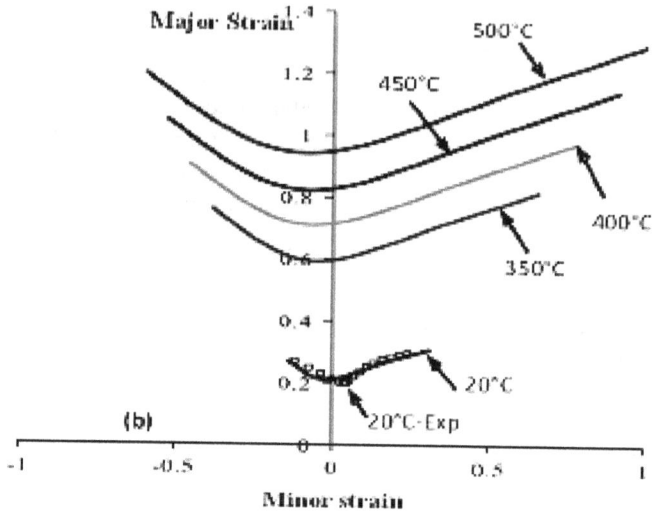

Fig. 8.45 Offset FLCs from room temperature (20 °C) to HFQ
conditions (high temperature) for AA6082. The computations were
carried out using $\alpha_1 = 0.4$, $\alpha_2 = -0.072$, $\alpha_3 = 0.05$, $\varphi = 4.0$ and $\Delta = 0.8$
(Mohamed, Lin, Foster, et al., 2014).

8.4.4 An examination of HFQ characteristics

a. Test rig

A die set was designed to investigate drawing depth and formability for
anisothermal hot stamping conditions. **Figure 8.46 (a)** shows the test rig
design. A test-piece with a central hole, as shown in **Fig. 8.46 (b)**, was
placed on the blankholder and during deformation was clamped between
it and the top plate, so that the outer edges were fixed in place while the
central region was stretched over a fixed hemispherical punch (the punch
diameter was 80 mm). Meanwhile, the top plate and blankholder
descended, thereby imposing a biaxial state of strain (i.e. radial and
circumferential) on the sheet. **Figure 8.46 (c)** shows a test-piece failed
mid-way between the central hole and clamped region.

Square test-pieces of AA6082 with dimensions of
170 mm × 179 mm × 2 mm, with central hole of diameter that was
varied, were used (**Fig. 8.46 (b)**). The hole allowed the material to flow
over the punch to a degree depending on the hole size and forming speed.

In addition, the evolution of the central hole diameter was used for qualitatively evaluating plastic deformation, verifying the modelling results and determining failure modes (Mohamed, Foster, Lin *et al.*, 2012). Here, the effect of hole size and forming speed on the formability of the material is investigated numerically and experimentally for AA6082 under conditions of HFQ.

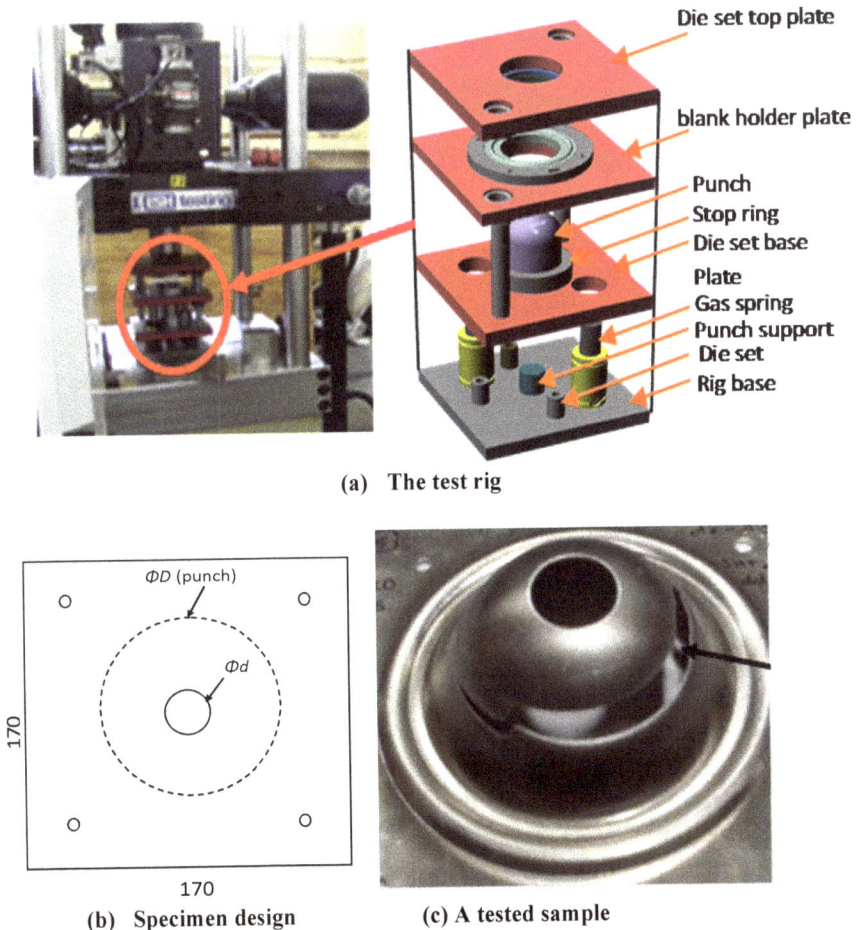

(a)   The test rig

(b)   Specimen design   **(c) A tested sample**

Fig. 8.46 (a) Hydraulic press and the formability test rig design; (b) testpiece with dimensions (in mm); and (c) a tested specimen (Mohamed, Foster, Lin *et al.*, 2012).

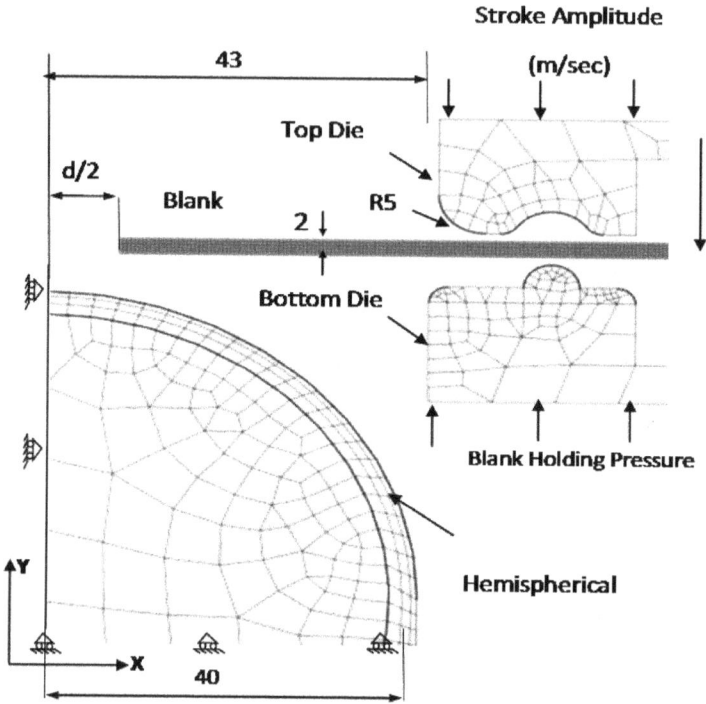

Fig. 8.47 Axisymmetric FE model with hemispherical punch and
testpiece with central hole (Mohamed, Foster, Lin *et al.*, 2012).

b. FE model

Figure 8.47 shows the FE model with boundary and loading conditions
for the simulation of the testing process shown in **Fig. 8.46**. The FE
model was used for studying the specimen's central hole design, so that
the formability and failure of the material could be evaluated for the hot
stamping condition. FE simulation of hemispherical cup forming at
elevated temperatures was conducted for AA6082 aluminium alloy using
the explicit FE code ABAQUS and coupled temperature-displacement
axisymmetric deformation mode. The sheet was meshed using
axisymmetric quadratic elements.

The unified viscoplastic damage constitutive equation set Eq. (8.9),
determined for AA6082, was used for the simulation. The heat transfer
between work-piece and die is related to contact pressure and clearance.
The heat transfer data used in the simulation are given by Foster,

Mohamed, Lin and Dean (2008). Lubrication and friction conditions for the hot stamping process have also been studied by Foster, Mohamed, Lin and Dean (2008) and the reported friction data were used in the simulation.

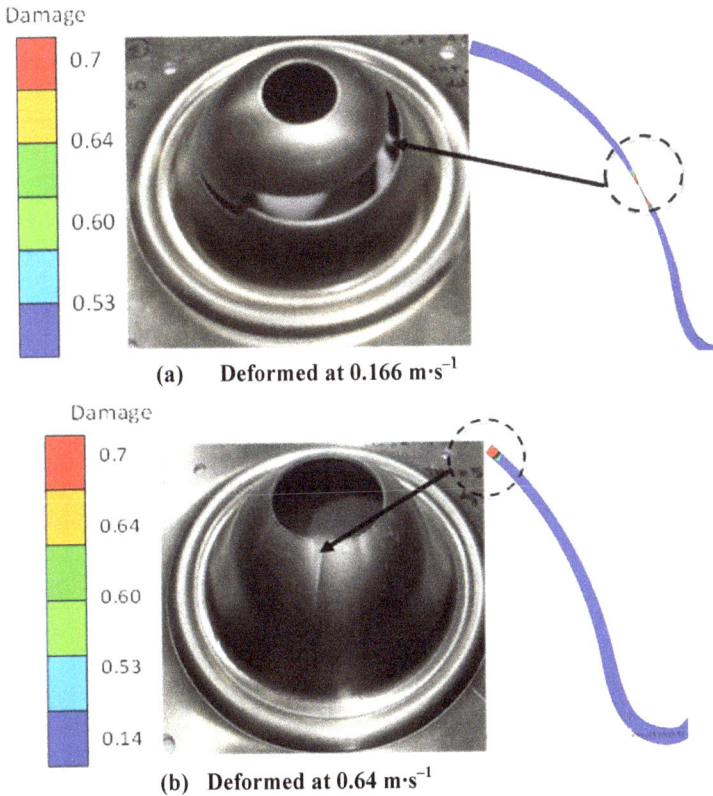

(a)   **Deformed at 0.166 m·s^{-1}**

(b) **Deformed at 0.64 m·s^{-1}**

Fig. 8.48 FE formability simulation and experimental results for the (a) low forming rate (0.166 m·s^{-1}), and (b) high forming rate (0.64 m·s^{-1}), showing contours of the damage parameter. The diameter of the central hole is 16 mm.

c. Simulation results

Two failure modes were observed in the hot stamping tests: (a) circumferential necking and subsequent tearing, occurring approximately halfway between the base and apex of the formed cup, and (b) radial necking and subsequent tearing, emanating from the central hole. **Figure 8.48** shows experimental results, together with FE predictions, for the two failure modes with a deformation temperature of 470 °C, punch stroke of 42 mm and forming rates of (**Fig. 8.48 (a)**) 0.166 ± 0.01 m·s^{-1} and (**Fig. 8.48 (b)**) 0.64 ± 0.01 m·s^{-1}, respectively. Diameters of the punch and the central hole on the specimen are 80 mm and 16 mm respectively for this case.

Figures 8.48 (a) and **8.48 (b)** also represent the predicted results of the damage value for the slow and fast forming rates, respectively. According to the theory, as the damage reaches 0.7, failure will take place. This can be considered as the initiation of tearing. From the figure, it can be seen that tearing is initiated by localised necking and the location of the failure is forming rate dependent. The developed viscoplastic damage equations and the FE process modelling method, together with the friction and heat transfer coefficients defined by Foster, Mohamed, Lin and Dean (2008), are able to predict the failure modes accurately.

A series of FE models, with different central hole sizes varying from 0 mm (no central hole) to 24 mm and a fixed punch size, were used for process simulation. Punch size was expressed by a diameter ratio ($\gamma = d/D$), where d is the sample central hole diameter and D is the punch diameter. γ is within the range of (0 to 0.25). Two forming rates were used in this study; fast forming rate (0.64 m·s^{-1}) and slow forming rate (0.166 m·s^{-1}).

Figure 8.49 shows the relation between punch stroke and diameter ratio of the formed cups for simulations at a forming rate of 0.64 m·s^{-1}. At the high speed, the time for heat transfer is very short and the material is close to isothermal forming conditions. In this case, the temperature effects on deformation are reduced.

Three different failure modes are observed in **Fig. 8.49**, which are related to diameter of the central hole, or the value of diameter ratio γ.

- **Mode 1:** The failure is a circumferential tear when the γ-value is within the range of 0 to 0.05. This is due to constraint of test-piece in the central region reducing material flow. Failure occurs in the stretched mid-way region, where close-to-plane strain deformation occurs.

- **Mode 2:** The failure location is changed to the central hole failure when the γ-value is within the range of 0.1 to 0.2 as constraint by the punch is less and hoop stretching occurs on the hole periphery.

- **Mode 3:** For the value of $\gamma > 0.2$, the punch could penetrate through the central hole without tearing it and only plastic deformation occurred.

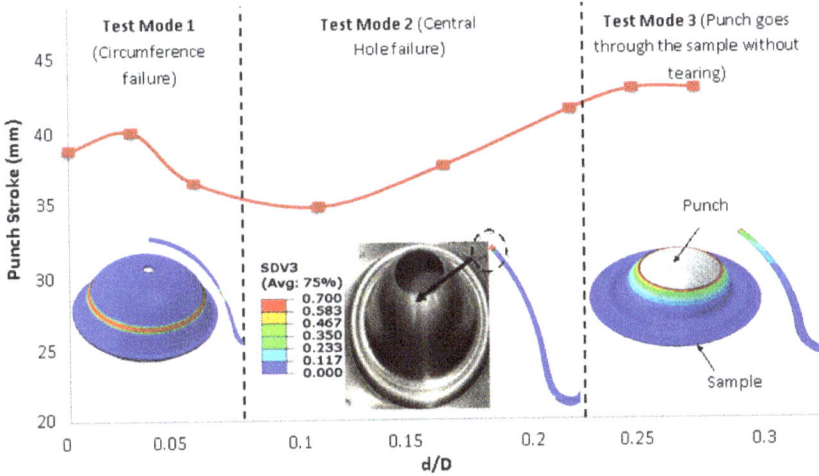

Fig. 8.49 Variation of failure mode of the specimens with a central hole at a forming rate 0.64 m·s^{-1}, temperature of 470 °C for AA6082 aluminium alloys. The solid curve represents the limit of the punch stroke, at which the failure or penetration through occurs (Mohamed, Lin, Foster *et al.*, 2014).

It can be seen from **Fig. 8.49** that if the central hole size is too small, then the effects of the hole on stress relief are not obvious. If the size of the hole is too big (relative to the punch size), the punch may penetrate through the hole during the deformation. This is related to the ductility of the material, which is a function of temperature and strain rate. Thus, to evaluate failure features at practical forming rates and temperatures, it is necessary to find an optimum design for the specimen, so that useful information for a hot stamping process can be captured.

8.5 Hot Rolling of Steel

Hot rolling is a common metal working process that is normally conducted above the recrystallisation temperature of the metal. The objectives of this operation include breakdown of a cast ingot into blooms or slabs for subsequent finishing into bars, plates or sheets, improvement of microstructure and removal of casting defects. In hot rolling of steel, the slabs are normally preheated and soaked at 1200 °C to 1300 °C depending on alloy composition. Normally a billet is rolled for about 10–20 passes. The temperature in the last finishing rolling typically is between 700 °C and 900 °C. In the case of continuous casting, metal is usually fed directly into the rolling mills at an appropriate rolling temperature.

When grains have been deformed sufficiently and are at a high enough temperature, dynamic recrystallisation (between the rolls) and static recrystallisation (between roll stands), can occur. This is followed by grain growth. This process helps to refine microstructure and prevent work hardening. Hot rolled metals generally have little directionality in their mechanical properties and little deformation-induced residual stress.

In hot rolling, it is important to be able to predict microstructural evolution and in this section an example is provided of using unified viscoplastic constitutive equations for that purpose.

8.5.1 *Unified viscoplastic constitutive equations*

Constitutive equations for viscoplasticity have been developed for steels under hot rolling conditions (Lin, Liu, Farrugia *et al.*, 2005; Liu, Lin, Dean *et al.*, 2005; Foster, Lin, Farrugia *et al.*, 2011). The equations enable a wide range of time-dependent phenomena to be modelled, such as strain hardening and stress relaxation, and in addition enable important time-dependent effects, such as strain rate, recovery and creep, to be modelled. A set of unified viscoplastic constitutive equations to model the evolution of recrystallisation, dislocation density, hardening and grain size, and to rationalise their inter-relationships and effects on viscoplastic flow, is presented here for steel under hot rolling conditions.

$$\dot{\varepsilon}^p = A_1 \sinh[A_2(\sigma - R - k)]_+ d^{-\gamma_4}$$
$$\dot{S} = Q_0 [x\bar{\rho} - \bar{\rho}_c(1 - S)]_+ (1 - S)^{Nq}$$
$$\dot{x} = A_0(1 - x)\bar{\rho}$$

$$\dot{\bar{\rho}} = (d/d_0)^{\gamma_d}(1 - \bar{\rho})|\dot{\varepsilon}^p| - c_1\bar{\rho}^{c_2} - [c_3\bar{\rho}/(1 - S)]\dot{S} \quad (8.10)$$
$$R = B\bar{\rho}^{\varphi}$$
$$\dot{d} = \alpha_0 d^{-\gamma_0} - \alpha_2 \dot{S}^{\gamma_3} d^{\gamma_2}$$
$$\sigma = E(\varepsilon^T - \varepsilon^p).$$

In the unified constitutive equations, if the normalised dislocation density $\bar{\rho}$ reaches a critical value $\bar{\rho}_c$, then, given sufficient time, recrystallisation occurs, which is controlled by the evolution equation of $\dot{S}$. The recrystallised volume fraction S varies from 0 to 1. The incubation time is controlled by the variable x, (varies from 0 to 1) and its rate is directly related to dislocation density $\bar{\rho}$. The isotropic hardening parameter R is also directly related to dislocation density. The exponent φ varies from 0.2 to 1.0, but normally it is given a fixed value of 0.5. However, for a C-Mn steel deformed at 1100 °C, it is chosen as 1.0, since the hardening effect of this material is significant during hot rolling.

The evolution of normalised dislocation density $\dot{\bar{\rho}}$ includes three terms. The first term models accumulation of dislocations due to plastic deformation and dynamic recovery, the second term models static recovery or annealing at high temperature, and the third term models the effect of recrystallisation on reduction of average dislocation density. The third term indicates that recrystallisation creates new, dislocation-free grains, which reduces overall dislocation density within the material. Large grain size obstructs grain rotation and thus increases rate of dislocation accumulation. This grain size effect is also modelled in the dislocation equation. The effect of recrystallisation on flow stress is modelled by isotropic hardening and grain size variables. Small grains contribute to grain rotation and thus grain boundary sliding is the dominant deformation mechanism, which reduces dislocation accumulation in plastic deformation and also enhances viscoplastic flow, $\dot{\varepsilon}^p$, at low stress. The sign $[...]_+$ indicates that if the values in the brackets are less than 0, then their corresponding rates are 0.

In the equation set, E is the Young's modulus and is equal to 100 GPa, whilst A_1, A_2, k, γ_4, Q_0, $\bar{\rho}_c$, N_q, c_1, c_2, c_3, d_0, γ_d, A_0, B, α_0, γ_0, α_2, γ_2 and γ_3 are material constants, which were determined by Lin, Liu, Farrugia *et al.*, (2005) from experimental data of a C-Mn steel reported by Medina and Hernandez (1996a; 1996b). The initial average grain size was 189 μm and the tests were carried out at 1100 °C. The determined material constants within the equations are given in **Table 8.9**.

Table 8.9 Material constants for Eq. (8.10) (Lin, Liu, Farrugia et al., 2005).

A_1 (s^{-1})	A_2 (MPa^{-1})	γ_4 (-)	Q_0 (-)	$\overline{\rho}_c$ (-)	N_q (-)	A_0 (-)
1.81×10^{-6}	3.14×10^{-1}	1.00	30.00	1.84×10^{-1}	1.02	40.96
c_1 (-)	c_2 (-)	c_3 (-)	d_0 (µm)	γ_d (-)	B(MPa)	φ(-)
16.00	1.43	8.00×10^{-2}	36.38	1.02	75.59	1.00
α_0 (µm)	γ_0 (-)	α_2 (µm)	γ_2 (-)	γ_3 (-)		
1.44	3.07	78.68	1.20×10^{-1}	1.06		

Computations are carried out for uniaxial loading cases with strain rates of 0.544 s^{-1}, 1.451 s^{-1}, 3.628 s^{-1} and 5.224 s^{-1}. The predicted/computed results for flow stress, recrystallised fraction, grain size and normalised dislocation density are shown in **Fig. 8.50**. It can be seen that, for a low strain rate, dynamic recrystallisation takes place at a low plastic strain. Physically, this shows that when the material is exposed for a long time under high temperature deformation, the combined thermo-mechanical effects encourage recrystallisation to take place at a low dislocation density. Mathematically, the introduction of the onset variable x enables this thermo-mechanical behaviour to be modelled.

Grain growth is not significant during high strain rate hot deformation. This is because of the short time span it takes for the strain to reach relatively high values, before recrystallisation occurs. However, once recrystallisation takes place, many new grains nucleate and the average grain size decreases quickly. High strain rates normally result in smaller grain size due to shorter grain growth time.

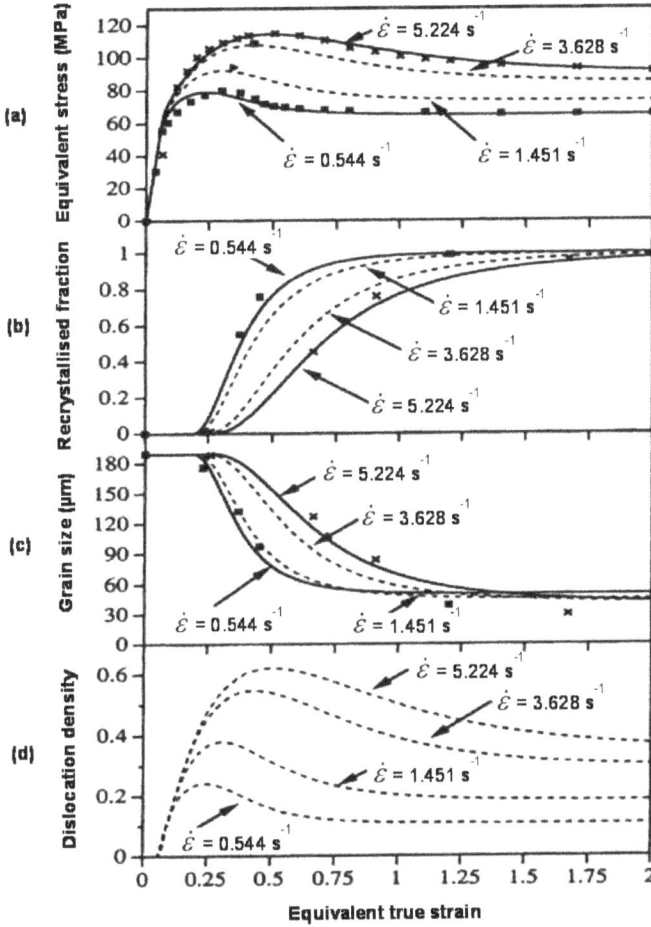

Fig. 8.50 Comparison of experimental data (symbols) (Medina and Hernandez (1996a) with computed (thick curves) and predicted (thin curves) results for different strain rates for (a) stress, (b) recrystallised fraction, (c) grain size and (d) predicted normalised dislocation density with equivalent true strain (Lin, Liu, Farrugia *et al.*, 2005).

Flow stress increases quickly with the increase of strain rate. In addition to strain rate effect, material hardening is directly related to the increment of the dislocation density, as discussed by Sandstrom, *and* Lagneborg (1975). It is argued that

$$\sigma = \alpha b G \sqrt{\rho} \,,$$

where α is a constant of about 0.5–1.0, b the Burgers vector and G the shear modulus.

For high strain rate deformation, dislocations accumulate more quickly since little time is available to allow recovery to take place (**Fig. 8.50 (d)**). This contributes to the high flow stress associated with high strain rate shown in **Fig. 8.50 (a)**. Once dislocation density reaches a critical level $\overline{\rho}_c$ for a strain rate, given sufficient time, recrystallisation takes place due to thermo-mechanical effects. This reduces dislocation density, as shown in **Fig. 8.50 (d)**, and grain size, shown in **Fig. 8.50 (c)**, which subsequently decreases flow stress, **Fig. 8.50 (a)**. The steady state flow stress is observed once the cycle of the dynamic recrystallisation is completed. This also results in a balanced state of material hardening (dislocation density **Fig. 8.50 (d)**), grain growth and refinement, **Fig. 8.50 (c)**. The relationship between flow stress and grain size for a steady state basically follows the empirical equation (Humphreys, 1999)

$$\sigma = Ad^{-m} \,,$$

where A is a parameter related to each particular material, Burgers vector and shear modulus. This flow behaviour can be seen in **Fig. 8.50**.

During hot deformation, if dislocation density reaches a certain level, dynamic recrystallisation may take place provided that the incubation time for the onset of recrystallisation is sufficient. This critical value of dislocation density, $\overline{\rho}_c$, corresponds to a critical strain, ε_c, for dynamic recrystallisation, which varies with strain rate, initial grain size and temperature. When a certain volume fraction of recrystallisation is reached, the flow stress drops due to grain refinement and reduction in dislocation density.

Figure 8.51 shows the variation of critical strain and peak stress with strain rate and initial grain size, using the determined constitutive equations. High values of critical strain for recrystallisation (**Fig. 8.51 (a)**) and lower peak stresses (**Fig. 8.51 (b)**) are obtained for the steel with small initial grain size. This is due to the deformation mechanism of grain boundary sliding and grain rotation, which is more likely to be dominant for a material with small grain size. The result is a lower rate of dislocation accumulation and thus lower flow stress.

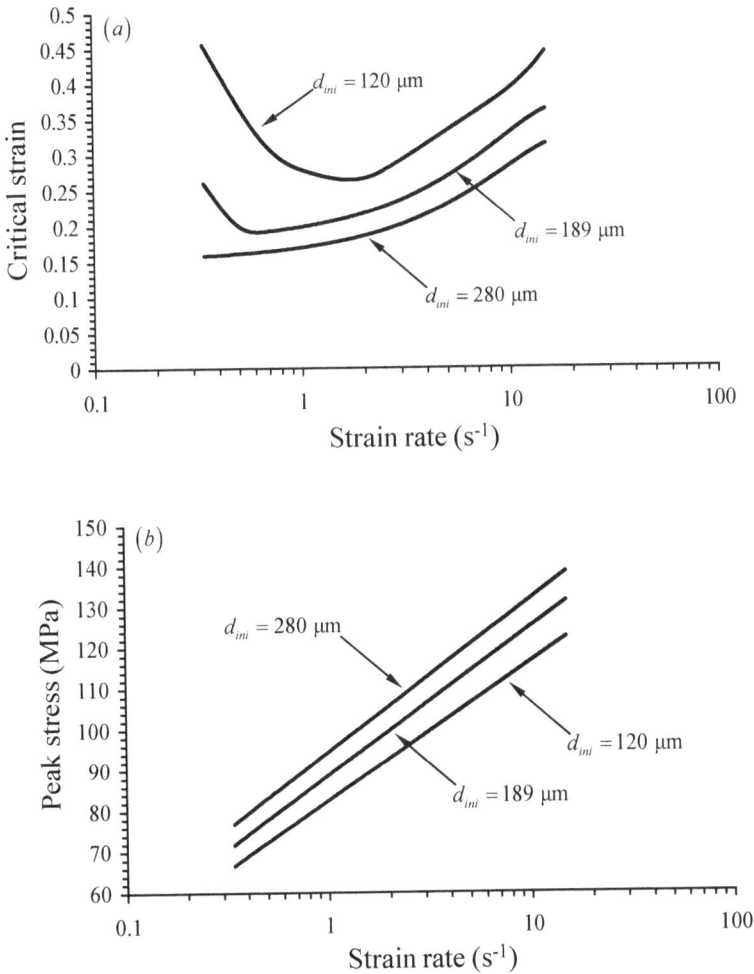

Fig. 8.51 Variations of (a) critical strain for recrystallisation and (b) peak stress with strain rate for different initial grain sizes d_{ini} (Liu, 2004).

If a low strain rate (e.g. 0.3 s^{-1}) is applied to the material with a given initial grain size, the accumulation rate of dislocations is low, due to static recovery (annealing). Thus a high value of critical strain is required to enable the dislocation density to be accumulated to the critical value $\bar{\rho}_c$ (Fig. 8.51 (a)). If strain rate is high (e.g. 10 s^{-1}), the critical strain is also high (**Fig. 8.51 (a)**). This is due to the incubation time required for recrystallisation to happen, although that time is shorter for a higher dislocation density. However, for steel deformed at a relatively low strain rate with a small initial grain size, grain boundary sliding and grain rotation may be the dominant deformation mechanism. This can slow down the accumulation rate of dislocations. Thus for a material with a given initial grain size, an optimum strain rate exists for the occurrence of the lowest critical strain. In addition, steel may be able to demonstrate its own "superplastic phenomenon", if a proper combination of initial grain size and deformation rate at a given high temperature is applied.

Multiaxial constitutive equations from equation set Eq. (8.10) are obtained by assuming von-Mises behaviour and by defining an energy dissipation rate potential, which has been detailed in earlier sections for the sinh-law type of equations (Lin, Liu, Farrugia *et al.*, 2005).

8.5.2 *FE model and simulation procedures*

Simulation of hot rolling is carried out at a constant temperature of 1100 °C. The FE model for two pass hot rolling is shown in **Fig. 8.52**. The initial thickness of the stock decreases from 50 mm to 30 mm after the first pass and reaches its thickness of 20 mm after the second pass. The viscoplastic deformation of the material is governed by the multiaxial constitutive equation set, which was generalised from equation set Eq. (8.10) and implemented in the FE solver ABAQUS/standard through the user-defined subroutine CREEP. To use an implicit numerical integration method, the gradients, $\partial \Delta \varepsilon_e^p / \partial \varepsilon_e^p$ and $\partial \Delta \varepsilon_e^p / \partial \sigma_e$, are calculated and implanted in the system for the step time control. Due to its symmetry, only a half of the rolling system is modelled. Eight-node quadrilateral plane strain elements are used to mesh the work-piece. Friction, as a surface interaction property, is related to the contact rollers and the top and left-hand side surfaces of the work-piece, by specifying a sticking friction coefficient. The detailed FE procedures for the multipass rolling simulation are given below (Lin, Liu, Farrugia *et al.*, 2005).

- **Stage 1:** The work-piece is moved into contact with Roller 1.

- **Stage 2:** The work-piece is rolled with a rolling velocity of 6.58 rad/s. It takes 0.92 s for the work-piece to pass entirely through Roller 1.

- **Stage 3:** The interpass time applied to the work-piece at a constant temperature of 1100 °C is 20 s, during which recovery, recrystallisation and grain size evolution continue to occur. Then the work-piece is forwarded to Roller 2.
- **Stage 4:** The work-piece is rolled with a rolling velocity of 3.74 rad/s with Roller 2.

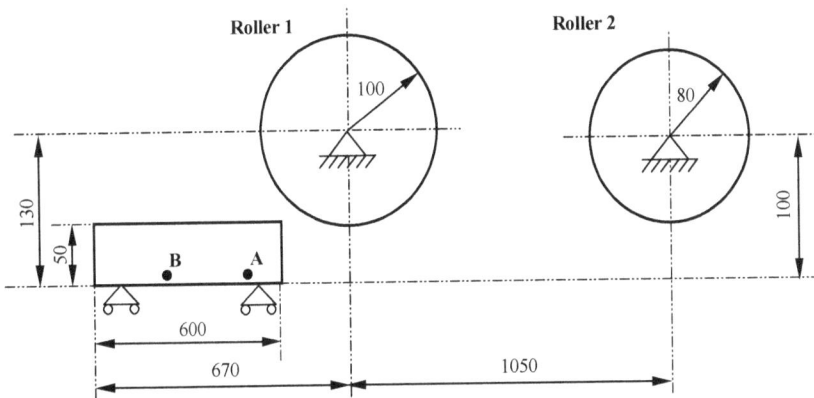

Fig. 8.52 FE model for the simulation of two pass rolling (Liu, 2004).

8.5.3 *Prediction results*

The FE analysis is carried out using the FE model and numerical procedures described above for a two pass rolling process. **Figure 8.53** shows field plots of effective stress, dislocation density, recrystallisation and grain size, for the first pass. The microstructure at the left end of the work-piece remains almost unchanged. Stress and strain levels change dramatically around the rolling region. After the steel is rolled, stress reduces to a very low level (residual stress) and the pattern remains constant except at the leading end where stress states are more complicated due to initial contact conditions between roller and work-piece.

Microstructural evolution exhibits a different behaviour. The normalised dislocation density increases immediately **(Fig. 8.53 (b))** from the original state once the material is fed into the rollers. The field pattern of the normalised dislocation density for the initially deformed part is very much similar to that of the effective stress (**Fig. 8.53 (a)**), since the increment of dislocation density is directly related to the plastic strain rate, as explained when the equation set was described.

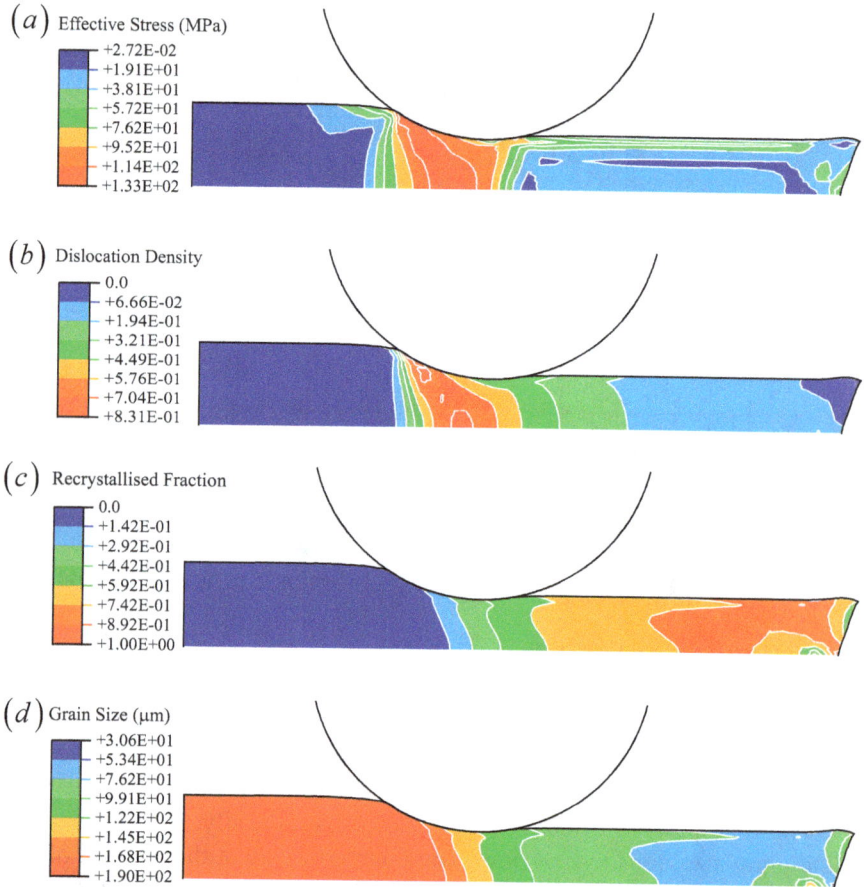

Fig. 8.53 Predicted distributions of (a) effective stress, (b) normalised dislocation density, (c) recrystallisation volume fraction and (d) grain size for the first pass (Lin, Liu, Farrugia *et al.*, 2005)

Dislocation density reaches its maximum value at almost the same time and in the same location as the occurrence of maximum effective stress (**Fig. 8.53 (a)**). After the dislocation density reaches this critical value, dynamic recrystallisation takes place. It can be seen clearly from the field plot, shown in **Fig. 8.53 (c)**, that recrystallisation does not start immediately the material enters the roller.

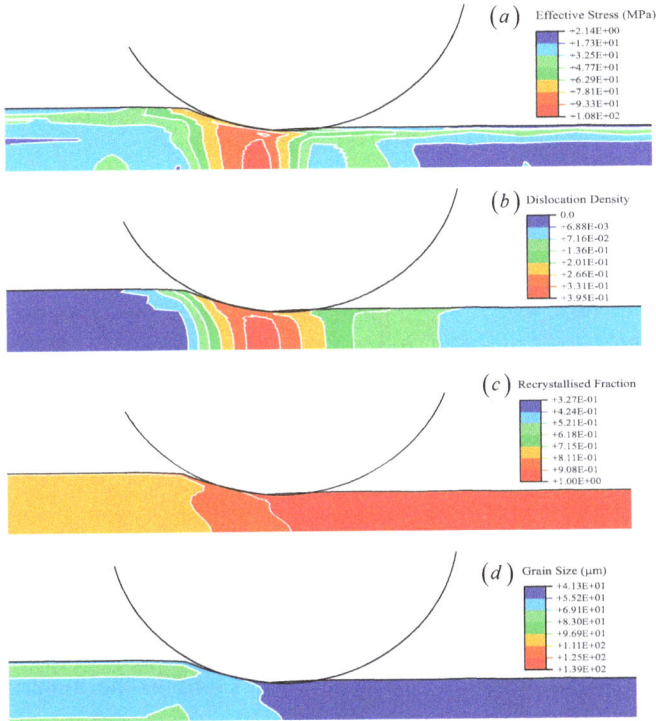

Fig. 8.54 Predicted distributions of (a) effective stress, (b) dislocation density, (c) recrystallised volume fraction and (d) grain size evolution for the second pass (Lin, Liu, Farrugia *et al.*, 2005).

There is an obvious delay due to the onset parameter control and the critical dislocation density accumulated from deformation. Recrystallisation (static) continues after the work-piece passes the roller (**Fig. 8.53 (c)**) due to the high dislocation density (≈ 0.4 **Fig. 8.53 (b)**). This results in the recrystallised volume fraction of the material increasing from about 0.4 just after the material passes the roller to about 0.8, over a very short time period. The continuous recrystallisation and static recovery cause dislocation density reduction after rolling, which can be seen from **Fig. 8.53 (b)**. Grain refinement takes place only when dynamic recrystallisation begins. The average grain size (**Fig. 8.53 (d)**) reduces to about 100 µm from 189 µm just after the first pass due to dynamic recrystallisation and continues to decrease to about 60 µm due to the following static recrystallisation.

Figure 8.54 shows field plots of the same parameters around the rolling area for the second pass with a lower rolling speed of 3.74 rad/s. It can be seen that after 20 s relaxation, the normalised dislocation density (**Fig. 8.54 (b)**) reduces to about zero due to recrystallisation and recovery. This minimises the hardening caused by plastic deformation in the first pass and increases formability of the steel. The stress level (**Fig. 8.54 (a)**) is lower in the second pass due to the low rolling velocity and smaller grain size. The average grain size (**Fig. 8.54 (d)**) reduces to about 45 µm after the second pass due to the continuous recrystallisation (**Fig. 8.54 (c)**). Thus the lower stress level in the second pass is partially due to the lower rolling speed and partially due to the further refinement of grains, which enables grain boundary sliding to take place more easily.

Figure 8.55 shows the variation of effective stress, dislocation density, recrystallisation and grain size at locations *A* and *B*, indicated in **Fig. 8.52** for the two pass rolling process. Both location *A* and *B* have passed the first roller; however for the second roller, location *A* has passed, but *B* has not. The recrystallised volume fraction increases quickly after the first pass but stops when dislocation density vanishes due to recovery and recrystallisation. This results in the material being partially recrystallised.

Recrystallisation continues at the second pass, once dislocation density has increased again, until the material is fully recrystallised. This is the starting point for the second cycle of recrystallisation.

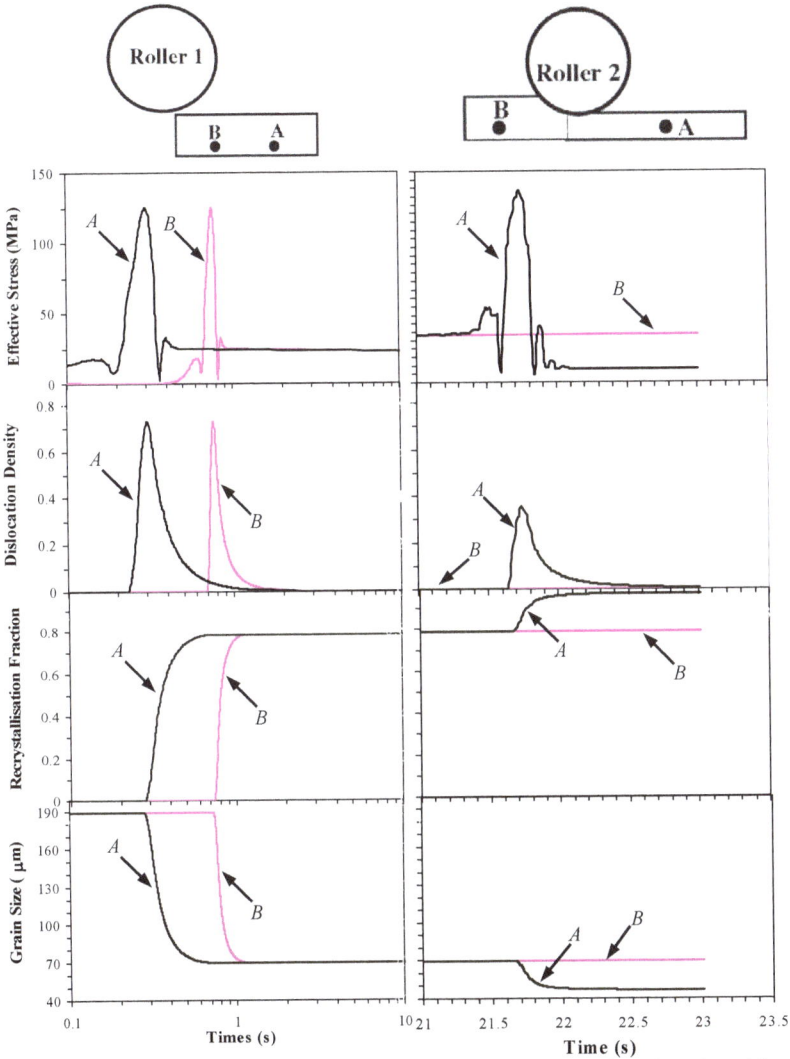

Fig. 8.55 Variation of (a) effective stress, (b) dislocation density, (c) recrystallised fraction and (d) grain size at locations *A* and *B*, indicated in Fig. 8.52, for the two passes rolling process. The computation is carried out with the rolling speeds: 6.58 rad/s and 3.74 rad/s for the two rolling passes (Liu, 2004).

The developed unified viscoplastic constitutive equations, integrated with an FE solver, enable detailed microstructural evolution, such as recrystallisation and grain size during and after hot rolling, to be modelled. For example, according to the evolution of recrystallisation, dislocation density recovery and grain growth between rolling passes can be determined as a function of interpass time. If the interpass time is too short, the recovery of dislocations may not be complete. This would result in high rolling forces and low formability of the material in subsequent passes. If the interpass time is too long, the excess static grain growth after recrystallisation may result in poor mechanical properties of the rolled steel. Thus, by the use of this equation set within an FE package, it is possible to control the microstructure of the material processed and to optimise hot rolling processes.

8.6 Modelling of Two-phase Titanium Alloys during Forging

Titanium alloys are used in significant quantities in gas turbines (**Fig. 8.56**), especially for compressor blades (Eliaz, Shemesh and Latanision, 2002), where the operating temperature is below approximately 400 °C. Hot extrusion and forging are the most popular shaping processes for the manufacture of titanium blades. During these forming processes, microstructural evolution can affect the viscoplastic flow of the alloys and subsequently influence the mechanical properties of the formed parts. Thus, unified viscoplastic constitutive equations are required to predict the interactive effect of microstructural evolution and viscoplastic-plastic flow.

Fig. 8.56 Section of a gas turbine (Eliaz, Shemesh and Latanision, 2002).

8.6.1   *Hot forming of Ti-6Al-4V gas turbine blades*

The hot forging of titanium alloy Ti-6Al-4V for gas turbine blades includes a number of operations and a typical example is shown in **Fig. 8.57**. Firstly, a cylindrical billet is extruded to produce a preform. The size and geometry of the cylinder depend on the type and size of the turbine blade to be manufactured. Then the billet is coated with glaze coating by dipping and drying at room temperature.

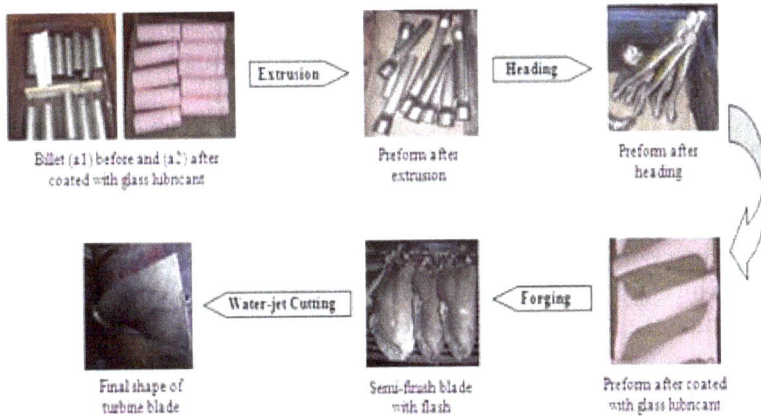

Fig. 8.57 A typical manufacturing process of gas turbine blades (Bai, 2012).

The billet is extruded at 920 °C first, then headed and twisted at the same temperature. Then the preform is again coated with the same glaze and is heated again to 920 °C in a furnace. A graphite lubricant is applied on the top and bottom dies soon before forging to minimise die wear and promote metal flow. The forged blade is removed from the bottom die after forging and cooled to room temperature. Finally the flash is removed by water-jet cutting or mechanical trimming.

In the hot extrusion, heading, twisting and forging processes, heat transfer between the work-piece and tools takes place and microstructural evolution occurs which affects the flow stress. The forming temperature is defined to control the correct ratio of α and β phases and their grain size for the $\alpha + \beta$ titanium alloy, which is also affected by the plastic deformation.

8.6.2 Deformation and softening mechanisms

In hot compression tests for Ti-6Al-4V, it was found that the flow stress reduces gradually with deformation (Bai, Lin, Dean et al., 2013). This phenomenon becomes less obvious as the deformation temperature increases. The softening behaviour of the material is considered to be due to globularisation of the lamellar alpha phase within beta grains (Stefansson and Semiatin, 2003).

Flow softening may be ascribed to deformation heating, texture changes and the microstructure morphology for hot-worked Ti-6Al-4V; however, globularisation of lamellar alpha is the dominant mechanism in flow softening of Ti-6Al-4V and transforms microstructure. Semiatin, Seetharaman and Weiss (1999) determined that the breakdown of the transformed microstructure during hot working at a temperature below the beta transus temperature plays a key role in the development of the equiaxed alpha microstructure.

The alpha phase is considerably harder than the beta phase; it behaves like a dispersion of hard particles in a soft matrix (Weiss and Semiatin, 1998). Since the alpha phase deforms less than the beta phase, strain concentrations may develop in the softer beta phase in the vicinity of the alpha phase, causing the formation of smaller highly disorientated sub-grains in comparison with the sub-grains produced in the remaining beta matrix.

At higher strains, sub-boundaries develop in the alpha phase followed by cusp formation and beta phase penetration along the alpha/alpha sub-boundaries. Weiss, Froes, Eylon et al. (1996) investigated the modification of lamellar alpha phase in Ti-6Al-4V by hot working. They found that the morphology change from alpha lamellae into low aspect ratio grains was by a break-up of the alpha lamellae, essentially in a two-step process: a formation of low and high angle alpha/alpha boundaries or shear bands across the alpha plates and then a penetration of beta phase to complete the separation. Also they found this break-up takes place during hot deformation and subsequent annealing, as shown in **Fig. 8.58 (a)**, which was obtained from backscattered SEM (Weiss and Semiatin, 1998). The initial microstructure was a lamellar-transformed microstructure.

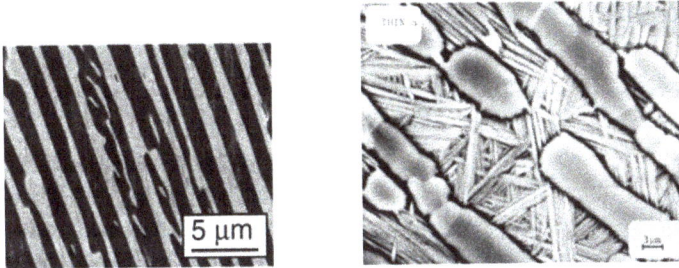

(a) Initial microstructure   **(b) Globularisation of secondary alpha phase**

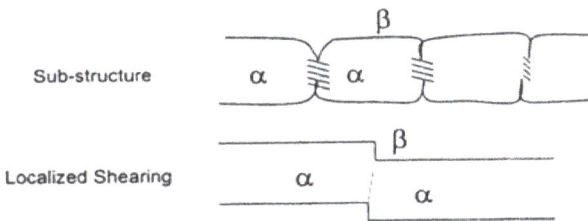

(c) Globularisation driven by the formation of sub-boundaries in shear bands within alpha lamellae (Semiatin and Furrer, 2008).

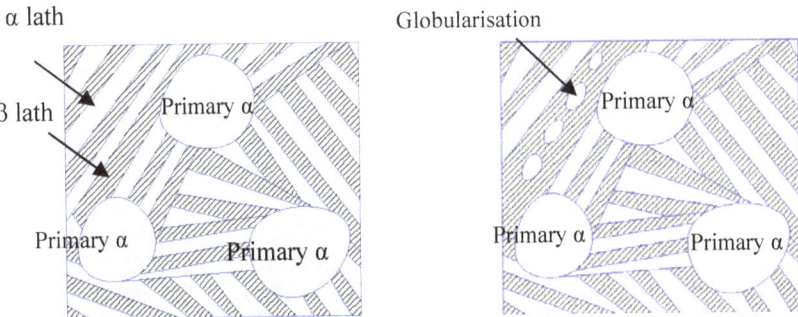

(d) Schematic of globularisation of secondary alpha phase in Ti-6Al-4V with primary alpha phase (Bai, Lin, Dean *et al.*, 2013).

Fig. 8.58. Evidence of globularisation in hot forming titanium alloy; (a) was obtained from backscattered SEM by Semiatin and Bieler (2001); (b) from channelling contrast SEM by Weiss and Semiatin (1998).

After hot working in the alpha and beta fields, globularisation takes place in the secondary alpha phase, which is evidenced in **Fig. 8.58 (b)** obtained from channelling contrast SEM. Observations by Weiss, Froes, Eylon *et al.* (1986) and Stefansson and Semiatin (2003) have revealed that the globularisation of colony alpha can occur by localisation of strain into intense shear bands within the alpha lamellae (**Fig. 8.58 (c)**), and it eventually leads to fracture and segmentation. Similarly, in a Ti-6Al-4V microstructure consisting of primary alpha phase and beta matrix, the beta matrix includes alpha laths and beta laths (**Fig. 8.58 (d)**) (Bai, Lin, Dean *et al.*, 2013). More deformation occurs in softer beta matrix in comparison with that in harder primary alpha grains. As a result, it can be assumed that the globularisation also occurs in the alpha laths in Ti-6Al-4V during hot working, to which softening during hot working is attributed (Bai, Lin, Dean *et al.*, 2013).

8.6.3 *Unified viscoplastic constitutive equations*

a. *Modelling of globularisation softening*

Beta strain induced globularisation, w, is proposed as a state variable in a set of constitutive equations for describing the softening mechanism of Ti-6Al-4V in hot working (Bai, Lin, Dean *et al.*, 2013). The beta phase strain controlled globularisation equation was formulated as

$$\dot{w} = c_w(1-w)\dot{\varepsilon}^{p,\beta}$$

$$\dot{\varepsilon}^{p,\beta} = \left(\frac{\sigma-R-k_\beta}{K_\beta(1-w)}\right)^n, \tag{8.11}$$

where the parameters, c_w and K_β, are temperature-dependent constants, k_β is the threshold stress for beta grains and R is the isotropic hardening. Globularisation of alpha laths is strongly related to the beta phase plastic strain $\varepsilon^{p,\beta}$. The extent of globularisation of alpha laths increases with the increase of beta deformation. The viscoplastic flow of the beta phase, $\dot{\varepsilon}^{p,\beta}$, shows that alpha phase plastic strain is associated with globularisation, w, which varies from 0 (the initial state) to 1 (the saturated state of globalisation). This condition is achieved by the first equation in Eq. (8.11).

b. *Adiabatic heating*

Many researchers (e.g. Khan, Sung-Suh and Kazmi, 2004) attributed flow softening of titanium alloys in hot working partially due to adiabatic heating at high strain rate. Ding, Guo and Wilson (2002) stated that adiabatic heating raises the temperature of the sample and also of a proportion of the softer beta phase. Normally, adiabatic heating is included in the set of constitutive equations by assuming 90% of the plastic work becomes heat (Khan, Sung-Suh and Kazmi, 2004). For hot forming, thermal softening from adiabatic deformation was effectively considered by converting the increment of temperature from the stress–strain curve using

$$\Delta T = \frac{\eta}{cd} \int_0^{\varepsilon_p} \sigma(\varepsilon_p) d\varepsilon_p \qquad (8.12)$$

where η, ($\eta = 0.9$), is the fraction of the heat dissipation caused by plastic deformation and c and d are specific heat and mass densities, respectively (Khan, Sung-Suh and Kazmi, 2004). The temperature increase is computed numerically. Here, the user-defined state variable of temperature rise follows the rate form as given by

$$\dot{T}_\varepsilon = \eta \frac{\sigma}{cd} |\dot{\varepsilon}_p|,$$

where T_ε is strain induced temperature rise. For isothermal conditions, the instantaneous temperature T equals the initial temperature T_0; in the cases for which heat is generated due to large plastic deformation, the instantaneous temperature T equals ($T_0 + \Delta T$, where ΔT ($\Delta T = \dot{T}_\varepsilon \times \Delta t$, where Δt is time increment in numerical integration) is the temperature rise as a function of deformation and deformation rate. From the equation, it can be seen that given a strain rate and the current stress, deformation induced temperature increase can be obtained through numerical integration.

c. *Phase transformation*

Titanium can have two crystal structures due to allotropic phase transformation: *hcp* alpha phase at lower temperatures and *bcc* beta phase at higher temperatures. For two-phase titanium alloys, the volume fraction of the alpha phase decreases as temperature increases, and complete transformation is obtained above the beta transus temperature (Fan and Yang, 2011). In this work, the Ti-6Al-4V alloy at room temperature, prior to hot working, consists of primary *hcp* alpha grain particles, with a dispersion of *bcc* beta phase around alpha grain boundaries, as shown in **Fig. 8.59**. The alpha phase transforms to beta phase starting at 873 K; the entire microstructure consists of transformed beta grains above 1268 K (Ding, Guo and Wilson, 2002).

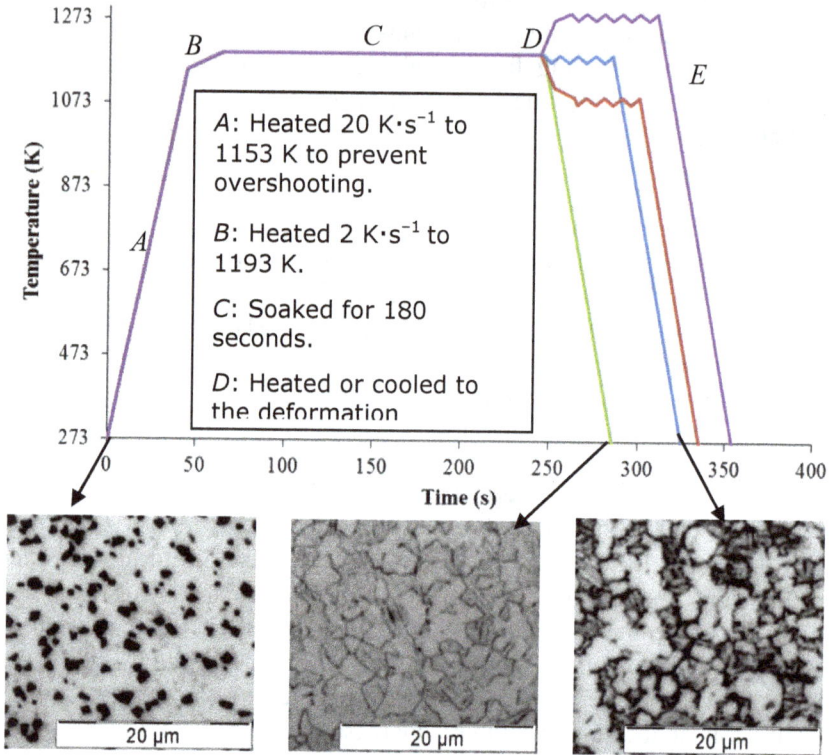

Fig. 8.59 A schematic showing the experimental programme and the corresponding microstructures at each stage (Bai, Lin, Dean *et al*, 2013).

A temperature profile for a hot compression test programme is shown in **Fig. 8.59**. The heating rate was constant but, to avoid overshoot, was reduced at a value of about 40 K below the specified deformation temperature. Take the deformation temperature of 1193 K for example. A sample was heated to 1153 K at a rate of 20 K·s^{-1}, and then heated to 1193 K at a heating rate of 2 K·s^{-1} to prevent temperature overshoot. After soaking at 1193 K for 180 seconds to ensure temperature uniformity, it was deformed at one of a specified number of strain rates and subsequently cooled to room temperature by water quenching. The microstructures of the material at the initial state, after soaking and after hot deformation, are also shown in **Fig. 8.59**.

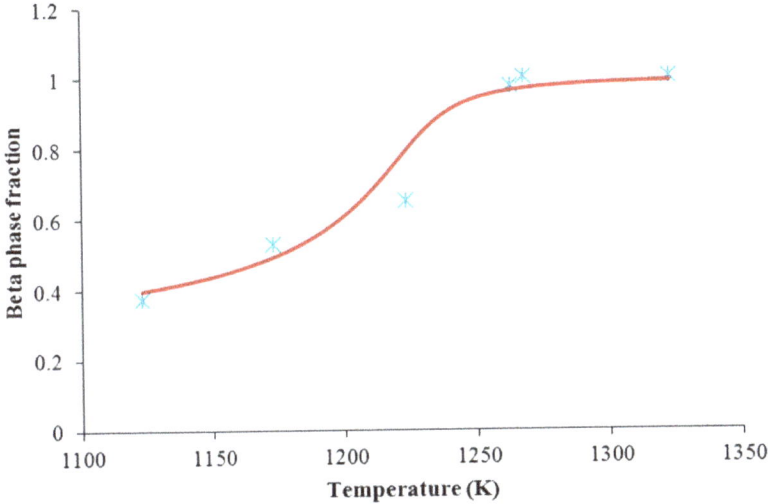

Fig. 8.60 Comparison of experimental (symbols, by Ding, Guo and Wilson (2002)) and computed (solid curves by Bai, Lin, Dean *et al.* (2013)) beta phase fraction–temperature relationships.

In non-isothermal hot forming processes, phase volume fractions change with temperature. Due to the fact that the *bcc* beta phase is softer, the proportion of the softer beta phase affects material flow. Therefore, it is important to model the phase transformation with temperature. The alpha-beta phase transformation can be described by an Avrami equation. Picu and Majorell (2002) obtained beta volume fraction directly from the phase diagram and used Eq. (8.13) to describe and predict the volume fraction of beta phase for all temperatures.

$$C_\beta(T) = \left(\frac{T}{1270}\right)^{10}, \qquad (8.13)$$

where $C_\beta(T)$ is the beta phase fraction as a function of temperature, and T is sample temperature.

Evolution of beta phase volume fraction is modelled using (Bai, Lin, Dean *et al.*, 2013),

$$\dot{f}_\beta = X \left(\frac{f_\beta}{0.5}\right)^\alpha (1 - f_\beta)^\gamma \dot{T}_T, \qquad (8.14)$$

where X and α are material constants and T_T is the current temperature. When the deformation takes place in an isothermal condition, in which case heat transfer can be ignored, $T_T = T_\varepsilon$. The equation was determined by Bai, Lin, Dean *et al.* (2013) using the experimental data generated by Ding, Guo and Wilson (2002). The fitted results are shown in **Fig. 8.60** and the values of the constants are listed in **Table 8.10**.

d. Formulation of unified viscoplastic constitutive equations

Bai, Lin, Dean *et al.*, (2013) developed a set of unified viscoplastic constitutive equations for the two phase material, which models evolution of dislocation density, dislocation hardening, temperature increase due to adiabatic deformation, phase transformation and beta strain controlled globularisation, to rationalise their inter-relationships and effect on plastic flow. For simplification, dynamic recrystallisation of the material at the initial deformation process is not considered in their work. The equations enable the dominant softening mechanism of the material for hot large deformation conditions to be modelled. By omitting the dynamic recrystallisation phenomena, a set of unified viscoplastic constitutive equations for modelling the softening mechanism is shown as below (Bai, Lin, Dean *et al.*, 2013):

$$\dot{\varepsilon}^{p,\alpha} = \left(\frac{\sigma - R - k_\alpha}{K_\alpha}\right)^n$$

$$\dot{\varepsilon}^{p,\beta} = \left(\frac{\sigma - R - k_\beta}{K_\beta(1-w)}\right)^n$$

$$\dot{\varepsilon}^p = \dot{\varepsilon}^{p,\alpha}(1 - f_\beta) + \dot{\varepsilon}^{p,\beta}f_\beta$$

$$\dot{\bar{\rho}} = A(1 - \bar{\rho})|\dot{\varepsilon}^p| - C\bar{\rho}^\delta$$

$$R = B\bar{\rho}^{0.5} \tag{8.15}$$

$$\dot{T}_\varepsilon = \eta\frac{\sigma}{cd}|\dot{\varepsilon}^p|$$

$$\dot{f}_\beta = X\left(\frac{f_\beta}{0.5}\right)^\alpha (1 - f_\beta)^\gamma \dot{T}_T$$

$$\dot{w} = c_w(1 - w)\dot{\varepsilon}^{p,\beta}$$

$$\sigma = E(\varepsilon^T - \varepsilon^p).$$

The state variables within the equation set represent individual physical phenomena of the material during hot forming. The alpha phase is harder than the beta phase; therefore the flow strain varies from phase to phase (Bai, Lin, Dean *et al.*, 2013). Thus, the strain rate of alpha phase, $\dot{\varepsilon}^{p,\alpha}$, and strain rate of beta phase, $\dot{\varepsilon}^{p,\beta}$, must be modelled separately.

- The plastic strain rate of alpha is a function of stress σ, isotropic hardening R and temperature-dependent yield stress k_α.
- The plastic strain rate of beta is a function of stress σ, globularisation-induced softening, w (varies from 0 to 1), isotropic hardening R and temperature-dependent yield stress k_β. In these two equations, $\varepsilon^{p,\alpha}$, $\varepsilon^{p,\beta}$, σ, R and w are state variables, n is material constant and k_α, K_α are temperature-dependent parameters for the alpha phase. From high temperature indentation tests on the two phase titanium alloy Ti-6Al-4V, it was found that the ratio of beta yield stress over alpha yield stress is approximately 0.8. As a result, an assumption has been made that for all values of elevated temperature, $k_\beta = 0.8k_\alpha$. Due to the fact that the main element in alpha phase and beta phase is titanium, the material properties for both phases are considered as similar and it has been assumed that $K_\beta = 0.9K_\alpha$.

The properties of two-phase materials depend on various factors including the volume fraction and morphology of the component phases, the anisotropic nature of these phases and the degree of preferred orientation. Each phase, during hot working, is treated as an isotropic, elastic-viscoplastic solid. The effect of the volume fraction of phases on material properties can be predicted with two simple assumptions and utilisation of the law of mixtures rule. The first, is that strain is the same in both phases (isostrain), known as the Voigt estimate, and the second, that stress is the same in both phases (isostress), known as the Reuss Estimate (Nemat-Nasser, 2004). The overall plastic strain rate is $\dot{\varepsilon}^p$, $\dot{\varepsilon}^{p,\alpha}$ is the plastic strain rate with regard to alpha phase, $\dot{\varepsilon}^{p,\beta}$ is the plastic strain rate with regard to beta phase, and f_β is the beta phase fraction, which varies with temperature.

Isotropic hardening, due to plastic deformation, R, is directly related to normalised dislocation density $\bar{\rho}$. The normalised dislocation density evolution $\dot{\bar{\rho}}$ is a function of total plastic strain rate in itself. A is a material constant and C and B are temperature-dependent parameters, which are defined using the classic Arrhenius relations and are listed below:

$$k_\alpha = k_{\alpha 0} \exp\left(\frac{Q_p}{kT}\right) \qquad\qquad K_\alpha = K_{\alpha 0} \exp\left(\frac{Q_a}{kT}\right)$$

$$C = C_0 \exp\left(-\frac{Q_c}{kT}\right) \qquad\qquad c_w = c_{w0} \exp\left(-\frac{Q_w}{kT}\right)$$

$$B = B_0 \exp\left(\frac{Q_b}{kT}\right) \qquad\qquad E = E_0 \exp\left(\frac{Q_e}{kT}\right).$$

The equations can be solved using a numerical integration method.

8.6.4 *Determined unified viscoplastic constitutive equations*

The experimental data generated from the test programme and shown in
Fig. 8.59, were used to determine the material constants contained in the
unified constitutive equations. Evolutionary programming (EP)
optimisation techniques for determining these material constants were
based on minimising the residuals between the computed target values
and experimental data. Details of the optimisation method and the
corresponding numerical procedure for this type of problem are
described in Chapter 7. Cao (2006) developed the solutions for
determination of a pair of data points for error assessment defined the
"error", which gives the "true" error in a fitting process and formulated a
universal objective function.

The optimisation system, OPT-CCE (Cao, Lin and Dean, 2008b)
contains three parts. Part one is to enable constitutive equations,
boundary values of the constants and the corresponding experimental
data to be input into the system. Part two is the EP optimiser. The
constitutive equations need to be solved numerically via a numerical
integration method. Differences between the model predictions and
experimental data are assessed using a well-defined objective function.
Part three outputs the optimised material constants and the fitted results
graphically.

Table 8.10 Determined constants for the set of unified viscoplastic constitutive equations for Ti-6Al-4V (Bai, Lin, Dean *et al.*, 2013).

$k_{\alpha0}$ (MPa)	Q_p (J/mol)	$K_{\alpha0}$ (MPa)	Q_α (J/mol)	n (-)	A (-)	C_0 (-)
6.57×10^{-3}	7.79×10^4	4.08×10^{-2}	7.11×10^4	2.8	10.0	1.62×10^{16}
Q_c (J/mol)	δ (-)	B_0 (MPa)	Q_b (J/mol)	η (-)	X (K^{-1})	α (-)
3.28×10^5	2.0	1.34	4.15×10^4	0.95	0.01	5.7
γ (-)	c_{w0} (-)	Q_w (J/mol)	E_0 (-)	Q_e (J/mol)		
1.81	7.09×10^2	6.70×10^4	7.57×10^4	1.14×10^3		

The procedure for determining the values of material constants within the constitutive equations was divided into two steps by Bai (2012). The first step was to determine the constants relating to phase transformation, which was modelled using Eq. (8.14). Experimental data of the beta phase fraction reported by Ding, Guo and Wilson (2002) is shown using symbols in **Fig. 8.60**. The constants X, α and γ within the phase transformation equation were determined using the experimental data and the EP based optimisation method by Bai, Lin, Dean *et al.* (2013), and listed in **Table 8.10**. The fitted numerical results are shown by solid lines in **Fig. 8.60**.

The second step was to determine the other constants within equation set Eq. (8.15) and their corresponding temperature-dependent variables. Seven selected experimental stress–strain curves from compression tests at different strain rates and temperatures were used for the work, and the results are shown with symbols in **Fig. 8.61** and **Fig. 8.62** for a temperature varying from 1093 K to 1293 K and strain rates from 0.1 s^{-1} to 10 s^{-1}. The temperature and strain rate values selected are those most appropriate for hot forming Ti-6Al-4V components in industry.

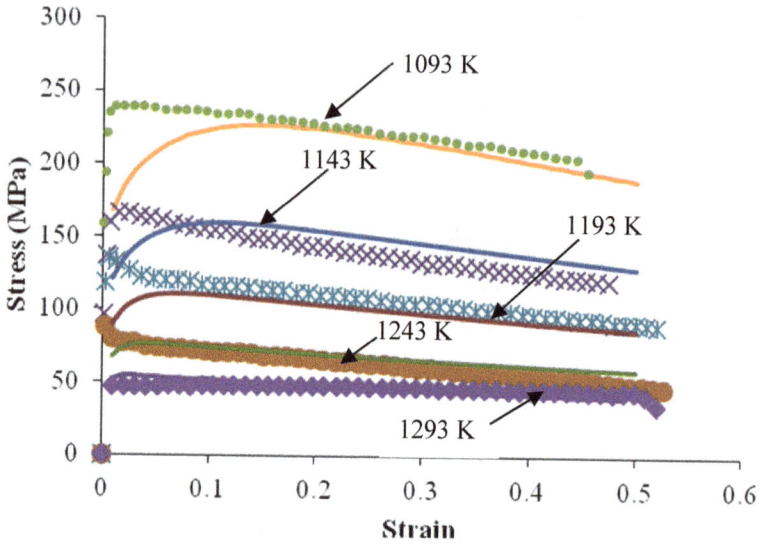

Fig. 8.61 Comparison of experimental (symbols) and computed (solid curves) strain–stress relationships for the compression tests with temperatures at strain rate 1.0 s^{-1} (Bai, Lin, Dean *et al.*, 2013).

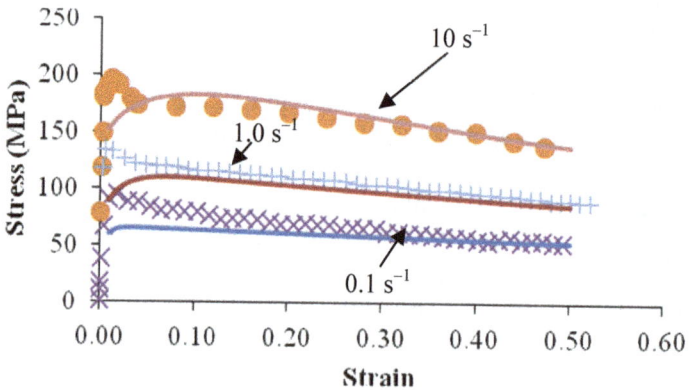

Fig. 8.62 Comparison of experimental (symbols) and computed (solid curves) strain–stress relationships for the compression tests with strain rates at temperature 1193 K (Bai, Lin, Dean *et al.*, 2013).

The technique, described above, for determining constants using EP-based optimisation was used for this step. The determined values of the constants are listed in **Table 8.10** and their fit with experimental values is shown by comparing solid lines with symbols in **Fig. 8.61** and **Fig. 8.62**. The general trend exhibited by softening mechanism of Ti-6Al-4V during hot forging is correctly described.

Using a similar method to that discussed in previous sections, the unified viscoplastic constitutive equations can be input into commercial FE solvers via user-defined subroutines for the modelling of extrusion and/or forging processes of Ti-6Al-4V alloy.

Chapter 9

Crystal Plasticity for Micro-forming Process Modelling

Plastic flow in a single crystal is anisotropic and the constitutive equations introduced in the previous chapters cannot be used to accurately model the plastic deformation of the material. Thus, constitutive equations are required to model the slip activity in the crystal directly and the rotation of individual grains (crystals) for polycrystalline structures.

The crystal plasticity theory was proposed a long time ago (Taylor, 1934) and has been used to understand/explain the deformation mechanism of single and polycrystal materials. Only in recent years has the theory been developed significantly and implemented into finite element systems to analyse deformation and failure of crystal structural materials in engineering applications (Dunne, Rugg and Walker, 2007; Dunne, Kiwanuka and Wilkinson, 2012; Huang, 1991).

Metal forming techniques have been developed in recent years for the forming of micro components, or metal components with micro features (Engel and Eckstein, 2002; Cao, Krishnan, Wang et al., 2004). In many cases, only a few grains exist across the minimum section of formed parts (Krishnan, Cao and Dohda, 2007) and thus crystal plasticity theories should be used to accurately predict the deformation and material flow features in micro-forming processes (Zhuang, Wang, Lin et al., 2012). Based on this need, crystal plasticity finite element (CPFE) methods have been developed for the modelling of micro-forming processes (Cao, Zhuang, Wang and Lin, 2009a). The details of deformation, material flow and texture evolution of materials cannot be well predicted without the use of CPFE micro forming simulation techniques (Wang, Zhuang, Balint and Lin, 2009b).

This chapter introduces the basic crystal plasticity theories, slip systems for crystals, constitutive equations and CPFE modelling methods. In addition, the integrated crystal plasticity FE modelling system, and grain structure generation facility – VGRAIN – will be briefly introduced. A few case studies are given at the end of the chapter.

9.1 Crystal Plasticity and Micro-forming

9.1.1 *Micro components and size effects*

In recent decades, the development towards miniaturisation of products and devices in industries such as electronics, optics, communications, etc. has increased the demand for metallic parts manufactured at the micro-scale. Such parts encompass a wide variety of geometries, materials, functionalities and production processes. Examples of micro-parts include screws, fasteners, connector pins, springs, micro-gears and micro-shafts. These are manufactured by employing a variety of manufacturing processes such as machining, folding, bending, stamping, drawing, moulding, lithography and forward/backward extrusion (Geiger, Kleiner, Eckstein *et al.*, 2001).

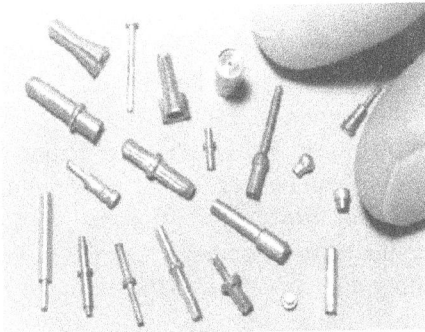

Fig. 9.1 Examples of micro-pins (Engel and Eckstein, 2002).

Some examples of extruded micro-parts are shown in **Fig. 9.1** (Engel and Eckstein, 2002). Forming is a particularly appropriate manufacturing technique for these parts, as they often have a complicated shape and machining would be time consuming and produce low yield. Process modelling plays an ever increasing role in these areas, for product design and for reducing lead time and manufacturing costs.

Extruded micro-pins are now used widely in electronics devices. They have diameters of the order of 100 μm to 2 mm. The quality of the formed micro-pins is affected by the grain size, grain orientation and grain distributions of the material and the geometrical defects cannot be captured using conventional continuum based FE forming simulation techniques.

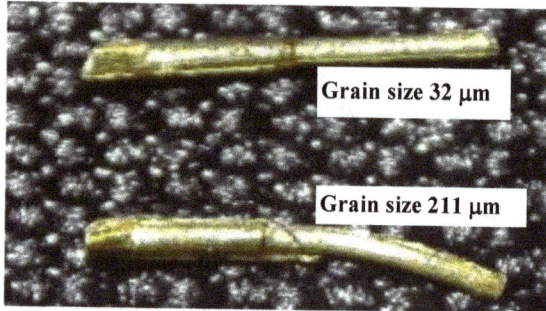

Fig. 9.2 Extruded micro-pins of 0.57 mm in diameter (Krishnan, Cao and Dohda, 2007).

Figure 9.2 shows micro-pins of 0.57 mm diameter extruded using the same material, but having undergone different heat treatments to produce different grain sizes (Krishnan, Cao and Dohda, 2007). It was reported that for material with a grain size of 32 μm, or about 16 ~ 18 grains across the extruded diameter, and with a deformation ratio of about 1.3, fairly straight micro-pins can be extruded and continuum FE analysis can be used for this prediction. However, for material with a grain size of 211 μm, about 2 ~ 3 grains across the diameter of the extruded micro-pins, uncontrollable bending and curvature of the extruded micro-pins usually can be observed experimentally for the same extrusion ratio (Krishnan, Cao and Dohda, 2007).

The aforementioned motivation has led researchers to perform numerical studies incorporating both the crystal plasticity material model explicitly representing the microstructure and finite element methods (Cao, Zhuang, Wang *et al.*, 2010). Zhuang, Wang, Lin *et al.* (2011) utilised the method to study local thinning and tearing of micro-tubes during hydroforming and it was shown by Wang, Zhuang, Balint *et al.* (2009) that this approach was capable of predicting necking in micro-films under tension.

9.1.2 *Crystal plasticity*

First, plastic flow in micro lengthscale is briefly introduced here. Yield phenomenon in crystalline metals is discussed. The reader is encouraged to refer to relevant books and publications related to crystal plastic deformation of metals for a more rigorous and detailed presentation of the subject.

a. *Micro-plasticity*

Metallic materials in their generally used form have a polycrystalline structure. That is their structure consists of an aggregate of crystals (called grains in the metallurgy context) with varying orientation and shape. In each grain, atoms are stacked in a regular manner whose structure depends on many different factors such as the metal itself, temperature, alloying elements etc.

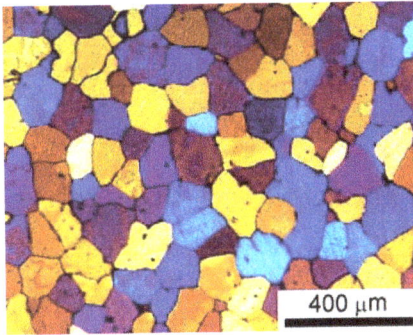

Fig. 9.3 Micrograph of ploycrystalline as-cast wrought-grade aluminium alloy (Quested, 2003).

A typical morphology that a polycrystalline aluminium alloy can adopt during solidification is illustrated in **Fig. 9.3**. The observed colours are dependent on the grain orientation and it clearly shows the orientation distribution in a polycrystalline material. This was achieved by viewing the oxide layer produced on the surface after etching under cross-polarised light, which produces orientation dependent colours (Quested, 2003).

J. Lin

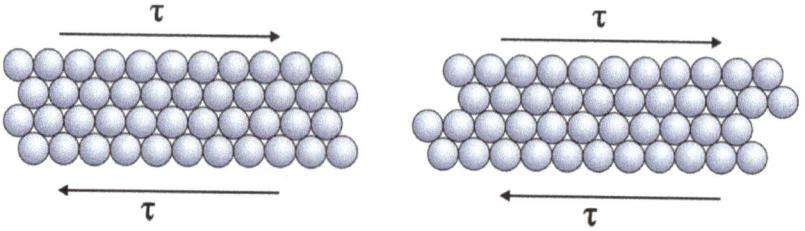

Fig. 9.4 Schematic representation of plastic deformation of crystals while preserving lattice structure (Dunne and Petrinic, 2005).

In crystalline materials, plasticity is caused by relative motion of planes of atoms (**Fig. 9.4**). This involves breaking and reforming of interatomic bonds. It should be noted that as shown in **Fig. 9.4**, crystalline slip is caused by shearing and unlike elastic deformation, does not change the volume. Furthermore the crystalline structure is preserved after the deformation and except for the extremities of the crystal the rest remains unchanged.

b. Crystal slip

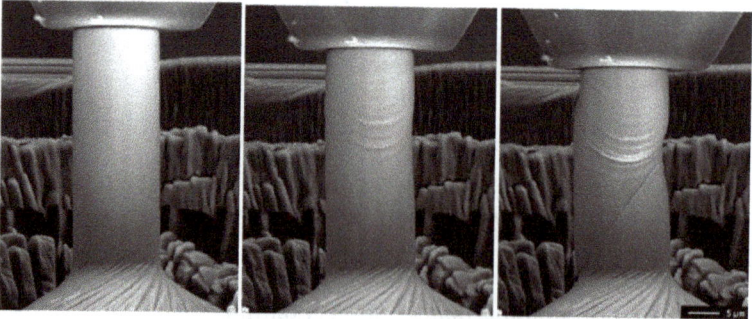

Fig. 9.5 SEM images showing the deformation for a single crystal micropillar sample compression test (Uchic, Dimiduk and Wheeler *et al.*, 2006).

Crystalline slip is the source of inelastic deformation in crystalline materials. Single crystal experiments carried out by Taylor (1934) and the work of Schmid and Boss (1935) during the first half of the nineteenth century laid out the foundation of the field of crystal plasticity. Advances in the experimental methodologies and availability of modern testing and measurement equipment such as SEM imaging and nano-indentation machines have made it possible to study more complicated aspects of this phenomenon. The work of Uchic, Dimiduk, Wheeler *et al.* (2006) on in-situ compression of single crystal micro-pillars, shown in **Fig. 9.5**, has significantly improved the understanding of the plastic flow. The influence of sample size on plasticity, which can be explained by the mean free path of dislocations as shown by Balint, Deshpande, Needleman *et al.* (2008), was also studied. The latter is in complete agreement with the Hall–Petch effect observed in polycrystalline materials (Hall, 1951; Petch, 1953).

The single crystal micro-pillar, shown in **Fig. 9.5**, has been subjected to compressive force beyond the yield threshold. The atomic planes on which slip has occurred (slip planes) are evident in the figure and the discrete nature of this phenomenon can be clearly seen. The process has been schematically illustrated in **Fig. 9.6 (a)**. The top surface of the sample has not been laterally constrained by the indenter tip and as a result of sliding of the slip planes it has moved in the horizontal direction under the compressive load. Had it been constrained in the horizontal direction, then the sample would have had to rotate in order to accommodate slipping which would have caused lattice orientation change.

c. Critical resolved shear stress

Critical resolved shear stress, τ_c, is defined as the shear stress level required for slip to take place on a slip system. The shear stress required for positive or negative slip in most crystals is equal with the exception of BCC metals (Hosford, 1992). The required condition may be expressed as

$$\tau_{ms} = \pm\tau_c , \tag{9.1}$$

where m and s refer to the *slip plane normal* and the *slip direction* respectively. This crystallographic yield criterion is known as *Schmid's law* and as the name suggests was first proposed by Schmid and Boas (1935) in order to explain the observed behaviour of single crystal deformation. Considering the uniaxial tension configuration, shown in **Fig. 9.6 (b)**, with the tensile direction oriented along the *x*-axis the resolved shear stress is given by

$$\tau_c = \pm\sigma_x \cos\lambda \cos\varphi , \tag{9.1a}$$

where λ is the angle between the slip direction and the tensile axis, and φ the angle between the tensile axis and the slip plane normal, as shown in **Fig. 9.6**. The *Schmid's law* is usually expressed as

$$\tau_c = \pm \sigma_x m_x$$

$$m_x = \cos \lambda \cos \varphi \,,$$

where m_x is known as the *Schmid factor*. This concept is used as a measure of grain orientation and the ease or difficulty of slip in that particular grain.

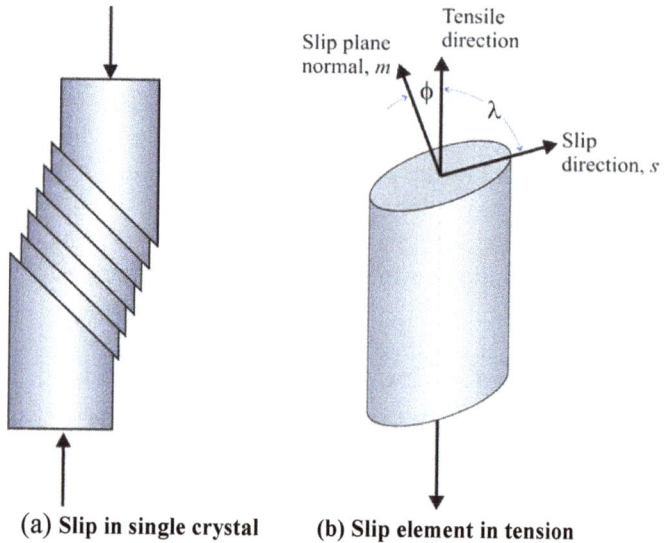

(a) Slip in single crystal   (b) Slip element in tension

Fig. 9.6 Schematic showing (a) the slip in a single crystal sample under unconstrained compression and (b) the slip element in uniaxial tension (Karimpour, 2012).

d. Dislocations

Plastic deformation in crystalline material takes place by breaking and reformation of inter-atomic bonds resulting in the rearrangement of the crystal. **Figure 9.7** shows dislocation presence in a stainless steel sample where each dislocation line is about 1000 atoms in length (Ashby and Jones, 2005). The theory of dislocations states that deformation in crystals occurs by the means of dislocation glide in the matrix.

Dislocation glide only involves local rearrangements of the lattice thereby explaining the low levels of yield strengths observed in experiments. It should be noted that the glide of an individual dislocation will result in a deformation equal to the Burgers vector, which is of the order of atomic spacing; hence large deformations require the glide of large numbers of dislocations present in the lattice.

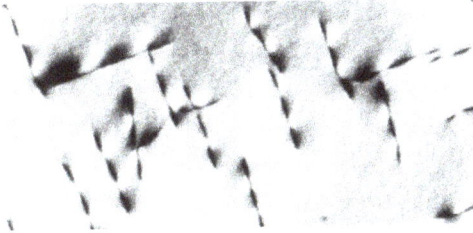

Fig. 9.7 Electron microscope picture of dislocation lines in stainless steel (Ashby and Jones, 2005).

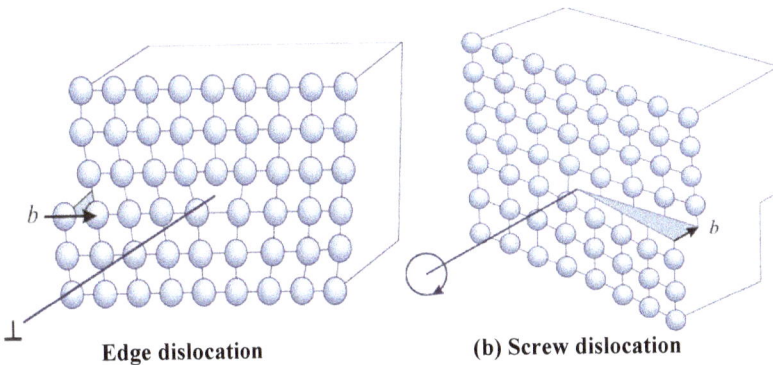

Edge dislocation   **(b) Screw dislocation**

Fig. 9.8 Representation of dislocations (Dunne and Petrinic, 2005).

There are two basic types of dislocations, namely *edge dislocation* (shown in **Fig. 9.8 (a)**) and *screw dislocation* (shown in **Fig. 9.8 (b)**). Any complex dislocation structure can either be explained by means of one (e.g. a dislocation line) or a combination of the two dislocation models (e.g. a dislocation loop).

As shown in **Fig. 9.8**, the main difference of the two models is that in an edge dislocation the Burgers vector *b* is perpendicular to the line of the dislocation whereas in the case of a screw dislocation the two are parallel. As a result, unlike an edge dislocation, the slip normal of a screw dislocation is not unique and can cause slip on any plane containing the dislocation line.

e. *Crystal structures and slip systems*

Experiments and crystallographic observations indicate that dislocation glide takes place preferentially on certain planes of atoms along specific directions. The slip planes and slip directions are usually those with the highest density of atoms and the closest distance between them. A unique combination of a slip plane and a slip direction is called a *slip system,* which clearly depends on the crystalline structure of the material. Slip systems are denoted as a slip direction followed by a slip plane expressed in Miller indices. Slip systems found in single crystals of some engineering materials are listed in **Table 9.1**.

Table 9.1 Slip system in some engineering materials (Karimpour, 2012).

Metal	Structure	Slip systems
Cu	FCC	$\langle 1\bar{1}0\rangle\{111\}$
Ni	FCC	$\langle 1\bar{1}0\rangle\{111\}$
γ-Fe	FCC	$\langle 1\bar{1}0\rangle\{111\}$
Ta	BCC	$\langle 11\bar{1}\rangle\{101\}, \langle 11\bar{1}\rangle\{112\}, \langle 11\bar{1}\rangle\{123\}$
α-Fe	BCC	$\langle 11\bar{1}\rangle\{101\}, \langle 11\bar{1}\rangle\{112\}, \langle 11\bar{1}\rangle\{123\}$
Ti	HCP	$\langle 100\rangle\{100\}, \langle 110\rangle\{100\}$
MnS	SC	$\langle 1\bar{1}0\rangle\{110\}$

In a face centred cubic (FCC) crystal, "closed-packed" planes on which slip occurs are located diagonally in the unit cell. The FCC unit cell and the corresponding slip systems are shown in **Fig. 9.9 (a)**. There are four slip planes, with three slip directions on each, which count for 12 slip systems in the family, and are written as $\langle 1\bar{1}0\rangle\{111\}$.

The unit cell of a simple cubic lattice has the simplest structure of all cubic crystal structures, with a single lattice point at each corner of the cube. The slip direction and the slip plane of such a crystal are illustrated in **Fig. 9.9 (b)** and the slip system family is expressed as $\langle 1\bar{1}0\rangle\{110\}$.

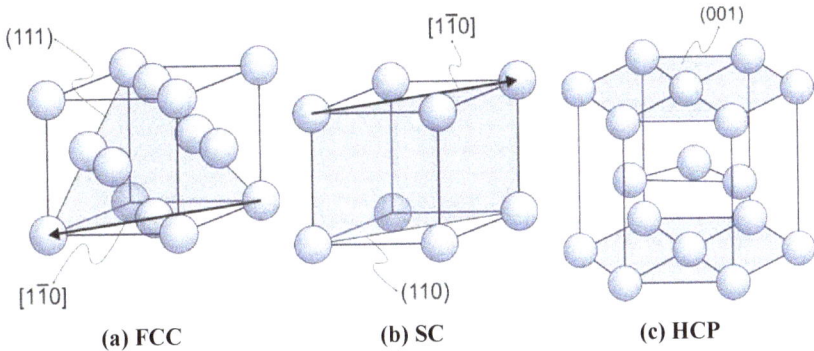

Fig. 9.9 Slip systems in (a) face centred crystals (FCC), (b) simple crystal (SC) and (c) hexagonal close-packed (HCP) lattices (Dunne and Patrinic, 2005).

Figure 9.9 (c) shows the hexagonal closed packed (HCP) unit cell. Slip systems lie on [001] planes and the slip direction have the general form of $\langle 100 \rangle$. The secondary family, $[110](100)$, is not considered as an independent slip system as it can be expressed in terms of $[100](100)$ and $[010](100)$ systems.

In the case of body centred cubic (BCC) crystals, there is no unique closely packed plane of atoms and there are multiple planes with similar density of atoms. Despite the non-uniqueness of the slip plane, $[11\bar{1}]$, which is the diagonal direction of the cube, is the only possible slip direction. The unique slip direction has been demonstrated in **Fig. 9.10** on the three possible slip planes. The slip direction combined with each slip plane forms a family of slip systems, making a total of three slip system families: $\langle 11\bar{1} \rangle \{101\}$, $\langle 11\bar{1} \rangle \{112\}$, and $\langle 11\bar{1} \rangle \{123\}$.

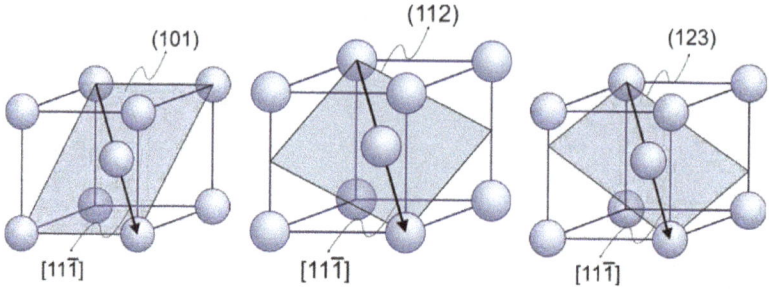

Fig. 9.10 Slip systems in a body centred cubic (BCC) lattice (Dunne and Patrinic, 2005).

9.2 Crystal Plasticity Constitutive Equations

The endeavour to incorporate micro-mechanical effects in continuum models of plastic deformation has led to the development of sets of viscoplastic constitutive equations that describe inelastic deformation in terms of crystalline slip. This was based on the pioneering work of Taylor (1938), who recognised that the plastic deformation of metals can be described based on the shearing of crystallographic slip systems.

9.2.1 *Crystal kinetics*

The basis of the crystal kinematics used in almost all crystal plasticity material models is the decomposition of the deformation gradient into the lattice and plastic parts that was proposed by Lee (1969). The complete set of equations describing the kinematics of crystalline slip was developed by Hill (1966), Rice (1971) and Hill and Rice (1972).

The total deformation gradient of finite strain from the reference frame to the current frame, F_{ij}, is defined by

$$F_{ij} = \frac{\partial x_i}{\partial X_j},$$

where X_j and x_i denote the reference and current particle positions, respectively. Here, tensor conventions for subscripts are adopted. All indices i, j, k and l are running from 1 to 3.

In crystal plasticity theory, a crystalline material is embedded on its lattice, which undergoes elastic deformation and rotation. The inelastic deformation of a single crystal is assumed here to arise solely from crystalline slip (i.e. deformation caused by grain rotation and twinning are not considered). The material flows through the crystal lattice via dislocation motion. The total deformation gradient F_{ij} is given by

$$F_{ij} = F_{ik}^* F_{kj}^P \qquad (9.2)$$

where F_{kj}^P denotes plastic shear of the material to an intermediate reference configuration in which the lattice orientation and spacing are the same as in the original reference configuration, and where F_{ik}^* denotes stretching and rotation of the lattice. These are shown in **Fig. 9.11**.

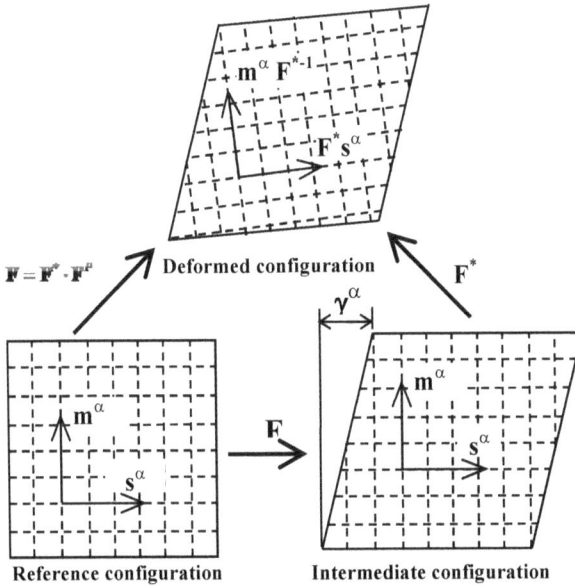

Fig. 9.11 Kinematics of elastic-plastic deformation of crystalline solid deforming by crystallographic slip.

The rate of change of F_{ij}^P is related to the slipping rate $\dot{\gamma}^\alpha$ of the αth slip system by

$$\dot{F}_{ik}^P \left(F_{kj}^P \right)^{-1} = \sum_\alpha \dot{\gamma}^\alpha s_i^\alpha m_j^\alpha , \qquad (9.3)$$

where the sum ranges over all activated slip systems and unit vectors s_i^α and s_j^α are the slip direction and the normal to the slip plane in the reference configuration, respectively.

The number of slip systems and their orientations depend on the crystal lattice, e.g. an FCC crystal contains four slip planes and each slip plane has three slip directions, which results in $\alpha = 1,2,...,12$. It is convenient to define the vector $s_i^{*\alpha}$, lying along the slip direction of the system α in the deformed configuration, by

$$s_i^{*\alpha} = F_{ik}^* s_k^\alpha .$$

A normal to the slip plane which is the reciprocal base vector to all such vector in the slip plane is

$$m_i^{*\alpha} = m_k^\alpha F_{ki}^{*-1} .$$

The velocity gradient in the current state is

$$L_{ij} = \dot{F}_{ik} F_{kj}^{-1} = D_{ij} + \Omega_{ij} , \qquad (9.4)$$

where the symmetric rate of stretching D_{ij} and the anti-symmetric spin tensor Ω_{ij} may be decomposed into lattice parts (superscript *) and plastic parts (superscript ^{P}) as follows:

$$D_{ij} = D_{ij}^* + D_{ij}^P$$
$$\Omega_{ij} = \Omega_{ij}^* + \Omega_{ij}^P ,$$

satisfying

$$D_{ij}^* + \Omega_{ij}^* = \dot{F}_{ik}^* F_{kj}^{*-1}$$
$$D_{ij}^P + \Omega_{ij}^P = \sum_\alpha \dot{\gamma}^\alpha s_i^\alpha m_j^\alpha . \qquad (9.5)$$

Finally, it is noted that the plastic parts of the rate of stretching and the rate of spin are given by the symmetric and skew part of Eq. (9.5). If it is defined that

$$P_{ij}^\alpha = \frac{1}{2}\left(s_i^{*\alpha} m_j^{*\alpha} + s_j^{*\alpha} m_i^{*\alpha}\right)$$

$$W_{ij}^\alpha = \frac{1}{2}\left(s_i^{*\alpha} m_j^{*\alpha} - s_j^{*\alpha} m_i^{*\alpha}\right),$$

then

$$D^P_{ij} = \sum_\alpha P^\alpha_{ij} \dot{\gamma}^\alpha$$

$$\Omega^P_{ij} = \sum_\alpha W^\alpha_{ij} \dot{\gamma}^\alpha .$$

It is worth noting that the plastic part of the rate of stretching, D^P_{ij}, is that part of the rate of stretching arising from slip in the current lattice direction $s^{*\alpha}_i$, in a plane where the current normal is $m^{*\alpha}_i$, while the plastic part of the rate of spin, Ω^P_{ij}, is that part of the rate of plastic spin resulting from the summation of rotation on each slip system.

9.2.2 *Crystal viscoplasticity constitutive equations*

For each grain, linear elasticity constitutive relations are given by the generalised Hooke's law,

$$\tau^{\nabla *}_{ij} = L_{ijkl} D^*_{kl}, \tag{9.6}$$

where L_{ijkl} the fourth-order stiffness tensor and D^*_{ij} the second-order symmetric rate of stretching of the lattice. $\tau^{\nabla *}_{ij}$ represents the Jaumann rates of Kirchhoff stress formed on axes that spin with the lattice

$$\tau^{\nabla *}_{ij} = \dot{\tau}_{ij} - \Omega^*_{ik} \tau_{kj} + \tau_{ik} \Omega^*_{kj}, \tag{9.7}$$

where $\dot{\tau}_{ij}$ is the material rate of Kirchhoff stress. The Kirchhoff stress τ_{ij} is defined as $(\rho_0/\rho)\sigma_{ij}$, where σ_{ij} is the Cauchy stress and ρ_0 and ρ are the material density in the reference and current states. On the other hand, τ^{∇}_{ij} is the Jaumann rate of Kirchhoff stress formed on axes that rotate with the material:

$$\tau^{\nabla}_{ij} = \dot{\tau}_{ij} - \Omega_{ik} \tau_{kj} + \tau_{ik} \Omega_{kj}. \tag{9.7a}$$

The difference between these two resulting from rotation caused by plastic part of the spin tensor rates is

$$\tau^{\nabla *}_{ij} - \tau^{\nabla}_{ij} = \sum_\alpha \left(W^\alpha_{ik} \tau_{kj} - \tau_{ik} W^\alpha_{kj} \right) \dot{\gamma}^\alpha. \tag{9.8}$$

Then Eq. (9.8) becomes

$$\tau^{\nabla}_{ij} = L_{ijkl} D_{kl} - \sum_\alpha \left(L_{ijkl} P^\alpha_{kl} + W^\alpha_{ik} \tau_{kj} - \tau_{ik} W^\alpha_{kj} \right) \dot{\gamma}^\alpha. \tag{9.9}$$

In crystal plasticity, plastic deformation is assumed to be caused solely by crystalline slip, which is driven by Schmid stress (or resolved shear stress), τ^α and is defined by

$$\tau^\alpha = m_i^{*\alpha} \tau_{ij} s_j^{*\alpha}, \tag{9.10}$$

where $m_i^{*\alpha}$ and $s_j^{*\alpha}$ are slip plane normals and directions for the αth slip system, respectively. The rate changes of this Schmid stress is given by Peirce, Asaro and Needleman (1982):

$$\dot{\tau}^\alpha = m_i^{*\alpha} \left(\tau_{ij}^{\nabla*} - D_{ik}^* \tau_{kj} + \tau_{ik} D_{kj}^* \right) s_j^{*\alpha}. \tag{9.11}$$

a. Crystal plasticity equations with self- and latent-hardening

The slipping strain rate $\dot{\gamma}^\alpha$ is assumed to be governed by the resolved shear stress τ_α given by a constitutive equation shown below:

$$\dot{\gamma}^\alpha = \dot{a} \left(\frac{\tau^\alpha}{g^\alpha} \right) \left(\left| \frac{\tau^\alpha}{g^\alpha} \right| \right)^{n-1}, \tag{9.12}$$

where $\dot{a}$ is the reference strain rate, n is the stress sensitivity parameter an g^α is the current strain hardened state of the crystal. Equation (9.12) is similar to the power-law viscoplastic constitutive equations. It calculates the plastic strain (shear strain) in the αth slip system, which is resulted by the shear stress τ^α in the direction. As the parameter n approaches infinity, this power-law approaches rate independent material, i.e. elastic-plastic problems. For most viscoplastic materials, the n value varies from 2 to 8.

The current hardened state g^α is defined by

$$\dot{g}^\alpha = \sum_\beta h_{\alpha\beta} \dot{\gamma}^\beta, \qquad \beta = 1,2,.....,12 \text{ for FCC crystals,}$$

where $h_{\alpha\beta}$ is the slip hardening moduli. Self- ($h_{\alpha\alpha}$) and latent- ($h_{\alpha\beta}$) hardening moduli are defined as

$$h_{\alpha\beta} = \begin{cases} h_0 \text{sech}^2 \left| \dfrac{h_0 \gamma}{\tau_s - \tau_0} \right| & \alpha = \beta \\ qh(\gamma) & \alpha \neq \beta \end{cases}, \tag{9.13}$$

where $\gamma = \displaystyle\sum_{\alpha=1}^{12} |\gamma^\alpha|$.

The parameter h_0 is the initial hardening modulus and τ_0 is the initial shear strength of the material; at $t = 0$, all g^α are equal to τ_0; τ_s is the threshold stress when plastic flow initiates; γ is the cumulative slip strain and q is a hardening factor. The material constants within the crystal plasticity constitutive equations are determined from experimental data.

Table 9.2 Material parameters for the crystal plasticity model.

$n\,(-)$	$\dot{a}\,(\mathrm{s}^{-1})$	$h_0\,(\mathrm{MPa})$	$\tau_s\,(\mathrm{MPa})$	$\tau_0\,(\mathrm{MPa})$
10.0	0.001	541.5	109.5	60.8

As an example, a set of the constants is listed in **Table 9.2** for a single crystal copper (Peirce, Asaro and Needleman, 1982). The high value of n is used here to reduce the viscoplastic behaviour of the material, as it is used in room temperature forming processes.

b. Slip-system-based hardening equations

Considering the form of constitutive equations introduced in Chapters 4 and 5 for macro-mechanics modelling, a set of crystal plasticity constitutive equations can be written as

$$\dot{\gamma}^\alpha = \dot{a}\left(\frac{\tau^\alpha - \tau_c^\alpha - R^\alpha}{K}\right)^n$$

$$\dot{\bar{\rho}} = A_1(1-\bar{\rho})|\dot{\gamma}|$$

$$\dot{\bar{\rho}}^\alpha = A_2(1-\bar{\rho}^\alpha)|\dot{\gamma}^\alpha| \tag{9.14}$$

$$R^\alpha = B\left[\chi\frac{\bar{\rho}}{N} + (1-\chi)\bar{\rho}^\alpha\right]^{\frac{1}{m}}$$

$$\dot{\gamma} = \sum_{\alpha=1}^{N}|\dot{\gamma}^\alpha|$$

with

$$\dot{a} = \begin{cases} +1 & \tau^\alpha > 0 \\ -1 & \tau^\alpha < 0 \end{cases}$$

$$\dot{\gamma}^\alpha = 0 \text{ when } |\tau^\alpha| - \tau_c^\alpha - R^\alpha \leq 0,$$

where $\dot{\gamma}^{\alpha}$ is the resolved shear strain rate for slip system α ($\alpha = 1,2,...,N$) and N is the total number of slip systems. $\dot{\alpha}$ is reference strain rate, τ_c^{α} represents the initial shear strength of crystal material for the αth slip system, R^{α} the material hardening due to dislocation development in that slip system and K the drag stress, which is similar to the general viscoplastic constitutive equations. The equations for $\dot{\bar{\rho}}$ and $\dot{\bar{\rho}}^{\alpha}$ describe the dislocation flow at overall crystal level and at individual slip system using normalised total dislocation density $\bar{\rho}$ and component dislocation density $\bar{\rho}^{\alpha}$. For both cases, the normalised dislocation density takes a value between 0 (the initial state) and 1 (the saturated state of a dislocation network) (Lin, Liu, Farrugia et al., 2005). n and m are material parameters.

The equation of R^{α} represents the hardening of individual slip system. The hardening is contributed by dislocations at both overall and individual slip system levels within a grain. By specifying the parameter $\chi = 1$, the equation becomes

$$R^{\alpha} = B\left[\frac{\bar{\rho}}{N}\right]^{\frac{1}{m}}.$$

The hardening is "isotropic" for each slip system. This is similar to "Taylor's hardening". When $\chi = 0$, the equation becomes

$$R^{\alpha} = B\left(\bar{\rho}^{\alpha}\right)^{\frac{1}{m}}.$$

It represents pure anisotropic hardening and each slip system hardens differently. That means the dislocation hardening in one slip system has no effect on the other slip system.

In this formulation, the traditional form of interaction matrix for the hardening evolution is not necessary. In the set of unified crystal plasticity constitutive equations Eq. (9.14), the interaction of dislocation hardening among different slip systems is rationalised by the *parameter* χ. The parameter χ can be a material variable and depends upon the crystal structures. The hardening of individual slip system can be considered as fully decoupled ($\chi = 0$) or fully coupled ($\chi = 1$). The χ value usually takes a value between 0 and 1 depending on materials.

9.3 Generation of Grain Structures

To carry out crystal plasticity FE (CPFE) analysis, one of the most important tasks is to generate grain structures to realistically represent the microstructural features of the material. Polycrystalline FE has given valuable insights into grain-to-grain interaction behaviour and local deformation mechanisms, acts as an effective utility to simulate micro-forming processes for small scale metal products and is capable of simulating local damage processes of components in service (Asaro and Rice, 1977). Since stress and strain are related to grain size, shape, orientation and their distributions, FE micro-mechanics simulations must be based upon a grain structure modelled within a FE/CAE computational environment. Therefore, building the FE model derived from realistic grain structures is critical for accurate simulations.

Representation of a grain structure can be obtained directly by mapping the metallographic observations to a FE model by either processing digitised images (Den Toonder, Van Dommelen and Baaijens, 1999) or processing SEM/EBSD data to reconstruct the grain structures (Barton, Bernier, Lebensohn *et al.*, 2009). These approaches are generally very time-consuming and laborious for practical applications (Zhang, Balint and Lin, 2011a). In addition to the experiment- and simulation-based methods, computational geometrical models have been applied in representing equivalent grain structures. In early studies, grain structures were commonly represented by simplified geometrical units, such as hexagons, cubes, rhombic dodecahedra and truncated octahedrons (Harewood and McHugh, 2006). However, in reality grains exhibit large variations in both shape and size, and more importantly, the morphological characteristics have strong influence on various mechanical behaviours such as strain localisation and micro-crack propagation.

Alternatively, Voronoi tessellation (VT) models have been traditionally applied in modelling polycrystalline grain structures for metallurgical applications (Boots, 1982) as they provide a natural solution to represent grain structures with non-uniform grain shapes. Since polycrystalline materials are originally formed by the nucleation and growth of grains, a VT describes the formation of a crystal aggregate growth process. Like a solidification process, grains in the Voronoi tessellation are spatially distributed and completely determined by initial nuclei/seeds and individual crystal growth velocities. Research into the utilisation of a Voronoi tessellation for modelling grain structures has been devoted to the development of sophisticated grain growth schemes and seed distribution control mechanisms. By designating different grain growth velocities within a given seed pattern, morphological features such as curved boundaries can be produced for particular simulation requirements (Mahadevan and Zhao, 2002), whilst, having defined a grain growth scheme, the grain size distribution properties of a VT is essentially decided by its seed lattice.

Efforts have been made to generate virtual grain structures to enable realistic microstructure of materials to be efficiently represented for CPFE analysis, particularly for micro-forming simulations (Zhuang, Wang, Lin *et al.*, 2012). The theory and algorithms for virtual microstructure generation have been updated over the years and the major contributions can be found in the work of Cao, Zhuang, Wang *et al.* (2009) and Zhang, Balint and Lin (2011a).

This section summarises the basic concept for the generation of virtual grain structures for CPFE analysis. Details related to the theories of virtual grain generation can be found in Zhang (2011), Zhang, Balint and Lin (2011a and b) and Zhang, Karimpour, Balint and Lin (2012a and b).

9.3.1 *Grain distribution and generation algorithms*

a. *2D Voronoi tessellation*

Polycrystalline materials can be formed by the nucleation and growth of grains. Provided that all the grain growth velocities are equal, the resultant structure can be described by a Voronoi tessellation. The initial nucleation site for each grain is referred to as its seed. A Voronoi tessellation is naturally analogous to the grain growth process, starting from the initialised seeds lattice (Boots, 1982). Growth ceases for a grain whenever it contacts a neighbouring grain. As a result, a Voronoi tessellation is uniquely determined by the designated seed lattice and growth velocities.

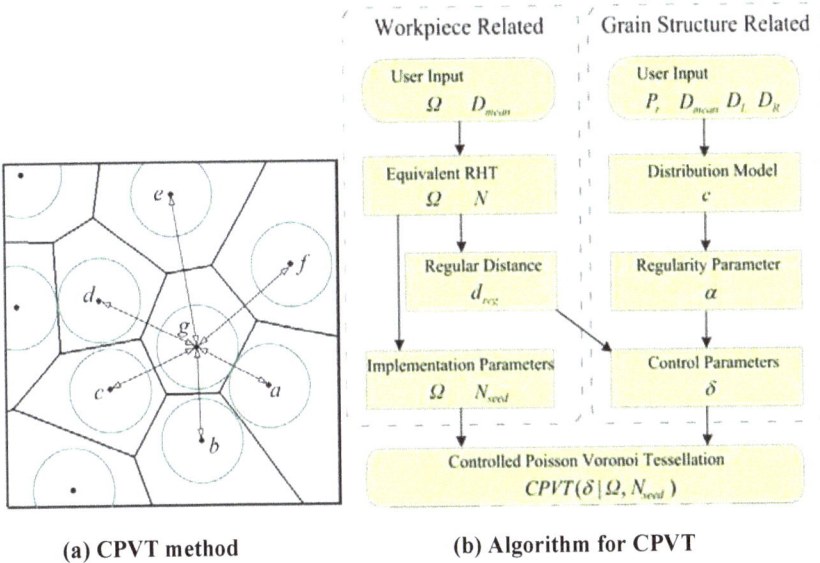

(a) CPVT method   (b) Algorithm for CPVT

Fig. 9.12 (a) Schematic of a controlled Poisson Voronoi tessellation (CPVT), where all seeds have identical diameter δ and the control parameter is equal to the diameter, and the seed distance $d(g,a) = \delta$ and $d(g,j) \geq \delta$, where $j = b,c,...,f$ and, (b) the integrated scheme with CPVT model for grain structure generation (Zhang, Balint and Lin, 2011a).

Given a plane of area A_0, and letting the seeds be produced in the area by sampling x- and y-coordinates independently from a uniform random generator, a corresponding tessellation can be achieved after a growth process. The resulting tessellation is generally referred to as a Poisson Voronoi tessellation (PVT). Note that the seeds are represented by points in a PVT. In contrast, assuming that each initial seed is a circle with diameter $\delta > 0$, and no seeds overlap each other in the seed lattice, then an alternative Voronoi tessellation can be produced by uniformly generating and growing the circular seeds. This random VT is called a controlled Poisson Voronoi tessellation (CPVT), where the circle diameter of a seed is used as a control parameter dictating the distances between it and its neighbours. The CPVT generation process proceeds as follows.

After the first seed is produced, each subsequent seed i is only accepted if it is greater than a minimum allowable distance δ_i from any existing seed, i.e., $d(i,k) \geq \delta_i$, where $k = 1,2,...,i-1$. Provided that all seeds have the same diameter δ, the distance between a newly generated seed i and an existing seed k satisfies the condition $d(i,k) \geq \delta$, $\forall\ k = 1,2,\ldots,i-1$. For example, as illustrated in **Fig. 9.12 (a)**, the new seed g would be accepted because the distances between g and the other adjacent seeds are all equal to or larger than δ.

The CPVT model aims to fully control the tessellation's regularity and its corresponding grain size distribution, and the control parameter is the primary link between the user-expected grain structure and the realised virtual grain structure. **Figure 9.12 (b)** shows the integrated system that employs the CPVT model to generate the grain structure based on the user input. The overall procedure has the following properties:

- work-piece-related input includes the size of a work-piece and the mean grain size, which are used to obtain the implementation parameters including the tessellation's domain Ω and the number of grains N_{seed};

- grain-structure-related input consists of four physical parameters (to be discussed later), which describe a set of higher order grain size distribution features, to determine the control parameter δ;

- in the process of deriving the regularity, two modules are involved: one maps from the physical parameters to a distribution parameter c, and the other maps from the distribution parameter to the regularity parameter α.

Now, the controlled Poisson Voronoi tessellation model can be denoted as $CPVT\ (\delta|\Omega,N_{seed})$, where under the evaluation of δ, the N_{seed} seeds are sequentially generated within the domain Ω. Details of the implementation will be discussed next.

b. Regularity of Voronoi tessellation

To evaluate the regularity of the tessellation, Zhu, Thorpe and Windle (2001) defined the regularity parameter α as

$$\alpha = \frac{\delta}{d_{reg}}, \tag{9.15}$$

where δ is the minimum distance among all neighbouring seeds in the tessellation, and d_{reg} is the distance between two adjacent seeds in the equivalent regular hexagonal tessellation (RHT), as shown in **Fig. 9.13**. According to the seed lattice of the RHT, the standard seed distance d_{reg} is calculated as

$$d_{reg} = \left(\frac{2 \cdot A_0}{\sqrt{3} \cdot N} \right)^{1/2} , \tag{9.16}$$

where A_0 is the area of the equivalent RHT and N is the number of hexagons in the RHT.

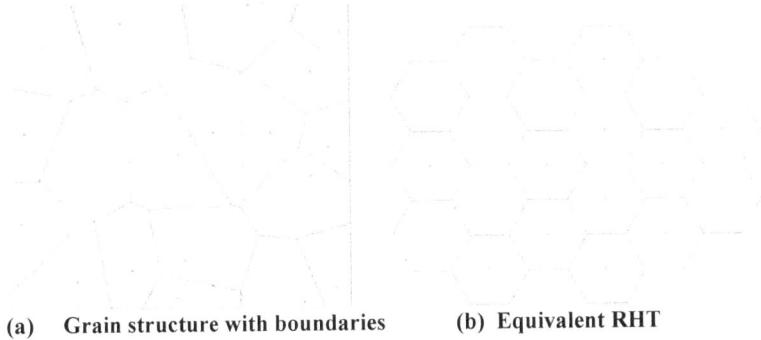

(a)   **Grain structure with boundaries**   (b) **Equivalent RHT**

Fig. 9.13 (a) The count of grain numbers and mean size with boundaries and (b) the illustration of the equivalent regular hexagonal tessellation (RHT) to the grain structure in (a) (Zhang, 2011).

Here the term "equivalent" means that a VT and its correlated RHT have an equal mean grain size D_{mean} and an equal domain area A_0. It is important to note that the equivalent RHT consists of space-filling hexagons, which partition a domain of area A_0 without overlap, therefore the number of grains in the RHT can be directly determined from the domain area, according to

$$N = \frac{A_0}{D_{mean}} . \tag{9.16a}$$

As a result, the number of grains in a VT can be determined by using the definition of its equivalent RHT, specified by user input.

In quantitative metallography, the number of grains is determined based on the heuristic rule that a grain cut by boundaries is counted as a half grain. However, in the proposed scheme, a VT is generated as follows:

- a domain is defined;
- the seeds are randomly created in the given area;
- the grain structure is generated.

Therefore, the final tessellation ignores the effect of seeds external to the domain, which exist in actual materials. For this VT scheme, the number of grains and seeds are approximately equivalent, i.e.

$$N_{seed} = N .$$ (9.17)

Therefore, based on the heuristic rule given by the above equation, the equivalent RHT can be used to determine the number of seeds required for the VT, together with the result of Eq. (9.16a).

c. Control parameter δ of the CPVT

The control parameter of a CPVT is defined by a minimum acceptable seed spacing (or seed distance), δ. The minimum seed spacing of a VT was employed to compute the regularity value α of a VT by Zhu, Thorpe and Windle (2001), whilst by prescribing an expected minimum seed spacing using the control parameter, a final VT has an approximate regularity α. This can be explained in that as randomly sampling a certain number of seeds, the minimum seed spacing would converge to the value of the control parameter. Also, when sampling a small number of seeds, the resultant VT tends to be more regular than the prescribed regularity, due to random error. However, the random error is small and allowable, when the number of grains is large.

Thus, the expected minimum seed spacing is taken as the control parameter to monitor the seed generation process. From the definition of the tessellation's regularity in Eq. (9.15), the control parameter δ can be determined by

$$\delta = \alpha \cdot d_{reg} .$$ (9.18)

(a)

(b)

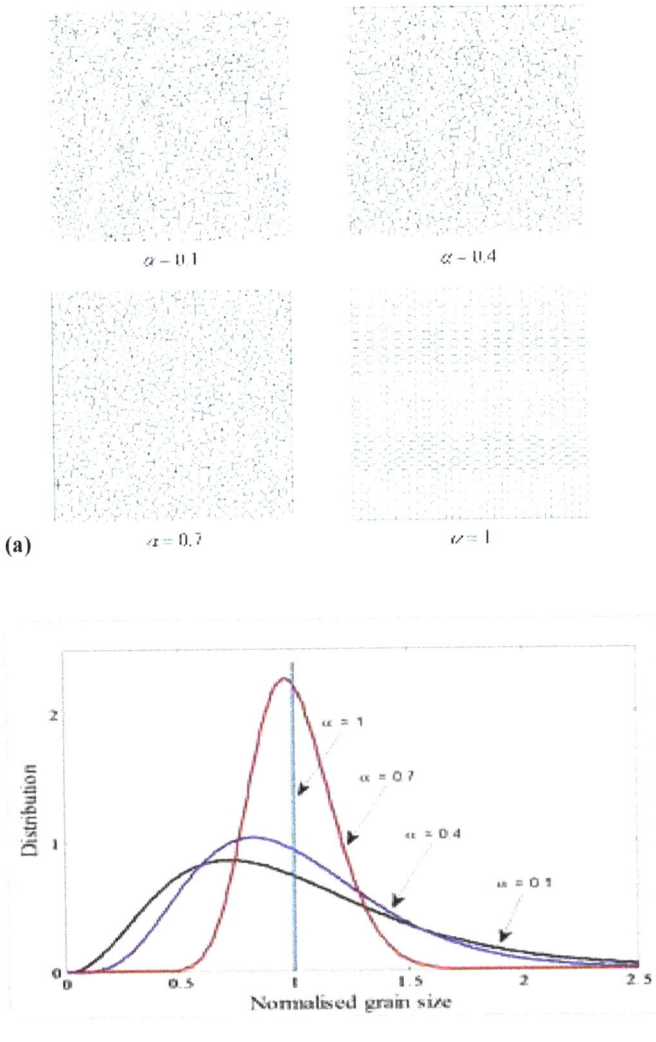

Fig. 9.14 Controlled Poisson Voronoi tessellations with different regularity parameters, (a) the grains generated and (b) the gamma distribution (Zhang, Karimpour, Balint and Lin, 2012a).

It can be seen that if $\alpha = 0$, then $\delta = 0$. and the resultant tessellation is a pure Poisson Voronoi tessellation, whilst letting the regularity $\alpha = 1$ gives $\delta = d_{reg}$, and the yielded tessellation in that case is exactly a regular hexagonal tessellation, as shown in **Fig. 9.13 (b)**. As the regularity α varies from 0 to 1, the tessellation becomes increasingly uniform. **Figure 9.14** shows a series of tessellations with varying regularity values from 0.1 to 1 (RHT).

In summary, implementation of the CPVT model $CPVT(\delta|\Omega,N_{seed})$ involves two parallel procedures: Determination of the implementation parameters $\{\Omega,N_{seed}\}$ and derivation of the control parameter δ. First of all, the number of seeds, N_{seed}, to be generated in a VT is determined by Eq. (9.17) and the standard seed distance d_{reg} is defined in Eq. (9.16). Secondly, with the assignment of a regularity value, a control parameter δ can be obtained from Eq. (9.18).

9.3.2 *Grain generation with physical material parameters*

In a CPVT, the regularity parameter α is used directly to specify the control parameter δ. However, physical measurements from quantitative metallography, such as the mean grain size and percentage of grains between the small and large sizes, are well known and widely used by engineers and material scientists. In such a circumstance, the regularity parameter is a good mathematical presentation, but it is not intuitive for practical use.

Two modules are to be developed to map the gap between the physical parameters and regularity parameter. The first module is the grain size distribution model linking the tessellation's regularity to its grain size distribution, and the other module is the mechanism to determine the distribution parameter from the given physical parameters.

a. *Distribution model*

Traditionally, the grain area distributions of Voronoi tessellations have been modelled by gamma distribution functions. Reported results include a three-parameter gamma distribution (Hinde and Miles, 1980) and a two-parameter gamma distribution (Kumar, Kurtz, Banavar *et al.*, 1992). If the experimental data is normalised by the mean grain area, then a one-parameter gamma distribution can also accurately fit the grain size distribution in term of grain area (Zhu, Thorpe and Windle, 2001; Zhang,

Balint and Lin, 2011a). The one-parameter gamma distribution function takes the form of

$$P_{x,x+dx} = \frac{c^c}{\Gamma(c)} x^{c-1} e^{-cx} dx \qquad x > 0, \tag{9.19}$$

where the distribution parameter $c > 1$, and $\Gamma(c)$ is the gamma function, defined as

$$\Gamma(c) = \int_0^\infty x^{c-1} e^{-cx} dx.$$

Note that the variance of the one-parameter gamma distribution is $1/c$. As the parameter c increases, the distribution becomes narrower, which is more suitable for modelling the tessellations having approximately similar grain sizes. In addition, there are two major advantages to using a one-parameter gamma distribution to describe the grain size distribution: only one parameter c is involved in the relation for the tessellation's regularity, and the mean value of this distribution is one, which is the normalised mean grain size.

In the work of Zhu, Thorpe and Windle (2001), a series of statistical tests have been performed to summarise the relationship between the fitting parameter c and the regularity α, where α ranges from 0 to 0.8. In the work of Ho, Zhang, Lin *et al.*, (2007) and Zhang, Balint and Lin (2011a), a descriptive model was proposed based on the statistical data of Mika and Dawson (1998), given by

$$\alpha(c) = A \cdot \left[z(c) - z_0 \right]^{k+nz(c)} \qquad c_0 \le c \le c_m, \tag{9.20}$$

where (Ho, Zhang, Lin *et al.*, 2007)

$$z(c) = \frac{c}{c_m}, \ z_0 = \frac{c_0}{c_m},$$

$c_0 = 3.555$, $c_m = 47.524$, $A = 0.74$, $k = 0.324$ and $n = -0.4144$.

Equations (9.20) and (9.18) provide a mechanism to assign the control parameter by prescribing an intuitive grain-size distribution model with a distribution parameter c rather than the regularity parameter α.

The next step incorporates a set of traditional physical parameters to replace the distribution parameter for regularity assignment, by which all abstract parameters are avoided when assigning the input to the CPVT model. Considering a set of physical parameters $\{D_L, D_{mean}, D_R, P_r\}$ from the quantitative metallography, where D_{mean} is the mean grain size, defined as $D_{mean} = \dfrac{1}{N}\displaystyle\sum_{i=1}^{N} D_i$, D_L and D_R are two specific grain size values and P_r is the percentage of the grains with size in the range of $[D_L, D_R]$ over the total number of the grains N , i.e.

$$P_r = \frac{1}{N}\sum_{i=1}^{N} m_i \text{ , where } m_i = \begin{cases} 1, & \text{if } D_i \in [D_L, D_R] \\ 0, & \text{Otherwise} \end{cases}.$$

It should be noted that grain size D henceforth represents an area in two dimensions as in this study, and a volume in three dimensions.

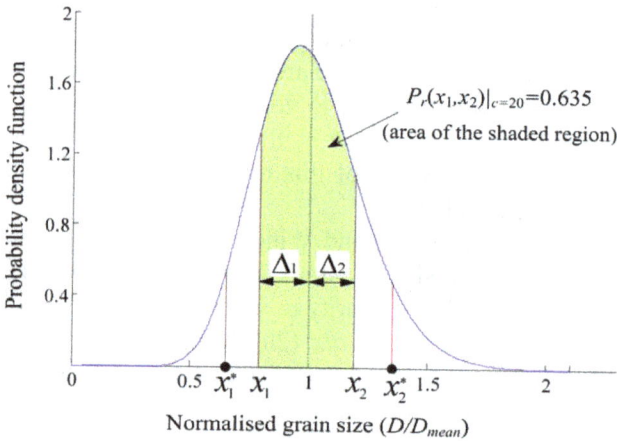

Fig. 9.15 Example of a one-parameter gamma distribution (Zhang, Balint and Lin, 2011a).

When modelling grain distributions by a one-parameter gamma distribution, the set of parameters conforms to the following equation:

$$P_r = \int_{x_1}^{x_2} \frac{c^c}{\Gamma(c)} x^{c-1} e^{-cx} dx , \qquad (9.21)$$

where the integral bounds x_1 and x_2 are

$$x_1 = \frac{D_L}{D_{mean}}, \ x_2 = \frac{D_R}{D_{mean}}.$$

Figure 9.15 presents an example of a one-parameter gamma distribution with $c = 20$, where the area of the shaded part between the lower limit D_L and the up limit D_R equals the percentage $P_R = 0.635$.

It can be observed that, by designating the set of physical parameters, the parameter c is correspondingly found by solving Eq. (9.21). However, there are still two major issues to be solved in order to utilise this scheme, which are:

(i)   conditions for which the parameter c is uniquely determined by Eq. (9.21), and

(ii) an efficient numerical algorithm to solve Eq. (9.21).

b. *Uniqueness and physical parameters*

Here uniqueness implies the uniqueness of the value of c obtained by solving Eq. (9.21). That is, given a set of physical parameters $\{D_L, D_{mean}, D_R, P_r\}$, there correspondingly exists a unique value of c characterising the one-parameter gamma distribution, describing the grain size distribution. Let

$$\begin{cases} x_1 = 1 - \Delta_1, & 0 < \Delta_1 < 1 \\ x_1 = 1 - \Delta_2, & 0 < \Delta_2 < 1 \end{cases},$$

where x_1 and x_2 are the integral limits of Eq. (9.21). For the symmetric case where $\Delta_1 = \Delta_2 = \Delta_3$, $\bar{x} = (x_1 + x_2)/2 = 1$, which is also the mean value of a one-parameter gamma distribution.

Lemma 1. There exists an interval $S^* = \left(x_1^*, x_2^*\right)$, such that for any interval $S = \left(x_1, x_2\right)$, where $x_1 < 1 < x_2$, if $S \subseteq S^*$, the implicit function

$$P_r(c) = \int_{x_1}^{x_2} \frac{c^c}{\Gamma(c)} x^{c-1} e^{-cx} dx \tag{9.22}$$

is strictly monotonically increasing, as the parameter c increases for any $c \geq 1$. Moreover, the interval $S^* = \left(x_1^*, x_2^*\right)$ can be estimated by

$$\begin{cases} x_1^* = 1 - \Delta_1^* \approx 1 - \dfrac{1}{\sqrt{c}} + O\left(\Delta^{3/2}\right) \\ x_2^* = 1 + \Delta_2^* \approx 1 + \dfrac{1}{\sqrt{c}} \end{cases} \tag{9.22a}$$

where the term $O\left(\Delta^{3/2}\right) > 0$.

Theorem 1. Given the constants x_1, x_2 and the percentage $P_r(x_1, x_2)\big|_c$, where $x_1 < 1 < x_2$, if the interval $S = [x_1, x_2] \subset S^* = (x_1^*, x_2^*)$ where x_1^* and x_2^* are determined from Eq. (9.22a), then the implicit function $P_r(c)$ in Eq. (9.22), defined over the domain $D(P_r) \subseteq [1, \infty]$ and range $R(P_r) \subseteq [0, 1]$, has the properties that

- $P_r(c)$ is a bijection;

- $P_r(c)$ has a continuous inverse $P_r^{-1}(c)$ on the range $R(P_r)$.

Lemma 1 and *Theorem 1* explain the existence of a valid range of physical parameters, where uniqueness is satisfied. Proofs of *Lemma 1* and *Theorem 1* are given by Zhang, Balint and Lin (2011a). This is realised by choosing a lower grain size limit D_L, a upper grain size limit D_R and a mean grain size D_{mean}, such that $S = [x_1, x_2] \subset S^* = (x_1^*, x_2^*)$. Then, the parameter c is uniquely determined by the percentage value P_r. Although the existence of such an interval of physical parameters that uniquely defines the c values has been proved, this interval, estimated using Eq. (9.22a), is not large enough for all practical applications. In the following, efforts are made to extend the effective interval S^*.

To estimate the interval S^* means to find a valid interval $S^* = (x_1^*, x_2^*)$ as large as possible such that $\partial P_r(c)/\partial c > 0$. Since $S^* = (x_1^*, x_2^*) = (1 - \Delta_1^*, 1 + \Delta_2^*)$, to estimate the valid interval S^* is equivalent to finding possible large values of both Δ_1^* and Δ_2^*. For simplicity and without loss of generality, only a symmetrical situation is considered, where $\Delta_1 = \Delta_2 = \Delta$ and the best estimation is specified by Δ^*, where Δ^* is the maximum $\Delta \in (0, 1)$ such that $\partial P_r(c, \Delta)/\partial c > 0$. Let the function $\varphi(c, \Delta)$:

$$\varphi(c, \Delta) = \frac{\partial P_r(c, \Delta)}{\partial c}$$

$$= \frac{c^c}{\Gamma(c)} \cdot \int_{1-\Delta}^{1+\Delta} x^{c-1} e^{-cx} \cdot \left[1 + \ln c - \psi(c) + \ln x - x\right] dx \tag{9.23}$$

Furthermore, from Eq. (9.23), it can be found that $\Delta_1^* < \Delta_2^*$. The following discussion for the selection of Δ^* is carried out with respect to three sub-intervals, i.e.

$$(0,1) = I_1 \cup I_2 \cup I_3 = \left(0,\Delta_1^*\right] \cup \left(\Delta_1^*,\Delta_2^*\right) \cup \left(\Delta_2^*,1\right] \tag{9.23a}$$

from *Lemma 1*, $\varphi(c,\Delta) > 1$ for any $\Delta \in I_1$ and $x \in (1-\Delta, 1+\Delta)$. The next discussion focuses on the value $\varphi(c,\Delta)$ over the last two intervals, I_2 and I_3.

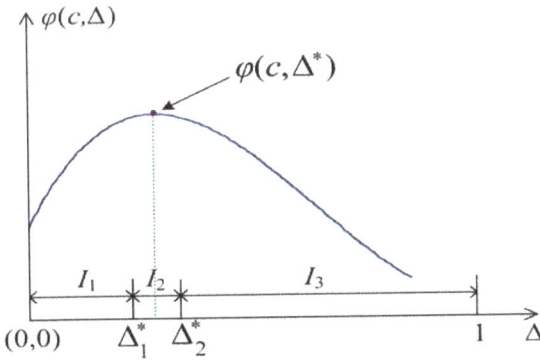

Fig. 9.16 Schematic of $\varphi(c,\Delta)$ over the intervals of $\Delta \in (0,1)$ and $c \in (1,\infty)$ is a constant value (Zhang, Balint and Lin, 2011a).

Lemma 2. There exists a point $\Delta^* \in \left(\tilde{\Delta}_1, \tilde{\Delta}_2\right)$, where Δ_1^* and Δ_2^* are given by Eq. (9.22a), such that for any $\Delta \in \left(0,\Delta^*\right)$, $\varphi(c,\Delta) > 0$. Moreover, the function $\varphi(c,\Delta)$ is strictly monotonically increasing for $\Delta \in \left(0,\Delta^*\right)$ and strictly monotonically decreasing for $\Delta \in \left(\Delta^*,1\right)$.

The proof of *Lemma 2* was given by Zhang, Balint and Lin (2011a). The shape function $\varphi(c,\Delta)$ is schematically illustrated in **Fig. 9.16**, where it can be observed that as Δ increases from 0 to Δ^*, $\varphi(\Delta)$ increases, starting from a positive value. But after Δ^*, as the variable Δ increases, the function value $\varphi\big|_{\Delta > \Delta^*}$ decreases correspondingly. Then

$$\min_{\Delta^* < \Delta < 1} \varphi(\Delta) = \lim_{\Delta \to 1} \varphi(\Delta). \tag{9.24}$$

In this situation, the asymptotic value is critical; that is, if $\lim_{\Delta \to 1} \varphi(\Delta) > 0$, then the interval S^* is such that $S^* = (0,1)$.

c. Algorithms for solving D_L on P_r and c-value

Given $x_1 = D_L/D_{mean}$ and $x_2 = D_R/D_{mean}$ within the interval given by Eq. (9.22), the function value $P_r(c,x_1,x_2)$ is monotonically increasing as c increases. On the other hand, the P_r value cannot be arbitrarily small due to the requirement of modelling grain size distributions; that is, a relaxation condition to determine the lower bound of P_r is derived based on $c \geq 1$. Therefore, the lower bound for the percentage $P_r(c,x_1,x_2)$ is given by

$$P_r\left(1, x_1, x_2\right) = e^{-x_1} - e^{-x_2}. \tag{9.25}$$

For the symmetrical case, i.e. $x_1 = 1 - \Delta$ and $x_2 = 1 + \Delta$, the lower bound is determined by

$$P_r\left(1, \Delta\right) = e^{\Delta-1} - e^{-\Delta-1}, \tag{9.25a}$$

where $\Delta \in [0, 0.999]$.

In the work of Ho, Zhang, Lin et al. (2007) and Zhang, Balint and Lin (2001b), the c value is determined by means of exhaustive enumeration starting from 1 with a prescribed incremental step, such as 0.001. Since each iterative step involves a series of computations including one numerical integration, this searching scheme is inefficient. To improve the solution procedure, an efficient gradient search method is proposed by Zhang, Balint and Lin (2011a). Given the four input parameters, D_L, D_{mean}, D_R and P_r, the following equation can be solved for c:

$$P_r = \int_{D_L/D_{mean}}^{D_R/D_{mean}} \frac{c^c}{\Gamma(c)} x^{c-1} e^{-cx} dx \triangleq F(c), \tag{9.26}$$

where the constant $\Gamma(c)$ is calculated by Eq. (9.19a). A Newton–Raphson method can be used to obtain the c value as follows. Let

$$f(c) = \int_{D_L/D_{mean}}^{D_R/D_{mean}} \frac{c^c}{\Gamma(c)} x^{c-1} e^{-cx} dx - P_r. \tag{9.26a}$$

Then

$$f'(c) = \int_{D_L/D_{mean}}^{D_R/D_{mean}} \frac{c^c}{\Gamma(c)} x^{c-1} e^{-cx} \left[1 + \ln(c) - \psi(c) + \ln(x) - x\right] dx \tag{9.26b}$$

and the iterative root finding procedure is

$$c_{i+1} = c_i - \frac{f(c)}{f'(c)},$$
(9.27)

where $i = 1, 2, \ldots$. Based on the discussion, the denominator is always non-zero. Since all the distributions are described by $c \geq c_0$, the searching process starts at $c = c_0$. This process continues until the prescribed tolerance ε is achieved, where the termination condition is formulated in terms of the successive change as

$$\left| c_{i+1} - c_i \right| \leq \varepsilon .$$
(9.28)

9.3.3 *Creation of 2D-VGRAIN system*

To facilitate grain structure generation and micro-mechanics modelling, an integrated system called VGRAIN was developed to build materials microstructures, where the CPVT model has been implemented to define the grain structures (Zhang, Zhu and Lin, 2012). The integrated process of defining the microstructure of the material is illustrated in **Fig. 9.17**. Two more modules have also been developed to include grain orientation and material properties assignment (Cao, Zhuang, Wang *et al.*, 2010). The orientation of each grain can be assigned based on a fixed texture, or be set by a random value from a random number generator based on a uniform distribution or a normal distribution. Grain orientations can also be defined according to measurements, such as EBSD. In the VGRAIN system (Zhang, Zhu and Lin, 2012), the generated grain structure together with the grain orientations can be directly imported into commercial FE codes, e.g. ABAQUS/CAE, for further preprocessing operations, such as meshing, boundary and loading conditions defined based on the simulation requirements. Examples of the application will be given in the last section of the chapter.

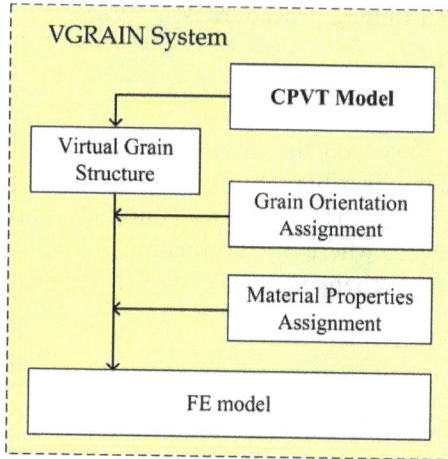

Fig. 9.17 The virtual microstructure generation system (2D-VGRAIN) for micro-mechanics modelling (Zhang, Balint and Lin, 2011a).

The CPVT model is the core mechanism for grain structure generation in the system shown in **Fig. 9.17**. The implementation procedure is as follows:

- Input general parameters to the model, including the domain Ω and the mean grain size, D_{mean}, lower and upper boundaries of grain size, D_L and D_R, and the percentage of the grains within the limits, P_r.

- Compute the distribution parameter c using the Newton–Raphson method with Eqs. (9.25)–(9.28);

- Derive the regularity parameter α from the empirical model using Eq. (9.19)–(9.20);

- Combine c and α with the obtained regular distance d_{reg} (Eq. (9.16)) to compute the control parameter δ (Eq. (9.18)).

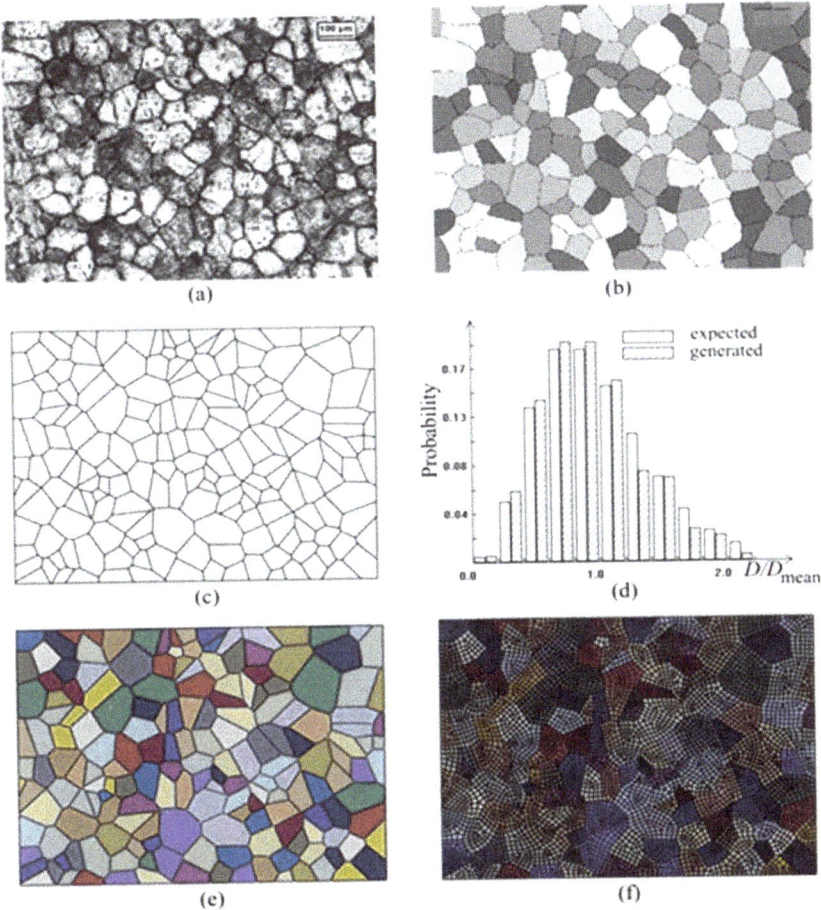

Fig. 9.18 Generation of grain structures. (a) Original microscopic image in a domain of 1000 μm × 730 μm. (b) Result after image processing - the grain structure was segmented into 179 grains. (c) The generated virtual grain structure. (d) Comparison of grain size distributions. (e) With the assignment of grain orientations. (f) With FE mesh (Zhang, Balint and Lin, 2011b).

- A seed lattice is generated from the model $\text{CPVT}\left(\delta|\Omega, D_{mean}\right)$ and the corresponding VT is used to define grain structure.

More detailed grain generation procedures are given by Zhang, Balint and Lin (2011a), and the use of the VGRAIN system was detailed in the user manual by Zhang, Zhu and Lin (2012). **Figure 9.18** shows an example of a 2D grain structure generated using the VGRAIN system

(**Fig. 9.18 (c), (e)** and **(f)**). By comparison with experimental observations (**Fig. 9.18 (a)** and **(b)**), very close agreement has been achieved as shown in **Fig. 9.18 (d)**. Grain orientations are assigned using the probability theory implemented into VGRAIN. Elements within grains are generated using ABAQUS/CAE. Loading and boundary conditions can also be applied using ABAQUS/CAE for CPFE analysis.

9.4 Creation of Cohesive Zones and 3D Grain Structures

Grain boundaries and interfaces between phases of materials (or different materials) are normally modelled using cohesive zones in FE analysis packages, such as ABAQUS. Interfaces normally have different mechanical properties from bulk materials. In this section algorithms for generating cohesive zones for grain boundaries are introduced. This is mainly based on the requirements of the FE code ABAQUS, where 4-sided elements are used to represent the cohesive zones. In addition, the algorithms for generating 3D grain structures are briefly introduced.

9.4.1 *Cohesive zones*

The mechanical properties of material interfaces (e.g. grain boundaries, inclusion and matrix interfaces) can be modelled using cohesive elements. To build an CPFE model for commercial FE/CAE codes, e.g. ABAQUS, it is required that

- each interface zone between two grains should be discretised with a single layer;
- cohesive zones including layers and the junctions of multi-layers can only be meshed by means of quadrilateral elements; and
- at least one of either the top or bottom edge of the cohesive element must be constrained to a grain element.

Obviously, meshing a cohesive layer boundary is straightforward, as illustrated in **Fig. 9.19**. However, at a junction of multiple layers, such as the triple junction ABC in **Fig. 9.19 (c)**, the part of the cohesive zone cannot be represented as a single mesh element and further procedures must be introduced to partition the junction into a few quadrilateral elements. As well as triple point junctions, there are a number of other special types of junctions. For example, **Fig. 9.19 (d)** depicts a linked triple point junction, where one edge of a cohesive layer is degenerated into a point A and hence two neighboured triple point junctions, i.e. EDA and CBA, are linked at the point A. Besides, **Fig. 9.19 (e)** also exemplifies an event of fivefold junction ABCDE. In CPFE simulations, there is a common fact that the more uniform is the mesh given, the more accurate the results will be. But the complicated junctions handicap the application of embedded cohesive zone models into Voronoi tessellations for crystal plasticity FE simulations.

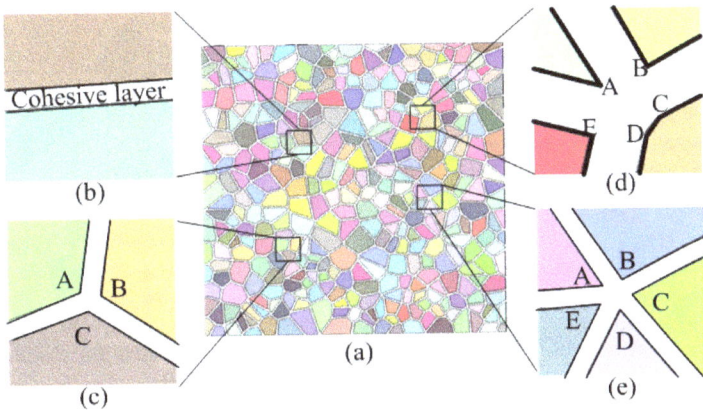

Fig. 9.19 The example of a VTclb and cases of cohesive junctions (Zhang, 2011).

a. *Voronoi tessellation with cohesive layer boundaries (VTclb)*

An offset method is introduced to transform ordinary Voronoi tessellations to non-zero thickness boundaries. The transformed VTs can be used to represent virtual grain structures and model the cohesive zones in applications of studying the grain boundary sliding, inter-granular cracking, propagation etc.

The geometrical properties of a Voronoi tessellation can be described as follows: given a domain $\Omega \subset \mathbf{R}^2$ with N seeds, $s_i \in \Omega$, $i=1,\ldots,N$, a tessellation divides the domain Ω into N polygons as grains. For a grain G_i, it contains the region that all points in this region are closer to the seed s_i than any other seeds. That is,

$$G_i = \left\{ x \in \Omega, \left\| x - s_i \right\| < \left\| x - s_j \right\| \right\}. \tag{9.29}$$

Obviously, all grains organised in this tessellation are convex and have straight edges. Furthermore, the region of boundary network in a VT is defined by

$$BN = \left\{ x \in \Omega, x \notin G_i, i = 1, 2, ..., N \right\}. \tag{9.30}$$

Since the grain boundaries are all formed by lines, the region BN has zero area.

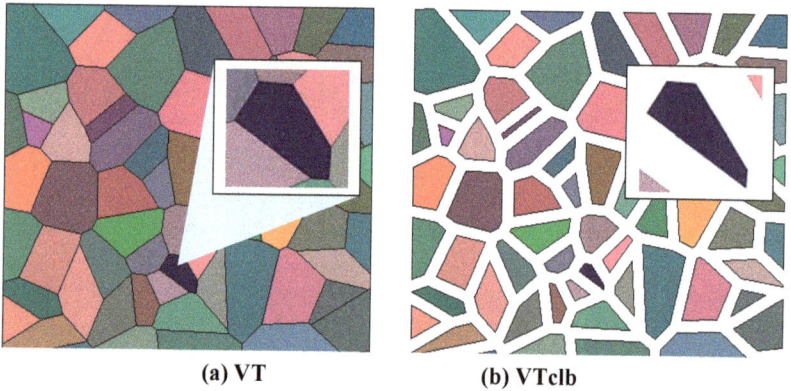

(a) VT   **(b) VTclb**

Fig. 9.20 Offsetting grain boundaries to represent cohesive layers. (a) A grain structure based on the VT representation; (b) a VTclb, where a degenerated grain was highlighted. It can be seen that one of the original boundaries vanished after offsetting (Zhang, 2011).

In order to incorporate a cohesive zone method (CZM) to formulate inter-granular traction-separation relations, cohesive interface elements need to be embedded along grain boundaries. As stated above, the grain boundaries in a VT consist of a network of lines. Naturally, in the presence of a VT, cohesive elements can be directly produced by replacing the original lines with a network of cohesive layers. **Figure 9.20 (b)** illustrates a VTclb, where grains are reconstructed by inward offsetting original grains from **Fig. 9.20 (a)** with a specific distance. The thickness of cohesive layers is generally thin, but, whatever the thickness is given, during generating the cohesive layers along grain boundaries there is a potential chance such that small grain edges could be vanished, due to the non-zero thickness of cohesive layers, as the highlighted grain in **Fig. 9.20 (b).**

b. *Offsetting of individual grains*

Generation of cohesive layers for a VT can be achieved through processing individual grains. An intuitive method of inward offsetting a convex polygon can be implemented by:

- creating offsets for all decomposed edges and then reconnecting the pieces according to original topological relations, i.e. the head to tail sequence of consecutive edges;

- tracking and removing loops (Held, 1991). However, tracking loops in this context needs complex heuristics to patch various counter-cases, which results in contradiction between efficiency and robustness.

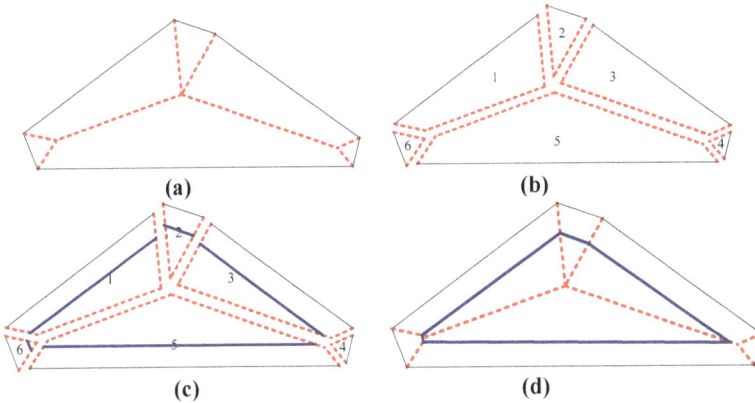

Fig. 9.21 Illustration of the structured offset method. (a) Generating the medial axis; (b) decomposing the grain into a set of sub-polygons; (c) structured offset boundaries; (d) formation of an inward-offset grain (Zhang, Karimpour, Balint and Lin, 2012).

A structured offset method is introduced by Zhang, Karimpour, Balint and Lin (2012) to produce inward-offset polygons and hence a complete VTclb. The term "structured" means that for each single edge, inward offset is restricted to a decomposed sub-polygon, which is divided by the polygon's medial axis. The medial axis of a simple polygon is a tree-like planar structure consisting of a set of line segments, as illustrated by the dot-lines in **Fig. 9.21 (a)**. A point on the line segments has the property that there is more than one closest point on the polygon's edges, i.e. there are at least two edges whose distances to the point are equal. Therefore, it can be easily obtained that each line segment of the medial axis is a bisector of the triangle of two edges. The algorithm to build the medial axis has been detailed by Aggarwal, Guibas, Saxe *et al.* (1987), which takes $O(n)$ time complexity where n is the edge number of the polygon.

c. Triple junction partition

In a real grain structure, grains are generally organised irregularly and hence the junctions are not uniform. There could be triple junctions where three boundary layers meet together, as shown in **Fig. 9.22**. It needs a general pattern to mesh them to quadrilateral elements.

Figure 9.22 (a) illustrates a technique to subdivide the junction. In this pattern, corner vertices, including A, B and C, are directly linked to curb the junction domain. In this way, the junction and boundary layers of two grains are separated into different domains. Since this junction is a triangle area, it should be further subdivided into quadrilateral elements. This can be achieved by constructing three perpendicular line segments to the triangle's sides from the centre of the triangle. Note that, this partition pattern involves free nodes, i.e., the intersection nodes m_1, m_2 and m_3. Although this can be fixed by introducing a linear multi-point constraint (MPC) relationship to constrain the degrees of freedom of the models, it could not provide any rationale for the choices of normal and tangential directions for these elements.

Figure 9.22 (b) provides an alternative scheme without introduction of any unnecessary free nodes. In this pattern, the junction region is initially cut by the line segments linking the centre and corner vertices, i.e. AO, BO and CO. In this way, the cohesive zone junction can be divided into three independent cohesive layers without isolating an extra junction as the first pattern does. Obviously, the ends of the layers are not regular. To minimise the effects of the irregular shape on meshing, particular procedures are required to initially mesh the ends of cohesive layers. In the initial meshing procedure, the first mesh element can be produced by cut the head of layer from a corner to the opposite edge, as the illustrated elements of AOBB', BOCC', and COAA'.

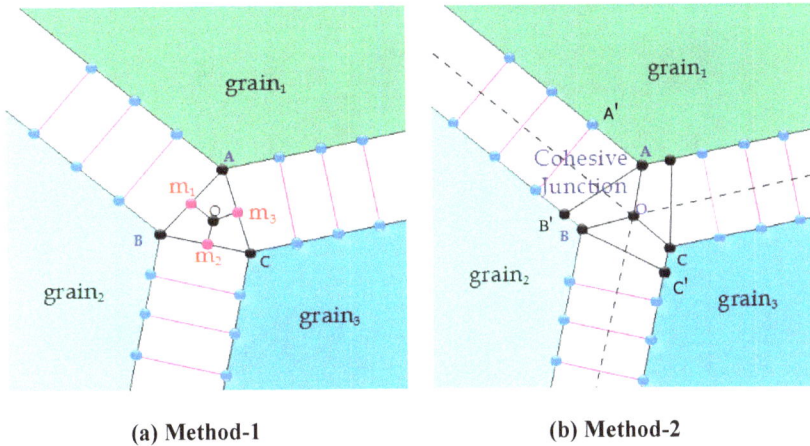

(a) Method-1   (b) Method-2

Fig. 9.22 Illustration of two junction partition methods of meshing cohesive zones for triple junctions (Zhang, Karimpour, Balint and Lin, 2012).

d. Natural geometric cases of grain junctions

Figure 9.23 (a) shows an example of a grain structure described by a VT with cohesive layer boundaries. Note that whilst VTs provide a natural representation of grain structures, they also introduce significant complexity in grain boundary connectivity. Therefore, a primary difficulty in developing a generic algorithm for automatically generating cohesive layers and partitioning the cohesive layer junctions is coping with the event of degeneration of small edges.

Furthermore, to build an FE model in commercial FE/CAE software, e.g. ABAQUS, which is typically required for industrial implementations, the following restrictions apply:

- each interface zone between two grains should be discretised with a single layer;
- cohesive zones including the layers themselves and their multiple junctions (e.g. triple points) can only be meshed by means of quadrilateral elements, since this shape naturally identifies the normal and tangential directions; and
- at least one of either the top or bottom edge of the cohesive element must coincide with a grain element.

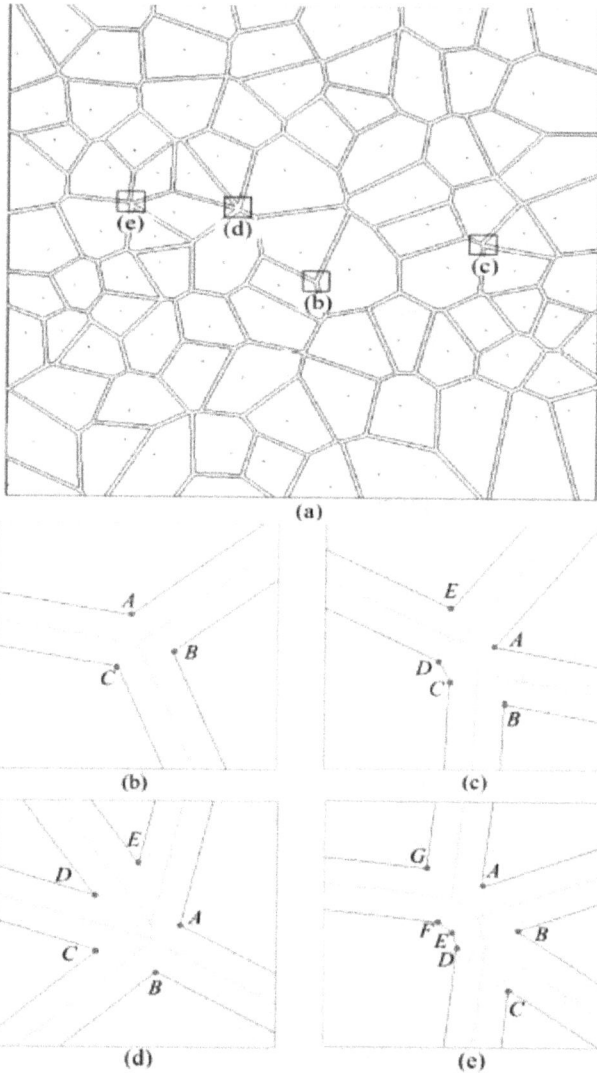

Fig. 9.23 An example of the VTclb scheme and specific cases of
cohesive junctions (Zhang, 2011).

At a junction of multiple cohesive layers, as shown in **Fig. 9.23 (b)–
(e)**, further procedures must be introduced to partition the cohesive zone
junction into quadrilateral elements. In addition to triple junctions, as
shown in **Fig. 9.23 (b)**, there are a number of special junction types. For

example, **Fig. 9.23 (d)** shows an occurrence of a fivefold junction ABCDE. **Fig. 9.23 (c)** depicts a linked triple junction, where one edge of a cohesive layer degenerates to a point A upon offsetting the boundaries, hence two neighbouring triple junctions, i.e. EDA and CBA, are linked to the point A.

Similarly, the junction in **Fig. 9.23 (e)** presents a double-linked triple junction. In FE simulations, it is a common fact that a more uniform mesh generally gives more accurate results, but the complicated junctions that may occur in a VTclb make it difficult to automatically generate uniform junction partitions.

To cope with general cases of grain boundary junctions, adjustments are normally required to convert general geometric cases to standard configurations. Four examples of the adjustments are listed in **Table 9.3**. The first is referred to as the *corner* case, in which the cohesive layer is split by a domain corner and the central line does not intersect the corner. In this case, the end is still formed by two edges, but, in contrast to the normal layer unit, in which the end point is the centre of a junction, the end point is the VT's domain corner. Therefore, for the purpose of uniformly deriving the data structure, the end point of the central line in this case is moved to the corner.

The *border* case is that of a cohesive junction cut by a border, where the related central lines cannot meet in the junction centre. Note that, although there is no centre point to be used to form the mesh element nodes, there are still two edges existing for each layer unit. Therefore, the central lines can be moved to the junction intersection, and the intersection point can then be used as the centre of cohesive junction.

The third case usually occurs at the VT's domain border, when the border cuts a cohesive layer unit such that only some part of the unit end remains. This case is referred to as the *two edges degenerated cut* case; that is, this type of layer unit is regarded as a cohesive layer unit that has its junction centre moved to the domain boundary, as shown in **Table 9.3**. The original triangle is relatively tiny, hence can be removed and the centre of the junction then moved to the other end of the central line segment.

The fourth is the case of *two edges degenerated*. The shape of the case might be non-convex, as shown in **Table 9.3**. Therefore, an additional procedure is required for generality that adjusts its shape by means of repositioning the two junction centres. If an end of a cohesive layer unit is found to be non-convex, or some inner angle is larger than a user-specified tolerance, the related junction centre O is moved to the centroid of triangle ABC.

Table 9.3 Special cases of cohesive layer units (Zhang, 2011).

	Natural geometric case	Adjusted geometric data
Corner		
Border		
Two edges degenerated cut		
Two edges degenerated		

Many other natural geometric cases need to be treated specifically. The details have been given by Zhang (2011); Zhang, Karimpour, Balint and Lin (2012a); and, Zhang, Karimpour, Balint *et al.* (2012). These are not introduced in the book.

9.4.2 *3D grain structures*

The Voronoi tessellation method introduced for 2D grain structure generation can be extended to the application of generating 3D grain structures, which follows the same procedures. The empirical results and rules relating to the mean area and mean perimeter of a 2D Poisson Voronoi tessellation (PVT), the mean volume and surface area of a 3D PVT, the number of vertices and edges of a typical grain, the length of a typical edge etc. should be calculated. In addition, a few higher level properties, such as the density of grain vertices or grain centroids, the expected edge length density and face area density also need to be determined.

a. *Regularity for 3D grain structures*

For a general three-dimensional Voronoi tessellation, a similar parameter proposed by Zhu, Thorpe and Windle (2001) can be used to assess the regularity of grain structures. The 3D regularity, α, is defined by the ratio of the minimum seed spacing of the tessellation to that of the equivalent regular truncated octahedral, or tetrakaidecahedral, tessellation.

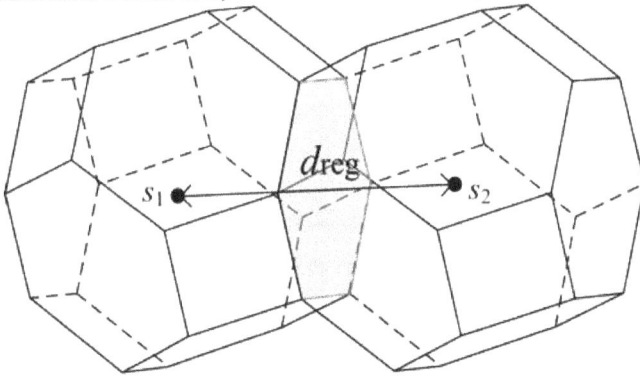

Fig. 9.24 Definition of the regular distance d_{reg}, where S_1 and S_2 are seeds from a body centred cubic (BCC) lattice (Zhang, Karimpour, Balint and Lin, 2012b).

A tetrakaidecahedral tessellation (TT) has been employed as the most regular tessellation, because a uniform tetrakaidecahedron with planar faces is a very good approximation to the so-called "Kelvin" polyhedron, which is a space filling polyhedron with minimum surface area. Note that a TT is a fully ordered 3D Voronoi tessellation, in which seeds are arranged in a body centred cubic (BCC) lattice. A tetrakaidecahedron has 14 planar sides; eight are regular hexagons and six are squares. As illustrated in **Fig. 9.24**, two adjacent grains in a TT are coincident at regular hexagonal faces and the seed distance must be equal to

$$d_{reg} = \frac{\sqrt{6}}{2}\left(\frac{D_{mean}}{\sqrt{2}}\right)^{1/3},$$

(9.31)

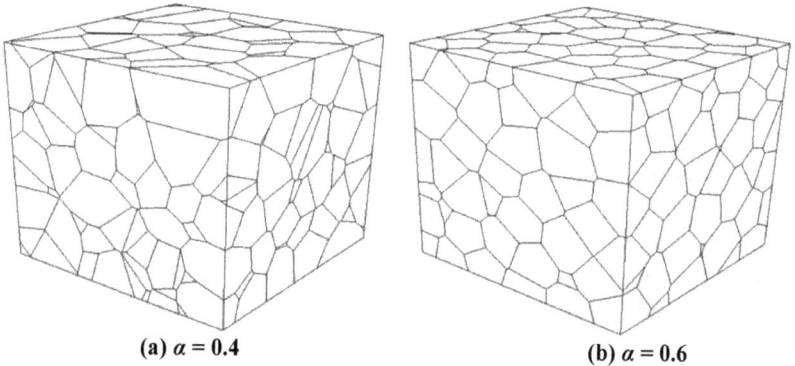

(a) $\alpha = 0.4$ (b) $\alpha = 0.6$

Fig. 9.25 Poisson Voronoi tessellations with different degrees of uniformity. (a) An irregular structure with a low value of α; (b) a more regular structure with a higher value of α (Zhang, Karimpour, Balint and Lin, 2012b).

Where D_{mean} is the mean grain size (in terms of volume) of the given VT. Therefore, the regularity parameter α can be defined as the same as the 2D case:

$$\alpha = \frac{\delta}{d_{reg}},$$

where δ is the minimum seed spacing in the given VT. Note that as the regularity α decreases from 1 to 0, the corresponding tessellation becomes more disordered, changing from a uniform TT to a fully-random tessellation, i.e. a Poisson Voronoi tessellation.

Figure 9.25 shows two 3D grain structures generated with different values of the irregular parameter α. It can be seen that more regular grains are generated with a higher value of α.

b. Grain size distribution

Given a Cartesian coordinate system and a cubic domain with volume V, seeds are placed in the cubic domain by generating x-, y- and z-coordinates sequentially from a uniform distribution random number generator. A constraint is introduced requiring that a subsequent seed is accepted only if the distances from it to the other existing seeds are all larger than or equal to the minimum seed spacing δ. This process is continued until N_{seed} points have been generated. Note that since seeds in the final seed lattice are controlled by a minimum distance, the final tessellation tends to have a slightly larger regularity than the specified value. Moreover, as the number of seeds N_{seed} increases, the regularity of a resultant VT asymptotically approaches the specified value. Similar to 2D case, the mean grain size can be expressed as

$$D_{mean} \approx \frac{V}{N_{seed}}.$$

This approximate mean grain size is used to calculate the seed distance in the equivalent TT according to Eq. (9.31).

An optimisation procedure can be employed to fit the grain size distribution for 3D grain generation using VT. This is based on the one-parameter gamma distribution, $P(x)$, as discussed in Section 9.3.2:

$$P(x) = \frac{c^c}{\Gamma(c)} x^{c-1} e^{-cx} dx \qquad x > 0, \tag{9.19}$$

where the variable x is normalised grain size, D/D_{mean}, the distribution parameter $c > 1$, and $\Gamma(c)$ is the gamma function, defined as

$$\Gamma(c) = \int_0^\infty x^{c-1} e^{-cx} dx.$$

The mean value of the distribution is one, hence 10 intervals given by

$$I_i = 0.3 \times [i, i+1],$$

where $i = 0, 1, \ldots, 9$ within a truncated domain $[0, 3]$, are used to evaluate the difference between the model and the statistical data. This range is defined according to the studies of Zhang, Karimpour, Balint and Lin (2012); for the most random case $\alpha = 0$ there are very few grains $D/D_{mean} > 3$.

Table 9.4 Optimisation results including the relations between the regularity α and the distribution parameter c, and the corresponding fitting errors defined by Eq. (9.32) (Zhang, Karimpour, Balint and Lin, 2012a).

α	0	0.1	0.2	0.3	0.4	0.5	0.6	0.7	0.8
c	5.156	5.209	5.446	6.210	7.702	10.609	16.031	28.971	48.187
$E(c)$	1.9E-5	1.9E-5	1.2E-5	1.5E-5	6.7E-5	1.1E-4	4.6E-5	2.1E-3	1.1E-2

The least squares error function is defined by:

$$E(c) = \sum_{i=1}^{n} \left[f_i(c) - \mu_i \right]^2 , \tag{9.32}$$

where $f_i(c)$ is the ideal probability in the interval I_i, i.e.

$$f_i(c) = \int_{0.3i}^{0.3i+0.3} P(x)dx ,$$

and μ_i is the frequency of the grains whose sizes are within the interval I_i. An optimisation algorithm (Hansen and Ostermeier, 2001) can be used to find the appropriate c values for corresponding regularity values. The optimisation results are presented in **Table 9.4**, where the second row lists the distribution values for the model fits and the third row list the related objective function values, defined by Eq. (9.32).

The data pairing between the regularity measurement, α, and the corresponding model parameter, c, presented in **Table 9.4**, reveals a nonlinear mapping. Equation (9.20), which is used for 2D VT, can also be used in 3D case with the different values of the optimised parameters:

$$c_0 = 5.156, \ c_m = 370, \ A = 2.4, \ k = 0.33 \ \text{and} \ n = 1.5 .$$

The final calibrated model and the $\alpha - c$ data pairs used in the calibration are plotted in **Fig. 9.26**. A good agreement is observed.

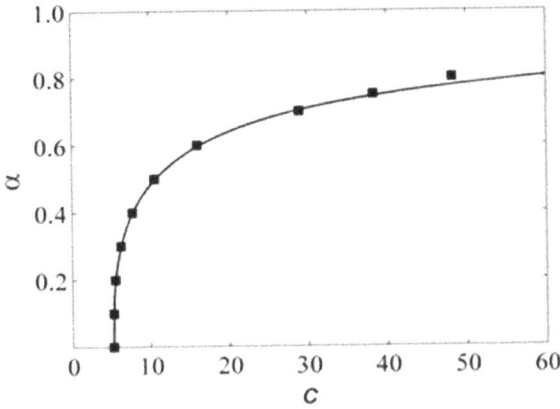

Fig. 9.26 The descriptive model, relating the regularity parameter α to the distribution parameter c of a one-parameter gamma distribution function (Zhang, Karimpour, Balint and Lin, 2012b).

c. Physical parameters

This is the same as the 2D case; the physical parameters consist of a mean grain size D_{mean}, a lower boundary of grain size D_L, an upper boundary of grain size D_R and the percentage P_r of grains within this range. Note that, the sizes D_{mean}, D_L and D_R are measured in terms of grain volume. The core mechanism in determining the distribution parameter c from the four physical parameters is solving the implicit Eq. (9.21) within the range defined in Section 9.3.2. The computational procedure is the same as that used in generating grain structures for 2D cases, which will not be detailed here.

9.5 VGRAIN System Development

The intention of this section is to give a brief introduction to the requirements of a virtual grain structure generation system for crystal plasticity FE analysis using a commercial FE code. VGRAIN, which was developed at Imperial College London (Zhang, Zhu and Lin, 2012) and linked with the FE preprocessor ABAQUS/CAE for CPFE analysis, is used here as an example to demonstrate the specific needs of a virtual grain generation system. The main theories for the system have been introduced in the previous sections, including the crystal plasticity constitutive equations in Section 9.2. Further readings for the theories include Cao, Zhuang, Wang and Lin (2010); Zhang, Balint and Lin (2011a); Zhang, Balint and Lin (2011b); Zhang, Karimpour, Balint and Lin (2012b), since not all details have been introduced in the book.

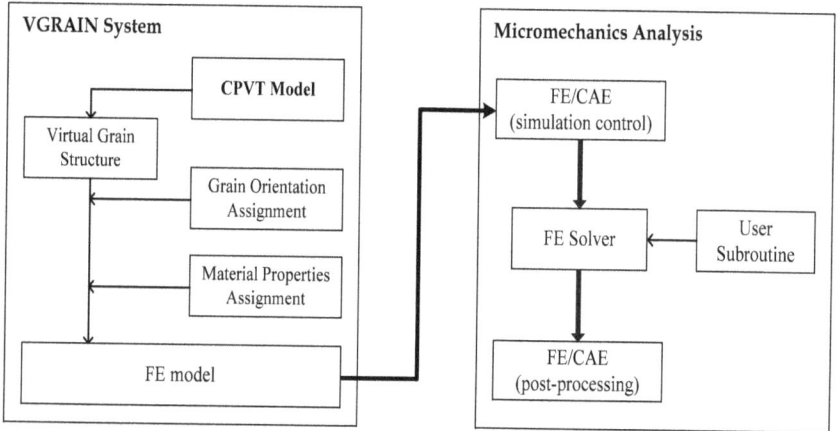

Fig. 9.27 Integration of the grain structure representation system and FE/CAE for micro-mechanics simulations (Zhang, Zhu and Lin, 2012).

9.5.1 *Overall system*

a. The system

A grain structure generation software system, VGRAIN, was developed based on the CPVT model described above to produce the virtual grain structures defined by the physical parameters. **Figure 9.27** illustrates the integration of the VGRAIN system with an FE platform. Apart from virtual grain structures, VGRAIN is capable of generating an entire materials model, including grain orientations and material properties assigned to the grains of a generated VT. The final virtual grain structure with material properties and cohesive junction partitions can be directly exported to commercial FE/CAE platforms, e.g. ABAQUS, by means of corresponding text scripts such as the python script for ABAQUS. Final meshing is then carried out using the commercial solver. Further preprocessing operations, such as meshing within grains, boundary constraints and loading conditions, can be performed within ABAQUS/CAE for crystal plasticity analysis.

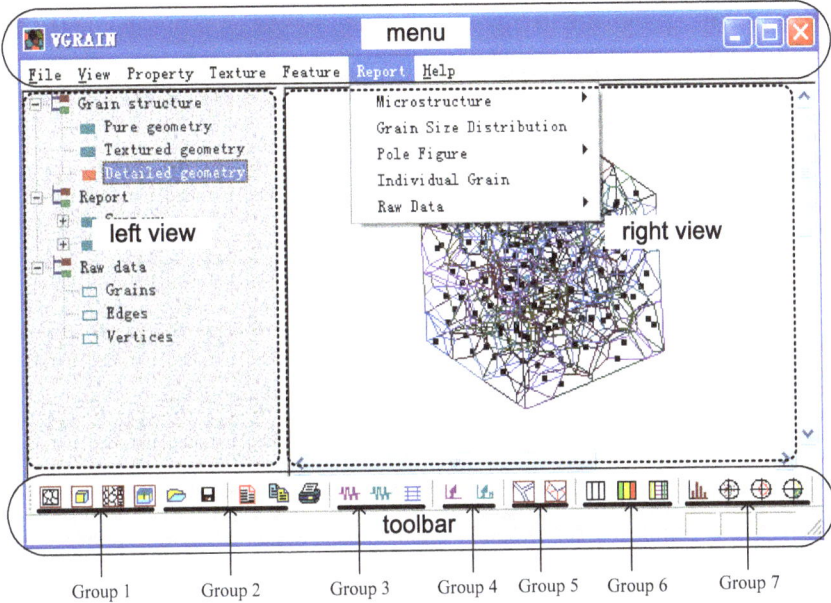

Fig. 9.28 Layout of the graphical user interface of the VGRAIN system (Zhang, Zhu and Lin, 2012).

b. The user graphic interface

Besides the grain structure modelling, the VGRAIN also acts as a stand-alone system allowing users to analyses of the morphological and orientation characteristics of the generated structures. The graphic interfaces for the VGRAIN software is designed with considerations of:

- providing sophisticated dialogues to assign the CPVT model parameters, configure the materials and texture generators, and specify the FE models to be exported;
- presenting sufficient information including graphs and spreadsheets to analyse generated grain structures.

The basic layout is presented in **Fig. 9.28**. The right view is a read-only region that allows users to access all the information relating to the current virtual grain structure, such as a variety of plots and spread sheets.

The menu bar and toolbar contains all the entities for user to operate this system. In the toolbar, there are mainly five groups of buttons to invoke corresponding dialogues for users to specify related parameters:

- *Group 1* contains the buttons to launch the dialogues for the CPVT models of 2D, 3D, 2D multi-zone and 3D multi-zone.

- *Group 2* performs save/open operations for the workspace and calls the dialogue to export the current virtual grain structure to a text file for commercial FE platforms.

- *Group 3* invokes the dialogues to assign material properties, according to the involved constitutive equations, for a generated grain structure.

- *Group 4* focuses on the orientation generators to assign grain orientations for a presented grain structure.

- Group 5 calls the cohesive zone model interface for users to specify the configurations of a CZ model and perform the junction partitioning.

Different from the other groups, *Groups 6* and *7* directly pass the user commands to the plot view (denoted as right view), providing related information such as simple geometrical grain structure plots, coloured structures with grain orientations, detailed grain structures, grain size histograms and a variety of pole figures. Note that the toolbar can be floating over the graphic user interface (GUI) or attached to any border, and here, to clearly present the graphic information, it was attached to the bottom.

In addition, the left view also provides the entities for users to request the information on the current model, which is to then be presented in the right view with plots or spreadsheets. Major queries include grain-structure-related plots, reports of statistics of grains, and direct access of grain morphological and orientation data.

c. *The architecture and input*

The system includes three major classes, which coordinate the inner workflow of the system: the CPVT class, the view class and the frame class, as shown in **Fig. 9.29**. The CPVT class contains all the modules for virtual grain structure generation. It receives the configuration data from corresponding dialogues and generates grain structures and other data according to the invoked commands. Moreover, it executes a series of drivers or functions to interact with external systems or devices:

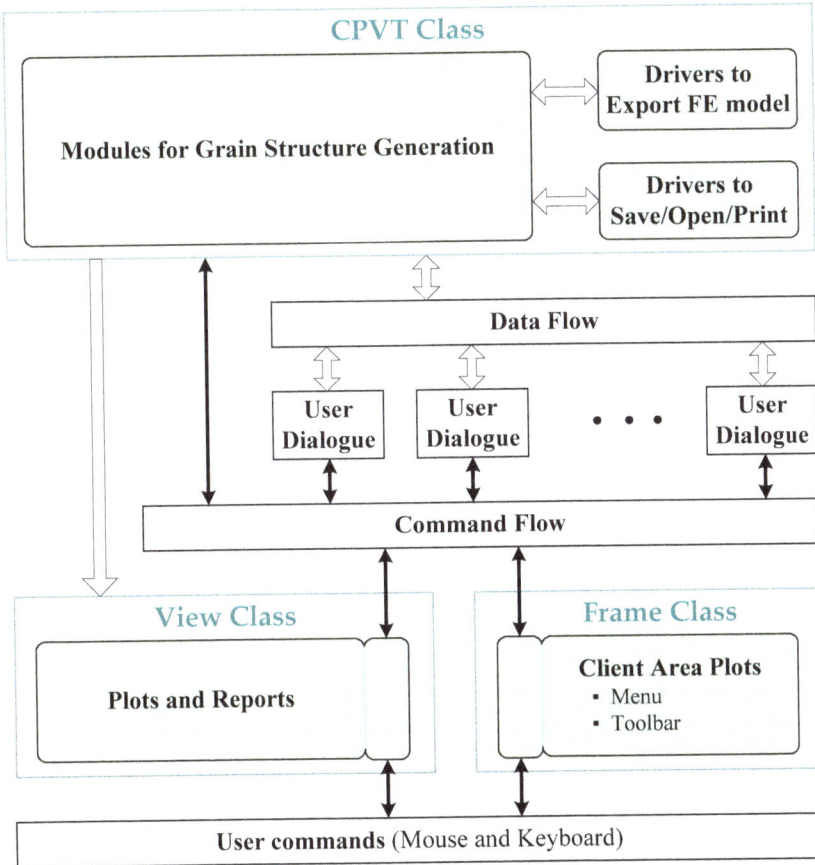

Fig. 9.29 An architectural view of the VGRAIN system. The command invoking flows are denoted by arrow lines and the data accessing relations are presented by block arrows (Zhang, Zhu and Lin, 2012).

- performing the save, open and print operations for the current workspace;
- exporting the current grain structure model into a script file for the specific FE/CAE platform.

Note that current version only focuses on the FE code ABAQUS and the script is based on the ABAQUS journal file, but it is easy to produce the corresponding scripts for other FE codes.

The view class maintains the left view command flow and also updates the right view plots or spreadsheets, while the frame class only executes the user commands that are invoked from the menus and buttons of the toolbar. The command flow and data flow of the system are illustrated in **Fig. 9.29**. Note that, based on this design, the GUI of the system receives user commands through the entities of the menu and the toolbar, and obtains the user data by the provided dialogues.

The VGRAIN system allows the user-input data to be interactively assigned or modified through user dialogues, with each parameter input dialogue relating to each module. The core modules implement the proposed CPVT models for generating a range of virtual grain structures. There are five grain structure generation modules: the 2D CPVT module, 2D multi-zone CPVT module, 3D CPVT module, 3D multi-zone CPVT module and the cohesive zone module. For each CPVT module, configuring one of the CPVT modules is performed by the corresponding user dialogue. As input parameters, Ω includes the vertices coordinates of the grain structure domain and the parameters of a mean grain size D_{mean}, a lower boundary of grain size D_L, a upper boundary of grain size D_R and the percentage P_r of grains within this range, which defines the characteristics of a grain structure. The detailed computational routine is given in **Fig. 9.30**.

Moreover, a grain orientation module and a material assignment module are given in the system. Three grain orientation generation mechanisms are attached to generate grain orientations including (Cao, Zhuang, Wang, *et al.* 2009)

- constant orientation assignment;
- uniform distribution random generator; and
- normal distribution random generator.

The user dialogues are provided for easily defining the orientation generators and accessing the orientation information.

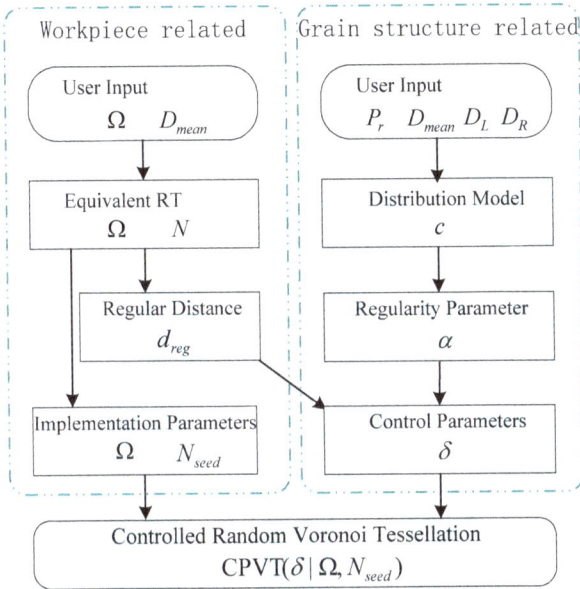

Fig. 9.30 Flow Chart of generating grain structures using the CPVT model (Zhang, Zhu and Lin, 2012).

The VGRAIN system is capable of studying the properties of the generated grain structures. A series of modules have been incorporated in the system, including:

- grain size distribution analysis – providing the histograms comparing distributions of the current generated grain structure and the user-specified;
- grain orientation analysis – providing the pole figures of the grain orientations assigned to the current grain structure;
- grain morphological analysis – giving statistics on the number of edges and vertices, and other topological relations; and
- spreadsheets – presenting the raw data on all the information of the grains.

9.5.2 *Grain structure generation*

Figure 9.31 presents the definition and generation of a grain structure with varying grain distributions, which simulate different heat treatment conditions of materials, such as a heat-affected zone of welds. As the user input, three sub-regions were defined and related physical parameters were provided. **Figure 9.31** gives a generated three-zone CPVT, which is purely the geometrical description.

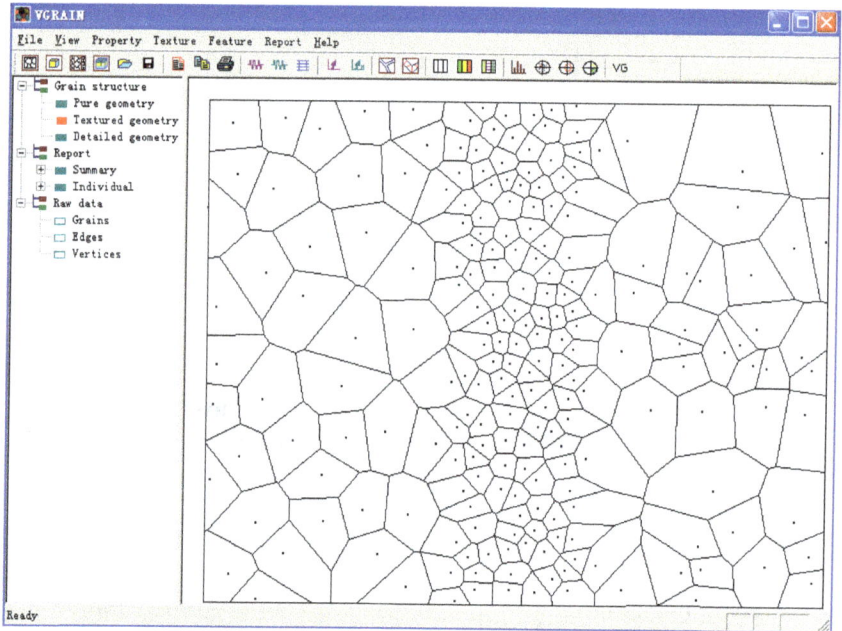

Fig. 9.31 The GUIs for generating grain structure (Zhang, 2011).

Grain orientations and material properties can be assigned to individual grains. **Figure 9.32** presents the grain structure model with complete geometry, grain orientations, grain boundary cohesive zones and material properties ready for CPFE simulations. Note that the colours in the figure denote the grain orientations.

The VGRAIN system has been linked with commercial FE codes, ABAQUS/CAE and DEFORM. Meshing, boundary constraints and loading conditions can be performed within the commercial FE codes. Further functions can be obtained from the user manual for VGRAIN (Zhang, Zhu and Lin, 2011).

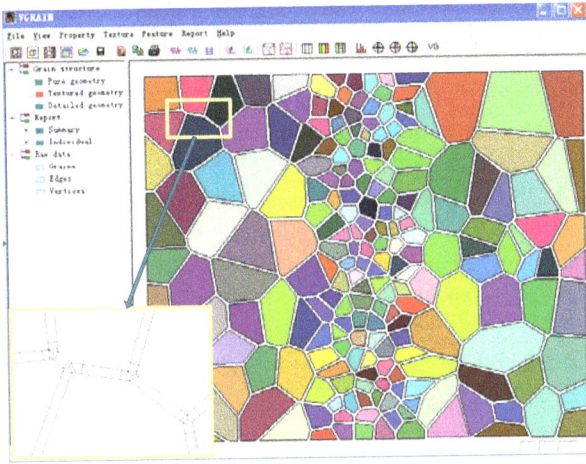

Fig. 9.32 Definition of grain orientations (colours) and the generation of cohesive zones (Zhang, 2011).

9.6 Case Studies for Micro-forming Process Modelling

In this section, a few examples are given to demonstrate the use of the crystal plasticity theories and grain generation facilities for CPFE analysis. In the following case studies, the same CPFE numerical procedure is used, which is shown in **Fig. 9.33** and based on the commercial FE code ABAQUS.

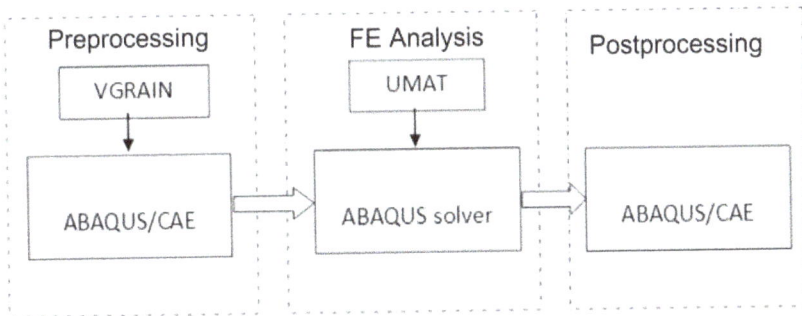

Fig. 9.33 Flow chart of an integrated numerical procedure for CPFE analysis (Cao, Zhuang, Wang *et al.*, 2009).

In pre-processing, the VGRAIN system generates virtual grain structures with their orientation information. ABAQUS/CAE is employed for further preprocessing, where a complete FE model with meshing, contact interaction, boundary and loading conditions is built. The crystal plasticity material model, which is introduced in Section 9.2, is implemented in ABAQUS via the user-defined subroutine VUMAT/UMAT to enable explicit/implicit crystal plasticity FE analyses to be carried out. The computational results are reviewed using ABAQUS/CAE as normal FE analysis (Cao, Zhuang, Wang *et al.*, 2009).

Table 9.5 Material constants for Eqs. (9.12)–(9.13).

n	$\dot{a}\,(s^{-1})$	$h_0\,(MPa)$	$\tau_s\,(MPa)$	$\tau_0\,(MPa)$
20.0	0.001	225.0	330.0	50.0

In the following simulations, 316L stainless steel (Harewood and McHugh, 2007) was used with material constants listed in **Table 9.5** for Eqs. (9.12)–(9.13) in Section 9.2. The Young's modulus and Poisson's ratio are 193 GPa and 0.34, respectively. A high value of n is used here to reduce viscoplastic behaviour of the material since it is being deformed at room temperature. It should be mentioned that 316L stainless steel has a face centred cubic (FCC) crystalline structure, which has its slip systems defined by the Miller indices <110>{111}.

9.6.1 *Necking in plain strain tension*

a. Localised necking

To study the general effects on deformation and necking localisation of grain orientation within a thin film structure, where only a few grains exist through the thickness of the material, a plane strain FE model is created. The overall dimension of the work-piece, the boundary and loading conditions are shown in **Fig. 9.34**. For this simulation, the mean grain size is estimated to be about 12.5 μm, where only about four grains are found across the thickness of the cross-section, on average. A displacement-controlled condition is applied on the right edge of the specimen and the lateral direction is free of constraint.

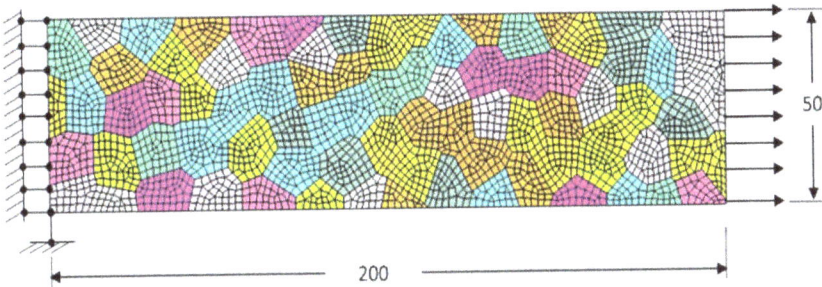

Fig. 9.34 Geometry and FE model (dimension in µm). The colours represent grain orientations, which were assigned randomly (Cao, Zhuang, Wang *et al.*, 2009).

CPFE analyses have been carried out using the 2D plane strain FE model. A displacement of 200 µm was applied at the right edge of the model. **Figure 9.35** shows the cumulative shear strain distributions superimposed on the deformed configuration. Overall, the material deformation is non-uniform due to each grain in the model having a different crystal orientation to the others, where their material behaviour is governed by the crystal plasticity constitutive equations described in Section 9.2. Significant localised necking can be observed at several locations along the length. This is thought to be the result of orientation mismatches among grains, as reported by Harewood and McHugh (2007).

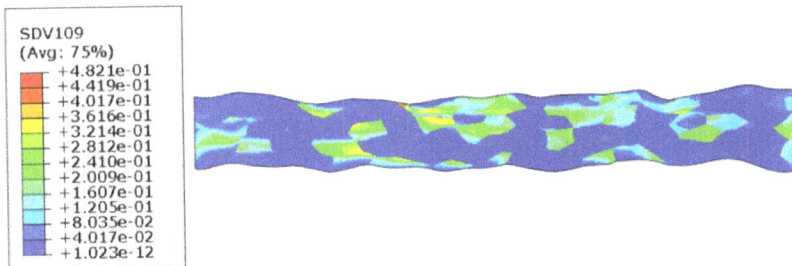

Fig. 9.35 Necking and accumulated shear strain distribution (Cao, Zhuang, Wang *et al.*, 2009).

Stress localisation and the coalescence of plastic strain, due to favourable grain orientations, can be attributed to the determination factors about the site of localised necking as well as eventual failure. Comparison of the {111} numerical pole figures in rolling direction is shown in **Fig. 9.36**. The concept of pole figure definition for crystal plasticity analysis can be found in the work of Cao, Zhuang, Wang *et al.* (2009) and Zhang, Zhu and Lin (2011). A set of 120 crystal orientations is used to represent the initial and deformed textures for this case. The first pole figure in **Fig. 9.36** is obtained before the deformation of the work-piece, corresponding to the assumed initial set of crystal orientations. It can be seen that the numerical representation shows the orientations to be randomly spread in the specified plane. After deformation of the work-piece, the orientations change, as shown in the second pole figure in **Fig. 9.36**.

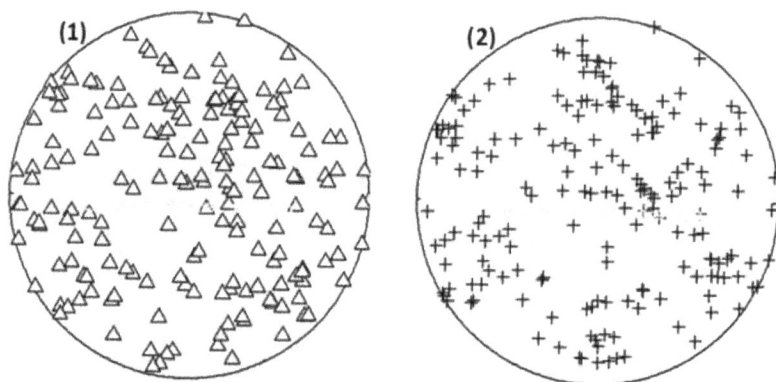

Fig. 9.36 Comparison of {111} numerical pole figures in rolling direction from (1) before the deformation; (2) after deformation in plane strain tension of microfilm (Cao, Zhuang, Wang *et al.*, 2009).

b. Grain size effect

To investigate the effect of grain size on necking and the exact spatial distribution of grain orientations on variations in the mechanical response under a tensional forming process, micro-film models with three, six and twelve grains through the width were used for the geometric dimensions of the FE model in **Fig. 9.34**. Each micro-film model was simulated with six different randomly generated grain orientation distributions to account for different microstructures (i.e. variations in the spatial distribution of grain shapes, sizes and orientations, with the average, maximum and minimum grain size fixed).

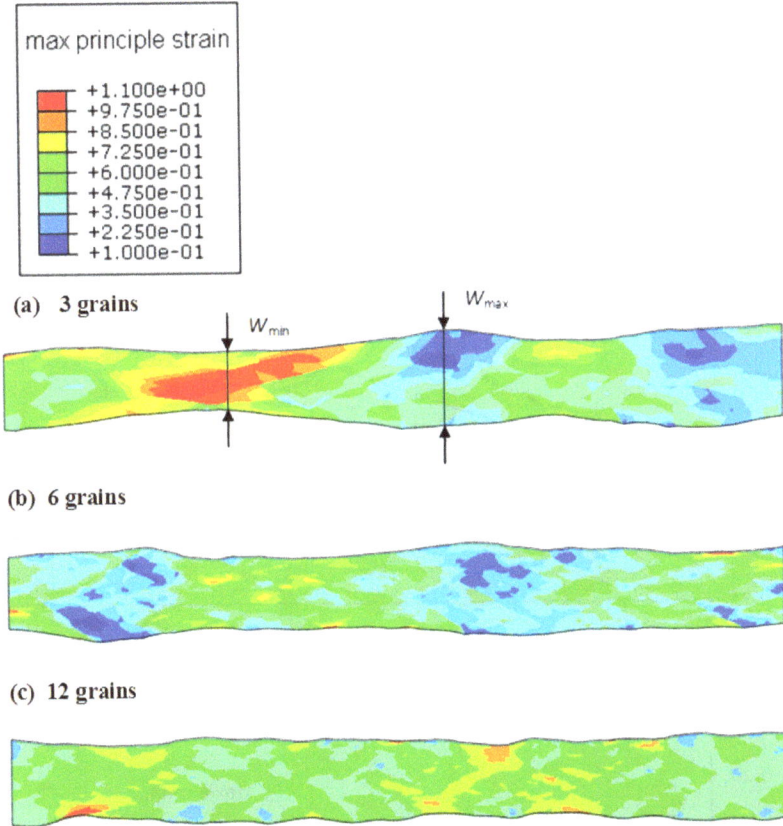

Fig. 9.37 Necking with superimposed maximum principle strain for micro-films with different number of grains through the width (Wang, Zhuang, Balint and Lin, 2009a).

Figure 9.37 shows the deformed shape of micro-films with superimposed contours of maximum principal strain for three, six and twelve grains through the film width W. Local thinning (or necking) can be observed for all the micro-films. Localised necking is observed to decrease as the number of grains through the width increases. Furthermore, the overall deformation is more uniform when the average grain size is small relative to the width. It can also be observed that there is a sharp strain gradient where necking takes place. The strain local to the neck is banded, indicating that the main mechanism of deformation is dislocation motion along slip planes.

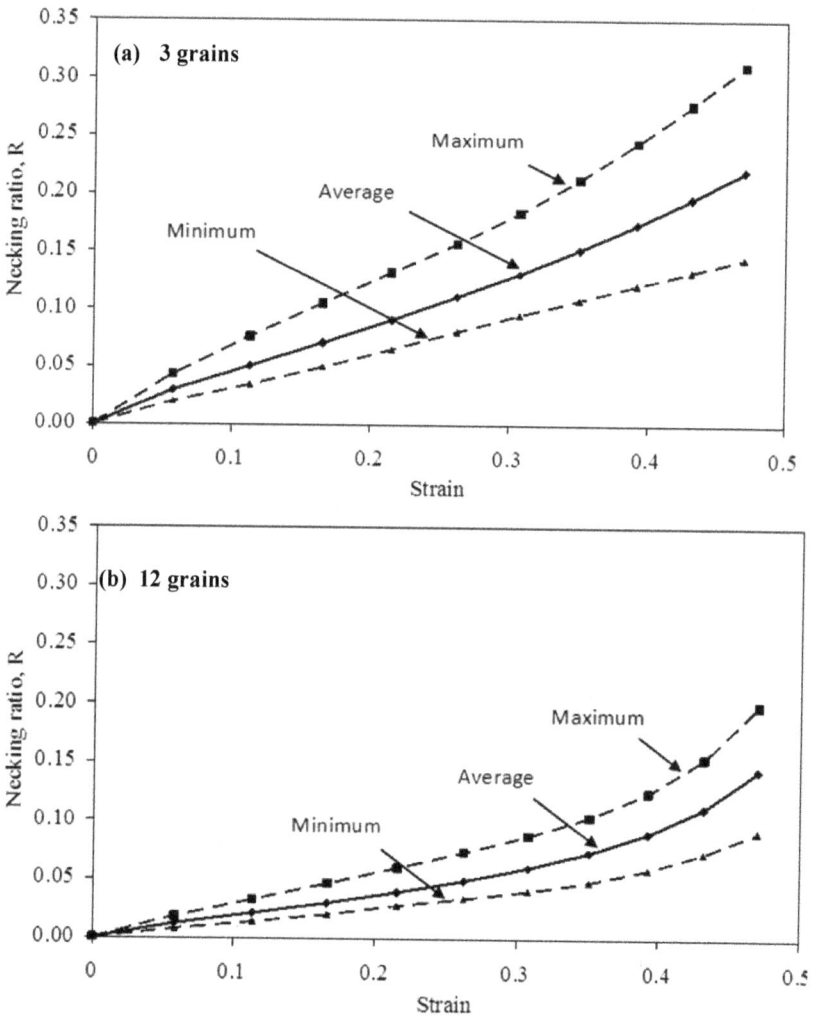

Fig. 9.38 Necking as a function of strain in micro-films under tension with (a) three grains through the width, showing approximately ± 50% scatter, (b) twelve grains through the width, showing approximately ± 40% scatter (Wang, Zhuang, Balint and Lin, 2009a).

It should be noted that conventional forming simulations fail to predict lateral thinning under a displacement controlled tensile loading with an elastoplastic or rigid-plastic FE model. The CPFE model accounts for activation of multiple slip systems, multiple grains and random grain orientations. Hence, necking is predicted by a CPFE model if the number of grains across the micro-film width is small enough.

To investigate the extent of necking quantitatively, the degree of necking is defined by

$$R = (W_{max} - W_{min})/W ,$$

where W_{max} and W_{min} are the maximum and minimum dimensions of the micro-film in the lateral direction as shown in **Fig. 9.37(a)**. For each FE model, the orientations were defined six times according to the normal distribution of probability theories, thus a total of 18 CPFE analyses were carried out.

Necking and its statistical scatter are shown in **Fig. 9.38**. The necking parameter R is plotted against applied strain for micro-films with three and twelve grains through the width W. It can be observed that necking increases monotonically with applied strain and there is significant scatter for both cases. The necking envelope is defined by approximately ±50% and ±40% of the average for the three- and twelve-grain cases, respectively. This degree of scatter indicates that the necking parameter defined here is a good measure of the local orientation mismatch between grains for the polycrystalline aggregate; grain orientations are assigned randomly from one simulation to the next, so considerable scatter in the necking should occur. Greater average necking is observed for the film with fewer grains through the width, as shown in **Fig. 9.37**.

The relationship between the critical strain and the ratio of micro-film width/average grain size is shown in **Fig. 9.39** for a range of necking ratios, $R = 3\%$, 6% and 12%. Here, for a microfilm with a certain number of grains through the width, the critical strain (ε_{cr}) is defined as the strain level when a specified amount of necking takes place, e.g. 3%. The necking ratio is a function of both strain and the ratio W/D_{mean}. If, for example, a necking ratio of $R = 3\%$ is acceptable for an engineering application, the corresponding axial strain is identified as the critical strain (Wang, Zhuang, Balint and Lin, 2009a). The critical strain for a given value of R increases as the ratio W/D_{mean} increases.

The critical strain also increases with the level of allowable localised necking. For a given level of allowable necking, macro-mechanics-based FE analysis techniques can be used for the application if the maximum strain is below the critical strain. If the maximum strain in the application is above the critical strain curve (see **Fig. 9.39**), the CPFE analysis method must be used. Otherwise, the error introduced to the analysis would be too large. Such a map as depicted in **Fig. 9.39** may serve as a general recommendation for when the size effect becomes significant and as a criterion for determining which simulation techniques to use in the forming of micro-film parts, or the application of crystal plasticity FE methods for particular applications.

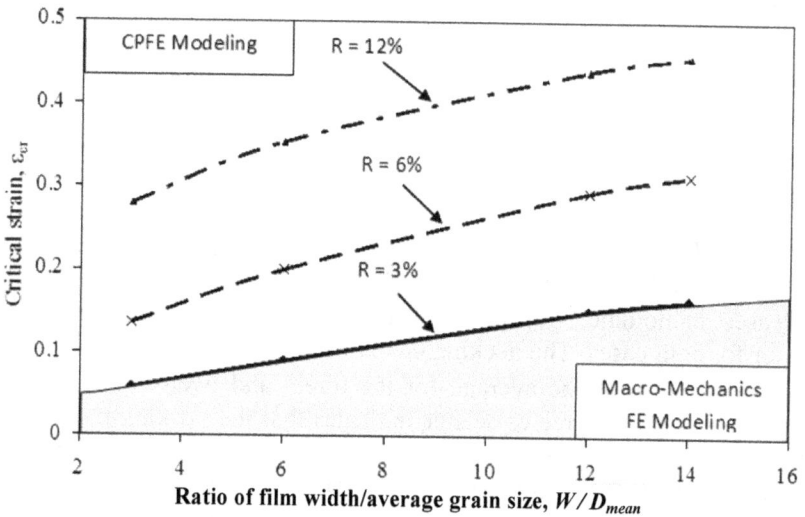

Fig. 9.39 Relationship between engineering strain and the ratio of film width to average grain size for different necking ratios (Wang, Zhuang, Balint and Lin, 2009a).

9.6.2 *Extrusion of micro-pins*

The quality of extruded micro-pins is affected by the grain size, grain orientation and grain distributions of the material and the geometric defects may be captured using conventional continuum-based FE forming simulation techniques. The established CPFE technique can be used for the process modelling. **Fig. 9.40(a)** shows the extruded micro-pin of 0.57 mm in diameter with the average grain size of 211 μm (Krishnan, Cao and Dohda, 2007). There are about 2 ~ 3 grains across the diameter of the micro-pin and uncontrollable bending and curvature of the extruded micro-pins are experimentally observed for the extruded pins (Krishnan, Cao and Dohda, 2007).

Grain size 211 μm

(a) **Extruded micro-pins (Cao, Krishnan, Wang, *et al.*, 2004)**

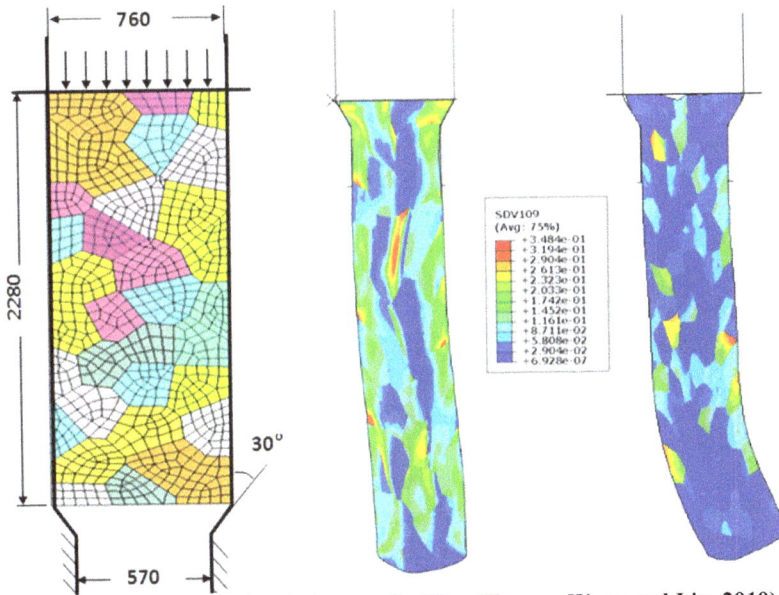

(b) **CPFE model and simulation results (Cao, Zhuang, Wang and Lin, 2010)**

Fig. 9.40 Comparison of micro-pins extruded (a) experimentally and (b) computationally. Dimensions of the FE model are in μm. The colour of the CPFE extruded micro-pins represents the shear strain.

Numerical investigations have been carried out using the developed integrated CPFE simulation system. Random distributed grains and their orientations with an average grain size of 211 μm were generated using the VGRAIN system (**Fig. 9.40(b)**). The dimensions of the work-piece and the die, shown in **Fig. 9.40 (b)**, are the same as those given by Krishnan, Cao and Dohda (2007), apart from the fact that a plane strain analysis model is created here for simplicity. The FE mesh was created using ABAQUS/CAE with quad-dominated elements (CPE4R).

A displacement of 2280 μm is applied on the extrusion punch. A friction coefficient of 0.1 between the die and the work-piece is used for the lubricated cold extrusion. The CPFE model is shown in **Fig. 9.40 (b)**, where the grain orientation is assigned randomly. Although the CPFE model is based on the plane strain description, which is a simplified model and different from the extrusion of circular micro-pins, similar features of geometric variations of physically and virtually extruded micro-pins can be observed.

Figure 9.40 (b) also shows the virtually "formed" micro-pins and the contour of cumulative shear strain distribution for two CPFE analysis results. The same FE model and grain structure are used for both FE analyses and the only difference is that the grain orientations were assigned to the grains using the same probability theories twice. Thus the grain orientations are different from the two CPFE models. It can be seen clearly that geometric errors for the extruded micro-pins are different. This could confirm the result obtained experimentally that if the ratio of the diameter of micro-pins and the grain size of the material is small, then uncontrollable bending and curvature of extruded micro-pins are the major geometric defects.

The CPFE analysis results demonstrate the validity of the developed CPFE tools in capturing the grain size and grain orientation effects in forming micro-pins. It can also be observed from the figure that the maximum cumulative shear strains occur locally along grain boundaries, which is induced by strong mismatches of orientations among grains. The uncontrollable curvature feature (**Fig. 9.40**) cannot be modelled if macro-mechanics based FE analysis is used. This clearly shows the grain size effect during the micro-forming process and CPFE needs to be employed to predict such feature.

Figure 9.41(a) shows the same size of micro-pins defined in **Fig. 9.40** and extruded with the same material, but with the smaller average grain size, 32 μm. In this case, there are about 20 grains across the cross-section of the micro-pin and straight pins were extruded (Krishnan, Cao and Dohda, 2007). The corresponding CPFE analysis was also carried out for the fine grain size material and the simulation result, as shown in **Fig. 9.41 (b)**, shows that the virtual extruded pin is straight (Cao, Zhuang, Wang and Lin, 2010). In this case, macro-mechanics simulation techniques can be used and the size effect on the geometry of formed parts is not that obvious.

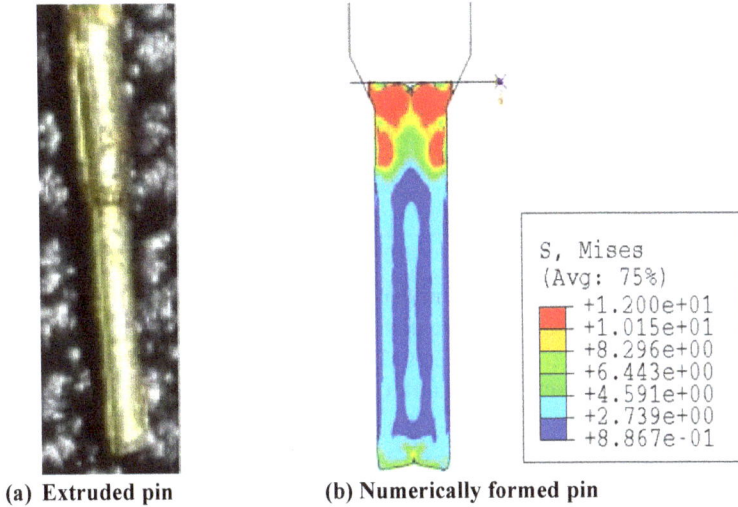

(a) Extruded pin   (b) Numerically formed pin

Fig. 9.41 Comparison of micro-pins extruded (a) experimentally (Cao, Krishnan, Wang, *et al.*, 2004) and (b) computationally (Cao, Zhuang, Wang and Lin, 2010) for the same dimensions of the pin shown in Fig. 9.40 with the average grain size of 32 μm.

9.6.3 *Hydroforming of micro-tubes*

a. Plane strain CPFE model

Figure 9.42 (a) shows some examples of hydroformed stainless steel micro-tubes having a final outer diameter of 1030 μm from a tube blank of 800 μm with wall thickness of 40 μm. Many tubes were formed using the same hydroforming parameters and it was found that rupture took place at different locations as a result of localised necking (Zhuang, Wang, Lin *et al.*, 2012). Further microstructure examination was carried out for the material and it was found that there were only about two grains across the thickness section of the tube, as shown in **Fig. 9.42 (b)**. Crystal plasticity FE analysis is used to investigate the random failure features for the forming of micro-tubes.

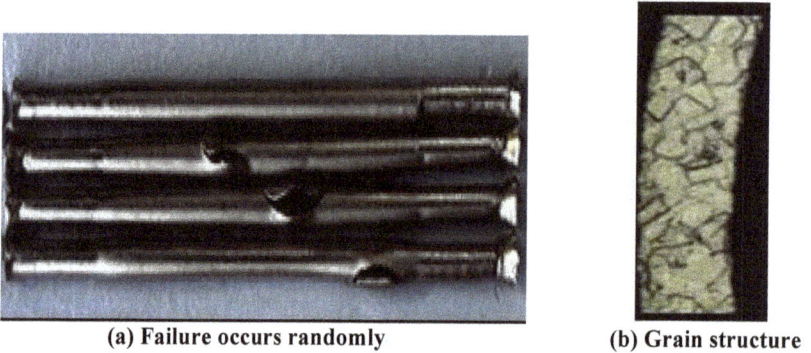

(a) Failure occurs randomly   (b) Grain structure

**Fig. 9.42 (a) Hydroformed micro-tubes with random failure locations
and (b) the typical grain size distribution across the section of the tube
(Zhuang, Wang, Lin *et al.*, 2012). The original wall thickness of the
tube is 40 μm and the outer diameter of the tube is 800 μm.**

The geometry and the FE model of the cross-section of a micro-tube
are shown in **Fig. 9.43**. In the CPFE model, a quarter section of the
micro-tube was considered with symmetry boundary conditions along
$X = 0$ and $Y = 0$. The minimum, average and maximum grain sizes of the
material were taken to be 25 μm, 30 μm and 40 μm, respectively, and
95% of the grains are within that range. The grains and their orientations
are generated using the VGRAIN system, and they are read into
ABAQUS/CAE for further mesh generation, boundary condition
specification and loading definition. The die is defined as a rigid part in
ABAQUS/CAE. The maximum applied pressure is 400 MPa; this high
pressure ensures that the work-piece is deformed sufficiently to come
into complete contact with the die. A friction coefficient of 0.1 is used
when the work-piece and the die are in contact during the forming
process. For simplicity, a 2D plane strain CPFE analysis was carried out
here.

It is worth mentioning that large-strain, polycrystal CPFE analyses
require considerable CPU time. A 2D simplification (e.g. plane strain)
reduces the computation time significantly and it is still possible to
capture the interesting features, such as localised thinning, failure etc., as
observed in hydroforming of micro-tubes. The CPFE process modelling
technique can be readily used for 3D hydroforming simulations if the 3D
grain structures can be constructed effectively for the initial metal tube.

The material model used here has been introduced in Section 9.6.1.

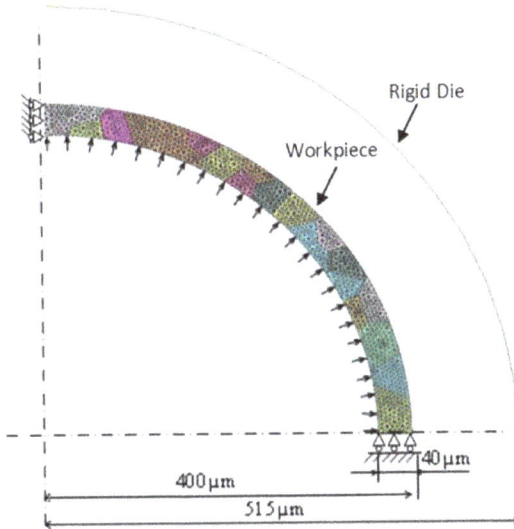

Fig. 9.43 CPFE model for tube hydroforming (Zhuang, Wang, Cao *et al.*, 2010). Grains and grain boundaries are shown in the figure; different colours indicate different orientations.

b. Grain orientation effects of single crystals

To investigate the localised thinning features associated with the grain orientation and the applied load, a single-crystal cross-section model (single grain, uniform orientation) was employed (Zhuang, Wang, Lin *et al.*, 2012). **Figure 9.44 (a)** shows a single FCC crystal with two sets of coordinate systems, the cubic crystal system and the global sample coordinate system, *X-Y-Z*. An assumption is made that the cubic crystal system may undergo only one rotation relative to the global system, around the *Z*-axis; this is depicted in **Fig. 9.44 (a)** where one of the potential slip systems, {111}<110>, is also indicated for reference. Hence, the effect of arbitrary rotations is not considered; the rotation defined by the one angle θ is sufficient to illustrate the effect of grain orientation on necking.

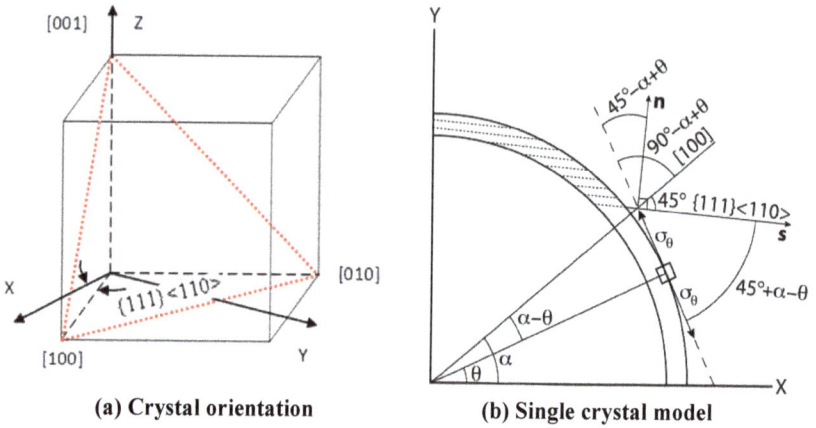

(a) Crystal orientation   **(b) Single crystal model**

Fig. 9.44. (a) FCC single crystal with one rotation relative to the micro-tube (global) coordinate system, about the Z-axis; (b) cross-section model of hydroforming a micro-tube with the slip direction represented by dotted lines; α gives the [100] direction, and θ a location on the tube in the global coordinate system (Zhuang, Wang, Lin *et al.*, 2012).

Figure 9.44 (b) illustrates the grain texture (in the global coordinate system) relative to a position on the tube specified by the angle θ. The direction of the hoop stress at position θ, σ_θ, the main driving force for forming the part, follows the tangent direction of the tube cross-section (thin tubes are considered here), whereas the crystal orientation remains unchanged over the tube cross-section. The <110>{111} slip system is indicated by the dotted lines, and the orientation of this slip system is specified relative to the [100] direction; the projection of the slip plane normal onto the XY-plane, **n**, and the slip direction, **s**, are indicated. It should be noted that the angle between the hoop stress direction and the projection of the slip plane normal onto the XY-plane is the same as that between the hoop stress direction and the slip plane normal itself, which is skewed in space relative to the XY-plane; writing

$$\mathbf{n} = \mathbf{N} - (\mathbf{N} \cdot \hat{\mathbf{K}})\hat{\mathbf{K}},$$

where $\mathbf{N}$ is the normal to the slip plane and $\hat{\mathbf{K}}$ is the unit vector in the Z-direction, it is clear that the scalar product between any vector in the XY-plane with **n**, hence the angle between them, is equal to the scalar product between that vector and $\mathbf{N}$. Therefore, calculations of Schmid factor are correct when using the projection of the normal, as done here. Henceforth, **n** is simply referred to as the normal to the slip plane.

The Schmid factor for crystal orientation α at location θ on the tube is $\cos(\pi/4 - \alpha + \theta)\cos(\pi/4 + \alpha - \theta)$. Thus, the location where plastic deformation occurs most readily on the <110>{111} system, the location of maximum resolved shear stress, occurs where $\theta = \alpha$, which gives $\tau^\alpha = \tau^{max} = \sigma_\theta/2$. The most difficult location to deform plastically is where $|\theta - \alpha| = \pi/4$, which gives $\tau^\alpha = \tau^{min} = 0$; in other words, the location most difficult to deform plastically is 45° from the location that is easiest to deform plastically. It should be noted that τ^α on the <110>{111} system at a certain location does not imply plastic deformation does not occur; other slip systems out of the plane may be activated instead, but plastic deformation will in all cases be less than it is at a location where τ^α is greater on the <110>{111} system. For example, if $\alpha = 45°$, then the greatest plastic deformation will occur at $\theta = 45°$ and the least at $\theta = 0°$ and $\theta = 90°$.

Since the crystal orientation is constant over the tube cross-section, the direction of the hoop stress varies relative to the slip direction and it can be envisaged that necking will occur at the position where the resolved shear stress is maximum, which will be at the location where the [100] and/or [010] directions are perpendicular to the local direction of the hoop stress. The only case, for the quarter model, in which both [100] and [010] are perpendicular to the local hoop stress direction is $\alpha = 0°$; for positive α, the only location is that indicated by the [100] direction, and for negative α, the only location is that indicated by the [010] direction; in the case of a polycrystal, there may be many locations where necking occurs in the quarter model. The quarter model with symmetry conditions assumes a somewhat unnatural symmetry of the necking locations since for any $\alpha > 0°$, the [010] direction is effectively neglected and the slip in the other quadrants occurs by symmetric reflections of that in the first quadrant about the global coordinate axes. However, the necking locations are separated such that they will not appreciably interact, and in any case, the quarter model assumption has no influence on the qualitative behavior demonstrated in the simulations, which is the purpose of this study.

Numerical investigations were carried out with a grain orientation of $\alpha = 60°$. A single crystal was used for the tube cross-section and was analysed using the loading and boundary conditions described above. **Figure 9.45** shows the deformed tube cross-section superimposed with contours of accumulated plastic shear strain. It can be observed, as described above, that the thickness of the formed tube varies due to the variation in the direction of the hoop stress.

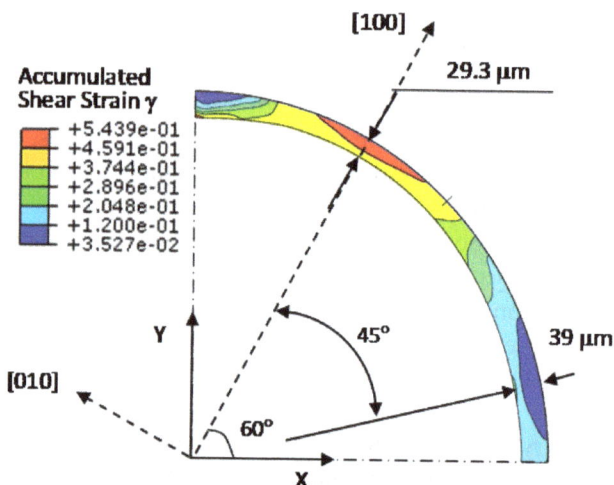

Fig. 9.45 Predicted localised necking for a single crystal structure with a grain orientation of 60° with respect to the sample (global) coordinate system (Zhuang, Wang, Lin *et al.*, 2012).

The necking ratio, which is defined as the minimum thickness divided by the maximum thickness, is 0.74 for the simulation shown in **Fig. 9.45**. The thickness of the formed part for the single crystal structure gradually varies from the minimum thickness section (29.3 μm) at the position indicated by the crystal axis [100] to the maximum thickness section (39 μm), where the slip direction <110>{111} (see **Fig. 9.44 (a)**) and the hoop stress are perpendicular (hence the resolved shear stress on the <110>{111} system is zero), which is the position 45° clockwise from the crystal axis [100]. The change in thickness for the formed tube is mainly due to the variation of the angle between the slip system and the local hoop stress direction, although the hardening of a material would affect the localised thinning behavior.

c. Random thinning locations in polycrystalline structures

Polycrystalline structures and grain orientations are generated using the VGRAIN system automatically as shown in **Fig. 9.43**. To simulate the deformation and thinning behaviour of hydroformed micro-tubes taken from the same piece of original material, the grain structures are generated twice using the same microstructure control parameters defined above, with grain orientations for both micro-tube models assigned randomly based on the orientation probability distribution within the VGRAIN system, characterised by the maximum, minimum and average grain sizes (Zhuang, Wang, Lin *et al.*, 2012). Grain structures and grain orientations may be different between the two CPFE models, but they are within the ranges of the material specification.

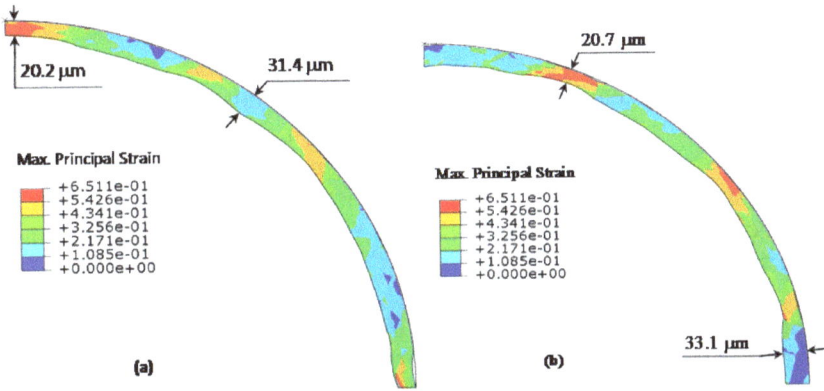

Fig. 9.46 Comparison of thinning features for microstructures generated twice by the VGRAIN system using the same control parameters (Zhuang, Wang, Lin *et al.*, 2012).

The simulation results for the hydroformed micro-tubes are shown in **Fig. 9.46**. It can be seen that the minimum and maximum values of the wall thickness of the formed tubes, shown in **Fig. 9.46 (a)** (20.2 μm and 31.4 μm, respectively) and **Fig. 9.46 (b)** (20.7 μm and 33.1 μm, respectively), are different for the two cases studied. The wall thickness of the two hydroformed micro-tubes having random grain orientations is non-uniform and difficult to predict. This is a result of the spatial variation of grain size, grain orientation and the constraints neighbouring grains with different orientations apply to one another; this cannot be captured using conventional macro-mechanics FE techniques. It can also be observed that localised thinning occurs at different locations due to the randomness of the grain orientations of the work-piece material, which is difficult to control in practice, unless using a material with sufficient fine grain size for the deformation ratio in the forming process.

9.6.4   Compression of micro-pillars

a.   CPFE model for micro-pillar compression

Figure 9.47 shows a tapered cylinder micro-pillar with a height of 180 μm, bottom radius of 30 μm and angle of 2°. A displacement of $U = 54$ μm (equivalent to 0.3 engineering strain) was applied on the upper face of the micro-pillar in all the simulations; the bottom face was fixed and the remaining faces were free of constraint. Two sets of simulations were conducted, one of which focused on the influence of the grain structure's regularity, i.e. the grain size distribution characteristics, on the compression deformation, and the other concentrated on the effect of grain size on the micro-pillar's inhomogeneous behaviour (Zhang, Karimpour, Balint and Lin, 2012b).

Equations (9.12)–(9.13) detailed in Section 9.2 are used here. The material constants were calibrated for a free cutting tool steel (Karimpour, 2012) for a temperature of 1100 °C and are listed in **Table 9.6**. Young's modulus and Poisson's ratio are 6.06 GPa and 0.3, respectively.

Table 9.6 Material constants for Eqs. (9.12)–(9.13) for a free cutting steel deforming at 1100 °C (Karimpour, 2012).

n	$\dot{a}(s^{-1})$	h_0 (MPa)	τ_s (MPa)	τ_0 (MPa)
3	10	33	150	23

b.   Grain structure regularity on deformation

In this set of simulations, three specimens with different grain structure regularities were generated. As shown in **Table 9.7**, the specimen R-1 is a purely random grain structure, i.e. $\alpha = 0$, the specimen R-2 has regularity $\alpha = 0.4$, and specimen R-3 has the most regular grain structure, $\alpha = 0.8$. The physical parameters used to generate the three grain structures are presented in **Table 9.7**, while the implementation parameters derived by the CPVT model are also given. Note that the grain size input to the CPVT model is in terms of grain volume, while the equivalent grain size assumed a spherical shape.

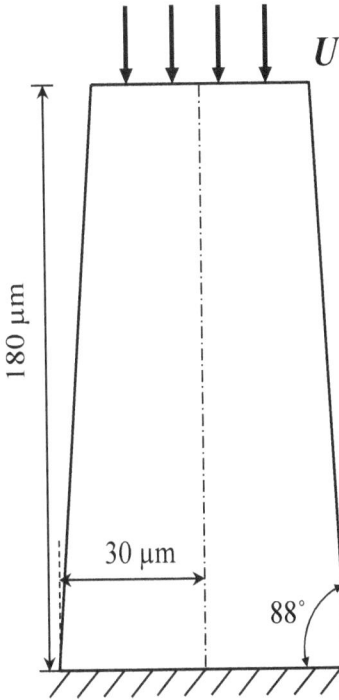

Fig. 9.47 Schematic diagram for the CPFE simulation of micro-pillar compression (Zhang, Karimpour, Balint and Lin, 2012b).

The undeformed micro-pillars with the specified regularities are shown in **Fig. 9.48**. It should be mentioned that repeated use of the same physical parameters using the CPVT model will result in slightly different grain structures, but all the virtual grain structures are statistically consistent with the specified physical parameters. The grain structures shown in **Fig. 9.48** also have crystallographic orientations, assigned by the VGRAIN system using a random number generator based on a uniform distribution.

Table 9.7 Physical parameters and corresponding grain structure properties for the two set of simulations (Karimpour, 2012).

Label	Physical parameters (μm^3)				Mean grain diameter† (μm)	CPVT model parameters			
	D_{mean}	D_L	D_R	P_r (%)	$\tilde{D}_{mean}$	d_{reg} (μm)	α	δ (μm)	N
R-1	14140	7070	21210	76.6	30	26.4	0.0	0	46
R-2	14140	9900	18380	60.3	30	26.4	0.4	10.5	46
R-3	14140	11310	16970	87.6	30	26.4	0.8	21.1	46
S-1	8180	5730	10630	60.3	25	22.0	0.4	8.8	78
S-2	4190	2930	5450	60.3	20	17.6	0.4	7.0	154
S-3	1770	1240	2300	60.3	15	13.2	0.4	5.3	361

† Assuming a spherical grain shape.

The deformed specimens shown in **Fig. 9.48** are demonstrations of regularity and grain size control in three dimensions, and compare the different deformation characteristics resulting from different grain size distribution properties. The three specimens have an identical mean grain size of 30 μm. The contours of accumulated shear strain show that the degree of plastic strain localisation is lower for sufficiently regular grain structures; the maximum value of γ is lower for the $\alpha = 0.8$ case, relative to the irregular grain structures, $\alpha = 0$ and $\alpha = 0.4$. The axial deviation is also lower for $\alpha = 0.8$ than it is for either $\alpha = 0$ or $\alpha = 0.4$, although there is inherent variability depending on the details of the grain orientation distribution.

Irregular grain structures have an inherently larger variability in deformation features, such as axial deviation in this case. For a given average grain size, a higher irregularity leads to a larger maximum grain size which may or may not dominate the deformation in a given cross-section depending upon its orientation, which is randomly assigned. In contrast, a regular grain structure has a more homogeneous grain size distribution, which precludes the occurrence of larger grains that may accommodate severe slip localisation if appropriately oriented, i.e. having a high Schmid factor for the boundary conditions.

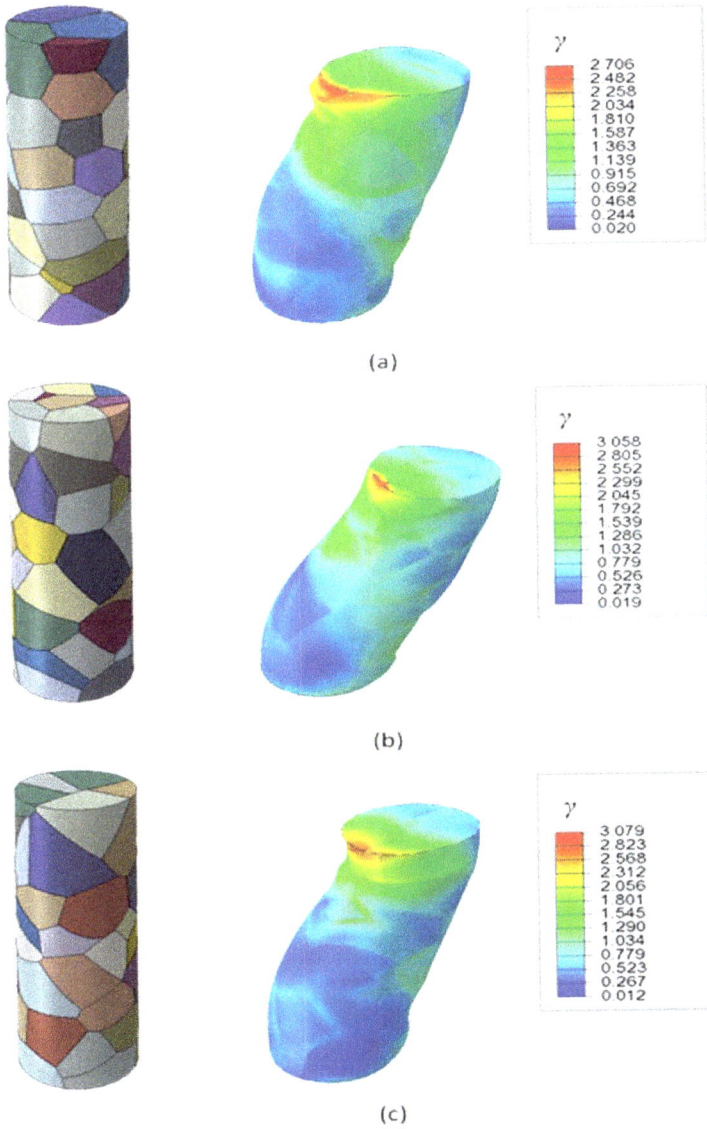

Fig. 9.48 CPFE simulations of micro-pillar compression for regular and irregular grain structures, for (a) $\alpha = 0.8$ (b) $\alpha = 0.4$ and (c) $\alpha = 0$ (Zhang, Karimpour, Balint, Lin, 2012b).

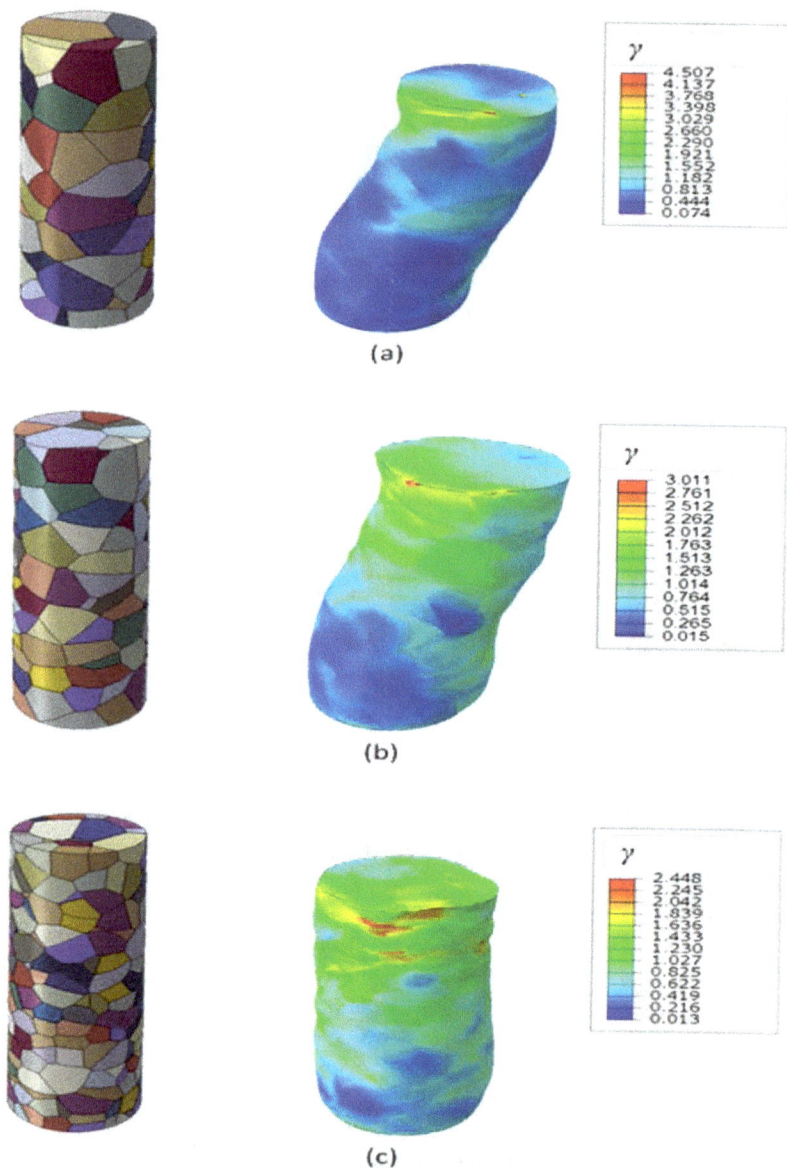

Fig. 9.49 CPFE simulations of micro-pillar compression for grain structures with different grain sizes for (a) $d = 25$ μm, (b) $d = 20$ μm and (c) $d = 15$ μm (Zhang, Karimpour, Balint, Lin, 2012b).

Figure 9.49 shows micro-pillar specimens with different mean grain sizes and the same regularity ($\alpha = 0.4$, moderately irregular); the normalised grain size distributions are statistically equivalent. Axial deviation of the deformed sample increases with increasing mean grain size for a given regularity, and a material with a smaller mean grain size exhibits a more homogenous deformation. This is clearly shown by the simulation in **Fig. 9.49 (c)** where the maximum value of γ is considerably lower and the cylinder has maintained its shape and not deviated axially after deformation. The grain size effects on deformation features are clearly shown in the CPFE simulation results.

Appendix A

Sets of Unified Constitutive Equations

For the convenience of the discussion in the book, especially in Chapters 7 and 8, we list a few sets of unified viscoplastic constitutive equations covering different applications. The dot "." within the equations indicates differentiation in terms of time, for example $\dot{\varepsilon}^p = d\varepsilon^p/dt$. Very brief descriptions are given to individual sets of equations. Units of individual equations are also given here for the convenience of discussion in the book, particularly in Chapter 7, to illustrate the unit problems in numerical integration and in the formulation of objective functions for optimisation.

A.1 SET I: Basic Elastic-viscoplastic Constitutive Equations

During hot/warm metal forming processes, the work-piece/material is deformed viscoplastically. To model the elastic viscoplastic deformation with the consideration of the strain hardening and strain rate hardening, the following set of constitutive equations can be used for a range of metals and alloys deforming at an elevated temperature without the consideration of recrystallisation, grain size evolution and damage due to plastic deformation (Lin and Dean, 2005). The formulation of the individual evolutionary equations is given in Chapter 5.

Plastic strain rate (s^{-1}): $\qquad \dot{\varepsilon}^p = \left[(\sigma - R - k)/K \right]^n \qquad$ (I.1)

Dislocation density rate (s^{-1}): $\qquad \dot{\bar{\rho}} = A(1 - \bar{\rho})\dot{\varepsilon}^p - C\bar{\rho}^{\gamma_0} \qquad$ (I.2)

Stress rate (MPa·s^{-1}): $\qquad \dot{\sigma} = E\left(\dot{\varepsilon}^T - \dot{\varepsilon}^p \right) \qquad$ (I.3)

where the isotropic hardening R is directly related to the normalised dislocation density $\bar{\rho}$, and can be expressed in an explicit form:

$$R = B\bar{\rho}^{1/2}.$$

are k, K, n, A, C, γ_0 and B are material constants to be determined from experimental data. The normalised dislocation density $\bar{\rho}$ varies from 0 (the initial state of the material) to 1.0 (the saturated dislocation network for the material), and is used to model the dislocation hardening, also known as strain hardening. The equation set has been determined for a micro-alloyed air-quenching steel in warm/hot forming conditions. Young's modulus, $E = 110$ GPa at a temperature of 850 °C. The material constants within the equations are listed in **Table A.1**.

Table A.1 List of material constants for Equation SET I (Cao, Lin and Dean, 2008a).

k (MPa)	K (MPa)	n (-)	A (-)	C (s^{-1})	γ_0 (-)	B (MPa)
0.34	51.67	1.31	10.49	0.25	1.01	200.18

A.2 SET II: Creep Damage Constitutive Equations

Continuum damage mechanics with two damage variables was verified and used to model tertiary creep softening in an aluminium alloy caused by grain boundary cavity nucleation and growth, and ageing of particular microstructures. The form of the constitutive equations is given as (Kowalewski, Hayhurst and Dyson, 1994):

Creep strain rate (h^{-1}):
$$\dot{\varepsilon} = \frac{A}{\left(1 - \omega_2\right)^n} \sinh\left(\frac{B\sigma(1 - H)}{1 - \phi}\right) \quad \text{(II.1)}$$

Primary hardening rate (h^{-1}):
$$\dot{H} = \frac{h_0}{\sigma}\left(1 - \frac{H}{H^*}\right)\dot{\varepsilon} \quad \text{(II.2)}$$

Ageing softening rate (h^{-1}):
$$\dot{\phi} = \frac{K_c}{3}\left(1 - \phi\right)^4 \quad \text{(II.3)}$$

Damage rate (h^{-1}):
$$\dot{\omega}_2 = D\dot{\varepsilon}, \quad \text{(II.4)}$$

where A, B, h_0, H^*, K_c and D are material constants, and n is given by

$$n = \frac{B\sigma(1 - H)}{1 - \varphi}\coth\left(\frac{B\sigma(1 - H)}{1 - \varphi}\right).$$

A state variable H is introduced to model the hardening of the material at the primary creep stage. Again, this is the same as the dislocation hardening even if they are expressed using different types of equations. H^* is the saturated value of the variable H. Equation (II.3) models the particle coarsening due to ageing, which softens the aluminium alloy, but does not lead to the failure of the material. More detailed descriptions for the creep damage constitutive equations are given by Kowalewski, Hayhurst and Dyson (1994). The equation set has been determined for an aluminium alloy at 150 °C and the material constants within the equations are listed in **Table A.2**.

Table A.2 Material constants for Equations SET II (Kowalewski, Hayhurst and Dyson, 1994).

$A\,(\mathrm{h}^{-1})$	$B\,(\mathrm{MPa}^{-1})$	$h_0\,(\mathrm{MPa})$	$H^*\,(\text{-})$	$K_c\,(\mathrm{h}^{-1})$	$D\,(\text{-})$
4.04e-15	0.11	2.95e4	0.11	1.82e-4	2.75

A.3 SET III: Unified Viscoplastic-damage Constitutive Equations

During warm metal forming processes, the work-piece/material is deformed viscoplastically. To model the elastic viscoplastic deformation and material hardening, the following set of constitutive equations is proposed (Foster, 2006) and determined for a micro-alloyed steel:

Plastic strain rate (s⁻¹):
$$\dot{\varepsilon}^p = \left[\frac{\sigma - (R+k)(1-D)}{K} \right]^n \cdot (1-D)^{-\gamma_1} \quad \text{(III.1)}$$

Dislocation density rate (s⁻¹):
$$\dot{\rho} = A(1-\rho)\left|\dot{\varepsilon}^p\right| - C\rho^{\gamma_2} \quad \text{(III.2)}$$

Isotropic hardening rate (MPa·s⁻¹):
$$\dot{R} = \frac{1}{2}B\rho^{-1/2}\dot{\rho} \quad \text{(III.3)}$$

Damage rate (s⁻¹):
$$\dot{D} = \beta\frac{1-\tanh\left(-\eta\dot{\varepsilon}_p\right)}{(1-D)^{\gamma_4}}\dot{\rho}^{\gamma_3} \quad \text{(III.4)}$$

Stress rate (MPa·s⁻¹):
$$\dot{\sigma} = E\left(\dot{\varepsilon} - \dot{\varepsilon}^p\right), \quad \text{(III.5)}$$

where $E = 110\,\mathrm{GPa}$ is Young's modulus, k, K, n, γ_1, A, C, γ_2, B, β, γ_3, η and γ_4 are material constants, which have been determined at three different strain rates $\dot{\varepsilon} = 0.01\,\mathrm{s}^{-1}$, $0.1\,\mathrm{s}^{-1}$ and $1.0\,\mathrm{s}^{-1}$ for a micro-alloyed quenchable steel at a temperature of 850 °C and are given in **Table A3**.

Table A.3 Material constants for Equations SET III (Cao and Lin, 2008).

k(MPa)	K(MPa)	n(-)	γ_1(-)	A(-)	C(-)
2.30e-3	89.90	4.55	2.99	2.19	4.26e-3
γ_2(-)	B(MPa)	β(-)	γ_3(-)	η(-)	γ_4(-)
0.38	2.60e2	0.11	0.77	3.30	7.63

A.4 SET IV: Mechanism-based Unified Viscoplastic Constitutive Equations

A set of mechanism-based unified viscoplastic constitutive equations is presented below to describe the kinetics of grain size, dynamic/static recrystallisation and dislocation density evolution, that enables the prediction of microstructure evolution of metals and alloys during and after hot deformation. It is particularly suitable for hot compressive forming applications, where the failure (or damage) of materials in the viscoplastic deformation is not an issue to consider. It has been used for hot rolling of steels by Lin, Liu, Farrugia *et al.* (2005).

Plastic strain rate (s^{-1}):

$$\dot{\varepsilon}^p = A_1 \sinh\left[A_2\left(\sigma - R - k\right)\right] \cdot d^{-\gamma_4} \tag{IV.1}$$

Recrystallisation rate (s^{-1}):

$$\dot{S} = Q_0 \cdot \left[x\bar{\rho} - \bar{\rho}_c\left(1 - S\right)\right] \cdot \left(1 - S\right)^{N_q} \tag{IV.2}$$

Onset of recrystallisation rate (s^{-1}):

$$\dot{x} = A_0\left(1 - x\right)\bar{\rho} \tag{IV.3}$$

Dislocation density rate (s^{-1}):

$$\dot{\bar{\rho}} = \left(\frac{d}{d_0}\right)^{\gamma_d}\left(1 - \bar{\rho}\right)\left|\dot{\varepsilon}^p\right| - c_1\bar{\rho}^{c_2} - c_3\frac{\bar{\rho}}{1 - S}\dot{S} \tag{IV.4}$$

Isotropic hardening rate (MPa s^{-1}):

$$\dot{R} = B\dot{\bar{\rho}} \tag{IV.5}$$

Grain size rate (μm·s^{-1}):

$$\dot{d} = \alpha_0 d^{-\gamma_0} - \alpha_2 \cdot \dot{S}^{\gamma_3} \cdot d^{\gamma_2} \tag{IV.6}$$

Stress rate (MPa·s^{-1}):

$$\dot{\sigma} = E \cdot \left(\dot{\varepsilon} - \dot{\varepsilon}^p\right) \tag{IV.7}$$

where $E = 100$ GPa is Young's modulus and A_1, A_2, k, γ_4, Q_0, $\bar{P}_c$, N_q, c_1, c_2, c_3, d_0, γ_d, A_0, B, α_0, γ_0, α_2, γ_2 and γ_3 are material constants, which are determined by Lin, Liu, Farrugia *et al.*, (2005) from experimental data of a C-Mn steel reported by Medina and Hernandez (1996a; 1996b). The initial average grain size is 189 μm and the tests were carried out at 1100 °C. The determined material constants within the equations are given in **Table A.4**. More discussion and the application of the equations for the modelling of hot rolling are given as case studies in Chapter 8.

Table A.4 Material constants for equation SET IV (Lin, Liu, Farrugia *et al.*, 2005).

A_1 (s^{-1})	A_2 (MPa^{-1})	γ_4 (-)	Q_0 (-)	$\bar{P}_c$ (-)	N_q (-)
1.81×10^{-6}	3.14×10^{-1}	1.00	30.00	1.84×10^{-1}	1.02

c_1 (-)	c_2 (-)	c_3 (-)	d_0 (μm)	γ_d (-)	A_0 (-)
16.00	1.43	8.00×10^{-2}	36.38	1.02	40.96

B (MPa)	α_0 (μm)	γ_0 (-)	α_2 (μm)	γ_2 (-)	γ_3 (-)
75.59	1.44	3.07	78.68	1.20×10^{-1}	1.06

A.5 SET V: Unified Viscoplastic-damage Constitutive Equations

A set of mechanism-based unified viscoplastic-damage constitutive equations developed by Liu (2004) and Lin, Foster, Liu *et al.* (2007) is given here for the modelling of the kinetics of grain size, dynamic/static recrystallisation, dislocation density and micro damage evolution. Particularly, the effect of grain size and deformation rate on grain boundary damage has been modelled in the equation set. These equations are useful for hot forming applications, where the failure and damage of the material is an important issue to investigate. The equation set has been used for the modelling of edge cracking in hot rolling of a leaded free cutting steel (Liu, 2004).

Plastic strain rate (s^{-1}):

$$\dot{\varepsilon}^P = A_1 \left[\frac{\sigma}{1 - D_{gb}} - (R + k) \cdot \left(1 - D_{pi}\right) \right]^{A_2} \cdot d^{-\gamma_1} \qquad (V.1)$$

Recrystallisation rate (s^{-1}):

$$\dot{S} = H \cdot \left[x\bar{\rho} - \rho_c (1 - S) \right] \cdot (1 - S)^{\gamma_s} \qquad (V.2)$$

Dislocation density rate (s^{-1}):

$$\dot{\bar{\rho}} = \left(\frac{d}{d_0}\right)^{\gamma_d} (1-\bar{\rho})|\dot{\varepsilon}^p| - c_1\bar{\rho}^{\gamma_2} - c_2 \frac{\bar{\rho}}{1-S}\dot{S} \tag{V.3}$$

Onset of recrystallisation rate (s^{-1}):

$$\dot{x} = A_3(1-x)\bar{\rho} \tag{V.4}$$

Isotropic hardening rate (MPa·s^{-1}):

$$\dot{R} = \frac{1}{2} \cdot B \cdot \bar{\rho}^{-1/2} \cdot \dot{\bar{\rho}} \tag{V.5}$$

Grain size rate (µm·s^{-1}):

$$\dot{d} = a_1 d^{-\gamma_3} - a_2 \cdot |\dot{S}| \cdot d^{\gamma_5} \tag{V.6}$$

Stress rate (MPa·s^{-1}):

$$\dot{\sigma} = E \cdot (\dot{\varepsilon}_T - \dot{\varepsilon}_p) \tag{V.7}$$

Total damage rate (s^{-1}):

$$\dot{D}_T = \dot{D}_g + \dot{D}_p \tag{V.8}$$

where:

$$\dot{D}_g = a_4 \cdot \eta \cdot \left\{ a_7 (1-D_g)\dot{\bar{\rho}} + \left[1/(1-D_g)^{n_3} - (1-D_g) \right] \cdot |\dot{\varepsilon}_p| \right\}$$

$$\eta = \exp\left(-a_5(1-d/d_c)^2\right)$$

$$d_c = a_6 \left(|\dot{\varepsilon}^p|\right)^{-n_1}$$

$$\dot{D}_p = a_9 \cdot \left[(1-D_p)\dot{\bar{\rho}} + a_8 \cdot (D_p d)/(1-D_p)^{n_5} \cdot \left(|\dot{\varepsilon}^p|^{n_6}\right) \right]$$

The total damage D_T is the sum of grain boundary (creep type) damage D_g and plasticity-induced damage D_p, and, the parameters η and d_c are used to express the effect of grain size and deformation on grain boundary damage, which have been introduced in Chapter 6. $E = 110$ GPa is Young's modulus. A_1, A_2, k, γ_1, H, $\bar{\rho}_c$, γ_s, d_0, γ_d, c_1, γ_2, c_2, A_3, B, a_1, a_2, γ_3, γ_5, a_4, n_1, a_5, a_6, a_7, n_3, a_9, n_5 and n_6 are material constants, which can be temperature-dependent as well. The equation set has been determined for a free-machining steel at 1273 K. The material constants within the equations are given in **Table A.5**. A detailed discussion of the equation set is given by Lin Foster, Liu *et al.* (2007).

Table A.5 Material constants for Equation SET V (Lin, Foster, Liu *et al.* 2007) .

A_1 (s^{-1})	A_2	k (MPa)	γ_1	H (s^{-1})	$\bar{\rho}_c$	γ_s	d_0 (μm)	γ_d
3.00	3.60	5.00	3.58	9.37e-2	0.10	0.50	31.67	6.00

c_1 (s^{-1})	γ_2	c_2	A_3	B (MPa)	a_1 (s^{-1})	a_2	γ_3	γ_5
8.20	1.07	10.07	6.46	146.89	7.48	1.80	5.89	1.82

a_4	n_1	a_5	a_6	a_7	n_3	a_9	n_5	n_6
0.41	0.17	6.0e-3	2.16	1.00	8.53	4.28e-2	8.53	1.08

A.6 SET 6: Temperature-dependent Unified Viscoplastic-damage Constitutive Equations

A set of unified viscoplastic damage constitutive equations for hot stamping of boron steel and aluminium alloy panels has been formed by Li, Mohamed, Cai, *et al.* (2011) and is given below:

$$\dot{\varepsilon}^P = \dot{\varepsilon}^o \cdot \left(\frac{\left| \frac{\sigma}{1-f_d} \right| - R - k}{K} \right)_+^{n_1} \cdot \frac{1}{\left(1-f_d\right)^{\gamma_1}} \tag{VI.1}$$

If $\left| \frac{\sigma}{1-f_d} \right| - R - k \leq 0$, $\dot{\varepsilon}^P = 0$; $\dot{\varepsilon}^o = \begin{cases} 1 & for \quad \sigma > 0 \\ -1 & for \quad \sigma < 0 \end{cases}$

$$\dot{\bar{\rho}} = A \cdot \left(1-\bar{\rho}\right) \cdot \left|\dot{\varepsilon}^P\right| - C \cdot \bar{\rho}^{n_2} \tag{VI.2}$$

$$\begin{cases} \text{Boron Steel}: \dot{f}_d = D_1 \cdot \dfrac{\sigma \cdot \left|\dot{\varepsilon}^P\right|}{\left(1-f_d\right)^{\gamma_2}} \\[4mm] \text{Al-alloys}: \dot{f}_d = D_1 \cdot f_d^{\gamma_2} \cdot \left|\dot{\varepsilon}^P\right|^{d_1} + D_2 \cdot \left|\dot{\varepsilon}^P\right|^{d_2} \cdot \cosh\left(D_3 \left|\varepsilon^P\right|\right) \end{cases} \tag{VI.3}$$

$$\sigma = E \cdot (1 - f_d) \cdot \left(\varepsilon^T - \varepsilon^P\right) \tag{VI.4}$$

where the material hardening R, $(R = B \cdot \bar{\rho}^{n_0})$ is directly related to dislocation density. In hot stamping processes, at the late stage of deformation, softening due to damage decreases the flow stress, which can be modelled based on void nucleation and growth mechanisms. The effective damage evolution is defined in Eq. (VI.4), where damage is 0 at the initial state of the deformation. The constants k, K, n_1, B, C, E, D_1 and D_2 are temperature-dependent parameters and A, n_2, γ_1, C, E, D_1 and D_2 are material constants. E is the Young's modulus. The following equations represent the Arrhenius formulation for temperature-dependent parameters, where κ is the universal gas constant and Q is the activation energy.

$$k = k_0 \exp\left(\frac{Q_k}{\kappa T}\right) \qquad B = B_0 \exp\left(\frac{Q_B}{\kappa T}\right) \qquad D_1 = D_{10} \exp\left(\frac{Q_1}{\kappa T}\right)$$

$$K = K_0 \exp\left(\frac{Q_K}{\kappa T}\right) \qquad C = C_0 \exp\left(-\frac{Q_C}{\kappa T}\right) \qquad D_2 = D_{20} \exp\left(\frac{Q_2}{\kappa T}\right)$$

$$n_1 = n_{10} \exp\left(\frac{Q_n}{\kappa T}\right) \qquad E = E_0 \exp\left(\frac{Q_E}{\kappa T}\right)$$

Table A.6 Material constants in viscoplastic-damage constitutive equations for Boron Steel and AA6082 (Li, Mohamed, Cai *et al.*, 2011).

Constants	A	n_2	γ_1	γ_2	d_1
Boron Steel	5.222	1.54	3.1	17.5	-
AA6082	13	1.8	0	1.2	1.0101
Constants	d_2	D_3	k_0 (MPa)	n_0	K_0 (MPa)
Boron Steel	-	-	12.4	0.4	30
AA6082	0.5	26.8	0.89	0.5	0.219
Constants	n_{10}	B_0 (MPa)	C_o	E_0 (MPa)	D_{10}
Boron Steel	0.0068	80	55500	1100	1.39e-4
AA6082	5	4.91	0.26	322.8191	10.32
Constants	D_{20}	Q_k (J/mol)	Q_K (J/mol)	Q_n (J/mol)	Q_B (J/mol)
Boron Steel	-	8400	8400	50000	8400
AA6082	5.49e-19	6679	27687.1	0	11625.8
Constants	Q_C (J/mol)	Q_E (J/mol)	Q_1 (J/mol)	Q_2 (J/mol)	κ (J/mol)
Boron Steel	99900	17500	10650	-	8.3
AA6082	3393.4	12986.7	6408.4	119804.6	8.3

The material constants within the equation sets are determined from experimental data over a range of strain rates and temperatures for a Boron steel and AA6082 respectively, and are listed in **Table A.6** (Li, Mohamed, Cai *et al.*, 2011).

References

Abramowitz, M. and Stegun, I.A. (eds.) (1972). *Handbook of Mathematical Functions with Formulas, Graphs, and Mathematical Tables*, Dover, New York.

Afshan, S. (2013). Micro-mechanical modelling of damage healing in free cutting steel, PhD thesis, Imperial College London, UK.

Aggarwal, A., Guibas, L., Saxe, J. *et al.* (1987). A linear time algorithm for computing the Voronoi diagram of a convex polygon, *Proceedings of the Nineteenth Annual ACM Symposium on Theory of Computing*, 39–45.

Ali, W.J. and Balod, A.O. (2007). Theoretical determination of forming limit diagram for steel, brass and aluminium alloy sheets, *Al-Rafidain Engineering*, **15**(1), 40–55.

Andersson, H. (1977). Analysis of a model for void growth and coalescence ahead of a moving crack tip, *Journal of the Mechanics and Physics of Solids*, **25**, 217–233.

Andrade, E.N.da C. (1914). The flow in metals under large constant stress, *Proceedings of the Royal Society London*, **90A**, 329–342.

Ashby, M.F. and Dyson, B.F. (1984). Creep damage mechanics and micro-mechanisms, *ICF Advances In Fracture Research*, Pergamon, Press, pp. 3–30.

Ashby, M.F. and Jones, D.R.H. (2005). *Engineering Materials 1: An Introduction to Properties, Applications and Design*, Elsevier Butterworth-Heinemann, Amsterdam.

Bäck, T. and Schwefel, H-P. (1993). An overview of evolutionary optimisation for parameter optimisation, *Evolutionary Computation*, **1**(1), 1–23.

Bai, Q. (2012). Development of a new process for high precision gas turbine blade forging, PhD thesis, Imperial College London, UK.

Bai, Q., Lin, J., Dean, T.A. *et al.* (2013). Modelling of dominant softening mechanisms for Ti-6Al-4V in steady state hot forming conditions, *Materials Science & Engineering A*, **559**, 352–358.

Bai, Q., Lin, J., Zhan, L., Dean, T.A., *et al.* (2012). An efficient closed-form method for determining interfacial heat transfer coefficient in metal forming, *Int. J. of Machine Tools & Manuf.*, **56**(1), 102–110.

Baker, T.J. and Charles, J.A. (1972). Deformation of MnS Inclusions in Steel, *Journal of the Iron and Steel Institute*, **210**, 680–690.

Balluffi, R. W., Allen, S. M. and Carter, W.C. (2005). *Kinetics of Materials*, Wiley, Hoboken, NJ.

Balint, D.S., Deshpande, V.S., Needleman, A. *et al.* (2008). Discrete dislocation plasticity analysis of the grain size dependence of the flow strength of polycrystals, *International Journal of Plasticity*, **24**, 2149–2172.

Banthia, V. and Mukherjee, S. (1985). On an improved time integration scheme for stiff constitutive models of inelastic deformation, *Journal of Engineering Materials and Technology*, **107**, 282–285.

Barton, N.R., Bernier, J.V., Lebensohn, R.A. *et al.* (2009). 'Direct 3D simulation of plastic flow from EBSD data', in Schwartz, Kumar, Adams *et al.* (eds), *Electron Backscatter Diffraction in Materials Science*, Springer, New York, pp. 155–167.

Bhandari, Y., Sarkar, S., Groeber, M. *et al.* (2007). 3D polycrystalline microstructure reconstruction from FIB generated serial sections for FE analysis, *Computational Materials Science*, **41**, 222–235.

Boots, B. (1982). The arrangement of cells in random networks. *Metallography*, **15**, 53–62.

Boyer, J.-C., Vidal-Sallé, E. and Staub, C. (2002). A shear stress dependent ductile damage model, *Journal of Materials Processing Technology*, **121**(1), 87–93.

Brust, F. W. and Leis, B. N. (1992). A new model for characterizing primary creep damage, *International journal of fracture*, **54**(1), 45–63.

Brozzo, P., De Luca, B. and Rendina, R. (1972). A new method for the prediction of formability limits in metal sheets, *Proceedings of the 7th Biennial International Deep Drawing Research Group Conference on Sheet Metal Forming and Formability*.

BSI, (2008). Metallic materials - Sheet and strip - Determination of forming-limit curves, Part 2: Determination of forming-limit curves in the laboratory, p. 27.

Brunet, M., Godereaux, S. and Morestin, F. (2000). Nonlinear kinematic hardening identification for anisotropic sheet metals with bending-unbending tests, *Journal of Engineering Materials and Technology*, **123**(4), 378–383.

Caraher, S.K., Polmear, I.J. and Ringer, S.P. (1998). Effects of Cu and Ag on precipitation in Al-4Zn-3Mg (wt. %), *Proceedings 6th International Conference on Aluminum Alloys (ICAA6)*, **2**, 739–744.

Cao, J. (2006). A study on the determination of unified constitutive equations, PhD thesis, University of Birmingham, UK.

Cao, J., Krishnan, N., Wang, Z. *et al.* (2004). Micro-forming: experimental investigation of the extrusion process for micropins and its numerical simulation using RKEM, *Trans. ASEM*, **126**, 642–652.

Cao, J. and Lin, J. (2008). A study on formulation of objective functions for determining material models, *Int. J. of Mech. Sci.*, **50**, 193–204.

Cao, J., Lin, J. and Dean, T.A. (2008a). An implicit unitless error and step-size control method in integrating unified viscoplastic/creep ODE-type constitutive equations, *Int. J. for Numerical Methods in Engineering*, **73**, 1094–1112.

Cao, J., Lin, J. and Dean, T.A. (2008b). User's manual for OPT-CCE (An Optimiser for Determining Constants in Constitutive Equations), *Report in Metal Forming and Materials Modelling Group*, Imperial College London.

Cao, J., Zhuang, W., Wang, S. *et al.* (2010). Development of a VGRAIN system for CPFE analysis in micro-forming applications, *Int. J. of Adv Manuf. Tech.*, **47**, 981–991.

Cao, J., Zhuang, W., Wang, S. *et al.* (2009). An integrated crystal plasticity FE system for micro-forming simulation, *J. Multiscale Modelling*, **1**, 107–124.

Chow, C.L. (2009). Anisotropic damage-coupled sheet metal forming limit analysis, *Int. J. of Damage Mechanics*, **18**(4), 371–392.

Cottingham, D.M. (1966). The hot workability of low-carbon steels, *Proceedings of The Conference on Deformation Under Hot working Conditions*, 146–156.

Cheong, B.H (2002). Modelling of microstructural and damage evolution in superplastic forming, PhD thesis, University of Birmingham, UK.

Cheong, B.H., Lin, J. and Ball A.A. (2000). Modelling of the hardening characteristics for superplastic materials, *Journal of Strain Analysis*, 35(3), 149–157.

Cheong, B.H., Lin, J. and Ball A.A. (2001). Modelling of the hardening due to grain growth for a superplastic alloy, *J. of Mat. Proc. Tech.*, 119, 361–365.

Chaboche, J.L. (1977). Viscoplastic constitutive equations for the description of cyclic and anisotropic behaviour of metals, *Bull. Acad. Pol. Sci. Ser. Sci. Tech.*, 25, 33.

Chaboche, J.L. (1987). Continuum damage mechanics: present state and future trends, *Nuclear Engineering and Design*, 105, 19–33.

Chaboche, J.L. and Rousselier, G. (1983). On the plastic and viscoplastic constitutive equations—part II: application of internal variable concepts to the 316 stainless steel, *J. Pressure Vessel Technol.*, 105(2), 159–164.

Chu, C.C. and Needleman, A. (1980). Void nucleation effects in biaxially stretched sheets, *Journal of Engineering Materials and Technology*, 102, 249–257.

Cockcroft, M.G. and Latham, D.J. (1968). Ductility and the workability of metals, *J. Inst. Met.*, 96, 6.

Cocks, A.C.F. and Ashby, M.F. (1980). Intergranular fracture during power-law creep under multiaxial stresses, *Metal Science*, 14, 395–402.

Cocks, A.C.F. and Ashby, M.F. (1982). On creep fracture by void growth, *Progress in Material Science*, 27, 189–244.

De Paoli, M. and Bennett, M. E-report: II. 4 Diffusive Mass Transfer, in *Microscopic Rheological Deformation of the Lithosphere*, [Online]. Available at: http://www.geosci.usyd.edu.au/users/prey/Teaching/Geol-3101/RheologyOne02/diffuse.htm.

Den Toonder, J., Van Dommelen, J. and Baaijens, F. (1999). The relation between single crystal elasticity and the effective elastic behaviour of polycrystalline materials: theory, measurement and computation, *Modelling and Simulation in Materials Science and Engineering*, 7(6), 909–928.

Deschamps, A., Solas, D. and Bréchet, Y. (1999). Modelling of microstructure evolution and mechanical properties in age-hardening aluminium alloys, *Proc. EUROMAT 99*, **3**, 121–132.

Deschamps, A., Solas, D. and Bréchet, Y. (2005). 'Modelling of microstructure evolution and mechanical properties in age-hardening aluminium alloys', in Bréchet, Y. (ed.), *Microstructures, Mechanical Properties and Processes - Computer Simulation and Modelling*, Wiley-VCH Verlag GmbH & Co. KGaA, Weinheim, FRG.

Dieter, G.E., Mullin, J.V. and Shapiro, E. (1966). Fracture of inconel under conditions of hot working, *Proceedings of The Conference on Deformation under Hot Working Conditions*, 7–12.

Ding R., Guo Z.X. and Wilson A. (2002). Microstructural evolution of a Ti–6Al–4V alloy during thermomechanical processing, *Materials Science and Engineering: A*, **327**, 233–245.

Djaic, R.A.P. and Jonas, J.J. (1972). Static recrystallisation of austenite between intervals of hot working, *Journal of Iron Steel Institute*, **210**, 256–261.

Doherty, R.D. (2005). 'Primary recrystallisation', in Cahn, R.W. *et al.*, *Encyclopedia of Materials: Science and Technology*, Elsevier, Amsterdam, pp. 7847–7850.

Doherty, R.D., Hughes, D.A., Humphreys, F.J. *et al.* (1997). Current issues in recrystallisation: a review, *Materials Science and Engineering*, **A238**, 219–274.

Dunne, F.P.E. and Hayhurst, D.R. (1992). Continuum damage based constitutive equations for copper under high temperature creep and cyclic plasticity, *Proceedings of The Royal Society A: Mathematical, Physical & Engineering Sciences*, **A437**, 545–566.

Dunne, F.P.E. and Petrinic, N. (2005), *Introduction to Computational Plasticity*, Oxford University Press, Oxford, UK.

Dunne, F.P.E., Rugg, D. and Walker, A. (2007). Lengthscale-dependent, elastically anisotropic, physically-based HCP crystal plasticity: application to cold-dwell fatigue in Ti alloys, *International Journal of Plasticity*, **23**, 1061–1083.

Dunne, F.P.E., Kiwanuka, R. and Wilkinson, A.J. (2012). Crystal plasticity analysis of micro-deformation, lattice rotation and geometrically necessary dislocation density, *Proceedings of the Royal Society A: Mathematical, Physical & Engineering Sciences*, **468**, 2509–2531.

Dyson, B.F. (1988). Creep and fracture of metals: mechanisms and mechanics, *Rev. Phys. Appl.*, **23**, 605–613.

Dyson, B.F. (1990). 'Physically-based models of metal creep for use in engineering design', in Embury, J.D. and Thompson, A.W. (eds), *Modelling of Materials Behaviour and Design*, The Minerals, Metals and Materials Society, pp. 59–75.

Dyson, B.F. and Loveday, M.S. (1981). 'Creep fracture in nomonic 80A under triaxial tensile stressing', in Ponter, A.R.S. and Hayhurst, D.R. (eds), *Creep in Structure*, Springer-Verlag, Berlin, pp. 406–421.

Dyson, B.F. and McLean, M. (1983). Particle-coarsening, σ_0 and tertiary creep, *Acta Metallurgica*, **30**, 17–27.

Dyson, B.F., Verma, A.K. and Szkopiak, Z.C. (1981). The influence of stress state on creep resistance: experimentation and modelling, *Acta Metallurgica*, **29**, 1573–1580.

Dyson, B.F. and McLean, D. (1977). Creep of Nimonic 80A in torsion and tension, *Metals Science*, **2**(11), 37–45.

Edington, J.W., Melton, K.N. and Cutler, C.P. (1976). Superplasticity, *Progress in Materials Science*, **21**, 63–169.

Eliaz, N., Shemesh, G. and Latanision, R.M. (2002). Hot corrosion in gas turbine components, *Engineering Failure Analysis*, **9**(1), 31–43.

Engel, U. and Eckstein, R. (2002). Micro-forming – from basic research to its realization, *Journal of Materials Processing Technology*, **125–126**, 35–44.

Estrin, Y. (1991). 'A versatile unified constitutive model based on dislocation density evolution', in *Constitutive Modelling: Theory and Application*, MD-Vol. 26/AMD-Vol. **121**, ASME, New York pp. 65–75.

Estrin, Y. (1996). 'Dislocation-density-related constitutive modelling', in Krausz, A.S. and Krausz, K. (eds), *Unified Constitutive Laws of Plastic Deformation*, Academic Press, USA.

Evans, R.W. and Wilshire, B. (1985). *Creep of Metals and Alloys*, The Institute of Metals, London.

Fan, X.G. and Yang, H. (2011). Internal-state-variable based self-consistent constitutive modelling for hot working of two-phase titanium alloys coupling microstructure evolution, *International Journal of Plasticity*, **27**, 1833–1852.

Faruque, M.O., Zaman, M. and Hossain, M.I. (1996). Creep constitutive modelling of an aluminium alloy under multiaxial and cyclic loading, *International Journal of Plasticity*, **12**(6), 761–780.

Ferragut, R., Somoza, A. and Tolley, A. (1999). Microstructural evolution of 7012 alloy during the early stages of artificial ageing, *Acta Materialia*, **47**, 4355–4364.

Fogel, D.B. (1991). *System Identification Through Simulated Evolution: A Machine Learning Approach to Modelling*, Ginn Press, Needham Heights, MA.

Ford, H. and Alexander, J.M. (1977). *Advanced Mechanics of Materials*, Ellis Horwood Ltd., England.

Foster, A.D., Dean, T.A. and Lin, J. (2012). Process for forming aluminium alloy sheet component, International Patent No.: WO2010/032002 A1, UK Patent Office.

Foster, A.D., Lin, J., Farrugia, *et al.* (2011). A test for evaluating the effects of stress-states on damage evolution with specific application to the hot rolling of free-cutting steels, *International Journal of Damage Mechanics*, **20**(1), 113–129.

Foster, A.D., Lin, J., Farrugia, D.C.J. *et al.* (2007). Investigation into damage nucleation and growth for a free-cutting steel under hot rolling conditions, *Journal of Strain Analysis*, **42**(4), 227–235.

Foster, A.D., Mohamed, M., Lin, J. *et al.* (2008). An investigation of lubrication and heat transfer for a sheet aluminium Heat, Form-Quench (HFQ) process, *Steel Research International*, **79**(11) II, 133–140.

Gaskell, J., Dunne, F.P.D., Farrugia, D.C.J. *et al.* (2009). A multiscale crystal plasticity analysis of deformation in a two-phase steel, *Journal of Multiscale Modelling*, **1**(1), 1–19.

Garett, R.P., Lin, J. and Dean, T.A. (2005). An investigation of the effects of solution heat treatment on mechanical properties for AA 6xxx alloys: experimentation and modelling, *International Journal of Plasticity*, **21**(8), 1640–1657.

Geiger, M., Kleiner, M., Eckstein, R. *et al.* (2001). Microforming, *CIRP Annals*, **50**(2), 445–462.

Gelin, J.C. (1995) 'Theoretical and numerial modelling of isotropic and anisotropic ductile damage in metal forming processes', in Ghosh, S.K. and Predeleanu, M. (eds), *Materials Processing Defects*, Elsevier, Amsterdam, pp. 123–140.

Gelin, J.C. (1998). Modelling of damage in metal forming processes, *Journal of Materials Processing Technology*, **80-81**, 24–32.

Ghosh, A.K. and Hamilton, C.H. (1979). Mechanical behaviour and hardening characteristics of a superplastic Ti-6AL-4V alloy, *Metallurgical Transactions A*, **10A**(6), 699–706.

Goods, S.H. and Brown, L.M. (1979). The nucleation of cavities by plastic deformation, *Acta Metallurgica*, **27**, 1–15.

Groover, M.P. (1996). *Fundamental of modern manufacturing*, Prentice-Hall International (UK) Ltd., London.

Gurson, A.L. (1977). Continuum theory of ductile rupture by void nucleation and growth: Part I – Yield criteria and flow rules for porous ductile media, *Journal of Engineering Materials and Technology: Transactions of the ASME*, **99**, 2–15.

Hall, E.O. (1951). The deformation and ageing of mild steel: III discussion of results. *Proceedings of the Physical Society. Section B*, **64**, 747.

Hansen, N. and Ostermeier, A. (2001). Completely derandomized self-adaptation in evolution strategies, *Evolutionary Computation*, **9**, 159–195.

Harewood, F.J. and McHugh, P.E. (2007). Comparison of the implicit and explicit finite element methods using crystal plasticity, *Computational Materials Science*, **39**, 481–494.

Hayhurst, D.R. (1972). Creep rupture under multiaxial states of stress, *Journal of Mechanics and Physics of Solids*, **20**, 381–390.

Hayhurst, D.R. (1983). 'On the role of creep continuum damage in structural mechanics', in Wilshire, B. and Owen, D.R.J. (eds), *Engineering Approaches to High Temperature Design*, Pineridge Press, Swansea, pp. 85–176.

Hayhurst, D.R., Dimmer, P.R. and Morrison, C.J. (1984). Development of continuum damage in the creep rupture of notched bars, *Philosophical Transactions of the Royal Society*, **A311**, 103–129.

Hayhurst, D.R., Dyson, B.R. and Lin, J. (1994). Breakdown of the skeletal stress technique for lifetime prediction of notched tension bars due to creep crack growth, *Engineering Fracture Mechanics*, **49**(5), 711–726.

Held, M. (1991). 'On the computational geometry of pocket machining', in *Lecture Notes In Computer Science*, Springer-Verlag, New York

Hill, R. (1950). *The Mathematical Theory of Plasticity*, Clarendon Press, Oxford.

Hill, R. (1966). Generalized constitutive relations for incremental deformation of metal crystals by multislip, *Journal of Mechanics and Physics of Solids*, **14**(2), 95–102.

Hill, R. and Rice, J.R. (1972). Constitutive analysis of elastic-plastic crystals at arbitrary strain, *Journal of Mechanics and Physics of Solids*, **20**(6), 401–413.

Hinde, A. and Miles, R. (1980). Monte Carlo estimates of the distributions of the random polygons of the Voronoi tessellation with respect to a Poisson process, *Journal of Statistical Computation and Simulation*, **10**(3), 205–223.

Ho, K.C. (2004). Modelling of age-hardening and springback in creep age-forming, PhD thesis, University of Birmingham, UK.

Ho, K.C., Zhang, N., Lin, J. *et al.* (2007). An integrated approach for virtual microstructure generation and micro-mechanics modelling for micro-forming simulation, *Proceedings of ASME MNC2007*, 203–211.

Holman, M.C. (1989). Autoclave age forming large aluminium aircraft panels, *Journal of Mechanical Working Technology*, **20**, 477–488.

Hosford, W.F. (1992). *The Mechanics of Crystals and Textured Polycrystals*, Oxford University Press, Oxford.

Hsu, E., Carsley, J.E. and Verma, R. (2008). Development of forming limit diagrams of aluminum and magnesium sheet alloys at elevated temperatures, *Journal of Materials Engineering and Performance*, **17**, 288–296.

Huang, Y. (1991). *A User-Material Subroutine Incorporating Single Crystal Plasticity in the ABAQUS Finite Element Program*, Harvard University Press, Cambridge, MA.

Humphreys, F.J. (1999). A new analysis of recovery, recrystallisation and grain growth, *Materials Science and Technology*, **15**, 37–44.

Humphreys, F.J. and Hatherly, M. (2004). *Recrystallisation and Related Annealing Phenomena*, Elsevier, Amsterdam.

Jeunechamps, P.P., Ho, K.C., Lin, J *et al.* (2006). A closed form technique to predict springback in creep age-forming, *International Journal of Mechanical Sciences*, **48**, 621–629.

Kachanov, L.M. (1986). *Introduction to Continuum Damage Mechanics*, Martinus Nijhoff, Dordricht.

Kachanov, L.M. (1958). *The Theory of Creep*, (English translation edited by Kennedy, A.J.), National Lending Library, Boston Spa.

Kalpakjian, S. and Schmid, S.R. (2001). *Manufacturing Engineering and Technology, 4ᵗʰ Edition*, Prentice-Hall International (UK) Ltd., London.

Karimpour, M. (2012). Modelling of interfacial problems at micro and nano scales in polycrystalline materials, PhD thesis, Imperial College London, UK.

Kaye, M. (2012). Advanced damage modelling of free cutting steels, PhD thesis, Imperial College London, UK.

Khan, A.S., Sung Suh, Y. and Kazmi, R. (2004). Quasi-static and dynamic loading responses and constitutive modelling of titanium alloys, *International Journal of Plasticity*, **20**, 2233–2248.

Kim, T.W. and Dunne, F.P.E. (1997). Determination of superplastic constitutive equations and strain rate sensitivity for aerospace alloys, *Journal of Aerospace Engineering*, **211**, 367–380.

Kim, S.B., Huh, H., Bok, H.H. *et al.* (2011). Forming limit diagram of auto-body steel sheets for high-speed sheet metal forming, *Journal of Materials Processing Technology*, **211**, 851–862.

Kocks, U.F. (1976). Laws for work hardening and low temperature creep, *Journal of Engineering Materials and Technology*, **98**, 76–85.

Kowalewski, Z.L., Hayhurst, D.R. and Dyson, B.F. (1994). Mechanisms-based creep constitutive equations for an aluminium alloy, *Journal of Strain Analysis*, **29**, 309–316.

Kowalewski, Z.L., Lin, J. and Hayhurst, D.R. (1994). Experimental and theoretical evaluation of a high-accuracy uniaxial creep testpiece with slit extensometer ridges, *International Journal of Mechanical Science*, **36**(8), 751–769.

Krajcinovic, D. (1989). Damage mechanics, *Mechanics of Materials*, **8**(2–3), 117–197.

Krishnan, K., Cao, J. and Dohda, K. (2007). Study of the size effect on friction conditions in micro-extrusion: Part 1: Micro-extrusion experiments and analysis, *Journal of Manufacturing Science and Engineering-ASME*, **129**, 669–676.

Kumar, S., Kurtz, S., Banavar, J. *et al.* (1992). Properties of a three-dimensional Poisson-Voronoi tessellation: a Monte Carlo study, *Journal of statistical physics*, **67**, 523–551.

Leckie, F.A. and Hayhurst, D.R. (1977) Constitutive equations for creep rupture, *Acta Metallurgica*, **25**, 1059–1070.

Lee, E.H. (1969). Elastic-plastic deformation at finite strains, *Journal of Applied Mechanics*, **36**, 1–6.

Lemaitre, J. and Chaboche, J.L. (1990). *Mechanics of Solid Materials*, Cambridge University Press, Cambridge, UK.

Lagos, M. (2000). Elastic instability of grain boundaries and the physical origin of superplasticity, *Physical Review Letters*, **85**(11), 2332–2335.

Lee, D. (1969). The nature of superplastic deformation in the Mg-Al eutectic, *Acta Metallurgica*, **17**, 1057.

Leroy, G. (1981). A model of ductile fracture based on the nucleation and growth of voids, *Acta metallurgica*, **29**(8), 1509–1520.

Li, N. (2013). Fundamentals of materials modelling for hot stamping of UHSS panels with gradied properties, PhD thesis, Imperial College London, UK.

Li, N., Mohamed, M.S., Cai, J. *et al.* (2011). Experimental and numerical studies on the formability of materials in hot stamping and cold die quenching processes, The 14th Int. ESAFORM Conf. on Material Forming, *AIP Conf. Proc.*, **1353**, 1555–1561.

Li, Z.H., Bilby, B.A. and Howard, I.C. (1994), A study of the internal parameters of ductile damage theory, *Fatigue and Fracture of Engineering Materials and Structures*, **17**(9), 1075–1087.

Li, D. and Ghosh, A.K. (2004). Biaxial warm forming behaviour of aluminum sheet alloys, *Journal of Materials Processing Technology*, **145**, 281–293.

Li, J., Lin, J. and Yao, X. (2002). A novel evolutionary algorithm for determining unified creep damage constitutive equations, *International Journal of Mechanical Sciences*, **44**(5), 987–1002.

Lin, J. (2003). Selection of material models for predicting necking in superplastic forming, *International Journal of Plasticity*, **19**(4), 469–481.

Lin, J., Ball, A.A. and Zheng, J.J. (1988). Surface modelling and mesh generation for simulating superplastic forming, *Journal of Materials Processing Technology*, **80/81**, 613–619.

Lin, J., Cao, J. and Balint, D. (2011). Development of determination of unified viscoplastic constitutive equations for predicting microstructure evolution in hot forming processes, *International Journal of Mechatronics and Manufacturing Systems*, **4**(5), 387–401.

Lin, J., Cao, J. and Balint, D. (2012). 'Chapter 7: Determining unified constitutive equations for modelling hot forming of steel', in Lin, J., Balint, D. and Pietrzyk, M. (eds), *Microstructural Evolution in Metal Forming Processes*, Woodhead Publishing Ltd., Sawston, UK, pp. 180–209.

Lin, J., Cheong, B.H. and Yao, X. (2002). Universal multi-objective function for optimising superplastic-damage constitutive equations, *Journal of Materials Processing Technology*, **125–126**, 199–205.

Lin, J. and Dean, T.A. (2005). Modelling of microstructure evolution in hot forming using unified constitutive equations, *Journal of Materials Processing Technology*, **167**, 354–362.

Lin, J., Dean, T.A., Foster, A.D. (2011). A method of forming a component of complex shape from aluminium alloy sheet, UK Patent No: GB2473298, UK Patent Office.

Lin, J., and Dunne, F.P.E. (2001). Modelling grain growth evolution and necking in superplastic blow-forming, *International Journal of Mechanical Sciences*, **43**(3), 595–609.

Lin, J., Dunne, F.P.E. and Hayhurst, D.R. (1999). Aspects of testpiece design responsible for errors in cyclic plasticity experiments, *International Journal of Damage Mechanics*, **8**(2), 109–137.

Lin, J., Dunne, F.P.E. and Hayhurst, D.R., 1998, Approximate method for the analysis of component under cyclic plasticity damage. *Journal of Strain Analysis*, **33**(1), pp55-65.

Lin, J., Dunne, F.P.E. and Hayhurst, D.R. (1996). Physically-based temperature dependence of elastic viscoplastic constitutive equations for copper between 20 and 500 °C, *Philosophical Magazine A*, **74**(2), 655–676.

Lin, J., Foster, A.D., Liu, Y. *et al.* (2007). On micro-damage in hot metal working Part 2: Constitutive modelling, *Journal of Engineering Transactions*, **55**(1), 1–18.

Lin, J. and Hayhurst, D.R. (1993a). The development of a bi-axial tension test facility and its use to establish constitutive equations for leather, *European Journal of Mechanics*, **12**(4), 493–507.

Lin, J. and Hayhurst, D.R. (1993b). Constitutive equations for multiaxial straining of leather under uni-axial stress, *European Journal of Mechanics*, **12**(4), 471–492.

Lin, J., Hayhurst, D.R. and Dyson, B.F. (1993a). A new design of uni-axial creep testpiece with slit extensometer ridges for improved accuracy of strain measurement, *International Journal of Mechanical Sciences*, **35**(1), 63–78.

Lin, J., Hayhurst, D.R. and Dyson, B.F. (1993b). The standard ridged uni-axial testpiece: computed accuracy of creep strain, *Journal of Strain Analysis*, **28**(2), 101–115.

Lin, J., Hayhurst, D.R., Howard, I.C. *et al.* (1992). Modelling of the performance of leather in a uni-axial shoe last simulator, *Journal of Strain Analysis*, **27**(4), 187–196.

Lin, J., Ho, K.C. and Dean, T.A. (2006). An integrated process for modelling of precipitation hardening and springback in creep age-forming, *International Journal of Machine Tools and Manufacture*, **46**(11), 1266–1270.

Lin, J., Liu, Y. and Dean, T.A. (2005). A review on damage mechanisms, models and calibration methods under various deformation conditions, *International Journal of Damage Mechanics*, **14**(4), 299–319.

Lin, J., Kowalewski, Z.L. and Cao, J. (2005). Creep rupture of copper and aluminium Alloy under combined loadings – Experiments and their various descriptions, *International Journal of Mechanical Sciences*, **47**, 1038–1058.

Lin, J., Liu, Y., Farrugia, D.C.J. *et al.* (2005). Development of dislocation based-unified material model for simulating microstructure evolution in multipass hot rolling *Philosophical Magazine A*, **85**(18), 1967–1987.

Lin, J., Mohamed, M., Balint, D. *et al.* (2014). The development of continuum damage mechanics-based theories for predicting forming limit diagrams for hot stamping applications, *International Journal of Damage Mechanics*, **23**(5), 684–701.

Lin, J., and Yang, J. (1999). GA-based multiple objective optimisation for determining viscoplastic constitutive equations for superplastic alloys, *International Journal of Plasticity*, **15**(2), 1181–1196.

Lin, J. Zhan, L. and Zhu, T. (2011). 'Chapter 8: Constitutive equations for superplastic forming simulations', in Giuliano, G. (ed.), *Superplastic Forming of Advanced Metallic Materials*, World Publishing, Woodhead, pp. 154–222.

Liu, Y., Lin, J., Dean, T.A. *et al.* (2005). A numerical and experimental study of cavitation in a hot tensile axisymmetric testpiece, *Journal of Strain Analysis*, **40**(6), 571–586.

Liu, Y. (2004). Characterization of microstructure and damage evolution in hot deformation, PhD thesis, University of Birmingham, UK.

Mahadevan, S. and Zhao, Y. (2002). Advanced computer simulation of metal alloy microstructure, *Computer Methods in Applied Mechanics and Engineering*, **191**, 3651–3667.

Maire, E., Bordreuil, C., Babout, L. *et al.* (2005). Damage initiation and growth in metals. Comparison between modelling and tomography experiments, *Journal of the Mechanics and Physics of Solids*, **53**, 2411–2434.

Marciniak, Z., Kuczynski, K. and Pokora, T. (1973). Influence of the plastic properties of a material on the forming limit diagram for sheet metal in tension, *International Journal of Mechanical Sciences*, **15**, 789–805.

McClintock, F.A. (1968). A criterion for ductile fracture by the growth of holes, *Journal of Applied Mechanics*, **35**, 363–371.

Mecking, H. and Knocks, U.F. (1981). Kinetics of flow and strain-hardening, *Acta Metallurgica*, **29**(11), 1865–1875.

Medina, S.F. and Hernandez, C.A. (1996a). Modelling of the dynamic recrystallisation of austenite in low alloy and micro-alloyed steels, *Acta Metallurgica*, **44**(1), 165–171.

Medina, S.F. and Hernandez, C.A. (1996b). General expression of the zener-hollomon parameter as a function of the chemical composition of low alloy and micro-alloyed steels, *Acta Metallurgica*, **44**(1), 137–148.

Meyers, M.A. and Chawla, K.K. (1999). *Mechanical Behavior of Materials*, Prentice Hall, Upper Saddle River, NJ, pp. 555–557.

Mika, D.P. and Dawson, P.R. (1998). Effects of grain interaction on deformation in polycrystals, *Materials Science and Engineering A*, **257**, 62–76.

Miller, A.K. and Shih, C.F. (1977). An improved method for numerical integration of constitutive equations of the work hardening recovery type, *Journal of Engineering Materials and Technology*, **99**(3), 275–277.

Miller, A.K. and Tanaka, T.G. (1988). NONSS: A new method for integrating unified constitutive equations under complex histories, *Journal of Engineering Materials and Technology*, **110**, 205–211.

Moffatt, W.G., Pearsall, G.W. and Wulff, J. (1976). *The Structure and Properties of Metals, Vol. 1*, John Wiley & Sons, New York.

Mohamed, M.S. (2010). An investigation of hot forming quenching process for AA6082 aluminium alloys, PhD thesis, Imperial College London, UK.

Mohamed, M.S., Foster, A.D., Lin, J., *et al.* (2012). Investigation of deformation and failure features in hot stamping of AA6082: Experimentation and modelling, *International Journal of Machine Tools and Manufacture*, **53**, 27–38.

Mohamed, M.S., Lin, J., Foster, A.D. *et al.* (2014). A new test design for assessing formability of materials in hot stamping, *ICTP 2014 Conference*.

Nakazima, K., Kikuma, T. andAsaku, K. (1968). Study on the formability of steel sheet, *Yawata Technical Report*, p. 264.

Needleman, A. (1972). Void growth in an elastic-plastic medium, *Journal of Applied Mechanics*, **39**, 964–970.

Nemat-Nasser, S. (2004). *Plasticity: A Treatise on the Finite Deformation of Heterogeneous Inelastic Materials*, Cambridge University Press, Cambridge.

Nicolaou, P.D. and Semiatin, S.L. (2003). An experimental and theoretical investigation of the influence of stress state on cavitation during hot working, *Acta Materialia*, **51**, 613–623.

Nieh, T.G., Wadsworth, J. and Sherby, O.D. (1997). Book Review: *Superplasticity in Metals and Ceramics*, pp. 32–48.

Omerspahic, D. and Mattiasson, K. (2007). Oriented damage in ductile sheets: constitutive modelling and numerical integration, *International Journal of Damage Mechanics*, **16**, 35–56.

Othman, A.M., Lin, J., Hayhurst, D.R. *et al.* (1993). Comparison of creep rupture lifetimes of single and double notched tensile bars, *Acta Metallurgica et Materialia*, **41**, 17–24.

Pearson, C.E. (1934). The viscous properties of extruded eutectic alloys of lead-tin and bismuth-tin, *Journal of the Institute of Metals*, **54**, 111–116.

Peirce, D., Asaro, R.J. and Needleman, A. (1982). An analysis of nonuniform and localized deformation in ductile single crystals, *Acta Metallurgica*, **30**, 1087–1119.

Petch, N.J. (1953). The cleavage strength of polycrystals, *The Journal of the Iron and Steel Institute*, **173**, 25–28.

Picu, R.C. and Majorell, A. (2002). Mechanical behavior of Ti-6Al-4V at high and moderate temperatures – Part II: constitutive modelling, *Materials Science and Engineering: A*, **326**, 306–316.

Pietrzyk, M., Cser, L. and Lenard, J.G. (1999). *Mathematical and Physical Simulation of the Properties of Hot Rolled Products*, Elsevier, Amsterdam.

Pilling, J. and Ridley, N. (1986). Effect of hydrostatic pressure on cavitation in superplastic aluminium alloys, *Acta Metallurgica*, **34**, 669–679.

Pilling, J. and Ridley, N. (1989). *Superplasticity in Crystalline Solids*, The Institute of Metals, London.

Politis, D. (2013). Process development for forging lightweight multi-material gears, PhD thesis, Imperial College London, UK.

Poole, W.J., Sæter, J.A., Skjervold, S. *et al.* (2000). A model for predicting the effect of deformation after solution treatment on the subsequent artificial ageing behaviour of AA7030 and AA7108 alloys, *Metallurgical and Materials Transactions A*, **31**(9), 2327–2338.

Press, W.H., Flannery, B.P., Teukolsky, S.A. *et al.* (2007). *Numerical recipes in C: the Art of Scientific Computing*, Cambridge University Press, Cambridge.

Quested, T. (2003). As-cast wrought-grade aluminium alloy micrograph, Department of Materials Science and Metallurgy, University of Cambridge, [Online]. Available at: http://www.doitpoms.ac.uk/miclib/micrograph_record.php?id=712 [Accessed 30 Oct 2011].

Rice, J.R. (1972). Inelastic constitutive relations for solids: An internal-variable theory and its application to metal plasticity, *Journal of the Mechanics and Physics of Solids*, **19**(6), 433–455.

Rice, R.J. and Tracey, D.M (1969). On the ductile enlargement of voids in triaxial stress fields, *Journal of the Mechanics and Physics of Solids*, **17**, 201–217.

Ridley, N. (1989). 'Cavitations and superplasticity', in *Superplasticity, AGARD Lecture Series No. 168*, Specialised Printing Services Limited, Essex, pp. 4.1–4.14.

Ringer, S.P. and Hono, K. (2000). Microstructural evolution and age hardening in aluminium alloys: Atom probe field-ion microscopy and transmission electron microscopy studies, *Materials Characterization*, **44**(1–2), 101–131.

Ritz, H. and Dawson, P.R. (2009). Sensitivity to grain discretization of the simulated crystal stress distributions in fcc polycrystals, *Modelling and Simulation in Materials Science and Engineering*, **17**, 1–21.

Rollett, A.D, Luton, M.J. and Srolovitz, D.J. (1992). Microstructural simulation of dynamic recrystallisation, *Acta Metallurgica*, **40**(1), 43–55.

Rossler, J. and Arzt, E. (1990). A new model-based creep equation for dispersion strengthened materials, *Acta Metallurgica et Materialia*, **38**, 671–686.

Sandstrom, R. and Lagneborg, R. (1975). A model for hot working occurring by recrystallization, *Acta Metallurgica*, **23**, 387–398.

Schmid, E. and Boas, W. (1935). Kristallplastizitaet mit besonderer Beruecksichtigung der Metalle. (Strukter und Eigenschaften der Materie, XVII), *Kristallplastizitaet mit besonderer Beruecksichtigung der Metalle*, ISSU.

Semiatin, S.L. and Bieler, T.R. (2001). The effect of alpha platelet thickness on plastic flow during hot working of TI–6Al–4V with a transformed microstructure, *Acta Materialia*, **49**(17), 3565–3573.

Semiatin, S.L. and Furrer, D.U. (2008). Modelling of microstructure evolution during the thermomechanical processing of titanium alloys, in *ASM Handbook, Volume 22A: Fundamentals of Modelling for Metals Processing*, pp. 536–552.

Semiatin, S.L., Seetharaman, V. and Weiss, I. (1999). Flow behavior and globularization kinetics during hot working of Ti–6Al–4V with a colony alpha microstructure, *Materials Science and Engineering A*, **263**, 257–271.

Shao, Z., Bai, Q., Lin, J. *et al.* (2014). Experimental investigation of forming limit curves and deformation features in warm forming of an aluminium alloy, In Press.

Shi, Z., Doel, T.J.A., Lin, J. *et al.* (2010). *Modelling Thermomechanical Behaviour Of Cr-Mo-V Steel, Joining of Advanced and Speciality Materials XII*, 2596–2607.

Shi, Z., Mohamed, M., Wang, L. (2012). Forming limit curves of AA5754 under warm forming conditions, *WAFT project Deliverable Report*, Imperial College London, UK.

Socha, G. (2003). Experimental investigations of fatigue cracks nucleation, growth and coalescence in structural steel, *International Journal of Fatigue*, **25**, 139–147.

Staub, C.J. and Boyer, C. (1996). An orthotropic damage model for visco-plastic materials, *Journal of Materials Processing Technology*, **60**, 297–304.

Stefansson, N. and Semiatin, S. (2003). Mechanisms of globularization of Ti-6Al-4V during static heat treatment, *Metallurgical and Materials Transactions A*, **34**, 691–698.

Weiss, I. and Semiatin, S.L. (1998). Thermomechanical processing of beta titanium alloys—an overview *Materials Science and Engineering A*, **243**, 46–65.

Taylor, G.I. (1934). The mechanism of plastic deformation of crystals. Part I. Theoretical, *Proceedings of the Royal Society of London. Series A*, **145** (855), 362–387.

Taylor, G.I. (1938). Plastic strain in metals, *Journal of the Institution of Metals*, **62**, 307–324.

Talbert, S.H. and Avitzur, B. (1996). *Elementary Mechanics of Plastic Flow in Metal Forming*, John Wiley & Sons, New York.

Thomason, P.F. (1990). *Ductile Fracture of Metals*, Pergamon Press, Oxford.

Tvergaard, V. (1990). Material failure by void growth to coalescence, *Advances in Applied Mechanics*, **27**, 83–147.

Tvergaard, V. (1981). Influence of voids on shear band instabilities under plane strain conditions, *International Journal of Fracture*, **17**, 389–407.

Tvergaard, V. (1982). On localization in ductile materials containing spherical voids, *International Journal of Fracture*, **18**, 237–252.

Tvergaard, V. and Needleman, A. (1984). Analysis of the cup-cone fracture in a round tensile bar, *Acta Metallurgica*, **32**, 157–169.

Tvergaard, V. (1985). Effect of grain boundary sliding on creep-constrained diffusive cavitation, *Journal of the Mechanics and Physics of Solids*, **33**(5), 447–446.

Tvergaard, V. and Needleman, A. (2001). 'The modified Gurson model', in Lematire, J. (ed.), *Handbook of Materials Behavior Models*, Academic Press, London, pp. 430–435.

Uchic, M.D., Dimiduk, D.M., Wheeler, R. *et al.* (2006). Application of micro-sample testing to study fundamental aspects of plastic flow, *Scripta Materialia*, **54**, 759–764.

Vetrano, J.S., Simonen, E.P. and Bruemmer, S.M. (1999). Evidence for excess vacancies at sliding grain boundaries during superplastic deformation, *Acta Materialia*, **47**(15–16), 4125–4129.

Vinod, K.A. (1996). Efficient and accurate explicit integration algorithms with application to viscoplastic models, *International Journal for Numerical Methods in Engineering*, **39**, 261–279.

Wang, L., Strangwood, M., Balint *et al.* (2011). Formability and failure mechanisms of AA2024 under hot forming conditions, *Materials Science & Engineering A*, **528**, 2648–2566.

Wang, S., Zhuang, W., Balint, D.S. *et al.* (2009a) A crystal plasticity study of the necking of microfilms under tension, *Journal of Multiscale Modelling*, **1**(3), 331–345.

Wang, S., Zhuang, W., Balint, D.S. *et al.* (2009b). A virtual crystal plasticity simulation tool for micro-forming, *Procedia Engineering*, **1**(1), 75–78.

Watcham, K. (2004), Airbus A380 takes creep age-forming to new heights, *Materials World*, **12**(2), 10–11.

Weiss, I., Froes, F.H., Eylon, D. *et al.* (1986). Modification of alpha morphology in Ti-6Al-4V by thermomechanical processing, *Metallurgical and Materials Transactions A*, **17**, 1935–1947.

Weiss, I. and Semiatin, S.L. (1998). Thermomechanical processing of beta titanium alloys—an overview, *Materials Science and Engineering A*, **243**(1–2), 46–65.

Wesley, A.S., Joshua, J.J., Timothy, A.M. *et al.* (2010). Effect of electrical pulsing on various heat treatments of 5xxx series aluminum alloys, *Proceedings of the ASME 2010 World Conference on Innovative Virtual Reality WINVR2010*, Ames, IA.

Williams, J.G. (1973) *Stress Analysis of Polymers*, Longmans Harlow, Essex.

Williams, J.G. (2013). Sir Hugh Ford, *Biographical Memoirs of Fellows of the Royal Society*, **59**, 145–156.

Wong, C.C., Dean, T.A. and Lin, J. (2003). A review of spinning, shear forming and flow forming processes, *International Journal of Machine Tools and Manufacture*, **43**, 1419–1435.

Yang, H. (2013). Creep age forming investigation on aluminium alloy 2219 and related studies, PhD thesis, Imperial College London, UK.

Yang, H., Davies, C.M., Lin, J. *et al.* (2013). Prediction and assessment of springback in typical creep-age forming tools, *Proceedings of IMechE, Part B: Journal of Engineering Manufacture*, **227**(9), 1340–1348.

Yao, X., Liu, Y. and Lin, G. (1999). Evolutionary programming made faster, *IEEE Transactions on evolutionary computation*, **3**(2), 82–102.

Zeng, Y.S., Yuan, S.J., Wang, F.Z. *et al.* (1997). Research on the integral hydrobulge forming of ellipsoidal shell, *Journal of Materials Processing Technology*, **72**, 28–31.

Zhan, L., Lin, J., Dean, T.A. *et al.* (2011). Experimental studies and constitutive modelling of the hardening of AA7055 under creep age forming conditions, *International Journal of Mechanical Sciences*, **53**, 595–605.

Zhan, L., Lin, J. and Dean, T.A. (2011). A review of the development of creep age forming: experimentation, modelling and applications, *International Journal of Machine Tools and Manufacture*, **51**(1), 1–17.

Zhang, P. (2011). A virtual grain structure representation system for micro-mechanics modelling, PhD thesis, Imperial College London, UK.

Zhang, P., Balint, D. and Lin, J. (2011a). An integrated scheme for crystal plasticity analysis: virtual grain structure generation, *Computational Materials Science*, **50**(10), 2854–2864.

Zhang, P., Balint, D. and Lin, J. (2011b) Controlled Poisson Voronoi tessellation for virtual grain structure generation: a statistical evaluation, *Philosophical Magazine*, **91**(36), 4555–4573.

Zhang, P., Karimpour, M., Balint, D. *et al.* (2012a). Cohesive zone representation and junction partitioning for crystal plasticity analyses, *International Journal for Numerical Methods in Engineering*, **92**, 715–733.

Zhang, P., Karimpour, M., Balint, D. *et al.* (2012b). Three-dimensional virtual grain structure generation with grain size control, *Mechanics of Materials*, **55**, 89–101.

Zhang, P., Karimpour, M., Balint, D. *et al.* (2012). A controlled Poisson Voronoi tessellation for grain and cohesive boundary generation applied to crystal plasticity analysis, *Computational Materials Science*, **64**, 84–89.

Zhang, P., Zhu, T. and Lin, J. (2012). *User's Manual for VGRAIN – (Virtual Grain Structure Generation System)*, Imperial College London, UK.

Zhang, P., Balint, D. and Lin, J. (2012). User's manual for OPT-CAF: the optimisation tool for calibrating creep constitutive equations, *Report in Metalforming and Materials Modelling Group*, Imperial College London, UK.

Zhao, K.M. and Lee, J.K. (2000). Generation of cyclic stress–strain curves for sheet metals, *Journal of Engineering Materials and Technology*, **123**(4), 391–397.

Zheng, M., Hu, C., Luo, Z.J. *et al.* (1996). A ductile damage model corresponding to the dissipation of ductility of metal, *Engineering Fracture Mechanics*, **53**(4), 653–659.

Zhou, M. and Dunne, F.P.E. (1996). Mechanism-based constitutive equations for the superplastic behaviour of a titanium alloy, *Journal of Strain Analysis*, **31**(3), 187–196.

Zhuang, W., Wang, S., Cao, J. *et al.* (2010). Modelling of localised thinning features in the hydroforming of micro-tubes using the crystal-plasticity FE method, *International Journal of Advanced Manufacturing Technology*, **47**, 859–865.

Zhuang, W., Wang, S., Lin, J. *et al.* (2012). Experimental and numerical investigation of localised thinning in hydroforming of micro-tubes, *European Journal of Mechanics A/Solids*, **31**(1), 67–76.

Zhu, H.X., Thorpe, S.M. and Windle, A.H. (2001). The geometrical properties of irregular two-dimensional Voronoi tessellations, *Philosophical Magazine A*, **81**, 2765–2783.

Zhu, A.W. and Starke Jr., E.A. (2001) Materials aspects of age-forming of Al-xCu alloys, *Journal of Materials Processing Technology*, **117**, 354–35.

Index

www.ingramcontent.com/pod-product-compliance
Lightning Source LLC
Chambersburg PA
CBHW070740220326
41598CB00026B/3709